Ion Exchange: Science and Technology

NATO ASI Series

Advanced Science Institutes Series

A Series presenting the results of activities sponsored by the NATO Science Committee, which aims at the dissemination of advanced scientific and technological knowledge, with a view to strengthening links between scientific communities.

The Series is published by an international board of publishers in conjunction with the NATO Scientific Affairs Division

A	Life Sciences	Plenum Publishing Corporation
B	Physics	London and New York
C	Mathematical and Physical Sciences	D. Reidel Publishing Company Dordrecht and Boston
D	Behavioural and Social Sciences	Martinus Nijhoff Publishers Dordrecht/Boston/Lancaster
E	Applied Sciences	
F	Computer and Systems Sciences	Springer-Verlag Berlin/Heidelberg/New York
G	Ecological Sciences	

Series E: Applied Sciences – No. 107

Ion Exchange:
Science and Technology

edited by
Alírio E. Rodrigues
Professor of Chemical Engineering
Faculty of Engineering
University of Porto
Portugal

1986 **Martinus Nijhoff Publishers**
Dordrecht / Boston / Lancaster
Published in cooperation with NATO Scientific Affairs Division

Proceedings of the NATO Advanced Study Institute on Ion Exchange: Science and Technology, Troia, Portugal, July 14-26, 1985

Library of Congress Cataloging in Publication Data

ISBN-13: 978-94-010-8445-1   e-ISBN-13: 978-94-009-4376-6
DOI: 10.1007/978-94-009-4376-6

Softcover reprint of the hardcover 1st edition 1986

Distributors for the United States and Canada: Kluwer Academic Publishers, 190 Old Derby Street, Hingham, MA 02043, USA

Distributors for the UK and Ireland: Kluwer Academic Publishers, MTP Press Ltd, Falcon House, Queen Square, Lancaster LA1 1RN, UK

Distributors for all other countries: Kluwer Academic Publishers Group, Distribution Center, P.O. Box 322, 3300 AH Dordrecht, The Netherlands

PREFACE

Ion exchange is a field in which cooperation between chemists and chemical engineers is highly desirable.This NATO Advanced Study Institute had as a primary objective to bring together chemists and chemical engineers.

The Institute was dedicated to the memory of Prof. Ted Vermeulen who was supposed to be our scientific advisor.

The lectures presented at the Institute are collected in this volume together with papers presented by F.Evangelista,A.Bunge and G.Grevillot.

The material is arranged in five sections. The Introduction contains a review of the contributions of Vermeulen's work to the field and an overview of ion exchange.Section 2 deals with the chemistry of ion exchange resins.Section 3 covers the chemical engineering aspects of ion exchange processes (thermodynamics,kinetics,modelling and modes of operation).Section 4 starts with a broad review of industrial applications of ion exchange followed by papers dealing with hydrometallurgy,reaction processes involving ion exchangers and zeolites.Finally section 5 deals with Alternative Processes to ion exchange.It includes ion exchange membranes,solvent extraction, liquid membranes and cementation.

I am very grateful to my coworker J.Loureiro for the collaboration as Institute Secretary.Cooperation of M.Hudson at several stages of the organization of the Institute is gratefully acknowledged.

The financial support of the NATO Advanced Study Institute Programme made this course possible.The participants made it all worthwhile.

October 1,1985
A.E.Rodrigues

NATO Advanced Study Institute

"Ion Exchange : Science and Technology"

Director: A.E.Rodrigues

Advisory Committee: M.Hudson (co-director) ,M.Streat and D.Tondeur

Lecturers:

S.Carrà,Politecnico Milano,Italy
C.Costa,Universidade do Porto,Portugal
P.Grammont,Duolite International,France
F.Helfferich,Penn.State University,USA
M.Hudson,University of Reading,UK.

G.Klein,Water,Thermal & Chemical Technology Center,University
        of Berkeley,USA
P.Meares,University of Exeter,UK
A.Myers,University of Pennsylvania,USA
S.Ortiz,Imperial College London,UK
A.Rodrigues,Universidade do Porto,Portugal
F.Ribeiro,Instituto Superior Técnico,Portugal
M.Streat,Imperial College London,UK
D.Tondeur,Laboratoire des Sciences du Génie Chimique,France
P.Wankat,Purdue University,USA
A.Warshawsky,The Weizman Institute of Science,Israel
J.Wesselingh,Delft Institute of Technology,Netherlands

TABLE OF CONTENTS

PART I
INTRODUCTION

THEODORE VERMEULEN'S CONTRIBUTIONS TO PROCESS DESIGN FOR SORPTION OPERATIONS

Gerhard Klein

University of California, Berkeley
Water Thermal and Chemical Technology Center
47th & Hoffman Blvd., Richmond, California 94804

ABSTRACT

A summary is given of Theodore Vermeulen's scientific contributions in the field of process design for sorption operations.

Where possible, enough background is presented for each item to make the pertinent advances more readily understandable. The salient ideas underlying them, and their theoretical and practical importance are discussed.

The topics are grouped as adaptations and extensions of H.C. Thomas' reaction-kinetic model of fixed-bed performance; relaxation of premises underlying this model; multivariant (multicomponent) systems, including ion exchange accompanied by chemical reaction; conceptual process designs; development of ion-exchange applications; handbook articles and reviews; and unpublished contributions. A comprehensive list of pertinent references is included.

INTRODUCTION

To put our topic in perspective, it must be mentioned that, impressive as Theodore Vermeulen's work in sorption-operation design is, it covers only part of his scientific activities, and that he has done comparably important work in the fields of applied kinetics and reactor engineering, and in applied fluid mechanics, in addition to having maintained interest in numerous other chemical-engineering topics.

A summary of what he has done alone or in cooperation with others spans a large part of the body of developments that today constitutes the field of sorption operations as they interest the chemical engineer.

Of the various approaches to providing such a summary that suggested themselves, the one that evolved owed its being selected largely to expediency. The area of Vermeulen's activities in the theory, design, and development of sorption operations was divided into a few major sections, and each is dealt with largely in chronological order, but not rigorously so. Handbook articles and reviews are discussed separately. Because of Vermeulen's protean creativity, his published contributions, even if only those restricted to sorption-operation design are considered, are so numerous that the list of references provided here, while reasonably comprehensive, cannot lay a claim to absolute completeness.

The question as to the role Vermeulen played in his cooperative work with others cannot be answered clearly. Those who have themselves been part of cooperative scientific work will know how difficult, if not impossible, it is in most cases to trace exactly the origin and the subsequent fate of an idea. Those of Vermeulen's co-workers whom it was possible to interview were unanimous in stating that his interaction with them was so lively that the individual roles of the participants in the development of a problem could not be delineated easily. Certainly, Vermeulen was an excellent, if always benevolent and supremely tactful, critic. He would examine a statement or experimental results from all imaginable angles, never tiring in the process himself, while often leading his partner, who tried to follow his flight of thought, to the brink of exhaustion. In this process, a number of subsidiary questions would be raised, which would subsequently be turned into fruitful research topics.

But, while he enjoyed – and indulged in – lively cooperation with students, colleagues, and professional researchers working under his direction, he conceived and developed numerous ideas entirely by himself, and this, due to his great modesty, with so little fanfare, that their importance was not always immediately evident. It was this modesty that in some cases kept hidden the enormous amount of work he had done in the process of developing some of his thoughts that could sometimes make it difficult to determine whether a statement of his was based on a fleeting thought and dealt with as something "just barely possible", or was the result of a long night's work, or of work every night during a six-months' period. In this connection, I remember having asked him once to convey to me the level of confidence he had in a statement of his, since I could not determine how much time I would be justified in verifying it. His answer in that case, "about 99.99 percent", saved me months of work.

The scope of the present summary has not permitted me to dwell on "Ted's" wonderful human qualities, which even over-shadowed his scientific and engineering merits, great as they were. Those who have known him will never forget him; to those who have not, it would be difficult to convey a true picture. I have alluded to these human qualities in capsule form (1984) at the Golden Jubilee Meeting of the American Institute of Chemical Engineers in Washington, D.C., in October 1983, which Professor Vermeulen had been eagerly looking forward to attending when his terminal illness struck him, and in the programming of the Adsorption and Ion Exchange section of which he had taken a major part.

ADAPTATION AND EXTENSION OF THOMAS MODEL

One of the earliest steps Professor Vermeulen took in his work on the design of fixed-bed operation was done in close cooperation with N.K. Hiester, who was then a Ph.D. candidate at the University of California in Berkeley. This work was based on the mathematical model of fixed-bed sorption performance that H.C. Thomas had published (1944, 1948). He had regarded the exchange of monovalent ion species as two opposing second-order reactions. By combining the differential equation descriptive of the rate law with the conservation relation, he had obtained a second-order linear differential equation of the simplest hyperbolic type which could be solved by available methods. The solution contains Bessel functions of zero order for purely imaginary argument and, with a knowledge of the forward and reverse reaction-rate constants, permits calculation of the concentrations in the fluid and exchanger phases as functions of time and axial distance.

Hiester and Vermeulen (1952) have expressed Thomas' solutions more conveniently in terms of the tabulated J-function and have normalized and nondimensionalized the independent variables to make them conform to practice in other branches of chemical engineering, thus following the fruitful unit-operations approach. The distance variable thus becomes a "number of transfer units", N, i.e., the product of a rate constant or mass-transfer coefficient and the residence time, reflecting the fact that a shallower bed of a fast-reacting sorbent is equivalent to a deeper bed of a more slowly reacting sorbent, and that, at a higher linear flowrate, a deeper bed will be required to do the same job as a shallower bed at a lower flowrate. Time is transformed into the product of N and a quantity proportional to the local contact time, yielding NT, which is considered a single variable. (These symbols correspond to those assigned to these variables in the latest papers.)

With these definitions, and with the assumption of the validity of the Thomas model, fixed-bed behavior can be expressed generally with a single explicit parameter, namely the constant separation factor for ion exchange, or the Langmuir equilibrium constant for Langmuir adsorption. The mathematical equivalence of these two processes had already been established by Thomas. For engineering prediction of effluent-concentration histories, the bed depth, in terms of the number of transfer units, must be set. For this case, Hiester and Vermeulen have provided plots of families of curves of effluent concentration vs. throughput parameter T, the latter being the number of ion equivalents of influent solution leaving the column, divided by the number of ion equivalents of exchanger in the bed. One such plot is required for each representative value of the equilibrium parameter.

To obtain the input value of N, the value of the mass-transfer coefficient or of the reaction-rate constant must be known. Following a suggestion already made by Thomas, it can be obtained from the slope of an experimental effluent-concentration history at its midpoint. Hiester and Vermeulen have elaborated this procedure for cases of external and internal diffusion control and provided a relation that permits determination of the rate-controlling diffusion step.

A further extension of Thomas' procedures was made for the exchange of ions of unequal valence by the simple expedient of using an effective selectivity coefficient, obtained by substituting the concentrations of the intersection of the isotherm with the diagonal of the diagram into the constant-separation-factor equation. The underlying transformation is equivalent to replacing a hypothetical third-order reaction by a second-order one. A somewhat more laborious but improved method is also provided for cases in which greater accuracy is required, or in which the isotherm for heterovalent exchange differs too greatly to be replaced by one for homovalent exchange. Here, effective separation factors are obtained from the isotherm for several of its segments, and the resulting segments of the effluent-concentration history are pieced together. The overall history is then shifted so as to satisfy the overall material balance.

Additional generalization of Thomas' results was achieved by Vermeulen and Hiester's work (1954, and Hiester and Vermeulen,

1948) on beds uniformly partially saturated initially with some of the exhaustant ion species (incomplete regeneration), and possibly also receiving a feed solution contaminated with regenerant ion. This development is based on the fact (probably not explicitly recognized by the authors at the time) that both the Langmuir adsorption and the constant-separation-factor ion-exchange isotherms are segments of hyperbolae, and that any part of such a segment, with its limits stretched so that they become equal to those of the full segment, becomes again a segment of a hyperbola, but of smaller curvature.

This stretching procedure is reflected in the transformations of variables proposed by Vermeulen and Hiester. The transformed variables, which were later found applicable also to multicomponent systems, were ultimately called "transition variables", as they relate to an individual composition change or transition, rather than to the entire system, which can exhibit several such transitions.

In the fundamental design of fixed-bed sorption columns, experimental data from laboratory or pilot plant must first be evaluated to determine rate, equilibrium, and capacity parameters. To minimize the laboratory work, these coefficients should then be correlated with the operating variables of interest in full-scale operation. In this connection, Hiester et al. (1956), including Vermeulen, developed interpretive techniques for fixed-bed sorption with nonlinear isotherms in which the rate is controlled by combined diffusion mechanisms. Mass-transfer coefficients were derived from the midpoint slopes of experimental breakthrough curves obtained under defined conditions and resistances were combined with the aid of a factor to correct for nonlinearity of the isotherm. A relationship was developed between the reaction-kinetic rate constant obtained from the midpoint slope and the individual diffusion coefficients. In further elaboration of this work, Vermeulen and Hiester (1959) presented an important plot showing the effect of Péclet number, distribution ratio, and diffusivity ratio on the height of a transfer unit.

The utilization of the Thomas model as discussed so far has been limited to beds saturated uniformly initially and receiving a feed of constant composition. In cyclic operation, exhaustion and regeneration steps alternate and, except after backwashing, presaturation of the bed can not be expected to be uniform. Interference between successive concentration waves may thus be expected.

For this case, two students under Professor Vermeulen's direction have developed mathematical approaches based on the Thomas model, and one of them has tested the results against effluent-concentration histories obtained from laboratory experiments.

In a preliminary study, LeMaguer (1967) provided rigorous mathematical solutions and computer programs for the restricted cases in which (for ion exchange) the separation factor for the exhaustion step was the reciprocal of that for the regeneration step, and in which the total solution normality, the flowrate, and the number of transfer units were the same for both steps.

Results of greater practical value were obtained by Pancharatnam et al. (1969), who was able to relax these restrictions considerably by using a superposition method to combine the cyclic histories of the exhaustion and the regeneration steps, and who utilized this procedure as a basis for economic-optimization calculations. Good agreement between experiment and theory was obtained for systems with isotherms of limited nonlinearity (separation factors ranging approximately from 0.5 to 2) and limited interference of the effluent-concentration histories for the individual steps, resulting in a small memory term.

In trace chromatography, the complication due to nearly completely overlapping effluent-concentration histories is mitigated by the approximate linearity of the effective portion of the equilibrium isotherm. This case has been treated by Vermeulen and Hiester (1952), again on the basis of the Thomas model.

RELAXATION OF PREMISES OF THOMAS MODEL

The continuity relation underlying the Thomas model does not take into account any form of longitudinal dispersion, such as may be caused by flow irregularities (channeling, fingering) or longitudinal diffusion. Neither does it, in its form adapted to particle or film diffusion, allow for nonlinear concentration gradients in the sorbent particles or in the Nernst film surrounding them. Relaxation of these limitations has been the concern of several papers by Vermeulen and some of his co-workers.

In adsorption beds with low flowrates, large sorbent particles, or both, axial dispersion may produce spreading of the effluent-concentration-history curve beyond that produced by ordinary mass-transfer resistances. This case was studied by Quilicy and Vermeulen (1969) for external mass transfer, solid diffusion, pore diffusion, or reaction rate as the controlling mechanism, and numerical results were calculated for constant-pattern performance governed by separation factors corresponding to equilibrium ranging from slightly to irreversibly favorable.

For adsorption, LeVan and Vermeulen (1984 a, b) have observed that the effluent-concentration histories for beds of large diameter are broader than those for beds of smaller diameter. They explained this behavior in terms of channeling accompanied by radial and axial dispersion. A number of dispersion units was defined, which, for laminar flow, was proportional to molecular diffusivity and independent of velocity. Experimental data were used to determine the value of a velocity-profile parameter capable of correlating data for various particle sizes and bed diameters.

Diffusion in the solid particle frequently is the rate-determining step in gel-type synthetic ion-exchanger particles. For this case, and for irreversibly favorable equilibrium, as well as for equilibrium favorable enough to result in constant-pattern behavior in a fixed bed, Vermeulen (1953) succeeded in finding an empirical expression to predict the corresponding saturation history. This expression was not only simple, but also fitted the linear-isotherm case, for which an exact solution existed, considerably better than results obtained with the model of Glueckauf and Coates (1947), who assumed the driving force for mass transfer to be proportional to the difference between the outer surface concentration and the mean concentration of the particle.

The key to this development was the employment of a quadratic instead of a linear expression for the dimensionless driving force for mass transfer, which was subsequently used successfully also for other cases.

Hall and others (1966), including Vermeulen, have studied pore- and solid-diffusion kinetics in fixed-bed adsorption under constant-pattern conditions. As this case was not soluble analytically for the pore-diffusion mechanism, they sought numerical solutions. Such solutions were also obtained for the solid-diffusion case; in both cases for a range of favorable selectivity coefficients. For irreversibly favorable equilibrium, combinations of pore diffusion and external mass transfer were also analyzed. Vermeulen and Quilicy (1970) later also developed an analytic driving-force relation for pore-diffusion kinetics.

MULTIVARIANT SYSTEMS

Single-component adsorption and binary ion exchange may be considered to be monovariant in that, at equilibrium, only one concentration variable can be set independently. By extension, in a system of variance n, n concentration variables can be set independently.

The bulk of early adsorption and ion-exchange publications has dealt with monovariant systems, not only because theory for handling systems of higher variance was not readily available, but also because many problems of practical interest are approximately monovariant in the sense that some of the components occur only in negligible concentrations, or that several components exhibit sufficiently similar selectivity so that their concentrations can be combined.

A classic example of the first type is trace chromatography, which, even if a considerable number of trace components is present, may be termed quasi-invariant; and an example of the second type is the softening of dilute hard waters, where the selectivity of the exchanger for calcium and magnesium is so high as compared to that for sodium, that these hardness ions may be lumped together, as has been done conventionally.

Around 1960, Professor Vermeulen formed the Ion-Exchange Group of the Seawater Conversion Laboratory of the University of California, at Richmond, one of whose first tasks consisted in developing an ion-exchange softening system as pretreatment of seawater or other brines for desalting by evaporation. In the course of this work, it was soon realized that the multicomponent cationic nature of such brines could not be ignored, and an effort was made to provide a multicomponent theory for fixed sorption beds.

Initially, this was limited to systems locally at equilibrium. For this case, in particular for bivariant systems with Langmuir equilibrium, Glueckauf (1945, 1946) had presented specific equations. Building on this work and applying it to ion exchange, the Group calculated a considerable number of bivariant and trivariant ion-exchange examples and abstracted from them the Slope and Alphabet Rules, which permit rapid prediction of the nature of a transition (self-sharpening or proportionate pattern) and of the order in which the concentrations of components present only in the feed or presaturating solution become zero (Klein et al., 1967). Exceptions to the validity of these rules were later pointed out by Helfferich (Helfferich and Klein, 1970). Methods for calculating "composition paths" were found, along which, in composition space, concentrations can vary in a local-equilibrium system, provided that the composition wave is of the simple-wave type. Specifically, equations were derived for composition velocities in heterovalent ion-exchange systems governed by the ideal mass-action law.

In calculations applied to seawater and seawater concentrates, such paths were found to be apparently identical to paths calculated with an integral material balance, such as corresponds to a concentration step instead of a simple wave. However, no matter how often the calculations were repeated, every time to greater accuracy, and how many times the method was checked, there always remained a minute discrepancy. The fundamental question therefore arose whether there was a difference between such paths in principle. It was Professor Vermeulen, who, after intensive, prolonged, and agonizing thought, finally established that such a difference did exist.

One of the numerous other ideas he contributed to this work, but to which he, himself, never attributed much importance, was that of representing bivariant equilibrium surfaces as families of contour lines of constant solid-phase concentrations on a triangular diagram with a basic fluid-phase-concentration grid, or vice versa. From the moment he proposed it, this method was accepted universally, as a matter of course, but until then, the much less clear method of Selke (1956) was the only one generally known and used.

A general property of systems of variance n is that the composition velocity at each point in composition space has n eigenvalues. In calculations for bivariant constant-separation-factor systems (Tondeur and Klein, 1965), a general topology of the composition-path grid emerged. One singular point was found, at which the two eigenvalues coincided - later called the "watershed point" by Helfferich (Helfferich and Klein, 1970). In relation to this point, the rule of constant intercept ratios was discovered, which soon turned out to lead to important further developments in the calculation of constant-separation-factor systems of any variance. (By "intercept" is meant the distance of the intersection of a composition path with the borders of the composition triangle from a corner of this triangle or from the watershed point.)

Much of the further development of these concepts was due to Helfferich (cf. Helfferich and Klein, 1970), who, with the aid of the H-function, greatly facilitated the quantitative calculation of higher constant-separation-factor systems. Several discussions of this subject took place with members of our Group, during which Vermeulen urged Helfferich to orthogonalize the composition-path grid. This, while at first appearing difficult, was eventually done, with the most fruitful results.

The more general validity of some of the findings in the local-equilibrium theory began to become apparent when Shiloh (1966) began to look into the treatment of fixed-bed ion exchange accompanied by chemical reaction; especially of bivariant systems. This work was continued and extended by Golden (1972; Golden et al., 1974) and applied to a partial-deionization process accompanied by precipitation (Page et al., 1975; Popper et al., 1963). Bivariant systems resulting from Donnan uptake accompanying ion exchange were treated by Arnold (1978), after many detailed discussions with Professor Vermeulen. Wheelwright et al. (1984) handled adiabatic adsorption accompanied by condensation, based on the BET equilibrium model.

In a two-pronged attack on the dynamic behavior of multicomponent ion-exchange columns, both multicomponent diffusion and the possibility of applying the Thomas model to individual transitions were studied by Clazie (Clazie et al., 1968) and Omatete (1971).

Not strictly a part of multicomponent studies, but complementing them through clarifying the concept of variance, was an investigation of weak-electrolyte exchangers interacting with solutions of strong or weak electrolytes, and the reflection of this interaction in local-equilibrium theory (Klein et al., 1978). Experimental measurements of relevant parameters were carried out subsequently Klein et al. (1982).

CONCEPTUAL PROCESS DESIGNS

Familiarity with the local-equilibrium theory has led to several ideas for improved processes, which, without that theory, probably would not have been conceived, and, had they been conceived, would have required inordinate time and effort to test experimentally. The available theory, on the other hand, permitted rapid calculations to predict column performance semiquantitatively, thus making it possible to assess the effect of changes in operating conditions.

One of the problems in fixed-bed sorption operations is the balance of exchanger- and regenerant utilization. Equilibrium favorable for the exhaustion step in many cases can imply equilibrium unfavorable for the regeneration step, thus leading to poor regenerant utilization. Equilibrium unfavorable for the exhaustion step is usually avoided in practice, but could lead to poor exchanger and good regenerant utilization. The latter can often be improved by reverse-flow regeneration. In this case, the optimum case would correspond approximately to a linear isotherm, where some lengthening of the transition would occur in each step, but less than for steps with unfavorable equilibrium in the usual cases. These considerations have led to the concept of the layered- and mixed-bed processes described below.

The layered-bed process (Klein and Vermeulen, 1975) is based on a fixed bed consisting of two layers of either cation or anion exchangers. The exchanger nearer the inlet end for the exhaustant exhibits equilibrium unfavorable for the exhaustant ion species, and the other exchanger, equilibrium favorable for this species. The direction of regenerant flow is opposed to that of exhaustant flow, so that, with respect to flow direction and favorableness of equilibrium, these two steps are qualitatively symmetrical.

When representative concentration contours are constructed in a time-distance diagram, they are seen to diverge from the active inlet point to the boundary between the two exchanger types, and then to converge again toward the outlet end. The desired result is a sharp effluent-concentration history for exhaustion, as well as high regenerant utilization. Success of the method is contingent upon the availability of a pair of exchangers with equilibrium characteristics suitable for the intended application.

Calculations of exchanger utilization as function of exchanger proportions and of equilibrium parameters were carried out to make possible the selection of suitable exchangers and to establish the ratio in which they should be used for particular applications. It was recognized, however, that exchangers corresponding to the most desirable combination of these variables will not always be available.

An analogous study, yielding quite similar results, was later made for beds containing the two resins as a mixture, rather than in separate layers (Klein et al., 1981).

In many applications of fixed-bed sorption, it is desired to remove one or more "sandwich components" from a fluid, in which they are present in minor concentrations; sandwich components being solute species for which the sorbent exhibits selectivities intermediate between those of the components present in relatively large concentrations.

In this case, assuming the sorbent to be initially saturated with the component held the least strongly, and in the absence of selectivity reversal, a frontal-analysis pattern will develop; i.e., near the inlet end, a zone of nearly uniform concentrations ("plateau zone") of all components will appear, followed by a plateau zone containing all components but but the one most strongly held; and so forth, each consecutive plateau zone containing all components present in the plateau zone upstream of it, except the most-strongly held of them.

In conventional operation, one could thus apply feed, and save the effluent until the undesirable component or components began to break through (appear in the effluent), and route the next effluent fraction, containing the undesirable components, to waste. The final state of the bed would then be that of saturation with the most strongly held component(s). To regenerate the bed, a considerable excess of the least strongly held component would then be required.

One of Professor Vermeulen's last major ideas concerned itself with an attempt to eliminate most of this regenerant makeup. He called the resulting process concept that of the "segmented bed". It was presented as a minor part of a report covering several topics (Klein et al., 1982), but the concept in its original form was entirely his own. Specifically, it was applied to the removal of a relatively low concentration of phosphate ion from water containing primarily sulfates and chlorides. The advantage of saving regenerant is obtained at the expense of an increased exchanger inventory.

While several variants of this process are possible, only an outline of the simplest of these can be given here.

Assume the solution to be made up of exchangeable ion species A, B, and C, and A to be the most strongly held, C the least strongly held, and B, the species which it is desired to remove, present in minor proportion.

If such a mixture is admitted into a single column initially in the C-form, the familiar frontal-analysis pattern will develop, as mentioned above.

The actual concentrations in the two upstream plateau zones will depend on the governing equilibrium relations and on the influent composition. However, under many conditions, the first upstream plateau zone will contain primarily Species A, and the next plateau zone, primarily Species B. Roughly speaking, the bed will thus contain an A-, a B- and a C-zone.

The idea thus suggested itself to regenerate only the relatively shallow B-zone, and to use C-effluent to partially regenerate the A-zone, completing the regeneration with additional regenerant.

This could be accomplished with appropriate arrangements for side streams, but in the idea as discussed here, it is done with three beds in series, or three segments of one bed, whence the term "segmented bed".

The two outside beds are of the same size, which will depend primarily on exhaustant composition and selectivity. The middle bed only need be large enough to hold most of the amount of B contained in the feed volume of one exhaustion cycle.

Throughout this discussion, we shall number the beds 1, 2, and 3, regardless of the flow direction. Suppose that direction to have been from Bed 1 to Bed 3, and the condition of the beds after the exhaustion step to be that as described above. Bed 1 will thus be mostly in the A-form, Bed 2, in the B-form, and Bed 3, in the C-form.

After regeneration of only Bed 2 back into the C-form, possibly with a relatively concentrated regenerant solution, exhaustant feed is applied to the former outlet of Bed 3 until the latter is filled primarily with A, and Bed 2, with B. The effluent from Bed 2 enters Bed 1, and regenerates it with a diffuse front into the C-form, the bed being large enough to prevent significant leakage of C into the effluent.

After this step, the middle bed is regenerated again, and exhaustant feed is applied to the bed series in the reverse direction. During this exhaustion step, A re-establishes a sharp front, and the final concentration profile becomes the reverse of that which existed after the previous exhaustion step.

Conceptually, the process can be simplified by considering the limiting case in which the amount of B becomes negligible. The middle bed can then be ignored, and the concentration front in an outside bed, when C displaces A on the exchanger, is a simple wave, which, in the next cycle, when the flow direction is reversed, approaches a shockwave. The bed can be so dimensioned that, under the local-equilibrium assumption, this front only travels far enough to be thermodynamically reversible at all times.

Even under the assumption of the validity of the local-equilibrium theory, the multicomponent nature of such systems makes them complex. However, the theory presented by Bailly and Tondeur (1980) for a related process should be applicable, as well as Helfferich's H-function method (cf. Helfferich and Klein, 1970), as implemented in a computer program (Klein et al., 1984).

APPLICATIONS

Electrochromatography may be described as a combination of electrophoresis and chromatography to enhance the separation of electrically charged species. In continuous operation, with a direct-current electric field at right angles to the direction of fluid flow, the presence of a sorbent medium can modify the separation beneficially or detrimentally as compared to simple continuous electrophoresis.

Such a process was tested for the separation of rare earths by Nady and Vermeulen (1970), also in the presence of complexing agents. One type of apparatus used was a cylindrical chromatograph with axial flow and radial current (Hybarger et al., 1963; Kavanaugh, 1963; and Masson, 1965).

Starting in 1961, the then Seawater Conversion Laboratory of the University of California in Berkeley began a series of brine-softening investigations. In the first of these, it was of interest to remove hardness cations from seawater as pretreatment for desalination by evaporation. A number of cation-exchange resins were screened for effectiveness in this task by measuring points on the isotherms for the exchange of sodium and the hardness ions, and rough criteria were developed for evaluating the results in terms of softening performance (Klein et al., 1964). Later, this work, in which the most widely used types of strong-acid cation exchangers were also found to be the most useful for the present purpose, was used as a basis for process design.

Fixed-bed operation with regenerant flow in direction opposite to that of the exhaustant was the process arrangement selected. This made it possible to regenerate the exchanger exclusively with brine rejected from the evaporator. The effect of flowrates and regenerant concentration was investigated in a series of multicyclic column runs, and limits of regenerant concentration established. The latter were imposed on the one hand because, if the regenerant was too dilute (the concentration factor too small), the amount of regenerant brine produced by the evaporator was insufficient, and on the other hand, if the regenerant was too concentrated, calcium sulfate would precipitate during regeneration (Klein et al., 1968). Optimization of the process was undertaken by Makar et al. (1970). Awareness of the possibility of precipitation of calcium sulfate during regeneration should later prove itself useful in the development of a softening process for high-sulfate agricultural drainage water, described in another article for the present NATO Advanced Study Institute (Klein, 1985).

As a method of rendering waters containing undesirably high concentrations of phosphates fit for reuse, Haselbach-Williams (1981) studied the possibility of employing weak-base exchangers both experimentally in column runs, and theoretically (Klein et al., 1982). This work drew on the Group's earlier activities in the field of local-equilibrium theory, on ion exchange accompanied by reaction, and on theoretical and experimental work on weak-electrolyte exchangers and local-equilibrium theory as applied to them (Klein et al., 1987 and 1982).

HANDBOOK ARTICLES AND REVIEWS

Throughout the years, Professor Vermeulen, through interest, industry, and his fabulous memory, has accumulated an encyclopedic knowledge of the field, which benefited a number of review- and handbook articles of his (Vermeulen, 1958; Monet and Vermeulen, 1959; Vermeulen, 1963, 1977, 1978; Vermeulen and Klein, 1971, Vermeulen et al., 1984), while he himself kept up-to-date by writing them.

He had harbored long-range plans for a textbook on the subject. While this, regrettably, was not to materialize, he has left a rich legacy, and a crowning, if succinct, summary of the field, and particularly of his and his collaborators' work, in Section 16 of the last three editions of Perry's Chemical Engineers' Handbook, of which he was the guiding spirit and chief contributor (Vermeulen et al., 1984). This Section also represents a general and up-to-date survey of the work of others in the field.

UNPUBLISHED CONTRIBUTIONS

As we have seen, Professor Vermeulen's published works are voluminous, to say the least. But, had he never published a single paper, he still would have made major significant contributions to science, as lecturer, teacher, student counselor, thesis advisor, research director, organizer and attendant of technical meetings, vociferous chairman, editor, reviewer, and invaluable "sounding board" for the ideas of his colleagues and associates. Unfortunately, some of the work he has been involved in is still buried in theses.

For decades, Professor Vermeulen, partly with my assistance, has been teaching graduate courses in sorption operations at the University of California in Berkeley, thus making this institution one of the relatively few where this branch of chemical engineering was cultivated. Preparation for these lectures took a great deal of time, especially because he never presented the course in the same form, but varied it and brought it up to date from year to year. Together with Hiester and myself, he gave short courses on the subject in the series sponsored by the American Institute of Chemical Engineers.

Professor Vermeulen has given a number of lecture series and papers as invited speaker in the United States and other countries; at Syracuse University in 1971, and at the South China Institute of Technology in 1981, to mention only some of the most recent ones.

Together, his activities created at Berkeley one of the few important centers of progress in sorption-operation design - a field to which Professor Vermeulen's untimely death has meant an irremediable loss.

ACKNOWLEDGMENTS

I acknowledge my deep gratitude to Ted Vermeulen for an inexpressibly enriching association spanning nearly thirty years - as the teacher, superior, colleague, and friend.

To the sponsoring agency of this meeting, to its director, and to the members of the Advisory Committee must go the sincere thanks of all those who have known Ted Vermeulen personally or through his works, for having dedicated the meeting to his memory. For myself, I am deeply appreciative to them for having entrusted me with presenting this outline of his activities in our field.

REFERENCES

Arnold, D.F., 1978. Ion-Exchange Column Performance Under Equilibrium-Controlled Donnan Uptake. M.S. thesis in Chemical Engineering, University of California, Berkeley.

Bailly, M. and D. Tondeur, 1980. Two-Way Chromatography. Chem. Eng. Science 36, 455.

Clazie, R.N., G. Klein and T. Vermeulen, 1968. Multicomponent Diffusion. Generalized Theory with Ion-Exchange Applications. U.S. Office of Saline Water Res. Dev. Prog. Rept. 326.

Glueckauf, E., 1945. Chromatography of Two Solutes. Nature 156, 205.

Glueckauf, E., 1946. Contributions to the Theory of Chromatography. Proc. Royal Soc. Series A 186, 35.

Glueckauf, E., and J.I. Coates, 1947. Theory of Chromatography. Part IV. The Influence of Incomplete Equilibrium on the Front Boundary of Chromatograms and on the Effectiveness of Separation. J. Chem. Soc., 1315.

Golden, F.M., 1972. Theory of Fixed-Bed Performance for Ion Exchange Accompanied by Chemical Reaction. Doctoral dissertation in Chemical Engineering, University of California, Berkeley.

Golden, F.M., K.I. Shiloh, G. Klein and T. Vermeulen, 1974. Theory of Ion-Complexing Effects in Ion-Exchange Column Performance. J. Physical Chemistry 78, 926.

Hall, Eagleton, Acrivos and Vermeulen, 1966. Pore- and Solid-Diffusion Kinetics in Fixed-Bed Adsorption under Constant-Pattern Conditions. Ind. Eng. Chemistry Fundam. 5, 212.

Haselbach-Williams, L., 1981. Phosphate Removal by Weak-Base Ion Exchange. M.S. thesis in Chemical Engineering, University of California, Berkeley.

Helfferich, F. and G. Klein, 1970. Multicomponent Chromatography. Marcel Dekker, New York.

Hiester, N.K. and T. Vermeulen, 1948. Elution Equations for Adsorption and Ion Exchange in Flow Systems. J. Chem. Physics 16, 1087.

Hiester, N.K. and T. Vermeulen, 1952. Saturation Performance of Ion-Exchange and Adsorption Columns. Chem. Eng. Progress 48, 505.

Hiester, N.K. and T. Vermeulen, 1954. Ion-Exchange and Adsorption-Column Kinetics with Uniform Partial Presaturation. J. Chem. Physics 22, 96.

Hiester, N.K., S.B. Radding, R.L. Nelson, Jr., and T. Vermeulen, 1956. Interpretation and Correlation of Ion Exchange Column Performance under Nonlinear Equilibria. AIChE J. 2, 404.

Hybarger, R.M., C.W. Tobias, and T. Vermeulen, 1963. Design Principles for Annular-Bed Electrochromatography. Ind. Eng. Chemistry Process Design Devt. 2, 65.

Kavanaugh, M.C., 1964. Angular-Bed Electrochromatography and Related Investigations. M.S. thesis in Chemical Engineering, University of California, Berkeley.

Klein, G., 1985. Fixed-Bed Ion Exchange with Formation or Dissolution of Precipitate. Paper presented at NATO Advanced Study Institute on Ion Exchange: Science and Technology. Troia, Portugal.

Klein, G., 1984. Dedication to Theodore Vermeulen. AIChE Sympos. Ser. 233, Vol. 80 (Flyleaf).

Klein, G. and T. Vermeulen, 1975. Cyclic Performance of Layered Beds for Binary Ion Exchange. AIChE Sympos. Ser. 152, Vol. 71, 69.

Klein, G., M. Villena-Blanco and T. Vermeulen, 1964. Ion-Exchange Equilibrium Data in the Design of a Cyclic Sea Water Softening Process. Ind. Eng. Chemistry Proc. Des. Develop. 3, 280.

Klein, G., D. Tondeur and T. Vermeulen, 1967. Multicomponent Ion Exchange in Fixed Beds. General Properties of Equilibrium Systems. I&EC Fundamentals 6, 339.

Klein, G., S. Cherney, E.L. Ruddick and T. Vermeulen, 1968. Calcium Removal from Sea Water by Fixed-Bed Ion Exchange. Desalination 4, 158.

Klein, G., N.J. Norem and T. Vermeulen, 1978. Studies on the Behavior of Carbonic Acid and its Salts in Fixed Beds of Weak-Acid Ion Exchangers. Proc. Ion-Exchange Symposium, Central Salt and Marine Chemicals Research Institute, Bhavnagar. G.T. Gadre, ed. p. 119.

Klein, G., T.J. Jarvis and T. Vermeulen, 1979. Fluidized-Bed Ion Exchange with Precipitation - Principles and Bench-Scale Development. In "Recent Developments in Separation Science, V." N.N. Li, ed. CRC Press, West Palm Beach, p. 185.

Klein, G., E. Lee and T. Vermeulen, 1981. Cyclic Operation of Mixed-Resin Cation or Anion Exchange Columns with Flow Reversal Between Half-Cycles. Paper presented at AIChE Annual Meeting, New Orleans.

Klein, G., J. Sinkovic and T. Vermeulen, 1982. Weak-Electrolyte Ion Exchange in Waste-Water Reuse. U.S. Int. Dept. Rept. OWRT 82/7.

Klein, G., M. Nassiri, and J.M. Vislocky, 1984. Multicomponent Fixed-Bed Sorption with Variable Initial and Feed Compositions. Computer Prediction of Local-Equilibrium Behavior. AIChE Sympos. Series 233, Vol. 80, 14.

LeVan, M.D. and T. Vermeulen, 1984 a. Channeling and Bed-Diameter Effects in Fixed-Bed Adsorber Performance. AIChE Sympos. Ser. 233, Vol. 80, 34.

LeVan, M.D. and T. Vermeulen, 1984 b. Effects of Channeling and Corrective Radial Diffusion in Fixed-Bed Adsorption Using Activated Carbon. In "Fundamentals of Adsorption", A.L. Myers and G. Belfort, edts. Engineering Foundation, New York. p. 305.

LeMaguer, M., 1967. Cyclic Operation of Fixed-Bed Ion-Exchange and Adsorption Columns. M.S. thesis in Chemical Engineering, University of California, Berkeley.

Makar, K.M., T. Vermeulen and G. Klein, 1970. Design and Cost for Ion-Exchange Softening for a Seawater Evaporation Plant. In "Ion Exchange in the Process Industries", Society of Chemical Industry, London, p. 174.

Masson, M., 1965. Angular-Bed Electrochromatography and Related Investigations. M.S. thesis in Chemical Engineering, University of California, Berkeley.

Monet, G.P. and T. Vermeulen, 1959. Progress in Separation by Sorption Operations - Adsorption, Dialysis, and Ion Exchange. AIChE Prog. Sympos. Ser. 55, No. 25, 109.

Nady, L. and T.R. Vermeulen, 1970. Electrochromatographic Separations of Rare Earths. Lawrence Radiation Laboratory University of California Report UCRL-19526.

Omatete, O.O., 1971. Column Dynamics of Ternary Ion Exchange. Doctoral dissertation in Chemical Engineering. University of California, Berkeley.

Pancharatnam, Klein and Vermeulen, 1969. Design Optimization Procedure for Cyclic Ion Exchange and Adsorption. U.S. Office of Saline Water Res. Dev. Progr. Rept. 477.

Popper, K., R.J. Bouthilet and V. Slamecka, 1963. Ion-Exchange Removal of Sodium Chloride with Calcium Hydroxide as Recoverable Regenerant. Science 141, 1083.

Quilici and T. Vermeulen, 1969. Axial-Dispersion Constant-Pattern Kinetics of Ion-Exchange and Adsorption Columns. U.S. Office of Saline Water Res. Dev. Prog. Rept. 476.

Page, B.W., G. Klein, F. Golden and T. Vermeulen, 1975. Mixed-Bed Ion-Exchange Desalting by the Calcium Hydroxide Process. AIChE Sympos. Ser. 152, Vol. 71, 121.

Selke, W.A., 1956. Mass Transfer and Equilibria. Chapter 4 in Ion Exchange Technology, F.C. Nachod and J. Schubert, edts. Academic Press, New York. p. 64.

Shiloh, K.I., 1966. Ion Exchange Accompanied by Side Reactions. Column Performance. M.S. thesis in Chemical Engineering, University of California, Berkeley.

Thomas, H,C., 1944. Heterogeneous Ion Exchange in a Flowing System. J. Am. Chem. Soc. 66, 1664.

Thomas, H.C., 1948. Chromatography: A Problem in Kinetics. Ann. N.Y. Acad. Sci. 49, 161.

Tondeur, D. and G. Klein, 1967. Multicomponent Ion Exchange in Fixed Beds. Constant-Separation-Factor Equilibrium. I&EC Fundamentals, 6, 351.

Vermeulen, T., 1953, Theory for Irreversible and Constant-Pattern Solid Diffusion. Ind. Eng. Chemistry 45, 1664.

Vermeulen, T., 1958. Separation by Adsorption Methods. In "Advances in Chemical Engineering", T.B. Drew and J.W. Hoopes, Jr., eds. Vol. II. Academic Press, New York. p. 147.

Vermeulen, T., 1963. Industrial Adsorption. In "Encyclopedia of Chemical Technology", 2nd. ed. Interscience, New York.

Vermeulen, T., 1977. Process Arrangements for Ion Exchange and Adsorption. Chem. Eng. Progress 73, No. 10, 57.

Vermeulen, T., 1978. Adsorptive Separations. In "Kirk-Othmer: Encyclopedia of Chemical Technology", Vol. 1, third edition. Wiley & Sons, New York.

Vermeulen, T. and N.K. Hiester, 1952. Ion-Exchange Chromatography of Trace Components. A Design Theory. Ind. Eng. Chemistry, 44, 636.

Vermeulen, T. and N.K. Hiester, 1954. Ion-Exchange and Adsorption Column Kinetics with Uniform Partial Presaturation. J. Chem. Physics 22, 96.

Vermeulen, T. and N.K. Hiester, 1959. Kinetic Relationships for Ion Exchange Processes. AIChE Progr. Sympos. Ser. No. 24, Vol. 55, 61.

Vermeulen, T. and E.H. Huffman, 1953. Ion-Exchange Column Performance. Hydrogen-Cycle Rates in Nonaqueous Solvents. Ind. Eng. Chemistry 45, 1653.

Vermeulen, T. and G. Klein, 1971. Recent Background Developments for Adsorption Column Design. AIChE Sympos. Ser. 67, No. 117, 65.

Vermeulen, T. and R.E. Quilici, 1970. Analytic Driving-Force Relation for Pore-Diffusion Kinetics in Fixed-Bed Adsorption. I&EC Fundamentals 9, No. 1, 179.

Vermeulen, T., M.D. LeVan, N.K. Hiester, and G. Klein, 1984. Adsorption and Ion Exchange. Section 16 in "Perry's Chemical Engineers' Handbook", sixth ed., R.H. Perry, D.W. Green, and J.O. Maloney, eds. McGraw-Hill, New York.

Wheelwright, S.M., J.M. Vislocky and T. Vermeulen, 1984. Adiabatic Adsorption with Condensation: Prediction of Column Behavior Using the BET Model. In "Fundamentals of Adsorption", A.L. Myers and G. Belfort, edt. Engineering Foundation, New York. p. 721.

ION EXCHANGE: PAST, PRESENT, AND FUTURE

Friedrich G.Helfferich

Department of Chemical Engineering
The Pennsylvania State University

University Park,PA 16802 USA

ABSTRACT

The history and evolution of ion exchange are viewed in the broader context of the development of science and technology in our time.

HISTORY

The history of ion exchange has been told and retold count-less times, so a quite brief synopsis may suffice here. Thorough authors have traced ion exchange back to Moses, who softened the bitter waters of Mara to make them potable for his flock in the desert 1). Another often quoted ancient reference is to Aris-totle's observation that the salt content of water is diminished or altered upon percolation through certain sands 2). The thread continues to Thompson and Way, two British soil scientist who, in 1850-1852, studied the same phenomenon in the more systematic fash-ion of their time 3,4). With progressing industrialization the focus then, around the turn of the century, shifted to plant-scale water softening, first with natural and later with synthetic inor-ganic ion exchangers 5,6). The next major evolution came in 1935 with the work of Adams and Holmes 7), two English scientists, whose chance discovery that a shattered phonograph record exhib-ited ion exchange properties led them to invent ion exchange re-sins, materials with in many respects superior properties. In the following years these were further developed chiefly by the I. G. Farbenindustrie in Germany (Wofatits). In World War II, ion ex-change contributed significantly to the Manhattan Project by pro-viding solutions to the vexing problems of separating rare earths elements and other fission products 8-10). Two important advances in the first decade after the war were the advent of more stable and reproducible ion exchange resin on styrene basis 11) and of strong-base anion exchangers (with quaternary ammonium groups) 12). Two later important advances include the commercial development of effective inorganic ion exchangers in the form of synthetic zeolites 13) ["molecular sieves," first synthesized by Barrer 14)] and of "macroporous" ion exchange resins 15,16). The latter opened up many applications under conditions under which the active groups of conventional resins are not accessible for lack of swelling. Today, a wide array of improved organic and inorganic ion exchangers with a great variety of properties is available for laboratory- and plant-scale applications ranging from chemical analysis to preparative separations, from catalysis to organic synthesis, from biomedical uses to decontamination and detoxification.

ION EXCHANGE IN SCIENCE AND TECHNOLOGY OF OUR TIME

Rather than dwell further on this saga so often told, we might step back and try to take a more philosophical view, to see how ion exchange relates to the evolution of our science and technology and how it reflects in some fashion problems mankind has faced and changes in thinking and attitude that have occurred over the ages.

In one narrow respect, ion exchange from the outset has been intimately related to water, one of the most indispensable and precious resources of our planet -- without water, no life. Water treatment, be it for human consumption or industrial use, has been and remains the mainstay of ion exchange, from Moses at Mara till today and tomorrow, a time in which pure, clean water has once again become a gift we can no longer take for granted, an essential resource we must take great care to preserve.

In another way, ion exchange may strike us as an exceptionally versatile and practical chip of our technology. If we follow our chroniclers, we see it was born in the Arabian desert out of sheer necessity, and ever since has adapted in some fashion to newly evolving interests and needs: in the century of Malthus to soil science and agriculture; in the time of industrialization to preparation of boiler feed water for generation of steam and power; in the dawn of nuclear technology to its separation problems; in the early days of space technology to the development of miniature electric power sources; in our present time to the growing needs to recover valuable resources, to protect the environment. I know of few disciplines that have been so broadly involved in the evolution of our science and technology, whose thrust and emphasis has so well kept attuned to the changing outlook and needs of the time.

In many instances, ion exchange has contributed directly or indirectly to the evolution and success of new disciplines in chemistry, engineering, and biology. A few examples may illustrate this. In biophysics and biochemistry, our present understanding of transport across living membranes owes much to the theoretical and experimental work on ion exchange membranes by pioneers such as Teorell 17), K. H. Meyer 18), and Schlögl 19). Much the same is true for today's dialysis apparatuses in biomedical engineering, such as the kidney machine. Prediction and success of many techniques of enhanced oil recovery rely heavily on a correct understanding of ion exchange between reservoir fluids and clays 20) and, not infrequently, of adjustment of the ionic composition of injected fluids by ion exchange. If I had to single out three inventions as having shaped more than any others the working-day environment of the scientist in our time, I would choose the chromatograph, the xerox machine, and the personal computer; while the contributions of ion exchange to plain-paper copying and the microchip are somewhat far-fetched and debatable, those to chromatography have been numerous and significant and are continuing.

Ion exchange is principally a tool for chemical separations, and it is to this field that it has made some of its most remarkable contributions -- contributions today too easily taken for granted. To regain perspective, let me go back to my high school days, more years ago than I care to count. At that time the separation of rare earths was considered to be the hardest imaginable chemical separation problem -- short of that of optical isomers, which, almost by definition, seemed impossible to achieve by nondestructive chemical means. With ion exchange, rare earths were separated in kilogram quantities, excellent yields, and spectroscopic purity by Spedding 21) only a decade later. Today, with a technique closely related to ion exchange and thanks to pioneers such as Davankov 22), quite reasonable optical isomer separation are carried out without much ado.

It is instructive to examine how progress in ion exchange has been acchieved over the years, what methods and approaches have led to its remarkable successes. Seen in this light, ion exchange will impress us as a thoroughly practical -- not to say pragmatic -- science. Moses acted on divine inspiration to solve a practical problem, with no knowledge of how or why his technique worked. Perhaps even more noteworthy is the fact that the scientific empirical basis of ion exchange, in soils and by Thompson and Way, was laid before Arrhenius (in 1887) had established the existence of ions. The most striking example of this practical, Edisonian approach to problems in ion exchange is from a much later time; it is the separation of the rare earth elements and deserves to be singled out for more comment.

The difficulty of separating the rare earth (and transuranium) elements stems from their identical structure of outer electron shells, resulting in very closely similar chemical and physical properties. One of the few properties to vary substantially from one element to the next is the atomic diameter and, as a consequence, the stability of complexes with chelating agents. This sparked the idea of using such agents, in combination with ion exchange, to accomplish separations: the stronger the chelation in solution, the less strongly a rare earth ion should be taken up by an ion exchanger. In the context of the Manhattan Project, several groups worked on this approach in parallel, largely with the same types of ion exchangers and chelating agents. All groups but one obtained very similar results, a pattern much like that of analytical chromatography, with "peaks" of individual rare earths traveling through the column at different speeds and flattening out more and more on their way. Today, we know this happens when the chromatographic sorbent prefers the substances to be separated to the eluting agent, and call it "elution development." One group, headed by Spedding at Ames, Iowa, worked under slightly different conditions and obtained a pattern different in kind: the slug of rare earths travelled through the column at high concentration, without any flattening out, and within it the elements arranged themselves in sequence so as to emerge in high concentrations, close-up, and with some overlap 21). This so-called "displacement development" occurs when the eluting agent is preferred and so pushes the substances to be separated in a piston-like fashion through the column. Obviously, elution development, which can produce any desired sharpness of separation -- at the expense of product concentration -- , is best suited for analytical purposes; in contrast, displacement development, which maintains high concentration, is the better technique for preparative separations. In ion exchange chromatography with complexing agents, the same exchanger and same agent can produce either one or the other type of development, depending on concentration and pH of the agent. In the Manhatten Project, excellent analytical and preparative separations were developed without a real understanding of cause and effect. The correct explanation was given only years later, based largely on the work of Glueckauf 23) in the 1950's.

While the Edisonian component in ion exchange is very strong, there are also examples for successes obtained with the complementary, deductive approach. In fact, the general topic of complex formation in ion exchange provides a whole chain of these. Instead of exploiting complex formation of ions with an agent in solution to accomplish effective separations, the complexing agent can be built into the ion exchanger so as to provide preference for ions which are most strongly complexed. Starting with

Skogseid 24) in 1947, this idea has been actively pursued and has led to a number of commercially available chelating resins 25) now widely used for special purposes (e.g., Dowex A-1, Chelex 100). In a way, on a more primitive level, even the much older weak-acid resins with carboxylate groups can be considered members of this general class as the bond formation of their groups with hydrogen ions is the cause of their strong preference for the latter over metal cations. (A similar argument, of course, can be advanced for weak-base ion exchangers.)

The next logical step in exploitation of this idea was taken a few years later and is one of the many applications going beyond actual ion exchange. The ion exchanger can act as carrier of a metal ion that forms complexes of different strengths with various ligands, and can so be used for separations of such ligands 26). The first "ligand exchange" separations of this type were success-fully designed on paper, based on tabulated complex stability con-stants of the components involved 27). Today, ligand exchange has become a standard chromatographic technique and is widely used for separations of amines, amino acids, etc., largely thanks to the comprehensive work of Walton 28). [Incidentally, ligand exchange also illustrates how hard it is in our time for scientists to keep abreast of innovations in fields only distantly related to their own. When ligand exchange had long been an established technique of chemical chromatography, it was reinvented by biochemists for their own purposes and rechristened "metal chelate affinity chroma-tography." 29)]

Chelating resins and ligand exchange were only the first and fairly simple and obvious extensions of the general idea of combi-ning ion exchange with chemical reactions such as complex forma-tion. Two further examples deserve to be singled out here briefly. These are the extension of the ligand exchange idea, principally by Davankov 22), to equip resins with optically active centers and so achieve separations of optical isomers, and the generalization of the idea of "reactive ion exchange" (RIEX) by Janauer 30), who systematically combined ion exchange with reactions of many differ-ent types (e.g., oxidation-reduction, neutralization, complex for-mation with neutral or charged ligands) to achieve highly selec-tive separations, particularly at trace level and for protection of the environment.

To be successful, a discipline of science and technology needs not only pragmatic and deductive input, it also needs luck. This ion exchange has had. The best example for this is in the earliest large-scale application, to water softening, that is, removal of calcium ions in exchange for sodium ions from hard water. The way the mass action law, based on ideal thermodyna-mics, works for any process with other than one-to-one stoichio-metry is that low concentration favors the state with greater number of species. Applied to the sodium-calcium exchange in water softening, the mass action law thus tells us that the ion exchanger's preference for calcium will increase with dilution of the solution (more cations in solution if these are monovalent sodium instead of divalent calcium ions). In consequence, from hard water as a very dilute solution, calcium ions are highly preferred and are very effectively taken up, even if there is considerable competition from sodium and potassium ions. Yet, this does not impede removal of calcium from the exchanger in regeneration, which is effected by concentrated brine and thus under conditions under which the strong preference for calcium ions has disappeared. We might say this operation is one of the few exceptions to the Third Law of Engineering, the Law of Inherent Malice of Matter: if nature can make it hard for us, it will. In this case, nature gives us a break, providing favorable conditions for both steps of the operating cycle. So, if for reasons that at the time were neither recognized nor appreciated, ion exchange was highly successful. Were it not for this freak of nature, ion exchange might never have acquired commercial significance, and this NATO Advanced Study Institute would not have been held.

ION EXCHANGE IN BROADER CONTEXT

With ligand exchange and the separation of optical isomers we have already touched on topics that go beyond strict exchange of ions. In many instances, the desired effect is a separation, removal, accumulation, reaction, or other action of neutral molecules, and no ion exchange occurs. In ligand exchange, in fact, ion exchange would lead to the loss of complexing ions from the carrier, and so would be detrimental and must be suppressed. This is not an exception. Indeed, the great majority of applications of ion exchangers -- and the most interesting ones! -- are those in which the ability of materials to exchange ions under other conditions is used for quite different purposes.

One of the most important such application of ion exchangers is as catalysts. Many reactions of organic chemistry are catalyzed by ions in solution, the hydrogen ion-catalyzed hydrolysis ("inversion") of sucrose to give glucose and fructose being just one example. The recognition that such ion-catalyzed reactions can be conducted with ion exchangers instead of dissolved electrolytes as the catalyst sources 31) dates back to the 1930's and was one of the main incentives for the I. G. Farbenindustrie to acquire the Adams and Holmes patents. Today, catalysis by ion exchangers has gained special importance in key processes to manufacture automotive gasoline additives to provide effective no-knock performance in the absence of tetraethyl lead 32) (objectionable for environmental reasons) or with various types of synthetic fuels 33) (e.g., "gasohol"). The largest-scale application of ion exchange materials today is in the field of catalysis, but in a process for which the ability of the catalyst to act as ion exchanger is almost coincidental: catalytic cracking of hydrocarbon feedstocks in the gas phase over alumosilicates (zeolites) 34).

Along similar lines, an interesting example illustrating the diverse potential of ion exchange for meeting new demands is the production of high-fructose liquors, e.g., from corn starch 35). Because of its strong sweetness, fructose has firmly established itself as a natural low-calory replacement for artificial sweeteners that are being phased out as suspected health hazards. Here, ion exchangers are can be used in no less than three functions: as catalysts (or catalyst carriers) for the isomerization of glucose to fructose, to separate these two sugars, and to deionize the various process streams 35).

The use of ion exchangers to conduct ion-catalyzed reactions is interesting in still another, more academic but also more fundamental respect. It allows for a translation of homogeneous catalysis in solution into heterogeneous catalysis by an added solid, the ion exchanger, with all the advantages of easy separation of product from catalyst. The translation into heterogeneous catalysis furthermore provides an opportunity to develop catalysts of higher selectivity and effectiveness. For example, the solid catalyst can be so constructed that it will admit small molecules but exclude large ones, or will admit molecules of certain shapes but exclude others; it will then selectively catalyze reactions of only those molecules which have access to its sites 36). To enhance the effectiveness of catalytic reaction of certain molecules, the catalyst can be equipped with groups that will specifically attract these molecules to the vicinity of the catalytic sites. It is this principle to which enzymes largely owe their extreme catalytic activity and selectivity, and we are just beginning to learn how to copy nature in this respect.

Based on ideas somewhat along these lines is another novel application of ion exchangers: as disinfectants 37). Here, resins are used that have high sorption affinity for bacteria and carry antibacterial agents such as quaternary ammonium ions as either functional groups or counterions.

If I have dwelled here so long on aspects related to reactions and catalysis, perhaps giving them coverage out of proportion with other and equally important applications beyond literal exchange of ions, it is because they lead naturally into a further field that bears much promise and has recently moved into the foreground of attention. This is the use of ion exchangers or like solids as templates in organic synthesis 39). Here, in principle, a solid equipped with appropriate functional groups is used to hold a reactant in place in the right configuration for the de-

sired reaction with another agent. This "solid phase peptide synthesis" 39) has great advantages over synthesis in homogeneous
solution in two respects: A higher specificity can be achieved,
and high conversion can be attained through excess of the second
reagent, which then is easily removed as it remains in solution
while the product is still held on the solid. For his invention
of this ingenious method and its application to the synthesis of
proteins in high yield and with exactly specified sequences of
amino acids, Merrifield was recently awarded the Nobel Prize for
Chemistry.

THE POLYMER AGE

In order to gain a true appreciation of the role of ion exchange in our present and future world, we might step back still
farther for a yet more sweeping perspective. Can we find an underlying theme in the ubiquitous encounters with ion exchangers --
at least if that term is interpreted broadly enough to include all
those applications in which literal exchange of ions is coincidental, absent, or even undesirable?

I well remember from my childhood days a prophetic cartoon.
It showed some strange characters, obviously scientists, excitedly
crowding around an exhibit in a museum setting, one of them even
wielding a huge magnifying glass. The object of their rapt attention was labelled, "Table Made From Wood." In the infancy of plas-
tics, to suggest that wood would be supplanted as furniture material was bold indeed and provoked uproarious laughter. Each time
I look at my formica kitchen table and counters I am reminded of
this cartoon.

Let us try to imagine how a historian a hundred thousand
years from now, perhaps from another galaxy, might reconstruct the
evolution of intelligent life on our planet. In his perspective,
our entire recorded history will be but a brief episode. From the
artifacts he finds and with his dating techniques he might classify the epochs of mankind predating his time as the Stone Age, the
Bronze Age, the Iron Age ... and the Polymer Age. The latter, it
seems, we are about to enter.

Few will doubt the future role of polymers in our world.
Perhaps we should reorient our thinking about ion exchange with
more attention to its relation to polymers. As the present survey
may have illustrated, "ion exchange" has become too narrow a classification as many of the most fascinating and promising uses en-
tail no exchange of ions. Rather, what we really mean today when
we talk about ion exchange as a field is polymers, organic or inorganic, with ionic or ionizable functional groups. Even the distinction that the functional groups should be capable of ioniza-

tion is sometimes blurred or not essential. In this context,
then, actual ion exchange in the literal sense is only part of a
larger, more interesting and more colorful picture, that of "reactive" or "functional" polymers, that is, polymers carrying
groups capable of chemical reactions or of inducing such reactions
-- one such reaction being electrolytic dissociation. All the
many applications of ion exchangers in operations other than ion
exchange then have their logical place, and our thinking along
these lines might help to clarify connections, even lead to further innovation.

APOLOGY

The examples given in this survey are a subjective and often
arbitrary selection. My apologies are offers to those many whose
outstanding contributions have not been acknowledged here.

1. Exodus 15:23-25.
2. Aristotle, Works, (Clarendon Press, London, 1927), Vol. 7, p.
   933b.
3. H. S. Thompson, J. Roy. Agr. Soc. Engl., Vol. 11 (1850), p.
   68.
4. J. T. Way, J. Roy. Agr. Soc. Rngl., Vol. 11 (1850), p. 313;
   Vol. 13 (1852), p. 123.
5. R. Gans, Jahrb. Preuss, geol. Landesanstalt, Vol. 26 (1905),
   p. 179.
6. F. Harm and A. Rumpler, 5th Intern. Congr. Pure Appl. Chem.
   (1903), p. 59.
7. B. A. Adams and E. L. Holmes, J. Soc. Chem. Ind. (London),
   Vol. 54, (1935), p. 1T.
8. E. R. Tompkins, J. X. Khym, and W. E. Cohn, J. Am. Chem. Vol.
   69 (1947), p. 2769.
9. B. H. Ketelle and G. E. Boyd, J. Am. Chem. Soc., Vol. 69
   (1947), p. 2800.
10. F. H. Spedding, E. I. Fulmer, T. A. Butler, E. M. Gladrow, M.
    Gobush, P. E. Porter, J. E. Powell, and J. M. Wright, J. Am.
    Chem. Soc., Vol. 69 (1947), p. 2812.
11. G. F. D'Alelio (General Electric Co.), US Patent 2,366,007
    (1944).
12. R. M. Wheaton and W. C. Bauman, I&EC, Vol. 43 (1951), 1088.
13. D. W. Breck, Zeolite Molecular Sieves: Structure, Chemistry,
    and Use (Wiley, New York, 1974), p. 95.
14. R. M. Barrer, J. Chem. Soc. (1948), pp. 127, 2158; (1950),
    p. 2342.
15. R. Kunin, E. Meitzner, and N. Bortnick, J. Am. Chem. Soc.,
    Vol. 84 (1962), p. 305.

16. Farbenfabriken Bayer AG, Ger. Patents 1,045,102 and 1,113,570 (1957).

17. T. Teorell, Proc. Soc. Exptl. Biol., Vol. 33 (1935), p. 282.

18. K. H. Meyer and J. F. Sievers, Helv. Chim. Acta, Vol. 19 (1935), p. 649.

19. R. Schlogl, Stofftransport durch Membranen, (Steinkopff, Darmstadt, 1964).

20. G. A. Pope, L. W. Lake, and F. Helfferich, SPE J., Vol. 18 (1978), p. 418.

21. F. H. Spedding and J. E. Powell, in F. C. Nachod and J. E. Schubert, Ion Exchange Technology (Academic Press, New York, 1950), Chap. 15.

22. V. A. Davankov, Pure Appl. Chem., Vol. 54 (1982), p. 2159.

23. E. Glueckauf, in Ion Exchange and Its Applications (Soc. Chem. Ind., London, 1955), p. 34.

24. A. Skogseid (Norsk Hydro-Electrsk Kv.), Norw. Patent 72,583 (1947).

25. L. R. Morris (Dow Chemical Co.), US Patent 2,875,162 (1959).

26. F. Helfferich, Nature, Vol. 189 (1961), p. 1001.

27. F. Helfferich, J. Am. Chem. Soc., Vol. 84 (1962), pp. 3237 and 3242.

28. H. F. Walton, in J. A. Marinsky and Y. Marcus, Ion Exchange and Solvent Extraction (Marcel Dekker, New York), Vol. 4. (1973), Chap. 2.

29. J. Porath, J. Carlsson, I. Olsson, and G. Belfrage, Nature, Vol. 258 (1975), p. 598.

30. G. E. Janauer, R. E. Gibbons, Jr., and W. E. Bernier, in J. A. Marinsky and Y. Markus, Ion Exchange and Solvent Extraction (Marcel Dekker, New York), Vol. 9 (1985), Chap. 2.

31. I. G. Farbenindustrie, Ger. Patents 877,744 and 878,348 (1944).

32. M. Voloch, M. R. Ladisch, and G. T. Tsao, Reactive Polymers, Vol. 4 (1985), in press.

33. W. Neier, in D. Naden and M. Streat, Ion Exchange Technology (Ellis Horwood, Chichester, 1984), p. 360.

34. J. E. Germain, Catalytic Conversions of Hydrocarbons (Academic Press, New York, 1969).

35. R. Karonen, K. Poutanen, Y. Y. Linko, and P. Linko, Food Process Eng., Vol. 2 (1980), p. 123.

36. P. B. Weisz and V. J. Frilette, J. Phys. Chem., Vol. 64 (1960), p. 382.

37. M. B. Kril, G. E. Janauer, G. Wilber, and B. Kresge, in D. Naden and M. Streat, Ion Exchange Technology (Ellis Horwood, London, 1984), p. 407.

38. J. M. J. Fréchet, Tetrahedron, Vol. 37 (1981), p. 663.

39. R. B. Merrifield, Biochem., Vol. 3 (1964), p. 1385.

PART II
CHEMISTRY OF ION EXCHANGE RESINS

COORDINATION CHEMISTRY OF SELECTIVE--ION EXCHANGE RESINS

Michael J. Hudson

Chemistry Department, University of Reading,
Whiteknights, P.O. Box 224, Reading, Berks RG6 2AD, UK.

INTRODUCTION

The principal potential for coordinating copolymers in hydrometallurgy lies in the fact that they should have a greater capacity
and selectivity for metals than current ion-exchangers [1-4]. Consequently, coordinating copolymers have potential in the removal
of precious metals from base metals [3]; treatment of waste waters
[5] treatment of effluent from nuclear installations [6] and novel
separations such as nickel from cobalt solutions [7]. In addition,
coordinating copolymers may act as catalyst supports; as membranes
and as models for biological systems [8].

Coordinating copolymers are copolymers with covalently bound
side chains which contain a donor atom that is able to form a
coordinate bond to a metal. Coordination is frequently accompanied by ion-exchange and these copolymers are known as specific
and selective ion-exchange resins [9]. One of the first of such
resins to be prepared was the dipicrylamine resin which is
specific for the potassium ion [10]. There have been a number
of reviews concerning special applications and features of these
resins [11-19] and many new areas are being studied [20]. However,
these reagents have not fulfilled their potential and part of the
reason for this is due [21] to a lack of appreciation of the
fundamental coordination chemistry of these resins. Consequently,
this article will deal with some aspects of the coordination
of metals to selective ion-exchangers and will also emphasize
the use of ion-selective electrodes, electron spin resonance and
spectra in a study of fundamental coordination chemistry. The
principles of coordination chemistry itself which are used in the
ensuing article are to be found in text books [22,23]. Ion-selec-

tive electrodes [24] and electron spin resonance [5] particularly of copper [26,27] have been discussed.

SOME COMMERCIAL SELECTIVE ION-EXCHANGE RESINS

A list of some of the commercial selective ion-exchange resins is given in table 2.1. The donor atoms which are listed in the third column are part of the functional (ligand) group in the first column. In the case where there is more than one donor atom it is possible that the donor atoms may not all be in use. The list is not exclusive and new resins are coming onto the market but it does illustrate the range of substances which are available. The iminodiacetate resin is produced by most companies. Copolymers of the type 6 (dithiocarbamates based on polyetheneimine) will be used to illustrate the coordination chemistry and the use of physico-chemical methods. There appears to be no commercial resin with donor N,S atoms nor is there a commercial resin with amine groups adjacent to an aromatic ring.

CLASSIFICATION OF SELECTIVE ION-EXCHANGE RESINS

It is convenient to classify the selective ion-exchange resins according to the donor atoms which are present in the pendant ligand. One such classification is that in figure 2.1. An important feature of this is the distinction which is made between the number of donor atoms present in each pendant ligand. The capacity and selectivity of each copolymer are related to whether the ligand group is mono-, bi- or polydentate. The numbers in figure 2.1 which appear below the donor atoms (S, N or O) refer to the listings of the commercial copolymers (table 2.1).

CAPACITY

The distribution coefficient (D) is a useful measure of capacity

$$D = \frac{\text{amount of metal(ion) per g. of dry resin}}{\text{amount of metal per cm}^3 \text{ of solution}} \qquad (1)$$

$$= [\bar{A}]/[A] \quad \text{for dilute solutions.} \qquad (2)$$

Capacity, like selectivity, is a function of resin cross-linking, ionic strength of solution, temperature, ions in solution, functional groups on the resin and the aqueous chemistry of the metal ion itself. For a selective ion-exchange resin the following usually lead to a high capacity.

1. Little or no cross-linking.

TABLE 2-1

Functional Group	Nature of Group	Donor Atoms	Trade Name	Company
1. -SH	Aromatic Thiol	S	IMAC TMR	Rohm and Haas
			ES 465	Duolite (RH)
$-OCH_2CH \cdot CH_2SH$ OH	Aliphatic Thiol	S	Spheron 1000	Lachema
2. $-CH_2N \big\langle \begin{smallmatrix} CH_2CO_2H \\ CH_2CO_2H \end{smallmatrix}$	Iminodiacetate	NO_2	IMAC SYN 101	
			ES 466	Duolite (RH)
			DOWEX A-1	Dow
			Chelex-100	Biorad
			Diaion CR20	Mitsubishi
			Ligandex 1	Reanal
			IRC-718 (XE-318)	Rohm and Haas
			Wofatit MC50	VEB
			TP 207	Bayer
			UR 10,20,30, 40,50	Unitika

	Structure	Type	Designation	Company
3.	$CH_2N-CH_2-P=O$ with OH, OH	Amino phosphonic acid	NO_2	ES467	Duolite(RH)
4.	(pyridine ring) CH_2N-CH_2 / CH_2 / $HC-OH$ / CH_3	Weak base	N_2	XF-43084 (XFS 43084)	Dow
5.	(pyridine ring) R / $-CH_2N-CH_2$	Weak base	N_2	XF-4195(6) (XFS 4195)	Dow
6.	$-NCS_2H$	Dithiocarbamic acid	NS_2	Misso ALM 525 Sumichelate Q-10R	Nippon Soda

7. $-CH_2NH-(C_2H_4NH)_n H$	Basic (polyamine)	N	CR–20	Mitsubishi
8. $-CH_2-(N-CH_2CH_2)_n-$ with H	imine/amine	N	CR–40	Mitsubishi BDH 50% aqueous
9. $-C_6H_5N$	weak base	N	CR–2	Sumitomo, Croda
10. $-S-C(=NH)-NH_2$	isothiouronium	N_2	Srafion NMRR Ionac SR–3	Ayalon Ionac
11. Cryptand 221 B	Cryptand	NO	Kryptofix 221B	Parish
12. 222B	Cryptand	NO	Kryptofix 222B	Parish

FIGURE 2.1

Coordinating Copolymers – Donor Atoms in one Ligand

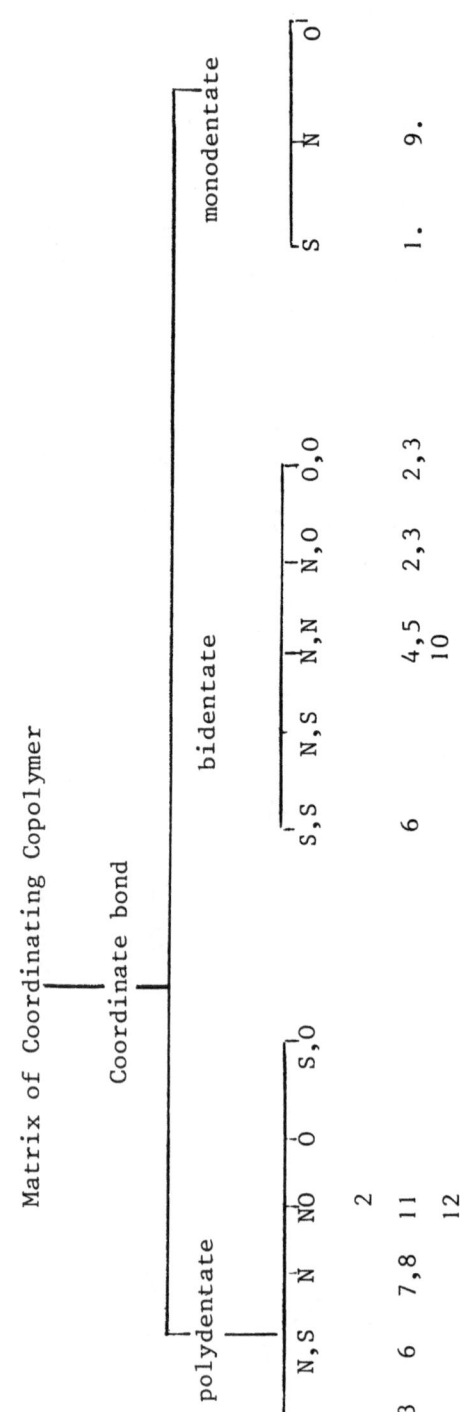

2. Flexible cross-links (e.g. alkyl) rather than rigid (e.g. aryl).

3. Maximum number of active functional groups on the copolymer.

4. Maximum number of active donor atoms per active functional group e.g. chelating rather than monodentate.

5. High metal to ligand affinity e.g. fast kinetics.

6. Strong metal-ligand bonds - slow rate of bond breaking.

7. Low coordination number of metal - particularly when extraction is required by highly cross-linked copolymers or one with rigid cross-links.

Selectivity

For two metals A and B which are in equilibrium with one selective ion-exchanger [28,29]

$$aB^b + bA^a \rightleftharpoons a\bar{B}^b + b\bar{A}^a \qquad (3)$$

$$K_{B/A} = \frac{[\bar{A}]^b[\bar{B}]^a}{[A]^b[B]^a} \qquad (4)$$

$[\bar{A}], [A]$ are activities (concentrations) in resin and solution. a, b are charges on the ions.

$K_{B/A}$ = selectivity coefficient [30]

For both standard ion-exchangers and selective ion-exchangers, selectivity depends on the nature of the copolymer (cross-linking, microporous, macroporous etc.) Several rules of thumb have been established [14] for those properties of the metal which influence the selectivity of standard cation-exchangers.

1. Selectivity increases [31] with the charge on the cation (e.g. for aquo-ions Th(IV)>La(III)>Ca(II)>Na(I).

2. Selectivity increases with decreasing *hydrated* radius of the cation e.g. approx. log $K_{M/H}$(Cs 0.8; Rb 0.7; K 0.55; Na 0.3, H, 0.0, Li - 0.05) respective ionic radii are Cs 1.5; Rb 1.55; K. 1.65; Na 2.2; H 2.45; Li 2.5Å).

3. Selectivity consequently increases with increasing polarising power (related to effective nuclear charge (radius) so that an ion with a high charge and a small radius (Al^{3+}) has a high polarising power.

For selective ion-exchange resins the charge and the hydrated radius play a role. In addition, however, a wide range of metal-ligand affinities are important.

INTRODUCTION TO SOME COORDINATION CHEMISTRY

The interaction between metals and selective ion-exchangers depends upon properties of the metal, the solution and the copolymer itself. With respect to the metal, it is necessary to be quite certain about the chemical species involved; the softness or hardness (A,B character) associated with the oxidation state in question; position in the Irving-Williams series; thermodynamic stability of complexes formed - formation constants of low and high molecular weight species; Ligand Field Stabilisation Energies; kinetics of metal complexes - rates of bond making and breaking; required stereochemistry of the metal; mechanisms of displacement reactions of metal-bound aquo groups. Capacity and selectivity are a function of these properties which are concerned with metal-ligand affinities.

METAL-LIGAND AFFINITIES

A. Lewis Acidity and Basicity

The thermodynamic data on metal-ligand affinities has been concerned with the stepwise displacement of coordinated water. The equations for these processes are

$$M(H_2O)_n + L \rightleftharpoons M(H_2O)_{n-1}L + H_2O \qquad (5)$$

$$M(H_2O)_{n-1}L + L \rightleftharpoons M(H_2O)_{n-2}L_2 + H_2O \qquad (6)$$

In practice the number of water molecules is frequently omitted. If the successive equilibrium constants are K_1 K_2 etc and overall stability constant is K. Then

$$K = K_1 . K_2 ..... K_n = \frac{ML_n}{[M][L]^n} \qquad (7)$$

For stable complexes K may be as high as 10^{40} [32] but for unstable complexes K may be less than 1. These thermodynamic stabilities do not refer to kinetics or to decomposition by hydrolysis, oxidation or thermal degradation. On the basis of a wide range of studies [32] the Lewis acidity of the first row dipositive ions is

$$Mn < Fe < Co < Ni < Cu > Zn \qquad (8)$$

This is the Irving-Williams series for kinetically labile divalent cations which can be explained by Ligand Field Theory [22,23].

B. Electronic Aspects

One generalisation concerning the relative stabilities of complexes is that there are hard or soft acids and bases [33] such that hard acids prefer to bond to hard bases and soft acids to soft bases. Hard acids and bases are usually small ions and not easily polarised. Soft acids and bases are generally large and easily polarised and form covalent bonds.

Hard Acids Na^+ H^+ Mg^{2+} Ti^{4+} Fe^{3+} Al^{3+} CO_2 SO_2

Soft Acids $Ag(1)$ $Pd(II)$ $Hg(I)$ Br_2 1,3,5-trinitrobenzene

Hard Bases NH_3 H_2O AcO^- F^-

Soft Bases C_2H_4, CO, R_3P, R_2S, I^-.

Most metallic soft acids are able to form π bonds with S and P. For a given metal, the lower oxidation states are 'softer' than the higher oxidation states (i.e. $Cu(I) > Cu^{2+}$; $Fe(II) > Fe^{3+}$; $Au(I) > Au(III)$.)

The platinum group metals (Ru,Os,Rh,Ir,Pd,Pt) are associated with 'soft' character as are the toxic metals (Hg(I), Hg(II), Cd(II)).

C. Entropy Changes

When a copolymer reacts with a metal there is bound to be a change in entropy. The change in Gibb's Function (free energy) is given by

$$\Delta G = \Delta H - T\Delta S \qquad (9)$$

For a favoured process, the $T\Delta S$ term should be large and positive. This may be viewed qualitatively as requiring an increase in disorder. For example, the formation constant for $[Cd(en)_2]^{2+}$ (en=ethenediamine) is 10^4 times greater than that of $[Cd(NH_2CH_3)_4]^{2+}$. The ΔH terms are comparable so that the $T\Delta S$ term contributes to the greater stability. The entropy change associated with the formation of a chelate is significantly more positive than for monodentates. The larger entropy increase associated with chelate formation is due to an increase in the total number of species in the system (2 molecules of en release four water molecules.) For this reason, if a scavenging copolymer is required, one with bidentate groups should be considered. For the formation of a metal-copolymer complex, however, the production of water molecules has to be balanced by the loss of free rotation in the polymer matrix and the pendant ligand group which is held in a restricted conformation. Nevertheless, the chelate effect does seem to be capable of extension to some copolymers. Whereas $K_1 > K_2$ etc for low molecular weight species, the reverse is true for copper with uncrosslinked poly(4-vinylpyridine) (successive log K values are ca. 1,2,3,4.5) [16]. Once the first metal-copolymer bond is formed subsequent coordination rapidly follows so that a 'polymer effect' may be regarded, in certain cases as an extension of the chelate effect.

STEREOCHEMISTRY

The preferred co-ordination number of metals in the first transition series may be related to the number of non-bonding d-electrons. In a selective ion-exchanger the metal tries to achieve the stereochemistries outlined in Table 2.2. This is counterbalanced by strain which is introduced into the copolymer which may produce novel stereochemistries. Later it will be shown that selectivity can be achieved when metals have different requirements for stereochemistry of the metals and steric hindrance and strain in the copolymer.

COVALENT BINDING

The ligands attached to metals may be placed in a series [34,35] which approximates to an order of decreasing covalent character in the metal ligand bond.

$$Br^-, \quad CN^-, \quad Cl^-, \quad SCN^-, \text{ oxalate, ethenediamine,}$$
$$NH_3, \text{ urea, } H_2O, F. \tag{10}$$

For different metal ions with a common set of ligands the order is

$$\tag{11}$$
$$Ir(III), \quad Rh(III), \quad Co(III), \quad Fe(III), \quad Cr(III), \quad Ni(II), \quad Mn(II)$$

This implies that the Ir-Cl bond is more covalent than the Mn-Cl bond. In selective ion-exchange copolymers the degrees of covalency of bonds may be related to rates of ligand exchange as a covalent bond, particularly strong ones, undergo slow rates of exchange. If rapid elution of a metal is required then strong covalent bonds may inhibit the rate of release of the metal by a copolymer in a simple displacement reaction. These

TABLE 2.2.

Number of non-bonding d-electrons	Shape Expected from Ligand Field Theory

d^0 Regular octahedron (K_2TiCl_6)

Regular tetrahedron $TiCl_4$

d^1 Slightly distorted octahedron $[Ti(H_2O)_6]^{3+}$

d^2 Slightly distorted octahedron $V(H_2O)_6^{3+}$

d^3 Regular octahedron $[Cr(NH_3)_6]^{3+}$

	Spin Free	Spin-paired
d^4	Square planar $CrCl_2.4H_2O$ distorted octahedron $CrSO_4.7H_2O$	Regular tetrahedron $[ReCl_4]^-$ Distorted octahedron $[Cr(CN)_6]^{3-}$
d^5	Regular octahedron $[FeF_6]^{3-}$ Regular tetrahedron $[FeCl_4]^-$	Regular octahedron $[Cr(NH_3)_6]^{3+}$
d^6	S. distorted octahedron $[Fe(H_2O)_6]^{2+}$	Regular octahedron $[Co(NH_3)_6]^{3+}$
d^7	S. distorted octahedron $[Co(H_2O)_6]^{2+}$ regular tetrahedral $[CoCl_4]^{2-}$	octahedron square planar square pyramid
d^8	regular octahedron $[Ni(H_2O)_6]^{2+}$	octahedron square planar square pyramid

---

d^9 Distorted (Jahn-Teller) octahedron: square planar

d^{10} Regular octahedron $[Zn(NH_3)_6]^{2+}$

      Regular tetrahedron $[Zn(acac)_2]^{2+}$

---

and related points will be illustrated in the subsequent sections.

FAJANS-TSUCHIDA SPECTROCHEMICAL SERIES

Analysis of a large number of spectra have indicated that ligands can be arranged according to the Crystal-Field and Ligand-Field Splitting (Δ) [22,23]. These are arranged in increasing splitting I^-, Br^-, Cl^-, NO_3^-, F^-, OH^-, $C_2O_4^{2-}$, H_2O, NH_3 pyridine, dipyridyl, NO_2^-, CN^-. An increase in the oxidation state from +2 to +3 doubles the value of Δ. A ligand high in the spectrochemical series may displace one which is lower. This displacement, and the change in oxidation state, is relatable to the elution of metals from the metal-copolymer compound as a metal may be displaced by a strong ligand especially after reduction with formaldehyde or sulphur dioxide.

LIGAND FIELD STABILISATION ENERGIES (LFSE)

This refers to the transition metals-particularly the first row elements. The splitting of orbitals as complexes are formed causes some orbitals to become more stable and other less stable than sperically symmetrical d^0, (d^5), d^{10} configurations. Weak ligands such as water can give spin-free complexes whereas strong ligands (see Table 2.2) such as CN^- give spin-paired complexes. When electrons are in the lower energy orbitals they contribute to the thermodynamic stability. An <u>approximate</u> comparison is given between high spin and low spin complexes in Table 2.3. For the configuration d^4 [Cr(II)] to d^7 [CoII)] the low-spin complexes are more stable. This implies that a metal in a spin-free complex on a copolymer may be eluted by a stronger ligand which forms a spin-paired complex.

The differences in LFSE partially explain why selective ion-exchangers react with one metal rather than another. Copper(II) (d^9) with an LFSE of $3/5\Delta$ is extracted by oximes in preference to zinc (no LFSE) as resultant metal-copolymer compound has a higher formation constant. The Irving-Williams order of stability [37] (see (8) can be explained by Ligand Field Theory [22,23] as can the mechanisms of substitution.

ELUTION

In cases where a metal needs to be eluted from a metal-copolymer complex, there are a number of general ideas which may be successful.

a) *Change of oxidation state of metal*-low oxidation states are

TABLE 2.3

Ligand (Crystal) Field Stabilisation Energies

Electronic Configuration	Spin-Free	Spin-Paired	Difference
0	0	0	
1	2/5	2/5	
2	4/5	4/5	
3	6/5	6/5	
4	3/5	8/5	5/5
5 Mn(II)	0	10/5	10/5
6 Fe(II)	2/5	12/5	10/5
7 Co(II)	4/5	9/5	5/5
8 Ni(II)	6/5	6/5	
d^9 Cu(II)	3/5	3/5	
d^{10} Zn(II)	0	0	

Values are approximate LFSE/Δ_o (o = octahedral)

for octahedral complexes.

softer than higher ones e.g. Elution of metal (Cd(II), Ag(I), from a thiol with oxidising agent followed by reduction of resin [38].

b) *Using a stronger ligand* stronger than that on the copolymer e.g. elution of metals by SCN^- from an N-oxide resin [39].

c) *Change of temperature* cobalt complexes may be tetrahedral at higher temperatures. Since the LFSE for tetrahedral complexes is lower than for octahedral elution is possible at a higher temperature.

INTERACTION BETWEEN METAL IONS AND SELECTIVE ION-EXCHANGERS

Introduction

The interaction between metal ions and selective ion-exchangers
is of interest in the fields of hydrometallurgy, membrane science
as well as bioinorganic chemistry [40]. Some of the types of
studies which can be made are illustrated using a polyfunctional
copolymer (PIED) which has bidentate (S,S) dithiocarbamate groups,
as well as monodentate primary, secondary and tertiary amine
groups. The techniques outlined here can be extended to other
metals and copolymers.

Dithiocarbamates have been used in analytical chemistry as
precipitating agents [41]. The dithiocarbamate group is a strong
ligand and there is an extensive chemistry of metal-dithiocarbamate
complexes [42]. The first example of a nickel(IV) compound on a
copolymer [3] has been reported using the copolymer PIED which has
been used to extract Rh(III) [2]; Cd(II) [5] and ^{106}Ru [6] and
these aspects will be considered in the section on hydrometallurgy.
Metals bound to dithiocarbamates can be placed in order of 'strength' of complex formation (Mn, weakest). It has been shown that

$$Mn, \ Zn, \ Fe(III), \ Cd, \ Pb, \ Cu, \ Ag \qquad\qquad (12)$$

it is possible for one metal which is higher in the series to
displace another which is lower in the series. Thus cadmium
may displace zinc and silver displace lead. The extension of
these ideas to the use of dithiocarbamates in copolymers has meant
that copolymers such as PIED can achieve some of the separations
which can be achieved with low molecular weight dithiocarbamates.
The poly(iminoethene) from which PIED was prepared was soluble
in water, had an average molecular weight of 40000 and is a
highly branched copolymer with an average ratio of primary,
secondary and tertiary groups of 1:2:1 [3]. An idealised representation of the repeating unit of the copolymer PIED is given
below. The PIED copolymer is zwitterionic above a pH of 4.5 but
below this the anionic acid group is protonated and the copolymer
becomes unstable [3]. Interestingly, the copolymer becomes more
stable [2] in the very strong hydrochloric acid solutions which
are required for precious metal refining. The dithiocarbamate
groups are attached to the primary amine groups [5]. Consequently
within the copolymer PIED there are residual primary amine groups,
secondary and tertiary amines as well as the dithiocarbamate
groups. The copolymer is soluble in basic solutions (pH>9.5) but
is regarded as insoluble at lower pH. The copolymer is uncross-
linked and the side chains confer another degreee of flexibility.

Conditional Stability Constants

The heterogeneous nature of the macromolecule; the random concentration of ligand groups; the effects of cross-linking and
strain in a copolymer as it binds to metals (and so on) means
that a simple extrapolation of the behaviour of low molecular

weight species to macromolecules is impossible without making
certain assumptions. With respect to the interactions of toxic
metals such as cadmium with the copolymer it is possible to
establish the value of conditional stability constants. These
are constants at a defined ionic strength. The following
assumptions are made [5]

a) The dithiocarbamato groups (and the amine groups) react with
cadmium without interacting with adjacent groups.

b) All the interactions between cadmium and the active
(coordinating) groups on the copolymer only involve complex
formation.

c) The functional groups on the copolymer react in the same
manner, regardless of their position.

The combination of a metal L with a copolymer can be considered
to be a series of stepwise reactions with the functional
groups [43,44]. This stepwise addition of the metal L can be
represented by the equilibrium below.

$$SL_{i-1} + L \rightleftharpoons SL_i \tag{13}$$

The conditional formation constant K_i which depends on ionic
strength and pH, is given by

$$K_i = \frac{[SL_i]}{[SL_{i-1}][L]} \tag{14}$$

where [ ] denotes activity (concentration). For a copolymer
it is necessary to consider a complex formation function $(\bar{n})$,
which is defined as the average number of metal atoms bound to
a functional group in the repeating unit.

$$\bar{n} = \sum_{i=1}^{n} \frac{n[SL_i]}{[SL_i]} \tag{15}$$

n = maximum number of binding sites.
n = can also be expressed [2] in an alternative way

$$\bar{n} = \frac{nK[L]}{1+K[L]} \tag{16}$$

where K = conditional formation constant of each particular
functional group if present alone with metal L. Equation (16)
can be rearranged as

$$\frac{\bar{n}}{[\bar{L}]} = nK - \bar{n}K \qquad (17)$$

In a plot of $\bar{n}/[L]$ against $\bar{n}$, the experimental data should follow
a straight line from which the conditional constant K and the
maximum number of bonding sites n can be calculated.

Since the macromolecule PIED. has more than one type of bonding
sites in the repeating unit because there are both the dithio-
cabamate and the amino groups, it is necessary to sum equation
(16) for each site

$$\bar{n} = \sum_{j=1}^{n} \frac{n_J K_j [L]}{1 + K_j [L]} \qquad (18)$$

in which n_j and K_j refers to the maximum of binding sites and to
the conditional constants, respectively, for the jth class of
binding sites. Assuming that the macromolecule has two different
types of metal bindings

$$\bar{n} = \frac{[L]_T - [L]_f}{[X]} = \frac{n_1 K_1 [L]_f}{1 + K_1 [L]_f} + \frac{n_2 K_2 [L]_f}{1 + K_2 [L]_f} \qquad (19)$$

in which $[L]_T$ = molar concentration of total metal added, $[L]_f$
= molar concentration of free metal ion after equilibriation
$[X]$ = molar concentration of dithiocarbamato groups on the
macromolecule.

Dividing by $[L]_f$ we obtain

$$\frac{\bar{n}}{[L]_f} = \frac{n_1 K_1}{1 + K_1 [L]_f} + \frac{n_2 K_2}{1 + K_2 [L]_f} \qquad (20)$$

There are quite separate sections for each active site in the
copolymer. In the Scatchard method $\bar{n}/[L]$ is plotted against
$\bar{n}$ and in favourable cases there are distinct regions in which
the magnitude of the conditional formation constant; the mole
fraction of each site occupied and a distinction between
chemisorption (bond making and breaking) and adsorption can be
made. In unfavourable cases, such as when the sites have equal
affinity for the metal, regions cannot be distinguished.

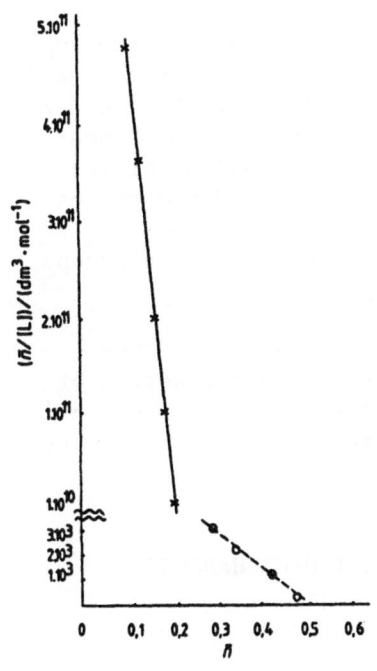

FIGURE 2.2.

Scatchard plot for the binding of Cd(II) at pH 5.5 and 0.1 M NaNO$_3$. [-NHCS$_2$$^-$] = 3.49. 10^{-3} mol dm^{-3}; $\bar{n}$ is the complex formation function.

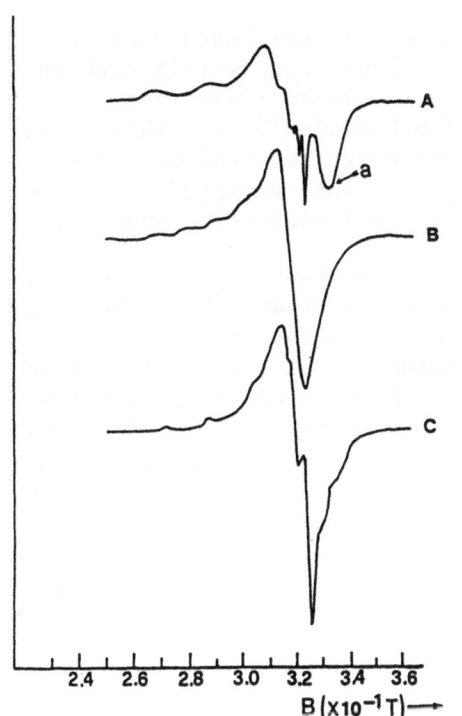

FIGURE 2.3

X-band
esr spectra of Cu-copolymers

A. 2-4% Cu IR-45CS$_2$H$_x$

B. 15% Cu-PIED.

C. 1-2% Cu-PIED.

All spectra recorded at 77 K.

When the conditional formation constants on different sites are roughly in the same order of magnitude a curve is obtained in which sites cannot be distinguished. With respect to the copolymer, PIED, there is a favourable case as one group is strong (S,S) chelating agent and the other is monodentate (N). In figure 2.2 there are two clear regions in which the analysis of the linear components allows the calculation of the conditional stability constants (K, is ca. 10^{11} and K_2 ca. 10^4 to 10^5). Moreover the mole fraction of sites is ca. 0.5 for both sites. The first formation constant is associated with the dithiocarbamate groups and the second with the N-containing groups.

Since cadmium has no absorption spectrum and no electron spin resonance signal it was decided to study the extraction of copper. Nevertheless the Scatchard Method is capable of extension to other metals and copolymers and promises to provide much useful information.

DETERMINATION OF ACTIVE SITES ON SELECTIVE ION-EXCHANGERS

Some Spectroscopic Studies

It is important to determine the sites at which metals are bound as this enables the selectivity and capacity to be explained for existing copolymers and aids the design of new copolymers.

With respect to the copolymer PIED, it was hoped that the copolymer would be selective for platinum group metals such as Rh(III) over base metals like Cu(II). In the event, the showed only small selectivity (Rh/Cu 8 at 4 m HCl). This fact, coupled with the observation that both metals could be eluted by dilute HCl from the high-loaded copolymer prompted a spectroscopic study (electron spin resonance and electronic spectra).

The basic problem was to determine whether copper (rhodium) is bound to S and/or to N atoms and to ascertain the numbers of each ligand atom which are in the coordination sphere of the metal. With respect to copper, the esr parameters have been related to the geometries of the coordination sphere of the metal for low molecular weight species [45,47] of known structure. It has been shown, for example, that $Cu(edtc)_2$ (edtc = diethyldithiocarbamate) has a CuS_4+S square-based pyramidal geometry. The esr parameters for this compound [48] are $g_{//}$ = 2.126, $g_\perp$ = 2,027 and $A_{//}$ = 147.5 G and similar parameters in other compounds are considered to be diagnostic of this type of stereochemistry [8]. Different stereochemistries and different ligand atoms give rise to different values and combinations of three parameters. When, for example, Cu(II) is doped into Ni(edtc)$_2$ in which the nickel is square-planar, the esr parameters for copper change to values which are attributable to a square-planar Cu(II) stereochemistry. Similarly the parameters for Cu(II), which is doped into Zn(edtc)$_2$ are ascribed to a distorted trigonal bipyramidal geometry associated with Zn(II). Some of the combinations of esr, and electronic spectral data which are thought to be associated with particular stereochemistries are discussed shortly. When copper is bound to a S-atom, the g-values are normally low. This is attributable π bonding effects [49] and so may also arise with phosphorus or aromatic nitrogens as donor atoms such as in poly(4-vinyl-pyridine).

Some esr spectra are given in Figure 2.3. For each of these spectra which were taken at liquid nitrogen temperature, the selective ion-exchanger was shaken with the copper solution. Two Cu(II) solutions were used. The first was copper(II) chloride acidified with one drop of HCl to prevent hydrolysis and the second was in 4 m HCl. For each solution a low-loaded sample was prepared by short contact time (10 secs) with a dilute solution (0.001 m) and a high-loaded sample with longer (2hr.) contact with a more concentrated solution (0.1 m). The samples were filtered, dried at 40-60° C (24hr) in vacuo, pulverised and dried for a further period (24hr). The samples were not hygroscopic after the second drying. The method allows active sites on the copolymer to be investigated. Different coordinations may be found in aqueous systems and further studies are in progress.

The esr spectra, which are illustrated in Figure 2.3, are interesting because they all give clear signals with well developed fine structure. This implies that the metal occupies one (or two) well-defined sites on the selective ion-exchanger. Coordination is not random. Each of the copolymers contained 2-dithiocarbamate ligands and there is no evidence for a [CuCl$_4$] ion in any of the metal-copolymer compounds. There was no evidence for lines at half-field strength [45] which implies that each Cu(II) is at a separate coordination site (there was also no evidence for a +1 to +2 oxidation state transition. One dithiocarbamate (IR45-CS$_2$H) was prepared by stirring IRA-45 with carbondisulphide in sodium hydroxide. The Cu(II) compound of this copolymer has parameters which are close to those for low molecular weight species ($g_{//}$ 2.135, $g_{\perp}$ 2.025 and $A_{//}$ 156). The diagnostic $g_{//}$ value is particularly close to that in Cu(II) doped zinc dithiocarbamate. This implies that the principal moiety is CuS$_4$+S. In the electronic spectrum there are bands at 16,000 cm^{-1} and 21,000 cm^{-1} (Figure 2.4) which supports this

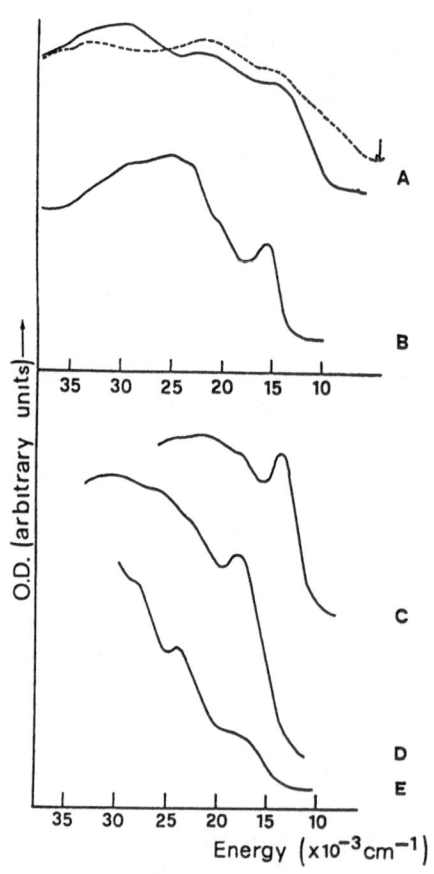

FIGURE 2.4

(After Ellis, Hudson and Tomlinson).

Electronic Reflectance Spectra

A.(.....) 15% Cu-PIED
   (-) 1-2% Cu-PIED
B. **ca.** 10% **Ni**-PIED
C. ca. 12% Co(III)-PIED
D. ca. 15% Rh(III)-PIED
Ec.ca. 15% Ir(III)-PIED

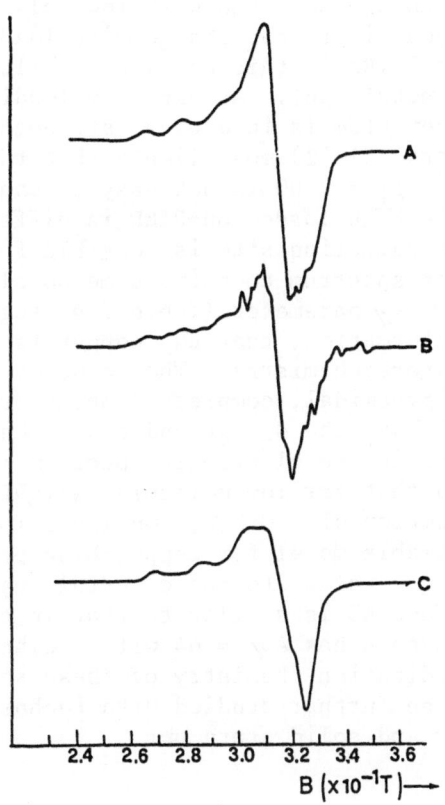

FIGURE 2.5

X-band esr spectra of Cu(II) - Rh(III) - PIED

coextracted

A. 5 mins.

B. 1 hour.

C. 24 hours.

assignment [50].

The Cu-PIED esr spectra are particularly interesting because they show features which are not seen with low molecular weight materials; how stereochemistry can change with loading; how the binding of one metal (Rh in this case) can influence the binding of the second metal (Cu). At very low-loading it is possible that the copper atom is in a novel stereochemical environment. It appears [51,52] most likely that the environment is CuN_2S_3 but the value $g_\perp = 2$ 00 is not easy to explain [8]. The esr spectrum of the high loaded Cu-PIED is different which means that a second coordination site is possible for the copper atom. However, the esr spectrum contains some novel features. In particular, the low $A_{//}$ parameter (Table 2.4) suggests, together with other information, that the copper is in a pseudo-tetrahedral (CuS_2N_2) stereochemistry. The value of $g_{//} > g_\perp > 2.02$ suggests that trigonalpyramidal, compressed octahedra and cisoctahedra [45], which all have the d_{z^2} ground state, can be discarded. The low-energy shoulder in the electronic spectrum (Figure 2.4) and a $g_{//}$ greater than that for low molecular weight (CuS_4+S) species supports the notion of a CuN_2S_2 configuration. This makes Cu-PIED a readily available model for copper-blue protein. It is possible to relate g values to the dihedral angle [52] and the value of 70° for Cu-PIED is similar to that in copper 'blue' proteins (e.g. plastocyanin has $A_{//} = 64$ with a dihedral angle of 80° [53]. The coordination chemistry of these selective ion-exchangers should be further studied with techniques such as Mossbauer Spectroscopy and solid state nmr .

AVAILABILITY OF ACTIVE SITES

Changes in Metal Ligand Atom Bond Strength

When a metal is bonded to the active sites in a copolymer the chains of the macromolecule move so that some, at least, of the stereochemical requirements of the metals can be accommodated. In certain cases, and particularly at high loadings, this results in a strain within the macromolecule such that the metal-ligand bonds are longer than in the low molecular weight species. Electronic spectra provide direct evidence of the reduction of the strength of a ligand-field which may result from this bond lengthening. This effect is particularly noticeable in MS_6 chromophores which require an octahedral arrangement of the ligands. The principal bands [54] in the same species are as follow ($x10^{-3}$ cm^{-1})

	d-d bands	Charge transfer and infra red ligand
Co(III)-PIED	14.82; 18.87	23.10; 37.04

Table 2.4 E.P.R. Spectra of Cu^{II}-loaded

Material	$g_{//}$	$g_{\perp}$	$A_{//}^{a}$	A^{a}
Cu^{II}-PIED (high-loaded)	2.236(3)	2.074(3)	104	-
" (low-loaded)	2.209(3)	2.065(3)	154(2)	8
Cu^{II}-IR45CS$_2^-$ (r.t.)	2.125(2)	2.025(2)	147(1)	15
" (77K)	2.135(2)	2.025(2)	156(1)	14
Cu^{II}-Rh^{III}-PIED (low-loaded)				
Cu^{II}-Rh^{III}-PIED				
high-loaded	2.206(2)	2.062	166(2)	-
after 1h loading	2.255	-	150	site a
	2.231	-	112	site b
	1.83	-	90	site c
after 5 min loading	2.262	-	140	

^{a} x 10^{-4} cm^{-1}

Rh(III)-PIED	19.40; 25.60	30.77; 39.2
Ir(III)-PIED	18.5; 23.80; 27.40	

Each of the bands lies in the range for all –S bonded species
[55]. However, they all appear at significantly lower frequenc-
ies than the corresponding low molecular weight compounds
[55,56]. They appear to be close to the values for dialkylphos-
phoro- and dialkylphosphine-dithioates which are known to give
weaker ligand-fields and longer M–S bond lengths [57,58]. This
appears to explain the ease of elution of the metals particularly
in the high-loaded samples. Moreover, the unusual stereochemistries mentioned above are, in effect, forced on the metal
and they also allow elution more readily than a low molecular
(unstrained) molecule. It appears to be a general rule
(particularly for metals of high coordination number) concerning

copolymers that they should be used well within their total capacity so that the metal is not too easily eluted. Conversely elution of the last traces of metal from a metal-copolymer might be difficult as the metal could be in an unstrained environment which has a maximum LFSE.

ACTIVE SITES DURING COMPETITIVE REACTIONS

In a competitive process in which two or more metals are being extracted it is likely that they will either compete for the same sites or the coordination of one will influence the sites which are available for another. Even if two metals are going for different sites (e.g. one for sulphur donor atoms and the other for nitrogen atoms) the coordination of one may restrict the freedom of the copolymer chains to rotate - particularly if the metal acts as a cross-linking agent between two polymer chains. As the copolymer becomes more loaded in certain cases it becomes more cross-linked and diffusion of the metal ions is severely restricted. Thus the particle size of a selective ion-exchanger is important. (See Hydrometallurgy section).

In a competitive reaction between Cu(II) and Rh(III) for sites on PIED the esr spectra were recorded for different loading times (5 min, 1 hr, 18 hr). The spectra are illustrated in Figure 2.5. After five minutes the esr spectrum is characteristic of a single Cu-PIED species which may be due to the faster up-take of copper than rhodium. Even so the esr parameters (Table 2.5) do not precisely coincide with a single high or low loaded Cu-PIED.

TABLE 2.5

Active Sites Available during Competitive loadings - esr signals of copper(II) on PIED when competing with rhodium.

Time	$g_{//}$	$g_{\perp}$	$A_{//}$	Comments
5 min.	2.262	-	140	like $g_{//}$ Cu-PIED high loaded
1 hour	2.255	-	150	like Cu-PIED (Low)
	2.231	-	112	like Cu-PIED (high)
	1.83	-	90	New
18 hour	2.206	2.062	166	like Cu-PIED (low)

A coordinated rhodium species may still be having some effect even at this early stage. After one hour there are sites which have parameters similar to the low and high loaded Cu-PIED species. Interestingly, however, there is still another new site which has

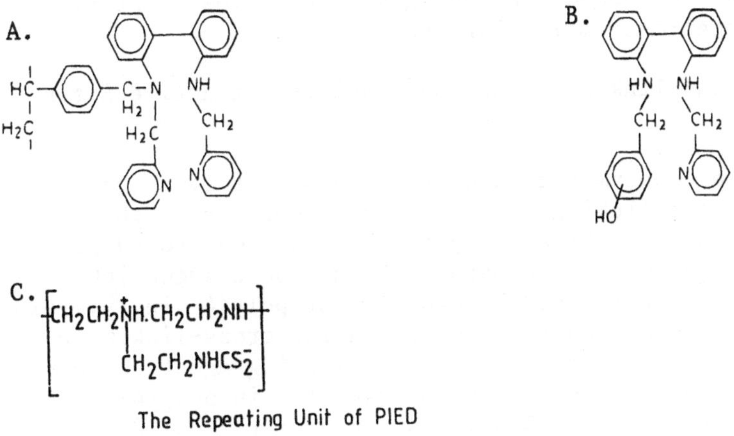

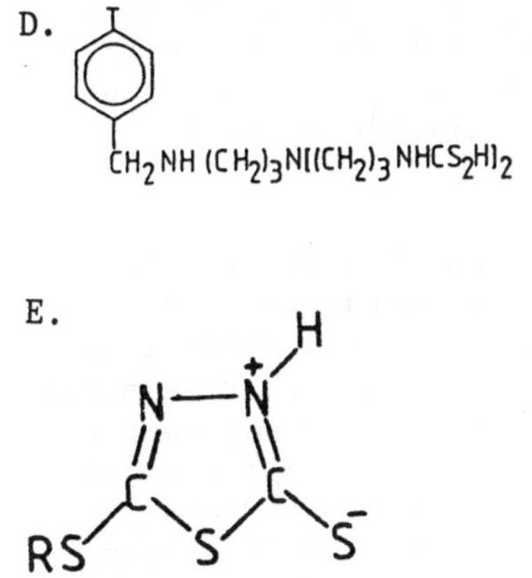

A.

B.

C.

$$\left[\begin{array}{c} CH_2CH_2\overset{+}{N}H.CH_2CH_2NH \\ CH_2CH_2NHCS_2^- \end{array}\right]$$

The Repeating Unit of PIED

D.

$$CH_2NH\,(CH_2)_3N[(CH_2)_3\,NHCS_2H]_2$$

FIGURE 2.6.

Some of the ligands mentioned in the text.

E.

unorthodox esr and electronic spectra. It appears that the copper may be forced into a compressed tetragonal or a trigonal bipyramidal geometry. The $g_{//} < g_{\perp} \ll 2.00$ is unusual as is the low $A_{//}$ value. The reflectance (electronic) spectra were superpositions of RhS$_6$ species and low-loaded Cu-PIED so there is no doubt that the rhodium is bonded to the copolymer. Further studies on competition for active sites is being undertaken.

Active sites can be put into a copolymer to enable a metal(Cu) which accepts four coordination to be separated from one, Fe(III), which requires 6-fold coordination. Thus the quadridentate ligand [59] (A in figure 2.6) has a high Cu(II) affinity but largely rejects Fe(III). Interestingly the related tridentate (B) ligand seems to have little affinity for either Cu(II) or Fe(III). Apparently, Fe(III) is unable to accommodate two of such

ligands which appear to sterically interfer with each other.
Further studies are needed concerning such examples of steric
hindrance in metal-macromolecular systems.

Influence of Cross-linking on the Availability of Active Sites (Capacity)

The ability of a copolymer to bind to metals is influenced by
the degree of cross-linking. Cross-linking influences the
rates of diffusion and the ability of the copolymer to adopt
conformations required by the metal. It has been shown [60]
that the rate of up-take and the capacity of poly(4-vinylpyridine
N-oxide) for metals decreases with increasing cross-linking by
divinylbenzene (DVB). For a given cross-linking agent, it has
been shown that the esr and electronic spectra in poly(4-vinyl-
pyridine) (PVP) for 4-12% DVB have esr parameters ($g_{//}$ 2.3,
$g_\perp$ 2.1 and $A_{//}$ 150±10) which are characteristic of a regular
(CuN_4) square-planar geometry. However, with higher cross-linking
[61] the esr parameters ($g_{//}$ 2.2, $g_\perp$ over 2.1 and $A_{//}$ 60-100)
approach those expected for a (distorted) tetrahedron. Active
sites with regular stereochemistry for a (square-planar) four-
coordinate complex do not appear to be available above 12%
cross-linking.

One of the criticisms which is levelled at selective ion-
exchangers is that they have slow rates of extraction. Part of
the reason may be associated with the need to find cross-linking
agents which are more flexible than divinylbenzene. This might
increase the rates of diffusion to the active site and confer
greater flexibility onto the copolymer. It is possible to use
[62,63] an experimental gel-type resin from Rohm and Haas
(XE 305) which has 2.4% cross-linking by long, flexible alkyl
chains. The functionaled XE 305 has acceptable rates and
capacities for base and precious metal extraction.

Another approach has been to use a longer side-chain to graft
the ligand onto the back-bone of the copolymer. Thus it has
been found [6] that longer side-chains (spacer-groups) allow the
metal to adopt a regular stereochemistry and the copolymer to
have a higher capacity for the metal. Thus the copolymer B in Fig-
gure 2.6 which has long oligo(ethene)oxide graft chains has a
regular square-planar copper $g_{//}$ = 2.242; $g_\perp$ = 2.070; $A_{//}$ = 150).
When n = 1, however, the copolymer has a low capacity for Cu(II)
and the available active site gives an irregular stereochemistry.

As further discussed in the hydrometallurgy section, uncross-
linked copolymers with dithiocarbamate groups have an acceptable
capacity for Rh(III) (0.6 mmoles Rh for PIED) (C in Figure 2.6)
and the rate of extraction is quite good ($t_{\frac{1}{2}}$ ca. 10 mins).

However, a copolymer such as IR-45 CS$_2$H (D in Figure 2.6) has little capacity for Rh(III) [8] or Ru [64] probably because the -NCS$_2$ groups are unable to achieve the conformation required by the metal.

It has been shown [65] that when uncross-linked poly(4-vinyl-pyridine) (PVP) binds to Cu(II) that the formation constants as successive aquo-ligands are displaced by donor N-atoms from PVP, increase. (K$_1$ ca. 10, K$_2$ = ca 100, K$_3$ ca 1000, K$_4$, ca 10,000) for low molecular weight species and copolymers with significant degrees of cross-linking that K$_1$> K$_2$>K$_3$> K$_4$ (For Cu-py aqueous K$_1$ ca 500 and K$_4$ 10). More studies are required to quantify the meaning of 'significant' in the last sentence. However, this observation for uncross-linked copolymers has been entitled "the polymer effect" and is considered to be an extension of the well known 'chelate effect' [23].

Since it was shown above [61] that distorted structures are obtained for copper extraction with cross-linking by DVB in PVP greater than 12%, it is possible that this percentage represents the highest degree of cross-linking for the 'polymer effect' to operate .

SOME ASPECTS OF KINETIC EFFECTS WITH SELECTIVE ION-EXCHANGERS

Much of the previous discussion has been concerned with thermo-dynamic effects. However, the selective ion-exchange have been crticised because of their slow kinetics (or slow rates of extraction). Several factors influence kinetics. These factors include the nature of the matrix e.g. gel, microporous, macro-porous or pellicular; the degree of cross-linking - percentag and rigidity; degree of functionalisation; hydrophilicity-nature of the ligand group; rates of ion-pair formation; rates of substitution of aquo-groups on the metal ion; mechanism of substitution; rates of elution; temperature; equipment design. The kinetics can be improved (see hydrometallurgy section) by decreasing the particle size [8]; decreasing the amount of rigid cross-linking [60]; increasing the degree of functionalisation; increasing the hydrophilicity of the resin e.g. cellulose matrix rather than a poly(styrene) one [66]; matching the affinity of the ligand for the metal; optimising the LFSE (Ligand Field Stabilisation Energy) of the ligand to promote a favourable mechanism [22 p.1185]. e.g. choose a strong ligand on the copolymer to displace the aquo-groups; increasing temperature - a 10° C rise in temperature may double the speed of displacement reactions - mobility of macromolecular chains increases; use equipment designed [67] for selective ion-exchangers.

There may be quite distinct differences in the rates of

extraction with a functional group according to whether the group is in a solvent extraction or a selective ion-exchanger. It was shown [1,68] that 1,3,4-thiadiazole-2-thiol-5-thioalkyl (E in Figure 2.6) may be used in solvent extraction (R = C_9H_{19}) or as a selective ion-exchanger (R = polymer). Both reagents are similar in that they have an affinity for Ag(I) ($t_{\frac{1}{2}}$ = ca. 2 mins). However, thiol groups have a higher affinity for Cu(I) than Cu(II) and, in order to extract copper, the thiol reduces Cu(II) whilst half of the thiol groups are oxidised to disulphides.

$$2RS^- \rightleftharpoons R-S-S-R + 2e \qquad (21)$$
$$2Cu^{2+} + 2e \rightleftharpoons 2Cu^+ \qquad (22)$$
$$2RS^- + 2Cu^+ \rightleftharpoons 2CuSR \qquad (23)$$

In order for reaction (21) to occur, two thiol groups must come together. This combination occurs rapidly with a solvent extraction reagent ($t_{\frac{1}{2}}$ for Cu is ca. 2 mins), but much more slowly in the copolymer. Consequently, the rate of extraction of copper is much slower ($t_{\frac{1}{2}}$ ca. 1 hour) (the capacity is half that for Ag(I) as half of the thiol groups are oxidised).

The rates of exchange of ligands depends upon the metal. Thus for a given type of complex, mechanism of substitution and oxidation state, the rate of substitution is Ni(II)>Pd(II)>Pt(II); Co(III)>Rh(III)>Ir(III). However, much depends on the oxidation state [69,70] of the metal and in strong chloride media the rates of substitution are Au(III)>Pd(II)>Pt(IV). Further examples will be quoted in the section on hydrometallurgy.

Metal complexes of divalent transition metal complexes with a high LFSE such as nickel may have slow rates of substitution of the aquo-groups. In principle, the rate of extraction of Mn(II) by a coordinating copolymer would be much greater than that of nickel. The slow rate of extraction of nickel by the pyridine-containing amine (5 in 2.1) may be largely due to the slow rate of substitution of nickel. However, metals with a high LFSE may have stable complexes which means that the overall thermodynamics may favour nickel extraction.

The nature of the ligand attached to the metal is important. For example, chloro-ligands attached to Rh(III) and Ir(III) are displaced at approximately the same rate [8]. However, a separation can be achieved. (D Ir/Rh. ca. 10) after five minutes with PIED because aquo groups attached to Ir(III) are more rapidly displaced than those attached to Rh(III) [8,70].

REFERENCES

1. Hudson M.J. and Shepherd M.J. 9 (1983) 223-234.

2. Hudson M.J. and Thorns J.F. Hydromet. 11 (1983) 289-295.

3. Giwa C.O. and Hudson M.J. Hydromet.8 (1982) 65-75.

4. Reichenberg D. in Marinsky J.A. (Ed.), Ion-Exchange, Vol.1. Dekker M. New York, NY 1966 p.227.

5. Hudson M.J. and Tiravanti G. Die Makromol. Chem. (1985) in press.

6. Hudson M.J. Leung B.K.O. Dyer A. and Keir D. Chem. Comm. 1984, 21, 1457-8.

7. Grinstead R.R. and Jones K.C. Chem. Ind. (1977) 637.

8. Ellis A.F. Hudson M.J. and Tomlinson A.A.G. Dalton (1985) in press.

9. Den Boef and Hulanicki A. Pure Appl. Chem., 55 (1983) 554.

10. Skogseid A. Norwegian Patent. 72.583 (1947) U.S. Patent, 2,592.350 (1952).

11. Millar J.R. Chem. Ind., (1957) 606.

12. Schmuckler G. Talanta, 4 (1963) 745.

13. Blasius E. and Brozio in Chelates in Analytical Chemistry, Vol.1, M. Dekker, New York, 1967, p.49.

14. Marcus Y. and Kertes A.S. Ion-Exchange and Solvent Extraction of Metal Complexes, Wiley Interscience, New York, 1969, 347.

15. Vernon F. Chem. Ind., (1977) 634.

16. Kaneko M. and Tsuchida E. Macromol. Rev., 16 (1981) 397.

17. Calmon C. Reactive Polymers 1 (1982) 3.

18. Lieser K.H. Pure Appl. Chem., 51 (1979) 1503.

19. Vernon F. Reactive Polymers 1 (1982) 51.

20. Warshawsky A. Selective Ion-Exchange Polymers, Angew. Makromol. Chem.

64

21.　Sahni K. and Reedijk J. Coord. Chem. Revs. 59 (1984) 1-139.

22.　Cotton F.A. and Wilkinson G. Advanced Inorganic Chemistry, Wiley, 1980, New York, USA.

23.　Greenwood N.N. and Earnshaw A.　Chemistry of the Elements, Pergamon, 1984, Oxford.

24.　Moody G.J. Thomas J.D.R.　Selective Ion-Sensitive Electrodes, Merrow, 1971, Durham, England.

25.　Banwell C.N. Fundamentals of Molecular Spectroscopy, McGraw-Hill, 1972, Maidenhead, England.

26.　Hathaway B.J. and Billing D.E. Coordin. Chem. Rev. 5 (1970) 143.

27.　Hathaway B.J. and Tomlinson A.A.G. Coordin. Chem. Rev. 5 (1970) 1.

28.　D'Alelio G.F. Hofman E.T. and Strazik W.F. J. Macromol. Sci. Chem. A6 (1976) 513.

29.　Luttrell G.H. (Jr.), More C. nd Kenser C.T. Anal. Chem. 43 (1971) 1370.

30.　Gregor H.P. Abolafia O.R. and Gottlieb M.H. J. Phys. Chem. 58 (1954) 984.

31.　Monk C.B. Electrolyte Dissociation, Academic Press, London 1961, p.270.

32.　Martell A. and Silten G.L. Stability Constants of Metal-Ion Complexes, Royal Institute of Chemistry, London 1964.

33.　Pearson R.G. J. Amer. Chem. Soc. 85 (1963) 3533.

34.　Jorgenson C.K. Acta Chem. Scand., 12 (1958) 1903.

35.　Jorgenson C.K. Absorption Spectra and Chemical Bonding in Complexes, Pergamon, 1962, New York.

36.　Fajans K. Naturwissenschaften 11 (1923) 165.

37.　Bjerrum J. and Jorgensen C.K. Rec. Trav. Chim. 75 (1956) 658.

38.　Hassan M.B. Ph.D. Thesis, Reading, 1986.

39.　Glaves L.R. Ph.D. Thesis, Reading, 1985.

40. Lonté R. Copper Proteins and Copper Enzymes, 1 CRC Press, Florida, 1984.

41. Delepine M. Bull. Soc. Chim. Fr. 3 (1908) 643.

42. Coucouvanis D. Progr. Inorg. Chem. 11 (197)) 234.

43. Mantoura R.F.C. and Riley J.P. Anal. Chim. Acta 78 (1975) 193.

44. Sposito G. Environ. Sci. Technol. 15 (1981) 396.

45. Hathaway B.J. and Billing D.E. Coordin. Chem. Rev. (1970) 143.

46. Hathaway B.J. and Tomlinson A.A.G. Coordin. Chem. Rev. 5 (1970) 1.

47. Hathaway B.J. Coordin. Chem. Rev. 41 (1982) 432.

48. Gregson A.K. and Mitra S. J. Chem. Phys. 49 (1968) 3696.

49. Palmer R.A. Tennant W.C. Dix M.F. and Rae A.D. J. Chem. Soc. (Dalton) (1976) 2345.

50. Brown D.B. Hall J.W. Scott M.F. and Hatfield W.E. Inorg. Chem. 16 (1977) 1813.

51. Addison A.W. and Rao T.N. and Sinn E. Inorg. Chem. 23 (1984) 1961.

52. Addison A.W. and Yokoi T. Inorg. Chem. 16 (1977) 1341.

53. Bereman R.D. Rao T.N. Reedjik J. van Rijn J. and Verschroor, J. Chem. Soc. (Dalton) (1984) 1349.

54. Thorns J.T. M.Sc. Dissertation, Reading, 1982.

55. Coucouvanis D. Progr. Inorg. Chem. 26 (1979) 236.

56. Tomlinson A.A.G. J. Chem. Soc. A (1971) 1409.

57. Lebedda J.D. and Palmer R.A. Inorg. Chem. 10 (1971) 2704.

58. Cavell R.G. Byers W. and Day E.D. Inorg. Chem. 10 (1971) 2710.

59. Melby L.R. J. Amer. Chem. Soc. 97 (1975) 4044.

60. Glaves L.R. Ph.D. Thesis, University of Reading, 1985.

61. Nishide H. Shimidzu N. and Tsuchida E. J. Appl. Polym. Sci. 27 (1982) 4161.

62. Shepherd M.J. Ph.D. Thesis, Reading, 1981.

63. Leung B.K.O. Ph.D. Thesis, Reading, 1985.

64. Sharp C. M.Sc. Dissertation, Reading/Salford, 1985 (details subject to confidentiality agreements).

65. Nishikawa H. and Tsuchida E. J. Phys. Chem. 79 (1974) 2072.

66. Hdradl J. Pevska J. Ilavsky M. and Stamberg J. J. Chromatography 125 (1976) 455. (Details from Inst. Macromol. Chem. Prague).

67. Brown C.J. "New Short Bed Ion-Exchange Technology for Metals Recovery", Paper, Conf. Soc. Mining Engineers (A.I.M.E.), Denver, Oct. 1984.

68. Hudson M.J. and Shepherd M.J. Hydromet. 1985 (in press).

69. Cleare M.J. Charlesworth P. and Bryson D.J. J. Chem. Tech. Biotechnol. 29 (1979) 210.

70. Griffith W.P. The Chemistry of the Rarer Platinum Metals, Wiley, New York, 1967.

MODERN RESEARCH IN ION EXCHANGE

Abraham Warshawsky

Department of Organic Chemistry
The Weizmann Institute of Science
Rehovot 76100, Israel.

INTRODUCTION

Ion exchange is now a very well established field. New developments in this area are sometimes viewed by the value to what
already exists, and not always for their own merit. This state of
affairs, typical of many mature areas of research, tends to reduce
enthusiasm for continuation and further exploration for new dis-
coveries, or for establishing better understanding of existing
knowledge. The purpose of this series of lectures is to consoli-
date the present advances in one area of modern research in ion
exchange, namely, selective ion exchange, pointing out, through
real examples taken from our own work as well as those of others,
to reasons for following certain directions, and trying to indicate the needs and challenges awaiting us. This course would,
therefore, be better defined as modern research in selective ion
exchange.

1. APPROACHES TO ION SEPARATION PROBLEMS.

A comprehensive approach to a particular separation problem should
consider the following: (a) Chemistry of a given ion under special
conditions, including understanding of the interaction of the reference ion, M^{n+} with all ligands, in solution, and also interactions of M^{n+} with other metalic ions, i.e. ligand exchange and
redox reactions. (b) Chemistry of the ionogenic group in the

resin phase. (c) Swelling behaviour of the macromolecular network under particular conditions of the separation scheme. (d) Kinetic phenomena in solution and resin phase.

The right match of all these factors is important to the design of a successful separation system.

1.1. Aqueous Chemistry of Metal Ions in Reference to The Chemistry of Ion Coordinating Resins

The reactivity of metal ions in solution may be related to the following: (1) electronic configuration of the ion, (2) ionic radii of metal ion, and (3) position on HSAB scale (hard-soft acid base scale) of metal ion and the ligands. Ions with electrons in s and p orbitals (alkali and alkaline earth) show a limited degree of variety in their chemical reactivity, whereas, ions with electrons in the d and f orbitals have complex behaviour.

The scale of HSAB allows a rough classification of ions and ligands divided into "hard" or "soft" categories. Hard bases (class A) are usually small in ionic radii with localized charges, whereas the soft bases (class B) are large in size with diffuse charges, and high polarizabilities. A rule of thumb is that "soft acids" tend to interact preferentially with "soft bases", and conversly, "hard acids" with "hard bases". Classification of selected ions and ligands by the HSAB scale is given in Table 1 (1).

Table 1   CLASSIFICATION OF LEWIS ACIDS

Hard	Soft
H^+, Li^+, Na^+, K^+	Cu^+, Tl^+, Hg^+
Be^{2+}, Mg^{2+}, Ca^{2+}, Sr^{2+}, Mn^{2+}	Pd^{2+}, Cd^{2+}, Pt^{2+}, Hg^{2+}, CH_3Hg^+, $Co(CN)_5^{2-}$, Pt^{4+}, Te^{4+}
Al^{3+}, Sc^{3+}, Ga^{3+}, In^{3+}, La^{3+}	Tl^{3+}, $Tl(CH_3)_3$, BH_3, $Ga(CH_3)_3$, $GaCl_3$, GaI_3, $InCl_3$
N^{3+}, Gd^{3+}, Lu^{3+}	RS^+, RSe^+, RTe^+
Cr^{3+}, Co^{3+}, Fe^{3+}, As^{3+}, Ce^{3+}	I^+, Br^+, HO^+, RO^+
Si^{4+}, Ti^{4+}, Zr^{4+}, Th^{4+}, Pu^{4+}	I_2, Br_2, ICN, etc.
UO_2^{2+}, $(CH_3)_2 Sn^{2+}$, VO^{2+}, MoO^{3+}	Trinitrobenzene, etc.
$BeMe_2$, BF_3, $B(OR)_3$	Chloranil, quinones, etc.
$Al(CH_3)_3$, $AlCl_3$, AlH_3	Tetracyanoethylene, etc.
RPO_2^+, $ROPO_2^+$	O, Cl, Br, I, N
RSO_2^+, $ROSO_2^+$, SO_3	$M°$ (metal atoms)
I^{7+}, I^{5+}, Cl^{7+}, Cr^{6+}	Bulk metals
RCO^+, CO_2, NC^+	CH_2, carbenes
HX (hydrogen-bonding molecules)	

Class A ions tend to form complexes with the following order of stabilities.

Class A (hard): $F^- > Cl^- > Br^- > I^-$
$O \gg S > Se > Te$
$N \gg P > As > Sb > Bi$

With class B metalic ions, this order is reversed. Of course, some borderline metal ions react with both classes.

Quantitative data for stability constants of metallic ions and organic or inorganic ligands is given in reference 2, and reference 3 (for macrocyclic ligands only).

On first approximation, examination of the electronic structure of the metal ion helps to screen the ions into two groups, "alkali alike ions" (AAI), and "transition metal alike ions" (TMAI). In the first group (AAI), steric considerations are predominant; in the second group, electronic considerations are predominant.

The complexation behaviour of AAI group may be best described under the terms of "host-guest complexation" coined by Cram, which means that complementary behaviour between ion (guest) and ligand (host) is required for the complexation process. For this to occur, the prima facia requirement is complementary sizes of ionic radii and ligand cavity. Illustration 1 shows the periodic table in steric perspective, pointing out the large differences in ionic radii of alkaline cations. Illustration 2 shows various macrocyclic ligands with varying cavity sizes, whereas illustration 3 shows the complementary behaviour of ion and ligand during complexation.

Stability constants (2) for several complexes of macrocyclic ligands and alkali metal ions are given in table 2. Usually this information is obtained in non-aqueous or mixed-aqueous solvents, and provides only partial information (3). Much more of what is needed, and usually not available, are stability constant for complexes in presence of varying concentrations of competing inorganic ligands, halides, nitrate, sulphate, phosphate, thiocyanate, etc.

In the second group (TMAI), electronic configuration, complex structure and stereochemistry are important. Crystal field splittings of the d orbitals of the central Mn+ ion in tetrahedral, octahedral, tetragonal and square planar complexes are shown in

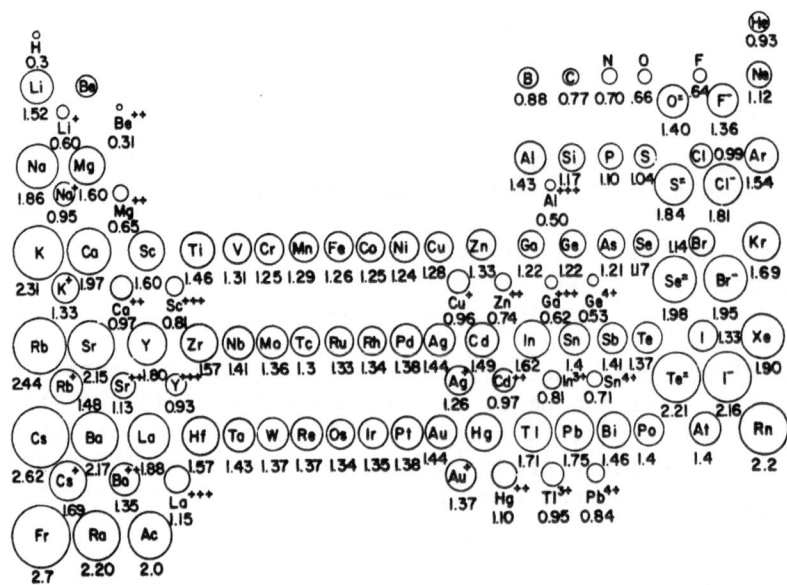

Illustration 1. Periodic table in steric perspective

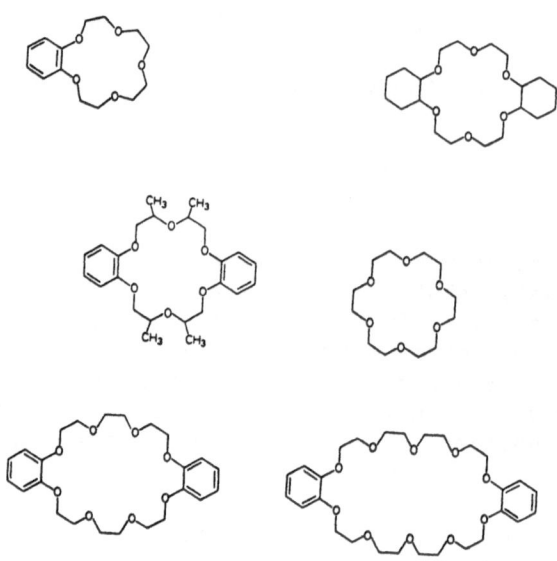

Illustration 2. Macrocyclic ligands with varying cavity sizes

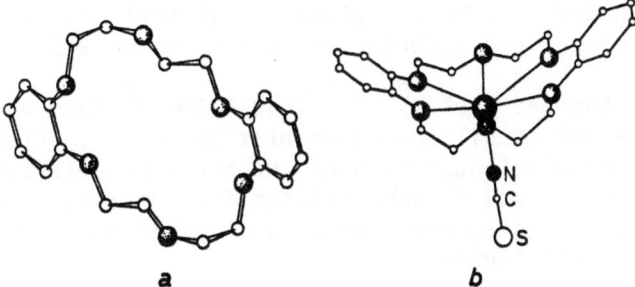

a **b**

I11. 3 Crystal structure and conformation of *a*) the free ligand dibenzo [18]-crown-6, and *b*) its RbNCS complex

Table 2. *Log* K, ΔH°, *and* ΔS° *values for the reaction,* $M^{n+} + L = ML^{n+}$, *in aqueous solution where* $L = dicyclohexyl$-18-crown-6, *isomer* A. $T = 25^\circ$.

M^{n+}	Cation ionic radius, A	Log K	ΔH° (kcal/mole)	ΔS° (cal/$^\circ$ K-mole)
K⁺	1.33	2.02	−3.88	− 3.8
Ag⁺	1.26	2.36	+0.07	+11.0
Rb⁺	1.48	1.52	−3.33	− 4.2
Tl⁺	1.40	2.44	−3.62	− 1.0
Sr²⁺	1.13	3.24	−3.68	+ 2.5
Hg²⁺	1.10	2.75	−0.71	+10.2
Pb²⁺	1.20	5.0	−5.58	− 3.9

illustration 4. The energy difference Δ (or Dq) between the degenerate d levels of the bare Mn+ ion and the stabilized d orbitals of the complex is indicative of the thermodynamic stability of the complex. Crystal field stabilization energies for d^n complexes (n designates number of electrons in the d orbital) of the common octahedral and square planar configurations, and for both weak and strong field ligands, are given in table 3.

It is observed that d^0, d^{10} ions (also d^5 ions at weak field ligands) are not stabilized by complex formation. Closer examination allows conclusions regarding the preferred stereochemistry of the ions, e.g. d^7 and d^8 ions benefit from complex formation of both octahedral and square planar configurations, particularly with strong field ligands.

The qualitative conclusions derived from examination of the theoretical data in table 3, may be corroborated with experimental observations for many systems. Thus, log K_T (total stability constant) for the first row metal ion complexes with ethylendiamine, the values of $Ca^{2+}(d^0)$, $Mn^{2+}(d^5)$ and $Zn^{2+}(d^{10})$ lie on a straight line, and deviate considerably in comparison to their neighbours (see illustration 5).

Table 3 CRYSTAL FIELD STABILIZATION ENERGY OF d^n COMPLEXES

System	Example	Octahedral		Square planar	
		Weak field	Strong field	Weak field	Strong field
d^0	Ca^{2+}, Sc^{3+}	0Dq	0Dq	0Dq	0Dq
d^1	Ti^{3+}, U^{4+}	4	4	5, 14	5,14
d^2	Ti^{2+}, V^{3+}	8	8	10, 28	10, 28
d^3	V^{2+}, Cr^{3+}	12	12	14,56	14,56
d^4	Cr^{2+}, Mn^{3+}	6	16	12,28	19,70
d^5	Mn^{2+}, Fe^{3+}, Os^{3+}	0	20	0	24,84
d^6	Fe^{2+}, Co^{3+}, Ir^{3+}	4	24	5,14	29,12
d^7	Co^{2+}, Ni^{3+}, Rh^{2+}	8	18	10,28	26,84
d^8	Ni^{2+}, Pd^{2+}, Pt^{2+}, Au^{3+}	12	12	14,56	24,56
d^9	Cu^{2+}, Ag^{2+}	6	6	12,28	12,28
d^{10}	Cu^+, Zn^{2+}, Cd^{2+}, Ag^+, Hg^{2+}, Ga^{3+}	0	0	0	0

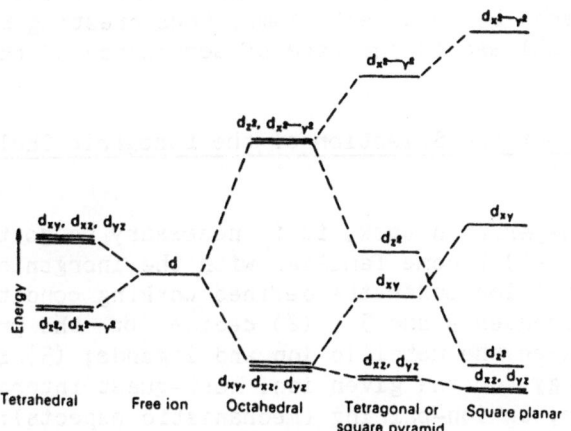

Ill. 4 *Crystal field splittings of the d- orbitals of a central ion in regular complexes of various structures*

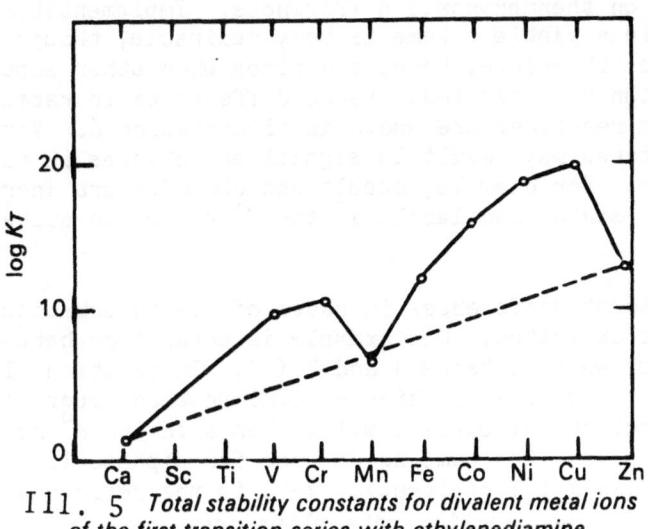

Ill. 5 *Total stability constants for divalent metal ions of the first transition series with ethylenediamine*

The inorganic complexes of the transition metal ions, formed in presence of anions by displacement of aqua$_2$complexes provide a special case. The stability of $[MX_4]^-$, $[MX_6]^{2-}$ type complexes (X=halide, thiocyanate) is so large that they tend to ion-pairing rather than to complex formation. Yet many times transition metal ions are sequestered by both mechanisms, thus creating many difficulties, as we shall see in the case of separation of the noble metals.

1.2. Guide Lines for the Selection of the Ionogenic Chelating Ligand.

Given a defined separation task, it is necessary to go through the following steps: (1) become familiar with the inorganic chemistry of the target metal ion under the defined working conditions (see for example, references 4 and 5); (2) decide on the predominant interactions between the metallic ion and ligands; (3) select a separation strategy for the given ion, host-guest interactions, chelate formation, or ion-pairing (mechanistic aspects); (4) consider the effect of the various separation strategies on the co-ions (selectivity problem) and (5) consider whether the various elution pathways are adequate (elution problem). This schematic approach will be further demonstrated.

1.3. Kinetic vs. Thermodynamic Separation, Non-Reversible Complexation.

The principles discussed in Section 1.1. are pertinent to separations based on thermodynamic differences. Implementation of these principles in a viable scheme is very desirable, though sometimes unobtainable, therefore, there are times when other separation principles can be exploited. Large differences in rates of ligand substitution reactions are shown in illustration 6. Variations in oxidation states may result in significant changes in rates of substitution. For example, cobalt and chromium are inert in the 3^+ oxidation state, and labile in the 2^+ oxidation state 1).

Significant differences in rates of ligand substitution reactions may be exploited. One example is separation between Pd(II) and Pt(IV) shown in schemes 1 and 2 (6). In solution, ligand substitution of chlorides by thiourea (Tu) proceeds much faster with the d^8 tetrachloropalladate complex than with the d^6 hexachloropalatinate complex. The same holds true for biphase liquid/liquid extraction system (S =solvent). Yet, in the resin phase (P =polymer), the reaction rate slows down, and separation becomes impossible. ---------------------

1) For comprehensive description of this topic see references 4 and 5

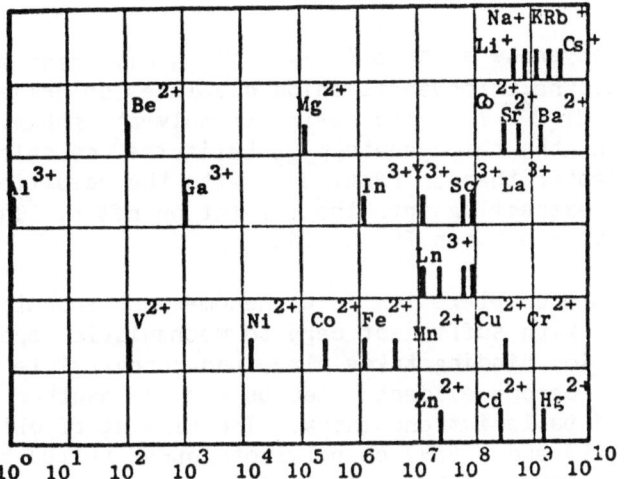

ill 6. Characteristic rate constants (sec^{-1}) for substitutions of inner sphere H_2O of various aquo ions.

| Organic phase | Organic phase | Aqueous phase |

$$\left\{\begin{array}{l} \circled{S}\text{-}\overset{+}{N}R_3\left[Pd\,Cl_4\right]^{2-}R_3\overset{+}{N}\text{-}\circled{S} \\ \circled{S}\text{-}\overset{+}{N}R_3\left[Pt\,Cl_6\right]^{2-}R_3\overset{+}{N}\,\circled{S} \end{array}\right\} \begin{array}{c}\xrightarrow[\text{Fast}]{Tu}\\ \xrightarrow[\text{Slow}]{Tu}\end{array} \left\{\begin{array}{l}\circled{S}\text{-}\overset{+}{N}R_3\,Cl^- \\ \circled{S}\text{-}\overset{+}{N}R_3\left[Pt\,Cl_6\right]^{2-}R_3\overset{+}{N}\,\circled{S}\end{array}\right\} \left[Pd(Tu)_4\right]^{2+}$$

| Resin phase | Resin phase | Aqueous phase |

$$\left\{\begin{array}{l}\circled{P}\text{-}\overset{+}{N}R_3\left[Pd\,Cl_4\right]^{2-}R_3\overset{+}{N}\,\circled{P} \\ \circled{P}\text{-}\overset{+}{N}R_3\left[Pt\,Cl_6\right]^{2-}R_3\overset{+}{N}\,\circled{P}\end{array}\right\} \xrightarrow[\text{Slow}]{Tu} \circled{P}\text{-}\overset{+}{N}R_3Cl^- \quad\begin{array}{l}\left[Pd(Tu)_4\right]^{2+}\\\left[Pt(Tu)_6\right]^{2+}\end{array}$$

__SCHEME 1__ $\circled{S}$= Solvent   Tu = $H_2N\overset{S}{\overset{\|}{C}}NH_2$ $\circled{P}$= Polymer

$$\boxed{\circled{P}\text{-}Tu + \left[Pd\,Cl_4\right]^{2-}}$$

$\circled{P}\text{-}Tu\longrightarrow\overset{|}{\underset{|}{Pd}}\text{-}\quad\xleftarrow{\quad}\overset{\times}{\xrightarrow{\quad}}\quad\circled{P}\text{-}\overset{+}{Tu}\left[Pd\,Cl_4\right]^{2-}\overset{+}{Tu}\text{-}\circled{P}$

Tu

$\circled{P}\text{-}Tu + \left[Pd(Tu)_4\right]^{2+}$

SCHEME 3

In the thioocyanate systems (scheme 2), a different mechanism operates. First, the thiocyanate anion displaces chloride in both $[PdCl_4]^{2-}$ and $[PtCl_6]^{2-}$ from the resin (or solvent) phase, and then, in solution, the thiocyanate(T$_c$) substitutes the chlorides in $[PdCl_4]^{2-}$ much faster than in $[PtCl_6]^{2-}$, with the resulting $[Pd(TC)_4]^{2-}$ back extracting into the solvent or resin. The net result is separation between Pt and Pd.

Non-reversible complexation is a phenomenon responsible for resin "poising". With sufficient care to mechanistic aspects, the conditions where ion binding takes place under reversible comple- xation terms will become evident. Let us examine another example from platinum and palladium chemistry. The binding of chloride complexes of Pd(II) and Pt(IV) on polyisothiourea (PITU) resins may proceed by an anion exchange mechanism or by ion coordination mechanism, as shown in scheme 3 (7). The existence of two differ- ent pathways for platinum group metals (PGM) binding was realized during simple continuous-absorption-elution tests (life tests) at 0.5 M HCl and 4 MHCl, shown in illustration 7.

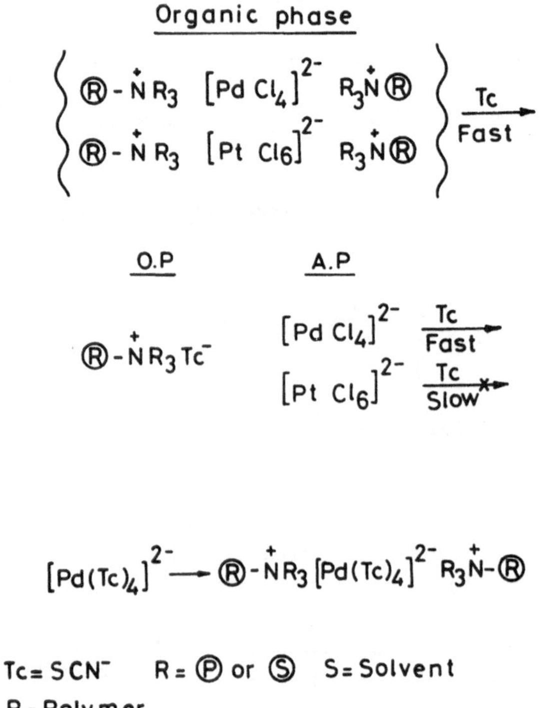

Organic phase

$$\left\{ \begin{array}{l} \circledR-\overset{+}{N}R_3 \; [Pd\,Cl_4]^{2-} \; R_3\overset{+}{N}\circledR \\ \circledR-\overset{+}{N}R_3 \; [Pt\,Cl_6]^{2-} \; R_3\overset{+}{N}\circledR \end{array} \right\} \xrightarrow[\text{Fast}]{T_c}$$

O.P	A.P

$\circledR-\overset{+}{N}R_3 Tc^-$ $[Pd\,Cl_4]^{2-} \xrightarrow[\text{Fast}]{T_c}$

$[Pt\,Cl_6]^{2-} \xrightarrow[\text{Slow}]{T_c} \times$

$$[Pd(Tc)_4]^{2-} \longrightarrow \circledR-\overset{+}{N}R_3 \, [Pd(Tc)_4]^{2-} \, R_3\overset{+}{N}-\circledR$$

$Tc = SCN^-$ $R = \circledP \text{ or } \circledS$ $S = Solvent$
$P = Polymer$

SCHEME 2

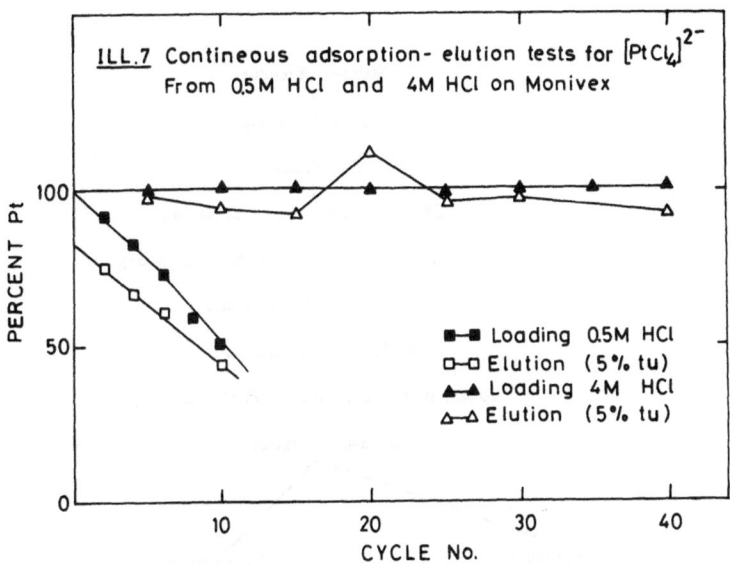

ILL.7 Contineous adsorption- elution tests for $[PtCl_4]^{2-}$
From 0.5M HCl and 4M HCl on Monivex

PERCENT Pt

■—■ Loading 0.5M HCl
□—□ Elution (5% tu)
▲—▲ Loading 4M HCl
△—△ Elution (5% tu)

CYCLE No.

The steady drop-off in the resin capacity at 0.5 M HCl indicated non-reversible behaviour.

1.4. Ion Exchange as Part of an Integrated Process

Ion exchange (IX) is today viewed as one unit process amongst several separation methods, and closely related to solvent extraction
(SX) and membrane separation (MS) processes. SX processes usually
offer better selectivity, larger mass output, lower reagent costs,
but great operating costs and considerable risks. MS processes
offer highest dynamic output efficiencies, excellent safety features, though highest "reagent" costs, and not very large selectivities. IX is usually a compromise between these extremes. The
best features of IX are performance in concentrating metals from
very dilute solutions with good selectivities. On the other hand,
large volumes of dilute eluant solutions are produced. Integrated
processes of IX and SX (and perhaps IX and MS) are therefore very
reasonable. The modern processing of PGM presents conditions for
integrated processing. The low abundance of the PGM (10 ppm) and
the large excess of base metals requires a selective preconcentra-
tion step, which collects all the PGM and rejects most of the
other elements. A schematic flow-sheet of such a process is shown
in scheme 4 (8).

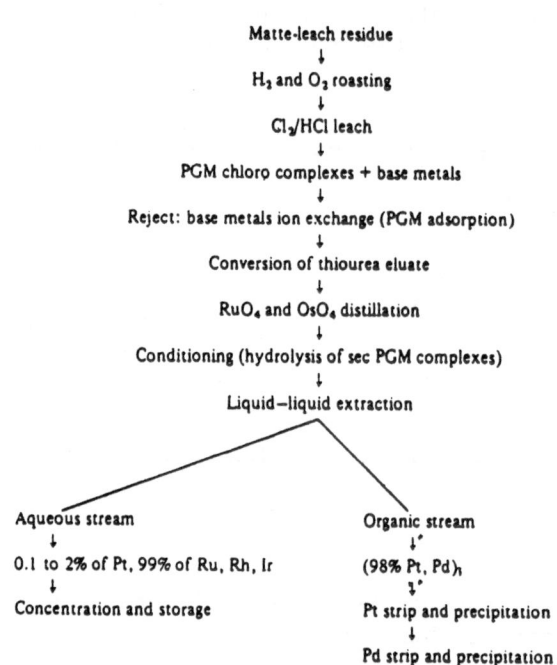

Scheme 4: Process flowsheet for the refining of PGM.

Scheme 5 describes the behaviour of PITU resins at various stages. First, the equilibria between the free base and conjugated acid forms of the resin, and then the adsorption of anionic complexes (at high acidicies) of the PGM ions, followed by elution with thiourea. This step converts the resin-bound anionic chloride complexes of the PGM to cationic thiourea complexes, and regenerates the resin. The effectiveness of these steps is shown in table 4.

Conversion of the thiourea PGM complexes back to chlorocomplexes is followed, as shown in scheme 5, by SX separation of the valuable metals, to produce high purity salts or metal sponges.

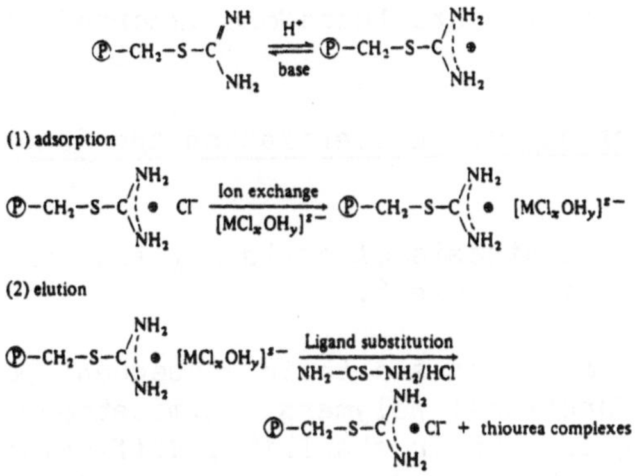

Scheme 5

METAL CONCENTRATION (ppm)	Pt	Pd	SEC PGM	Ag	Fe	Ni	Cu
Resin Feed	197	160	256				
Aqueous feed	8400	3800	660	70			
Aqueous barren	1.6	0.1	20	<0.1			
Loaded resin	71800	35600	4693				
Eluted resin	196	150	283				
% Recovery in Eluate	overall 99.5					0	
Purity	>99.9%				36	13	80
Conversion of Eluate , %	99.998	99.997	99.95				

SEC PGM = Rh, Ru, Ir.   Conditions: Absorption 2 M HCl
                         Elution: 5% thiourea

Table 4

2. STRATEGIES IN SYNTHESIS OF COORDINATING AND CHELATING RESINS

The main difference between synthesis of low molecular weight compounds and polymeric analogues is in product isolation and purification steps, which make exact characterization of polymers much harder. Consideration of synthetic strategies towards coordinating and chelating resins are therefore important in planning separation schemes.

2.1. Comparison of Direct Polymerization and Functionalization Methods

The main routes in synthesis of chelating and coordinating polymers are described in scheme 6.

Synthesis of monomers and their subsequent polymerization is a direct way to functional polymers. Limitations concerning synthesis of monomers, their stability, difficulties in polymerizations of several classes of monomers, and difficulties in achieving bead shaped particles of sufficient strength and porosity, considerably reduce the attractiveness of this route. The second method of templated ion synthesis promises better selectivity in resins, and is essentially a variation of the first, with very limited experience at the present time. The most common method is modification of preshaped polymers.

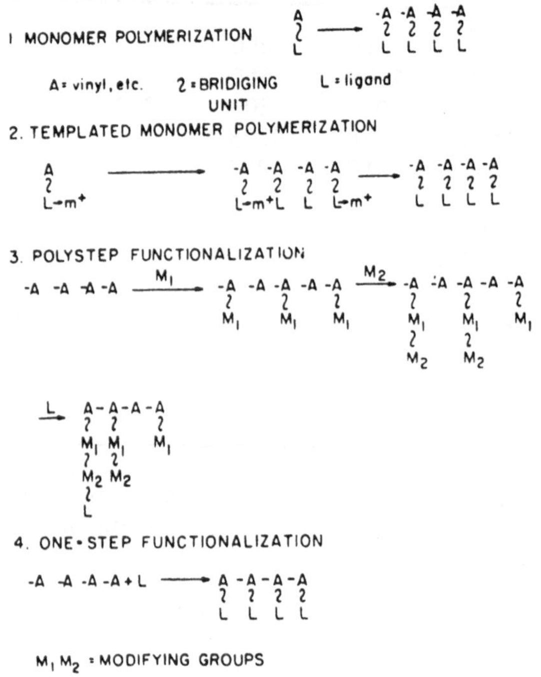

Scheme 6

The advantages of commercial availability of bead-shaped particles are confidence that the properties of the modified functional polymer will be closely related to the starting polymer.The ease of performing reactions on polymeric bead particles has prompted many researchers to perform multi-step consequential modification reactions. Assembling the ligand in several steps,(neglecting to note that each step "prints an error" which may not be corrected) causes the final polymer to include,in addition to the desired ligand, remnants of the intermediates. This obstacle contributed to the general disappointment from multi-step methodology. The last method of one-step functionalization tries to overcome the difficulties described above by pre-assembling the whole ligand, followed by one-step attachment. Limitations of this approach will be pointed out later.

2.2. Study and Control of Synthetic Procedure: Parallel Investigation of Insoluble Polymers and Analogous Hydrophobic Ligands

Choosing the preferred ligand, deciding on synthetic strategy and functionalization method does not guarantee an ion-selective resin. The ion complexation patterns of the resin may not fulfill our expectations, and there may be few clues to explain this behaviour. One way to overcome such difficulties is by planning, in advance, parallel investigations of synthesis and ion complexation of analogous hydrophobic ligands. The following paragraphs on copper-selective ligands and solvent impregnated resins will elaborate this point further.

2.3. The Special Case of Copper-Selective Resins

The conversion of classical copper-refining from pyrometallurgical processing to extractive metallurgy using chelating hydroxyoxime reagents (9) prompted a search for copper-selective resins (I) (scheme 7).

Examination of the K_1 values 2) for transition metal complexes of various ligands shows copper (II) to form the most stable complexes in the series, with the following ligands: ethylene imine (table 5), aromatic ethylene imines (table 6) and hydroxyoxime or 8-hydroxyquinoline (table 7).

----

2) Rigid polymeric networks tend to form 1:1 resin-to metal complexes. Comparison of the K_1 values is therefore pertinent to real situations. With gel-resins and other flexible systems, comparison with both K_1 and K_2 is advised.

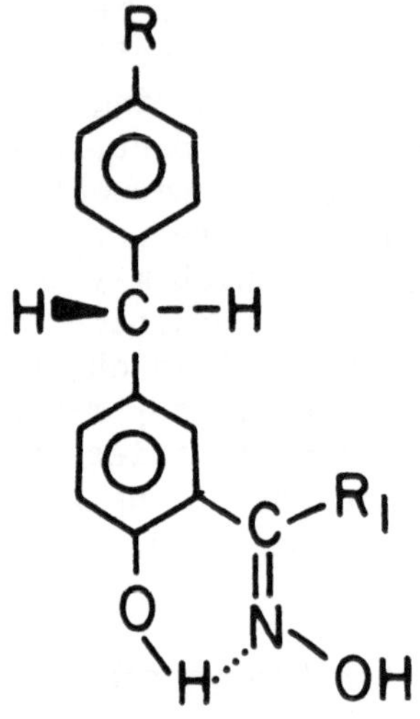

I   R = POLYSTYRENE

Scheme 7

Table 5

K^1 values for ethyleneimine ligands

	NH_3	$H_2N\ NH_2$	$H_2N\ NH\ NH_2$	$H_2N\ NH\ NH\ NH_2$
Mn (II)		500	10^4	10^5
Fe (II)		2×10^4	1.7×10^6	10^8
Co (II)	100	8×10^5	2×10^8	10^{11}
Ni (II)	500	4×10^7	6×10^{10}	10^{14}
Cu (II)	12×10^3	5×10^{10}	10^{16}	10^{20}
Zn (II)	200	8×10^5	10^9	10^{12}

Theoretically, incorporation of either of these ligands
should provide copper-selective resins. Indeed, different

Table 6

K^1 values for aromatic ethyleneimines

		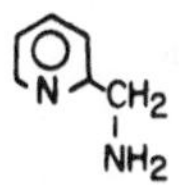 CH2–NH2	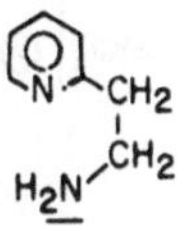 H2N–CH2	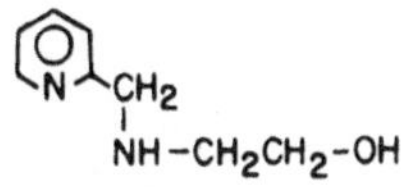 CH2–NH–CH2CH2–OH	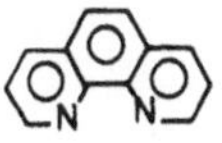 (bipyridyl)
Fe (II)	5	—		—	$10^{5.9}$
Co (II)	12	$10^{5.3}$		$10^{5.3}$	10^{7}
Ni (II)	50	$10^{7.1}$	$10^{5.2}$	$10^{7.1}$	10^{8}
Cu (II)	300	$10^{9.5}$	$10^{7.3}$	$10^{9.6}$	$10^{8.8}$
Zn (II)	10	$10^{5.2}$	—	$10^{5.2}$	$10^{6.3}$

research groups attempted to synthesize all three types (10–14).
Ours elected a unified approach to chelating liquid extraction
reagents and resins (scheme 8).

Table 7

K^1 values for salicyloxime and oxine (Conditions: 0.1 M
$NaClO_4$–dioxane–water)

	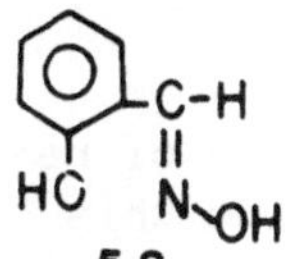 salicyloxime	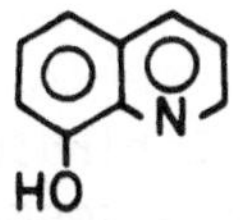 oxine
Mn (II)	$10^{5.8}$	$10^{7.3}$
Fe (II)	$10^{9.4}$	$10^{13.7}(Fe^{3+})$
Co (II)	$10^{6.4}$	$10^{9.6}$
Ni (II)	$10^{6.9}$	$10^{10.5}$
Cu (II)	$10^{12.6}$	$10^{13.3}$
Zn (II)	$10^{6.3}$	$10^{8.5}$

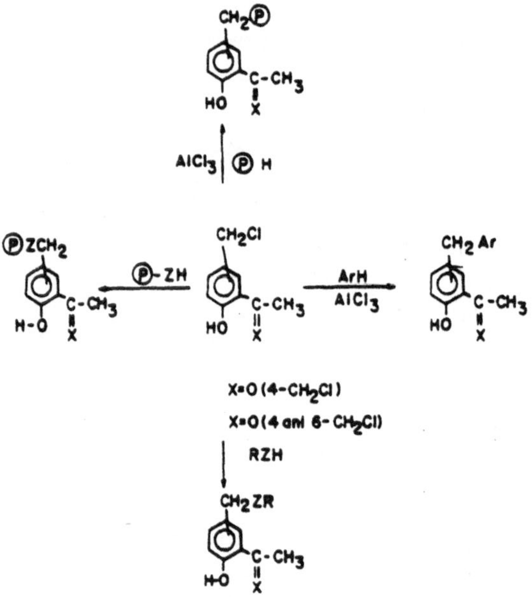

Scheme 8

The liquid extraction reagents, structurally related to the commercial hydroxy-oxime Lix-64 type reagent, showed, as expected,
excellent selectivity for copper (II) over other transition metal
ions (15,16).

The analogous resin (I) was prepared via Friedel-Crafts alkylation of polystyrene with 4-chloromethyl-2- acetyl phenol (scheme
8).

The metal binding properties of resin I were appallingly
poor, and for a considerable time this could not be explained.
Studies on the copper (II) liquid-liquid extraction properties of
the reagents 1A and 1B and of solvent impregnated resins (SIR)
models, described in the next section, confirmed the excellent
selectivity of hydroxyoximes for Cu(II). On the other hand, irreversible network molding observed for Friedel-Crafts alkylations
with 5-chloromethyl-8-hydroxyquinoline, have indicated the unsuit-
ability of the synthetic route described in scheme 8. Consequently, attachment of the same ligands to polymer carrying amino-
methyl groups, resulted in three polymers with good selectivity
for copper ions and generally satisfactory properties (scheme 9).

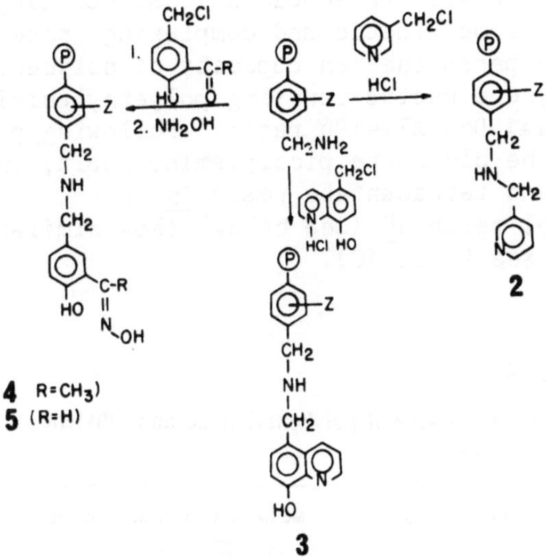

Scheme 9

Table 8

Comparison between hydroxyquinoline polymers **3** (NH$_2$ bridge) and **16** (CH$_2$ bridge)

POLYMER		$\bar{Cu}$	$\bar{Fe}$	pH
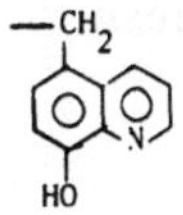	**16**	0.51	-	**4**
		-	0.57	**3**
		0.24	0.33	**2**
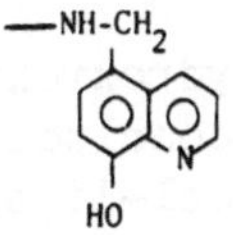	**3**	0.08	-	**4**
		-	0.12	**3**
		0.12	<0.01	**2**

Table 8, gives the relation between the spacer groups $[(CH_2)$ and $(NHCH_2)]$ resin capacity for Cu(II) and Fe(II), showing that the hydrophylic spacer with three-dimensional mobility $(N-CH_2)$ is responsible for the good kinetic and complexing properties of resins. Table 9 compares the ion capacity of polybenzylamine itself with polystyrene resins carrying polyethylenimine, picoly-lamine and commercial Dow XF-4196 resin. Following predictions (tables 5 and 6), the bidentate picolylamine resin, the tridentate XF-4196 (no. 13), and tetradentate resin 15 or 2(imidazolyl)pyridyl resin 14 (see below) show similar selectivity for copper (II). (see table 10).

Table 9

Ion capacity of several polybenzylamine and ethylene-imine polymers

POLYMER	NO	METAL ON POLYMER (mmol/g)			
		Cu	Fe	Ni	Zn
⟨O⟩-CH₂NH₂		0.1	0.12	1.0	0.78*
-CH₂ / HN-CH₂ (phenyl)	2	0.60	<0.001	0.54	0.33
-CH₂CH₂OH / HN-CH₂ (pyridyl)	XF-4196 / 13	0.70	0.001	0.39	0.23
HN-N-N-N-NH	12	0.73	0.19	0.27	0.51

CONDITIONS: pH=2, $[M^{n+}]_0$=0.05M, $[NaHSO_3]$=0.05M

Ion capacity and distribution data for hydroxyoxime and oxine resins shown in tables 11 and 12 allow us to conclude that pre-diction of ion exchange selectivity, using model-ligands, followed by appropriate synthetic strategies lead to tailor-made selective ion exchange resins for copper. Similar sequence of logical steps will lead to other selective resins.

Table 10. Comparison between several picolinic type resins

Resin	Ref.	Polymer No.	mmole/Gram Resin				
			Cu	Fe	Ni	Co	Zn
℗-CH₂ / HN-CH₂ N⬡	14		0.60	<0.001	0.54		0.33
℗-CH₂ / CH₂-N-CH₂ / CH₂ OH N⬡	10	13	0.70	0.001	0.39		0.23
℗-CH₂-N⬡N N⬡	12	14	0.83	0.09	0.09	0.01	
℗-CH₂-O NH HN N⬡	11	15	1.16	0.004	0.48	0.002	

2.4. Solvent Impregnated Resins. Aid in Understanding Diffusion Processes and Ion Selection Phenomena

The striking difference in the copper complexation behaviour of hydroxyoxime reagents **6A** and **6B** and analogous resins **I** can be attributed to several reasons. Hydrophobicity of the resin; insufficient mobility of the resin-bound ligand; much weaker complexation constants in the resin phase, etc. In order to sort out the various possibilities, it is necessary to prepare a model which resembles the resin phase, yet allows significant mobility of the ligand, a situation existing in solvent impregnated resins (SIR). Such systems were suggested as a compromise between synthetic limitations, availability of high porosity adsorbents, and the need for selective resins. Their properties are summarized in a recent comprehensive review for several ligand systems (17,18). The extractants ("solvents") are incorporated in the polymeric phase by impregnation, or alternatively, included in the monomer phase as additives and then polymerized. In both cases, the resulting SIR polymers (the first version) or Levextrel (the second version, commercial products by Bayer, AG) show very high reagent mobility and ion complexation efficiency. Concerning our case of copper-selective hydroxyoxime resins, reagents **6A** to **6E** were prepared from 4-chloromethyl-2-acetylphenol and 4-chloromethyl-2-formyl phenol (scheme 10) (15,16).

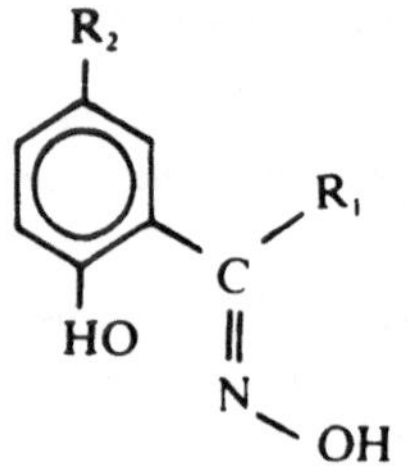

Reagents

6A: $R_1 = CH_3$ $R_2 = -CH_2-C_6H_4CH_3$
6B: $R_1 = H$ $R_2 = -CH_2-C_6H_4CH_3$
6C: $R_1 = CH_3$ R_2
  $= CH_2-O-CH_2CH_2-OCH_2CH_2-OC_4H_9$
6D: $R_1 = H$ R_2
  $= CH_2-O-CH_2CH_2-OCH_2CH_2-OC_4H_9$
6E: $R_1 = H$ $R_2 = CH_2-N(C_4H_9)_2$
LiX 65 N: $R_1 = C_6H_5$ $R_2 = C_9H_{19}$

Scheme 10

Table 11

Ion capacity of hydroxyquinoline and hydroxyoxime polymers

POLYMER	NO	METAL ON POLYMER (mmol/g)				
		$\overline{Cu}$	$\overline{Fe}$	$\overline{Ni}$	$\overline{Co}$	$\overline{Zn}$
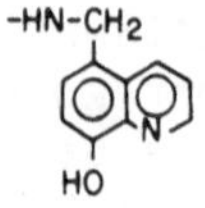 (hydroxyquinoline structure)	3	0.74	0.005	0.011	0.007	0.013
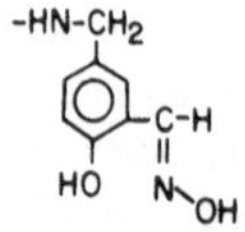 (hydroxyoxime structure)	5	0.36	0.005	0.004	0.003	0.005

COMPLEXATION: $[M^{n+}]_o = 0.1M$, pH=2, $[NaHSO_3] = 0.15M$

ELUTION: 3N H_2SO_4

The studies on copper complexation dependency on the impregnation method (table 13), oxime concentration on the polymer (table 13) and kinetic data showed the impregnation method to be extremely significant for impregnation of hydrophobic reagent of type 6A/6B. Removal of all the diluent from the polymer phase considerably reduced the reagent complexation efficiency. A wet impregnation method allowing retention of diluent in the resin phase, or impregnation of hydroxyoxime reagents with a hydrophilic side chain produced SIR resins with very good copper-complexation efficiencies. This, together with the linear dependency of complexation on oxime concentration in the polymer phase (illustration 8) pointed to the high mobility of the ligand in the polymer phase which enabled formation of 1:2 copper-to-ligand complexes (scheme 11).

Furthermore, the rate curves for complexation (illustration 9) and elution (illustration 10) showed both complexation and elution to be controlled by external driving forces. The complexation and elution rates increased together with copper and hydrogen ion solutions respectively (illustration 10). The high rates of complexation and elution reactions pointed to the complete reversibility of the copper complexation in the resin phase.

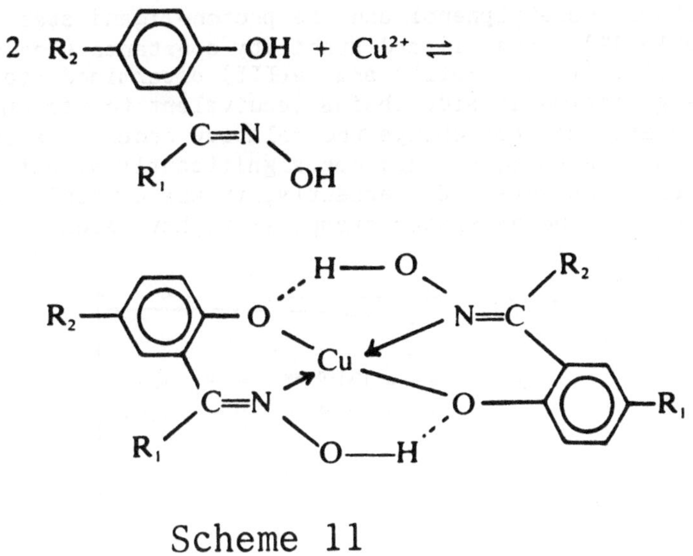

Scheme 11

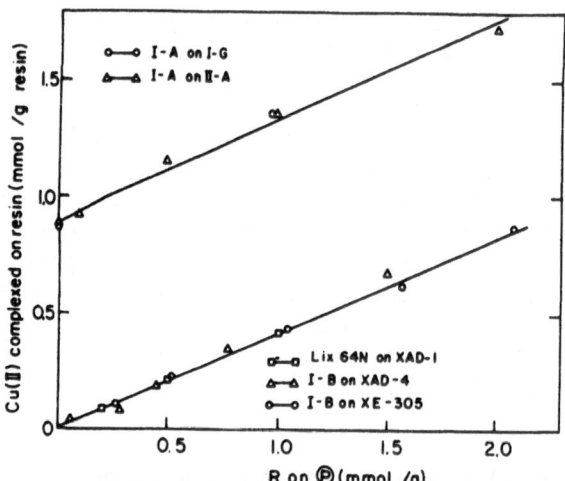

Ill. 8: Copper complexation dependency on oxime concentration on various polymeric supports.

The results obtained with the SIR hydroxyoxime extractants indicated the importance of tailoring the properties of the side group to match the requirementsof the polymer phase. Yet, it was necessary to prove that the side group itself does not significantly change the complexing properties of the main ligand (the hydroxyoxime).

Analytical grade hydroxyoximes were prepared by reaction of 4-chloromethyl-2-acetylphenol and the proton-ligand stability constants (table 14), metal ligand stability constants (log K_1) for Cu(II), Co(II), Ni(II), Fe(II) and Fe(III) determined (table 15) (16). The variation in side chains (equivalent to various spacer groups in resin) did not change the relative order of stability constants for the various ions; nor significantly affected the values of the constants. Consequently, it was possible to choose an ether or an amine as spacer group, as we have seen in the previous section.

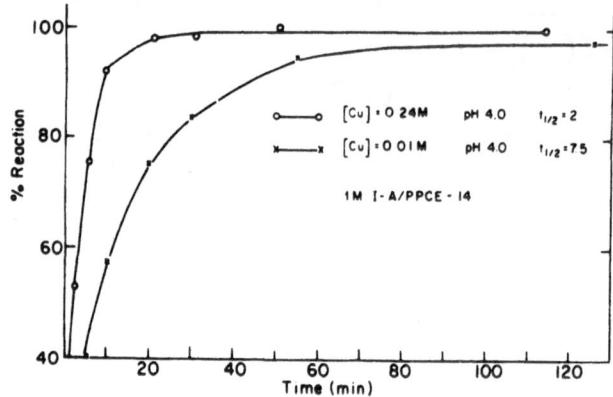

Ill. 9; Copper complexation rate dependency on solution concentration.

Table 12
Distribution factors for several transition metal ions on hydroxyoxime, hydroxyquinoline and ethyleneimine polymers

POLYMER	NO	$\overline{Cu}$	$\overline{Fe}$	$\overline{Ni}$	$\overline{Co}$	$\overline{Zn}$	pH
	3	4190	1.3	1.8	0.3	0.4	2
	5	2012	0.7	0.5	0.5	0.7	3.8
	12	666	7.4	6.2	2.1	4.4	4.0

CONDITIONS: $\left[M^{n+}\right]_0 = 0.01M$ $\left[NaHSO_3\right] = 0.03M$

Table 13 Dependency of Copper Complexation[a] on the Impregnation Method

Reagent (R)	Polymer (P)	SIR resin reagent concentration (mmol R/g P)	Impregnation method[b]	Copper concentratation on polymer (meq/g)	Reagent complexation efficiency (%)
I-B	XAD-2	0.35	A	0.098	28
I-B	XAD-2	0.65	B	0.680	100 ± 5
I-B	XAD-2	0.65	D	0.660	100 ± 5
Lix-64N	XAD-2	0.26	A	0.033	13
Lix-64N	XAD-2	0.26	D	0.240	92
Lix-64N	XAD-4	0.26	A	0.033	13
Lix-64N	XAD-4	0.26	B	0.26	100 ± 5
I-C	XAD-2	0.48	A	0.51	100 ± 5
I-D	XAD-2	0.41	A	0.42	100 ± 5

a. 0.05M CuSO₄ in pH = 4 buffered solution, 24 h.
b. A = dry method; B = wet method; D = 15% modifier, dry method.

Table 14 Proton ligand stability constants

R=		
	(anti)	12.05
	(anti)	12.20
	(anti)	12.09
-O(CH₂CH₂O)₂C₄H₉	(anti)	11.79
O C₈H₁₇	(anti)	12.05
N-(C₈H₁₇)₂	(anti)	12.30
N-(C₈H₁₇)₂	(syn)	10.26

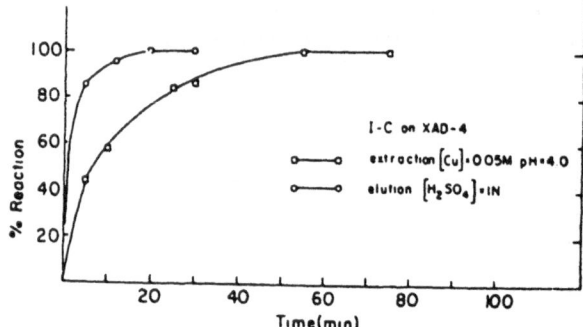

I11. 10 Comparison between rates of copper extraction and elution.

3. MATRIX EFFECTS IN SYNTHESIS AND SEPARATION

Early in the development of ion exchange resins, it was discovered that variations in resin matrix can significantly alter both rates of ion diffusion and selectivities (19). The resin matrix was varied by changing the nominal concentration of the crosslinking agent (usually divinylbenzene isomeric mixture). It is well recognized that the nominal D.V.B. content of the resin does not necessarily reflect the real D.V.B. content of the resin. Moreover, the rates of polymerization of the three positional isomers of D.V.B. vary.

The aspect of ion exchange relationship to real crosslinking agent concentration needs further investigation. At the present time, we wish to draw attention to other aspects concerning matrix effects in synthesis and their influence on separation mechanisms, as were revealed in several studies.

3.1. Matrix Effects vs. Ion Exchange Phenomena

In the previous section, it was shown how the mode of attachment of the functional group changed the rate and degree of Cu(II) binding by hydroxyoxime resins. This point is to be further demonstrated, borrowing from the extensive work of Egawa and colleagues at the Kumamato University in Japan, who paid much attention to variations in resin performance eminating from differences in reaction conditions (20,21).

The suspension polymerization of acrylonitrile and divinylbenzene (described in scheme 12), is followed by aminolysis with hydroxylamine to amido oxime resin RNH (7). Table 16 shows how parameters related to matrix structure, such as apparent density, pore volume, surface area and average pore radius, change with reaction medium. Table 17 provides the same parameters in rela-

Table 15. Metal ligand stability constants

R=		Fe^{+++}	Fe^{++}	Cu^{++}	Co^{++}	Ni^{++}
(benzene ring)	(anti)	13.36		10.87		
(toluene, CH_3)	(anti)	13.30	12.30	11.11		
(CH_3 — xylene — CH_3)	(anti)	13.66	12.55	11.03	8.55	7.28
$-O(CH_2CH_2O)_2C_4H_9$	(anti)	13.00	12.23	11.01	8.30	
$O\ C_8H_{17}$	(anti)	13.32	12.57	10.95		
$N-(C_8H_{17})_2$	(anti)	13.30	12.29	10.95		
$N-(C_8H_{17})_2$	(syn)	11.49		8.54	7.58	6.08

tion to composition of monomer mixture and the porogen (toluene), concluding that increasing the D.V.B. content beyond 10% does not increase the surface area. With 10% D.V.B., the major factor shaping pore volume and surface area is the volume of the porogen (see also illustration 11) (20).

The drastic effect on total ion exchange capacity is shown in illustration 12. Here, the significance of variation in aminolysis conditions (CH_3OH vs. H_2O) are also indicated. The uranium adsorption (illustration 13) of the resin is related to D.V.B. content, aminolysis media, reaching a maximum at 12% D.V.B.

The strong relation between rate of uranium adsorption and D.V.B. content of the RMH resin is shown in illustration 14.

Matrix effects may be understood by comparing ion binding of model ligands against oligomeric analogs and against insoluble chelating resins carrying varying concentrations of the chelating ligand. The aminodiacetic/EDTA type ligands were studied extensively. Let us compare bis(carboxymethyl)- iminomethylene derivative of oligomeric styrene 8 ($\bar{P}n-6-8$) (PEDA), iminodiacetic acid

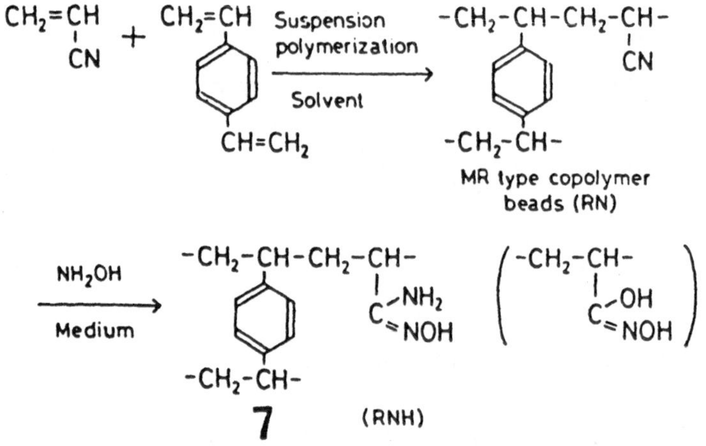

Scheme 12

Table 16 Effect of kinds of solvent on the pore structure and the adsorption of uranium

Solvent	Apparent density (g/cm³)	Pore volume (ml/g)	Surface area (m²/g)	Average pore radius (Å)	U adsorbed (μg)
Benzene	0.50	0.378	92.2	153	1.6
Toluene	0.47	0.491	121.0	168	2.1
p-Xylene	0.42	0.673	125.0	195	2.0
Chlorobenzene	0.59	0.370	80.3	181	1.4
Cyclohexane	0.38	0.963	23.7	410	1.0
Isooctane[a]	0.44	0.722	17.1	535	0.8
Chloroform	0.67	0.048	0.85	138	0.7
Carbon tetrachloride	0.45	0.566	108.0	158	1.9
Dichloroethane	0.67	0.074	8.03	123	0.9

DVB : 16.2 mol%, Solvent : 80 vol% ; a) 50 vol%.

(IDA), the analog benzyliminodiacetic acid (-BIDA) with iminodi-

Table **17** Effect of divinylbenzene and toluene on the pore structure of macroreticular copolymer

Composition of monomer		Toluene	Properties of copolymer			
DVB[a] (mol%)	AN[a] (mol%)	(vol%)[b]	Apparent density (g/cm³)	Pore volume (ml/g)	Surface area (m²/g)	Average pore radius (Å)
5.3	91.0	80	0.37	1.165	43.4	268
10.7	81.7	80	0.43	0.726	126.0	188
16.2	72.3	80	0.47	0.491	121.0	168
21.9	62.6	80	0.53	0.353	120.0	142
27.6	52.8	80	0.55	0.190	93.4	131
16.2	72.3	20	0.69	0.038	5.0	122
16.2	72.3	40	0.66	0.119	22.7	148
16.2	72.3	60	0.52	0.260	68.1	150
16.2	72.3	80	0.47	0.491	121.0	168
16.2	72.3	100	0.41	0.729	128.0	197
16.2	72.3	120	0.38	1.034	105.0	247

a) DVB=Divinylbenzene, AN=Acrylonitrile.
b) (Toluene/Monomer)×100%.

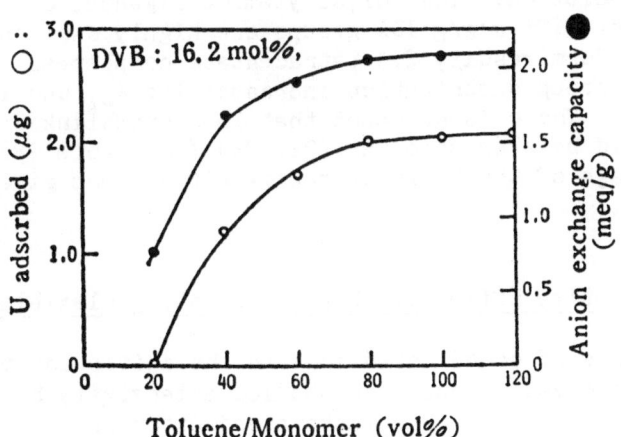

III. 11 Effect of toluene during the preparation of chelating resin on the adsorption of uranium

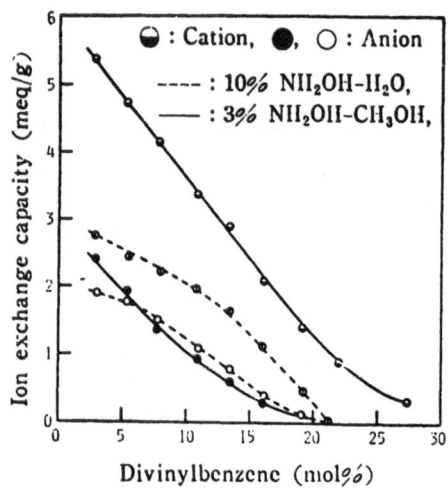

III. 12 Effect of reaction medium and cross-linking on
the ion exchange capacity

acetate polyacrylamides (10,11), and its model bis(carboxymethyl)-
iminomethylene N-isobutyramide (NBu-IDA). The log K_{MA} and log
K_{MA2} for Cu^{2+} and Zn^{2+} for the oligomeric styrene (91% substitution) are considerably higher than the corresponding values for
parent ligand IDA and analog benzyl-IDA (BIDA). Those values are
taken as maximal values for an iminodiacetate macromolecular
ligand in solution. The polyacrylamide ligands, containing
0.6 meg/g or 1.95 meg/g IDA groups are highly swollen gels, and
therefore, form readily 2:1 tetradentate complexes. Increasing
functional group distribution increases log K_{MA} and log K_{MA2} for
all metals. There is no doubt that in a crosslinked matrix this
effect would be even stronger (22, 23) (see table 18). Comprehensive coverage of the topic of matrix effects was given by Davankov
et al (24).

3.2. Cooperative Effects in Ion-Coordinating Flexible Networks

The influence of coordinated ions in the initiation mechanisms of
vinyl polymerization has been studied extensively by Kimura, Inaki
and Takemoto (25,26). Mikulaskova and Citovicky (27) have shown
that grafting polypropylene with styrene in a redox polymerization
system including the ligand TET (triethylenetetramine) and various
metal ions, is dependent on the chelate stability constant, as
well as on the redox potential of the solvated ion (26), as shown
in table 19.

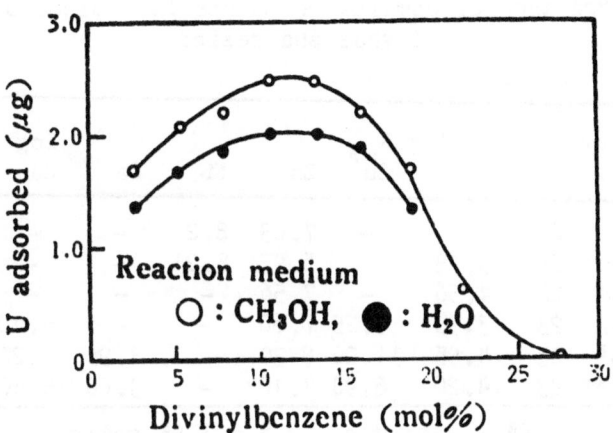

III. 13. Effect of cross-linking on the adsorption of uranium

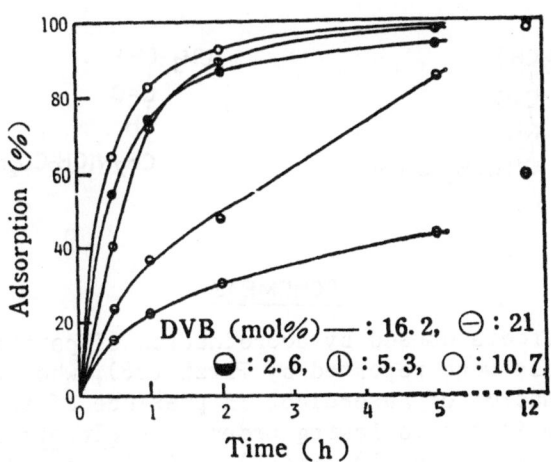

III. 14 Effect of cross-linking on the adsorption rate of uranium

Table 18. Comparison of complex constants for iminodiacetate (IDA) ligands and resins

No.	Polymer type	rf.	log K_{MA} Ca	Cu	Zn	Ni	log K_{MA2} Ca	Cu	Zn	Ni
	IDA	22	2.6	–	7.03	8.2	–	–	5.14	8.21
9	B–IDA	22	3.13		6.97	8.21	–	–	–	–
8	PS–IDA	22	4.56	–	7.56	–	–	–	6.28	6.43
	N–Bu–IDA	23	3.95	9.25	7.60	–	–	5.10	4.60	–
10	PAA–IDA*	23	4.95	11.20	8.30	–	3.95	6.20	6.40	–
10	PAA–IDA**	23	4.20	8.90	7.10	–	3.00	5.50	4.90	–

** = 0.6meg/g   * = 1.95 meg/g

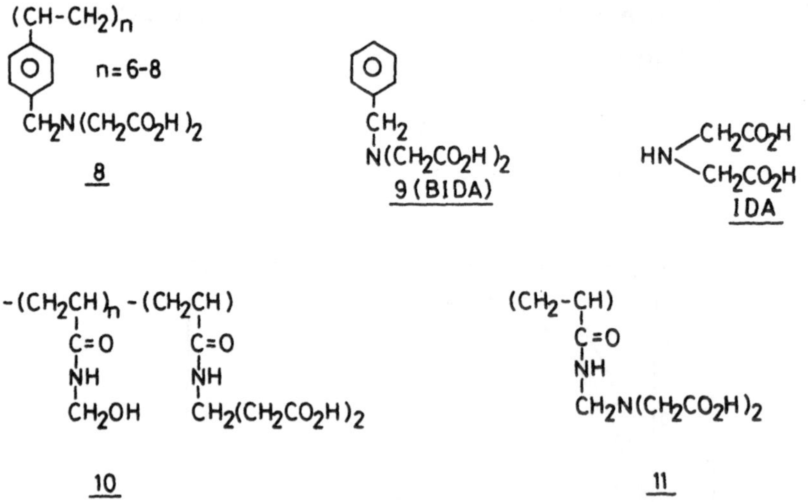

SCHEME 13

Kinetic effects caused by coordination of participating monomeric phenols were reported by Pizzi (28), who studied the rate of condensation of resorcinol in presence of several cations, and placed them in the following order: accelerating: Pb(II), Zn(II), Cd(II), Ni(II)>Mn(II), Mg(II), Cu(II), Co(II)>Mu(II), Fe(III)>>Be(II), Al(III), Cr(III), Co(II), retarding.

The retarding ions tend to form stable and kinetically inactive complexes.

The role of cooperative effects in shaping selectivity of flexible macromolecular networks was recently emphasized in sev-

Table 19. Stability constants and redox potential of TET styrene-
grafted polyethylene

M^{n+}	Complex Mn^+:TET	log K	E(volt)	% styrene in polypropylene
Mn^{2+}		8.2	-1.51	5.5
Fe^{2+}	2:3	12.1	-0.77	48.2
Co^{2+}		14.6	-1.84	10.6
Ni^{2+}		19.9	-	0.8
Cu^{2+}	1:1	10.6	0.17	0.5

eral publications. Nishide, Tsuchida and coworkers (scheme 14)
were able to crosslink polyvinylpyridine (29) in presence of templating metal ions, and thus induce selectivity for the given ion,
as shown in illustration 15, for complexation of Cu^{2+} or Fe^{3+} by
crosslinked PVP resins pretemplated by Cu^{2+} [DBQP(Cu(II)] or Fe
(III)[DBQP(Fe(III)]. In a related study, 1-vinylimidazole was
copolymerized in the presence of Ni(II), Cd(II) and Zn(II), and
then further crosslinked by X-ray irradiation in the presence of
1-vinylpyrrolidone. After removal of the templating ion by acid
elution, the resin absorbed better than the non-templated copolymer (30).

A dramatic effect of ion coordination during Friedel-Crafts
alkylation of styrene-divinylbenzene copolymers with
tris(5-chloromethyl-8-hydroxyquinoline) Aluminium complex was
observed by Warshawsky and Kalir (31), and is shown in illustra-
tions 16 and 17. The reaction carried out in solution (with fully
swollen XE-305 polymer) yields the normal round-shaped beads
(illustration 16), but in the dry impregnated state, the progress
of alkylation is noted by irreversible change of the beads to the
twisted form (illustration 17). A similar but reversible change
in the flexible matrix of XE-305 takes place when
tris(8-hydroxyquinoline) is impregnated. Since the oxine ligand
cannot be covalently bound to the resin, treatment with a solvent
restores the round shape of the bead (scheme 15).

Wulff (32) and Mosbach (33, 34) extended the term "template
polymerization" (26) to "host-guest polymerization" (33). According to this strategy (scheme 16), a monomer mixture containing a
large proportion of crosslinking units is polymerized in the pres-

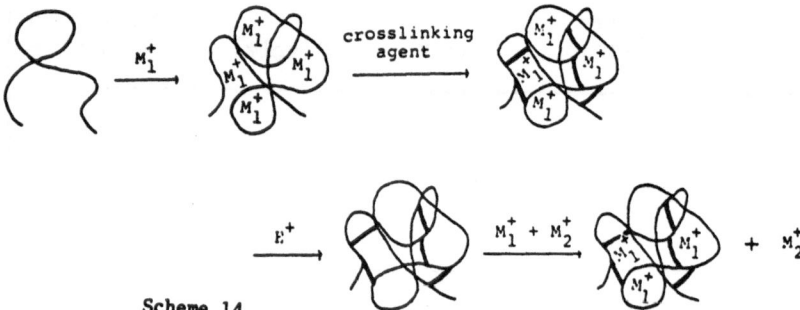

Scheme 14

ence of a free substrate, which acts as a templating center during polymerization. In the words of Arshady and Mosbach "--this is simply a mixing procedure, and no chemical attachment to the monomeric units is required. The monomers are, however, chosen in such a way as to have non-covalent binding abilities (i.e. ionic, hydrogen and hydrophobic, charge transfer, etc) complementary to those of the guest template ---, a host-guest relationship materializes during the polymer formation process and an _imprint_ of the guest (template) molecules is developed within the polymer matrix."

The examples by Wulff, using 4-nitrophenyl-α-D-mannopyranoside and 4,6-di-o-(4-vinyl-phenylboronate), and by Mosbach, using rhodanine blue and Safranine O, and the acrylate monomers N, N'-methylenediacrylamide, N, N'-1,4-phenylenediacrylamide and 3,5-bis(acryloylamido)benzoic acid show considerable selectivity for the binding of the pretemplated guests 3).

_____

3)The method practised by Mosbach closely resembles the preparation of Levextrel resins. Is it possible to prepare ion-specific Levextrel resins by pretemplating?

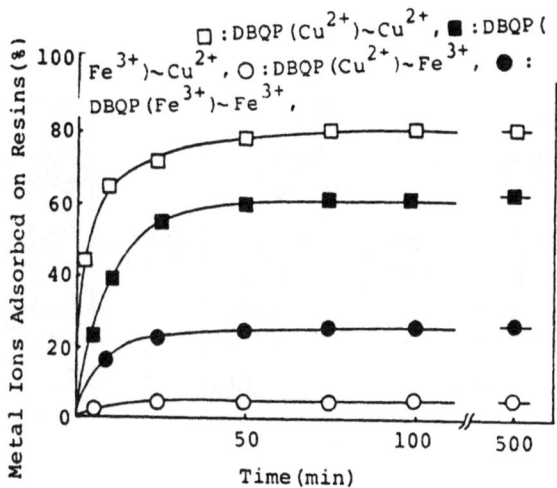

Ill. 15. Complexation between Resins and Cu^{2+}, Fe^{3+}

It is interesting to note that flexible networks with a low amount of crosslinking (such as XE-305) can be modulated by small ions and molecules, while highly crosslinked networks can be modulated in the presence of large organic molecules.

3.3. Matrix Structure and Ion Binding Mechanism. Isothiouronium Resins in Separation of Group VIII Elements

The importance of understanding the chemical behaviour of ions in solutions and identifying the exact ionic entity that is to be separated, was pointed out in section 1. The importance of understanding the exact interaction between the ionic entity and the polymeric ligand, and how the matrix structure participates in this interaction will be discussed here. Two systems have received consistent attention, poly(vinylpyridines) (PVP) and iso-thioureapolystyrene (PITU). Tsuchida and colleagues (35) concluded that polymeric ligands tend to form, due to high local concentration of ligand units, stronger complexes than their monomeric analogs i.e. the measured complex dissociation rate k' is much lower than the association rate constant k. In their detailed investigation, nineteen copolymers of styrene and 4-vinylpyridine (PSP_X, where X increased in steps of five, from 5 to 95) were prepared, and competitive complexation with two cobalt complexes, benzyl[bis(dimethylglyoximato)] Co(III) [BzCo(III) (DMG)] and [N,N'-ethylenebis(salicylideneiminato)] Co(II) [Co(II)

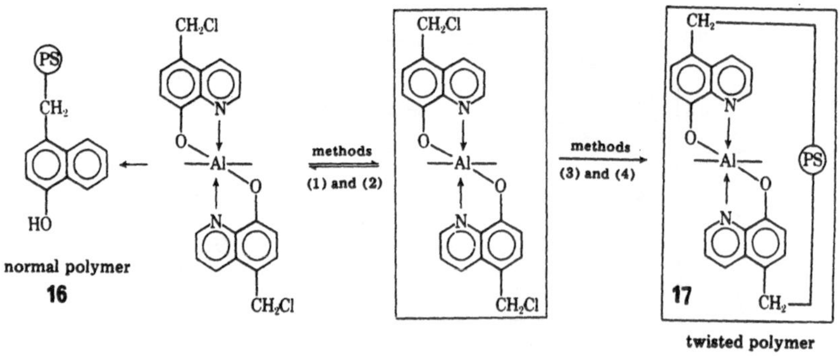

Scheme 15

salen] were measured by stopped-flow rapid scanning spectroscopy. For the complex formation reaction with BzCo(III) (DMG)$_2$, an unusual bell-shaped curve, relating to equilibrium constants(K values)and X, the percent 4-VP units in the PSP$_X$ copolymers was measured. Such a cage effect, when X is close to unity, is shown in illustration 18.

The interpretation of the complexation behaviour of PSP$_X$ polymers (soluble polymers) are in terms of reactions occurring on the "surface" of the macromolecular domain and "inside" the domain in three cases, (1) extremely "extended" structure, (2) extremely "shrunk" structure, and (3) intermediate shrunk structure. The apparent complex formation reaction is given by the equation

$$Kapp = \frac{pK_S + QK_D}{rK'_S + SK'_D}$$

S and D denote surface and domain. p,q,r,s are probability constants p+q=1, r+s=1. The p and r constants relate to high figures of X in PSP$_X$ (shrunk or rigid structures); and Q and S constants, the high figures of styrene (extended or flexible) structure.

Moving from macromolecules in solution to crosslinked insoluble polymers, there is a great decrease in ligand mobility, and

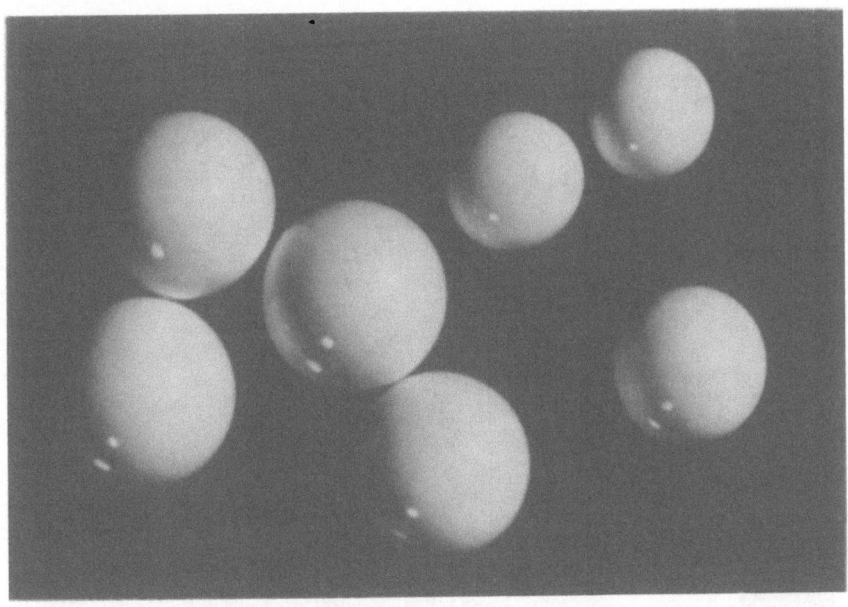

Ill. 16˙ Round-shaped polymer 2, typical methods (1) and (2), prepared from XE-305 (×10).

spacing the functional ligand away from the matrix is of considerable importance. Nishide et al. (36) prepared by chloromethyla-
tion of 20% D.V.B. copolystyrene, and further reaction with spacing glycols, the polymers 20-24 (scheme 17). With the spacers
showing two order increase in the value of the stability constants
(table 20).

Weak donors, such as polyvinylpyridine, may complex metal
ions through two mechanisms, ion coordination (IC) and ion
exchange (IX). Due to the high reversibility of both reactions,
they are hardly distinguishable in the PVP system. The polyiso-
thiouronium resins (PITU) present a very different situation which
is also of great practical significance.

The gel-type PITU resin, Srafion MRR, developed by Schmuckler and colleagues (37) showed a definite tendency for coordinating PGM ions through bonding with sulfur or nitrogen, as indicated
from the curve relating ion exchange capacity with hydrochloric
acid molarity (illustration 9). Postulating a matrix effect to
explain this behaviour of Srafion MRR, Warshawsky et al. (7) have
developed a resin derived from macroreticular flexible network,
Amberlite XE-305, with similar behaviour to Srafion in the low
activity range, binding PGM ions through IC mechanism (scheme 3).
However, in the high acidity range, good swelling allows Monivex
to bind PGM ions reversibly by IX mechanisms.

Ill. 17: Round-twisted polymer 3, prepared by method (3) from XE-305 (×10).

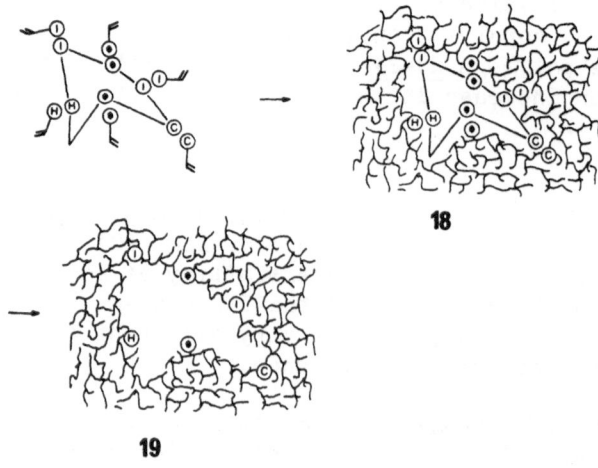

18

19

Scheme **16** Preparation of substrate-selective polymers by host-guest polymerization. Inter-actions: II, ionic; φφ, hydrophobic; HH, hydrogen bond; CC, charge transfer

Further work by Pope and Boyenes have indicated the impor-tance of hydrogen bonding in forming benzylisothiouronium and hex-

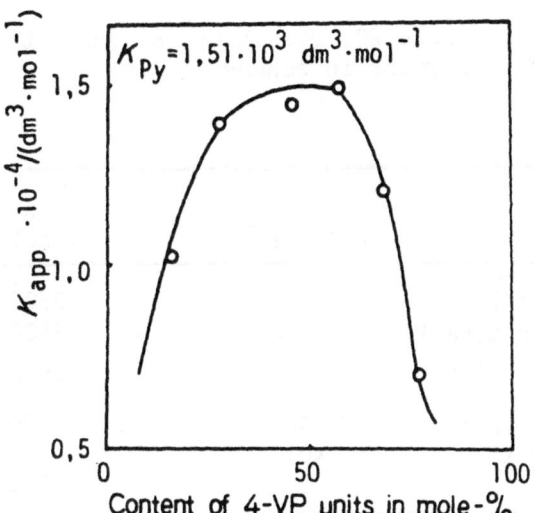

$K_{Py} = 1.51 \cdot 10^3$ dm$^3 \cdot$ mol^{-1}

Content of 4-VP units in mole-%

Ill. 18 Relationship between the content of 4-pyridylethylene (4-VP) units in PSP and K_{app}

achloroplatinate ionic pair complexes (38). Green and Hancock (37), attempting simulated synthesis of flexible networks (resins 4E, 4M and 404) and gel-type networks (4G) under identical and controlled conditions, have reconfirmed the dramatic difference between flexible network and gel-networks, as shown in illustration 20, for relationship between % Pt on resin and platinum concentration in solution 4).

Comparing PITU resins with polyvinylpyridium resins derived from the same matrices, there was no difference in the complexation of hexachloroplatinate (see illustration 21).

A reasonable explanation offered for this unexpected result is that $2[Hpy]^+[PtCl_6]^{2-}$ interaction is much stronger than the matrix effect, and the cooperative effect for monodentate complexation is small. With the PITU resins, the electrostatic interaction is weaker and the cooperative effect for bidentate complexation is large.

----

4) The ananomalous high concentration of Pt on XE-305 type PITU resin is explained by possible binding of monovalent $PtCl_5^-$ complex.

Table 20. Stability (K) and rate (k)
constants for polymers in scheme 17

Polymer	k $\times 10^4 (sec^{-1})$	K $(1/mol^{11})$	Polymer No.
Spacer type II	7.2	2500	20
Long spacer type V	116	3400	21
Long spacer type bipyridyl IV	11	4700	22
Graft type VIII	10	9100	24
Poly(4–VP) IX	<0.1	–	–
Pyridine		3.2	–

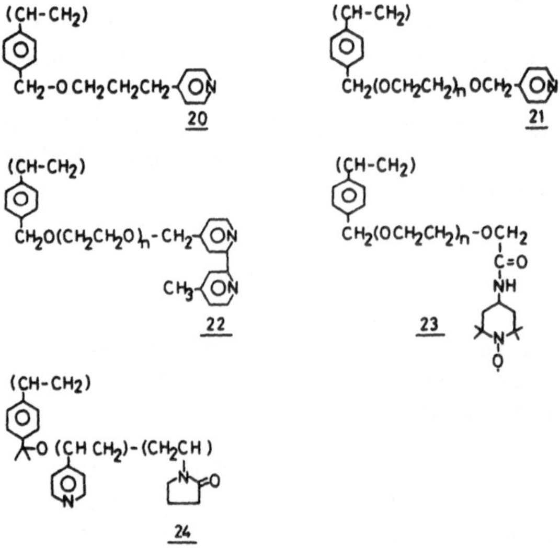

SCHEME 17

The swelling capacity of PITU resin is a direct factor
responsible for mechanistic differences in their behaviour, and
correlation between toluene regain and selectivity between
$[PtCl_6]^{2-}$ and chloride (illustration 22) confirms this experimentally.

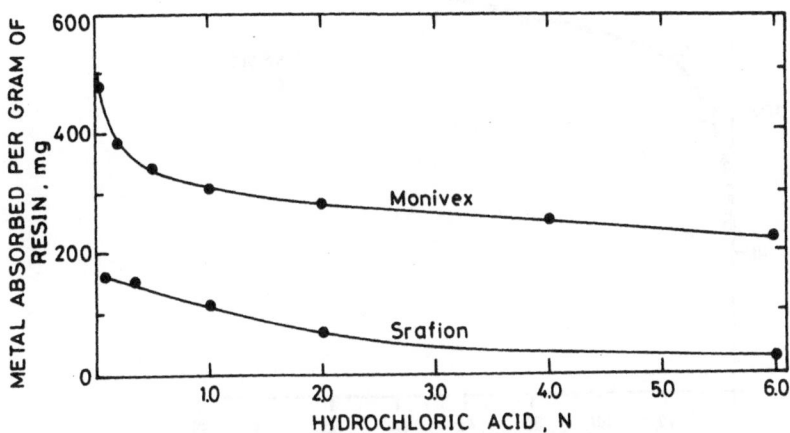

ILL.**19** The dependence on acid concentration of platinum
adsorption SRAFION NMRR and Monivex resins

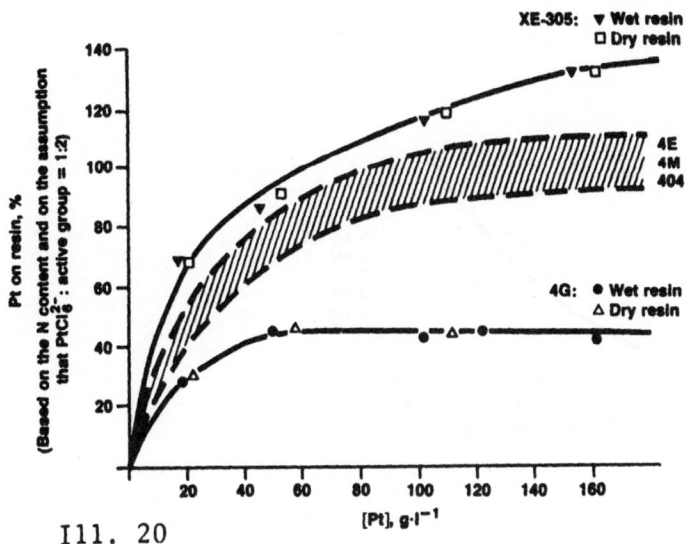

Ill. 20

. Equilibrium data for Pt chlorocomplex on isothiouronium resins.

108

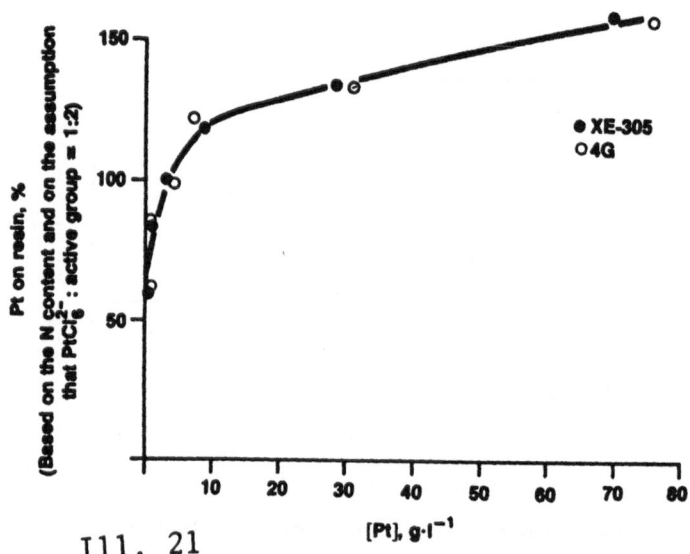

Ill. 21

Equilibrium data for Pt chlorocomplex on pyridinium resins.

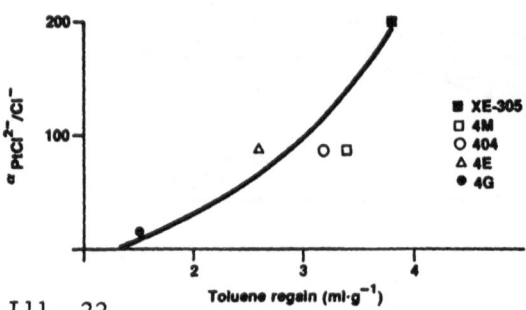

Ill. 22

. The effect of toluene regain of the matrix on the separation factor ($\alpha_{PtCl_6^{2-}/Cl^-}$ = $[\overline{PtCl_6^{2-}}][Cl^-]/[PtCl_6^{2-}][\overline{Cl^-}]$) for isothiouronium resins ([Pt] = 0.23 mol dm^{-3}, [Cl$^-$] = 4 mol dm^{-3}, units for concentration in resin phase: mmol g^{-1}).

3.4. Synthesis of Coordinating Polymers by Network Molding via Cooperative Effects. Polymeric Crown Ethers and Pseudocrown Ethers

In the previous section, we related mechanism of ion binding to matrix structure. In the case of flexible matrices, it is possible to use the matrix flexibility and ion templating of the ligand to synthesize new ligands which are part of the network structure itself. This is the case of polymeric pseudocrown ethers (40) (their synthesis is shown in scheme 18).

Open chain crown ethers were immobilized on polymers by Yanagida et al. (41) by polymer modification, and by Manecke et al. (42) (by polymerization of suitable monomers) and have shown catalytic behaviour in nucleophilic displacement reactions typical to "naked" anions, thus exhibiting regular cation complexation effects. Yanagida, studying blocked oligoethylene glycols (43) determined that the extraction effect is low with PEG-400 (n=9, $\%E$=6) : PEG-600 (n=14; $\%E$=14), but increase steadily to 100% extraction with PEG-4000 (n=90; $\%E$=100). The extraction patterns for PEG-4000 (n=23) are similar to 18-crown-6 and best with soft anions $SCN^->I^->>Br^-$, Cl^-, NO_3^-. On the other hand, polymeric psuedocrown (PPCE) do not bind cations, but rather anionic complexes, similar to amine-type anion exchange resins. Illustration 23 shows the ion coordination profiles for PPCE-14 in relation to chloride molarity (HCl or NaCl) for Au(III), Fe(III), Zn(II), Hg(II) and Cu(II), and illustration 24 shows complexation efficiency of several PPCE polymers (n=5,7,10,14) showing same trend of matching ionic diameters of complex and ligand cavity.

3.5. Thermoregulated Ion Complexation Effects – A Combination of Molecular and Macromolecular Effects

Thermoelution of ion exchange resins advanced by Bolto and his colleagues at the CSIRO in Melbourne (44) is a very appealing concept, since it allows a clean regeneration of resins. It is based on differences in water dissociation at various temperatures which influence the acid-base equilibria of caged polyelectrolytes, incorporated in the resin (44). Temperature-related differences in ion exchange equilibria are responsible for separations by "thermofractionation", particularly parametric pumping (45).

Thermodynamic properties of interactions between guest ions and host molecules are usually measured carefully, but rarely exploited practically. Only when the ligands are immobilized on a suitable matrice is it possible to take advantage of subtle differences. One example is the weak complexation of alkali salts by polymeric crown ethers. The enthropic factor relating to the

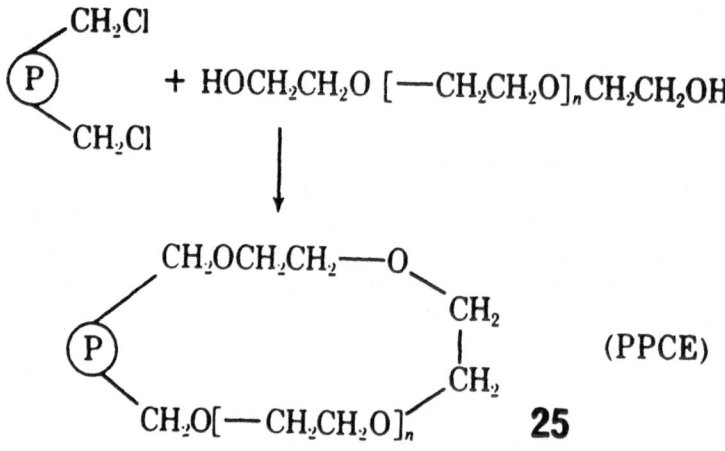

Scheme 18

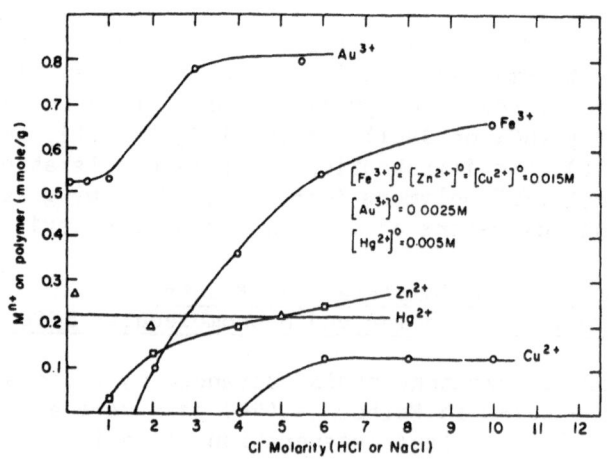

Ill. 23. Complexation efficiency for
PPCE-14 and anionic complexes of
several transition metals.

interaction of a small ion with a large host (macrocyclic crown

ether) coupled with another enthropic factor caused by the vicin-

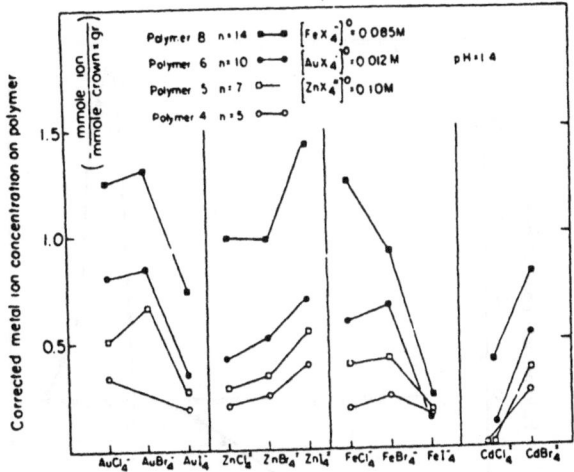

Ill. 24. Complexation profiles for
anionic halide complexes and several
$P(PCE)_n$ polymers (n=5,7,10,14).

ity of the macrocycle to a polymer backbone chain, result in complexation constants which are low compared to the complexation
constants of similar ligands in homogeneous phase. Consequently,
the IX complexation reaction

$$[P]-(CE)+MX \rightleftharpoons [P]-(CE)MX \qquad T_2 > T_1;$$
where [P]=polymer, CE=crown ether, M=alkali metal, and X=halide,

is shifted to the left when $T_2-T_1=30-40°C$. The polymeric crown
ethers described in scheme 19, synthesized from a porous network,
allow rapid ion diffusion, select alkali cations by the regular
ion-cavity matching priniciple, but by increasing the polymer
temperature to 60°C, a spontaneous decomplexation is observed
(illustration 25) (46-48).

Scheme 19

112

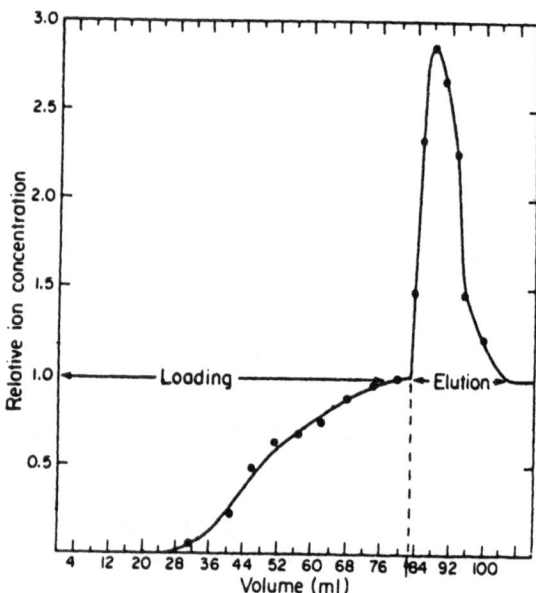

Ill. 25: Spontaneous ion elution from polymeric crown ether-6 (N-
K-41) by thermal shock at 60 °C

4. References

1. Cotton, F.A. and G. Wilkinson. Advanced Inorganic Chemistry, A
    Comprehensive Text, 4th Ed. (New York, Wiley & Sons 1980).
2. Christensen, J.J. and R.M. Izatt. Handbook of Metal-Ligand
    Heats and Related Thermodynamic Quantities, 3rd Ed. (New York,
    Marcel Dekker Inc. 1983).
3. Martell, A. and L.G. Silen. Stability Constants of Metal Ion
    Complexes. Chemical Society Special Publications, No. 17
    (1964), and No. 25 (1969).
4. Cattalini, L. Mechanisms of Square Planar Substitution. Inor-
    ganic Chemistry Series One, MTP Series, vol. 9 (London, But-
    terworth, 1972).
5. Langford, C.H. and V.S. Sastri. Mechanism and Steric Course of
    Octahedral Substitution. Inorganic Chemistry Series One, MTP
    Series, vol. 9 (London, Butterworth, 1972).
6. Warshawsky, A. Ion Exchange Separation of Platinum and Palla-
    dium by Nucleophilic Displacement with Thiocyanate or
    Thiourea. Separation and Purification Methods   (1983) **vol 12**
    37-48.

7. Warshawsky, A., M.M.B. Fibeberg, P. Mihalik, T.G. Murphy and
    Y.B. R as. The Separation of Platinum Group Metals (PGM) in
    Chloride Media by Isothiournium Resins. Separation and Purifi-
    cation Methods   (1980) 209-265. **vol 9**.
8. Warshawsky, A. Integrated Ion Exchange and Liquid-Liquid
    Extraction Process for the Separation of Platinum Group Metals
    (PGM). Separation and Purification Methods 11 (1982/3) 95-130.
9. Whewell, R.J. and C. Hanson. Metal Extraction with Hydroxyox-
    imes. Series on Ion Exchange and Solvent Extraction, vol. 8
    (New York, Marcel Dekker Inc. 1981).
10. Jones, K.C. and R.R. Grinstead. Properties and Hydrometalurgi-
    cal Applications of Two New Chelating Ion Exchange Resins.
    Chemistry and Industry (1977) 637-641.
11. Melby, L.R. Polymers for Selective Chelation of Transition
    Metal Ions. Journal of American Chemical Society, vol. 97
    (1975) 4044-4051.
12. Hancock, R.D., R.F. Bond and B.R. Green. Selective Ion
    Exchange Resins with Pyridyl-Imidazolyl Chelating Functional
    Groups. U.S.A. Patent No. 4202944, 1980.
13. Warshawsky, A. Selective Ion Exchange Polymers, Die Angewandte
    Makromolecularie Chemie, vol. 109/110 (1982) 171-196.
14. Warshawsky, A., A. Deshe, G. Rossey and A. Patchornik. Func-
    tionalization of Polystyrene II. Synthesis of Chelating
    Resins by Alkylation of Polybenzylamine. Reactive Polymers,
    vol. 2 (1984) 301-314.
15. Warshawsky, A. A Unified Approach to Chelating Phase Transfer
    Reagents. Proceedings of the International Solvent Extraction
    Conference, ISEC-77, Toronto, September 1977. Canadian Insti-
    tute of Mining, Special Volume (1979) 48-51.
16. Warshawsky, A., A. Deshe and Y. Shai. Transition Metal Chelat-
    ing Phase Transfer Agents from 3,4-Disubstituted Benzyl Hal-
    ides. Israel Journal of Chemistry (1985) in press.
17. Warshawsky, A. Solvent Impregnated Resins. Ion Exchange and
    Solvent Extraction, vol. 8 (New York, Marcel Dekker Inc. 1981)
    chapter 3.
18. Warshawsky, A. and A. Patchornik, Impregnated Resins. Metal
    Ion Complexing Agents Incorporated Physically in Polymeric
    Matrices. Israel Journal of Chemistry, vol. 17 (1978) 307-315.
19. Helfferich, F.G. Ion Exchange (New York, McGraw-Hill, 1962).
20. Egawa, H., H. Harada and T. Nonaka, Studies on Selective
    Adsorption Resins for Uranium in Sea Water. Nippon Kagaku Kai-
    shi (1980) 1767-1772.
21. Egawa, H., Y. Jogo and H. Maeda. Studies on Selective Absorp-
    tion Resins. Preparation of Chelating Resins Containing Mer-
    capto Groups from Poly(Glycidylmethacrylate) Beads and Absorp-
    tion of Metal Ions on Them. Nippon Kagaku Kaishi (1979)
    1759-1766.

114

22. Yamada, M., M. Takagi and K. Ueno. Synthesis and Chelating Behaviour of Bis(Carboxymethyl)Iminomethylene Derivative of Oligomeric Styrene. Journal of Coordination Chemistry, vol. 10 (1980) 257-262.

23. Lecat-Tillier, C., F. Lafuma et C. Quivoron. Synthese de Polyacrylamides Porteurs de Groupements Iminodiacetate et Etude de Leur Pouvoir de Complexation Vis-a-Vis de Quelques Cations. Metalliques European Polymer Journal, vol. 16 (1980) 437-445.

24. Davankov, V.A., S.V. Rogozhin and M.P. Tsyurupa. Influence of Polymeric Matrix Structure on Performance of Ion Exchange Resins. Ion Exchange and Solvent Extraction, vol. 7 (1977) 29-81.

25. Kimura, K., Y. Inaki and K. Takemoto. Vinyl Polymerization by Metal Complexes 30. On the Initiation Mechanism of Vinyl Polymerization by the System Copper (II) Chelate of Poly(Vinylalcohol)/Carbon Tetrachloride: Spin Trapping and Gelation Studies. Macromolecure Chemistry vol. 178 (1977) 317-328.

26. Akashi, M., H. Takada, Y. Inaky and K. Takemoto. Functional Monomers and Polymers. XXXXI. Template Polymerization of Methacryloyl-type Monomers Containing Nucleic Acid Bases. Journal of Polymer Science, Polymer Chemistry Edition, vol. 17 (1979) 747-757.

27. Mikulasova, D. and P. Citovicky. Redox Polymerization Systems for the Preparation of Grafted Polypropylene. I. Grafting of Polypropylene with Styrene in the Presence of Triethylenetetramine Metal Chelates. Chemiki Zvesti, vol. 27 (1973) 263-267.

28. Pizzi, A. Phenolic Resins by Reactions of Coordinated Metal Ligands. Journal of Polymer Science: Polymer Letters Edition, vol. 17 (1979) 482-492.

29. Nishide, H., J. Deguchi and E. Tsuchida. Selective Adsorption of Metal Ions on Crosslinked Poly(Vinylpyridine) Resin Prepared with a Metal Ion as a Template. Chemistry Letters (1976) 169-174.

30. Kato, M., H. Nishide and E. Tsuchida. Complexation of Metal Ion with Poly(1-Vinylimidazole) Resin Prepared by Radiation-Induced Polymerization with Template Metal Ion. Journal of Polymer Science, Polymer Chemistry Edition, vol. 19 (1981) 1803-1809.

31. Warshawsky, A. and R. Kalir. Twisted Polymers: Structural Reorganization in Macromolecular Network Through Metal Ion Coordination. Journal of Applied Polymer Science, vol. 24 (1979) 1125-1137.

32. Wulff, G. and I. Schulze. Enzyme-Analogue Built Polymers IX; Polymers with Mercapto Groups of Definite Cooperativity. Israel Journal of Chemistry, vol. 17 (1978) 291-297.

33. Arshady, R. and K. Mosbach. Synthesis of Substrate Selective Polymers by Host-Guest Polymerization. Makromolecular Chemistry, vol. 182 (1981) 687-692.

34. Andersson, L., B. Sellergren and K. Mosbach. Imprinting of Amino Acid Derivatives in Macroporous Polymers. Tetrahedron Letters, vol. 25 (1984) 5211-5214.

35. Shigeehara, K., A. Yamada, H. Sano and E. Tsuchida. Study on the Complex Formation of Cobalt Chelates with Polymer Ligands. Makromolecular Chemistry, vol. 181 (1980) 1823-1840, and references therein.

36. Nishide, H., N. Shimidzu and E. Tsuchida. Chelating Resin: Pyridine Derivatives Attached to Poly(Styrene)Beads with Spacer Groups. Journal of Applied Polymer Science, vol. 27 (1982) 4161-4169.

37. Koster, G. and G. Schmuckler. Separation of Noble Metals from Base Metals by Means of a New Chelating Resin. Analytica Chimica Acta, vol. 38 (1967) 179-184.

38. Green, R.B. and R.D. Hancock. The Role of Matrix Effects on Selectivity in Ion Exchange Resins. Hydrometallurgy, vol. 6 (1981) 353-363.

39. Pope, L.E. and J.C.A. Boyenes. Crystal structures of S-Benzyl-isothiouronium Hexachloroplatinate (IV) and Tetrachloroaurate (III). Journal of Crystallographic Molecular Structure, vol. 5 (1975) 47-58.

40. Warshawsky, A., R. Kalir, R. Deshe, H. Berkovitz and A. Patchornik. Polymeric Pseudocrown Ethers. 1. Synthesis and Complexation with Transition Metal Anions. Journal of American Chemisty Society, vol. 101 (1979) 4249-4258.

41. Yanagida, S., K. Takabashi and M. Okahara. Solid-Liquid Three Phase Catalysis of Polymer-Bound Acyclic Poly(Oxyethylene) Derivatives. Applications to Organic Synthesis. Journal of Organic Chemistry, vol. 44 (1979) 1099-1103.

42. Hiratani, K., P. Rueter and G. Manecke. Preparation and Catalytic Behaviour of Polymers with Pendant Oligoethylene Groups (Polymers of Non-Cyclic Crown Ethers). Israel Journal of Chemistry, vol. 18 (1979) 208-213.

43. Yanagida, S., K. Takahashi and M. Okahara. Metal Ion Complexation of Non-Cyclic Poly(Oxyethylene) Derivatives. I. Solvent Extraction of Alkali and Alkaline Earth Metal Thiocyanates and Iodides. Bulletin of the Chemical Society of Japan, vol. 50(1977) 1386-1390.

44. Bolto, B.A. and D.E. Weiss. The Thermal Regeneration of Ion Exchange Resins. Ion Exchange and Solvent Extraction (New York, Marcel Dekker Inc. 1977) **vol 7**.

45. Grevilliot, G. and D. Tondeur. Silver-Copper Separation by Continuous Ion Exchange Parametric Pumping. Ion Exchange Technology. (Chichester, Ellis Horwood Publishers, 1984) 653-660, and references therein.

46. Warshawsky, A. and N. Kahana. Temperature-Regulated Release of Alkali Metal Salts from Novel Polymeric Crown Ether Complexes. Journal of American Chemical Society, vol. 104 (1982) 2663-2664.

47. Kahana, N., A. Deshe and A. Warshawsky. Synthesis of Polymeric Crown Ethers and Thermoregulated Ion Complexation Effects. Journal of Polymer Science, Polymer Chemistry Edition, vol. 23 (1985) 231-253.

48. Warshawsky, A., A. Deshe and N. Kahana. Thermoregulated Ion Complexation Effects in Polymeric Crown Ethers. 2. Polymeric Sulfon- amidobenzo-18-crown-6. Reactive Polymer (1985) in press.

PART III
CHEMICAL ENGINEERING ASPECTS OF
ION EXCHANGE PROCESSES

THERMODYNAMICS OF ION EXCHANGE:PREDICTION OF MULTICOMPONENT EQUILIBRIA FROM BINARY DATA

Alan L.Myers and Sue Byington

Department of Chemical Engineering
University of Pennsylvania
Philadelphia,PA 19104 USA

INTRODUCTION

The elementary principles of ion exchange are covered in the excellent monograph by Helfferich [5].Solid materials such as zeolites and resins are used to capture ions from solution by adsorption. However,the requirement of electroneutrality means that the solid must release one ion equivalent for every one captured.A typical cation exchange is:

$$2NaX(s) + CaCl_2(aq) = CaX_2(s) + 2NaCl(aq) \qquad (1)$$

where NaX is a solid ion exchanger containing exchangeable Na^+ ions. Two sodium ions are released for each calcium ion captured as the cation exchanger,originally in the Na^+ form,is converted to its Ca^{2+} form.The degree of conversion,and therefore the amount of calcium adsorbed,increases with the original concentration of the calcium salt according to the law of mass action.

The reversible nature of ion exchange enables the solid to be regenerated and reused by treating the Ca-saturated solid with a concentrated solution of NaCl.

The ion (cation or anion) released by the ion exchanger to the bulk solution is called the counter ion,which may be exchanged with other ions of like charge in the bulk solution.The oppositely charged ions which are not exchanged are called co-ions.

The basic principles of multicomponent ion exchange in fixed beds have been worked out for the case of constant separation factor [15]. The question of whether separation factors can be correlated or predicted has been the subject of numerous studies[3,4,5,7,8,9,10,11, 16].Several of these investigations have focused on the question of whether equilibria for three cations (or anions) can be predicted from data on the three constituent pairs. Conclusions were, in general, positive. However, the approach is impractical for systems higher than ternaries: a ternary system requires data on 3 binaries, a quaternary system on 6 binaries, etc. One objective of this paper is to study the possibility of predicting multicomponent (N ions) exchange from (N-1) isotherms for the constituent pairs.

The second objective of this paper is to describe some applications of thermodynamics to ion exchange. Most of the previous work on this subject has been devoted to the interpretation of ion exchange as a chemical reaction [1,5,14] as suggested by Eqn.(1). This paper introduces a new approach in which ion exchange is treated as a phase equilibrium using standard procedures developed for solution thermodynamics. Surface effects are accounted for by introducing surface excess variables similar to those used to measure adsorption from liquid mixtures on solids.

EXPERIMENTAL METHODS

A binary ion exchange involves four components: a liquid solvent containing two dissolved electrolytes and the solid. The liquid solvent is usually water and the electrolytes often have a common co-ion. There are then five species : water, two counter ions, one co-ion, and the resin. However, these five species constitute a four-component system because of the requirement of electroneutrality. For four components (C=4) and two phases (P=2) at equilibrium, the Gibbs phase rule predicts four degrees of freedom (F=C+2-P). Convenient experimental variables are temperature, pressure, solution normality and ionic fraction of one of the counter ions. The effect of pressure is small and is usually ignored. The usual experimental procedure is to hold the temperature and solution normality fixed while varying the ionic fraction.

In general, the counter ions carry different charges and it is convenient to express their concentration in terms of the equivalent ionic fraction. For counter ions A and B carrying charges z_A and z_B and having solution molalities m_A and m_B, the equivalent ionic fraction is defined by:

$$x_A \equiv \frac{z_A m_A}{z_A m_A + z_B m_B} \tag{2}$$

This definition has the advantage that the number of equivalents which can be exchanged is the same for both counter ions. Let the equivalent ionic fraction of A in the solid phase be represented by $\bar{x}_A$. An ion-exchange isotherm is the equilibrium relation between $\bar{x}_A$ in the solid and x_A in the solution. For example, Figure 1 shows isotherms for Cu^{2+}/Na^+ exchange on Dowex 50-X8 at ambient pressure and for various total normalities $(CuCl_2+NaCl)$ in the bulk solution. Note the effect of the total solution concentration upon the equilibrium.

The resin shows a strong preference for copper at low concentration, while at high concentration sodium is preferentially adsorbed.At a total normality of 2.0,the resin is neutral and shows no preference for either ion.

A more sensitive parameter for measuring the effectiveness of ion exchange is the relative separation factor defined by [5]:

$$S_{A,B} \equiv \frac{\bar{x}_A x_B}{\bar{x}_B x_A} \qquad (3)$$

This quantity is also called selectivity [3] or selectivity coefficient [1]. Figure 2 is the same data as Figure 1,but illustrates with greater clarity how the separation effectiveness varies with solution normality or concentration.The relative separation factor decreases by more than two orders of magnitude as the solution normality increases.Even at constant normality,the separation factor is a strong function of counter-ion composition, especially in the region of low ionic fraction of copper where a separation process normally would be carried out.

The usual experimental procedure for determining the ionic fractions in Eq.(3) is to separate the solid from the solution after equilibrium with a feed of known concentration.Then the dry solid containing the two counter ions is washed with excess solution containing a third counter ion and the elutant is analyzed.However,it is difficult to separate the solid from the solution without disturbing the equilibrium.

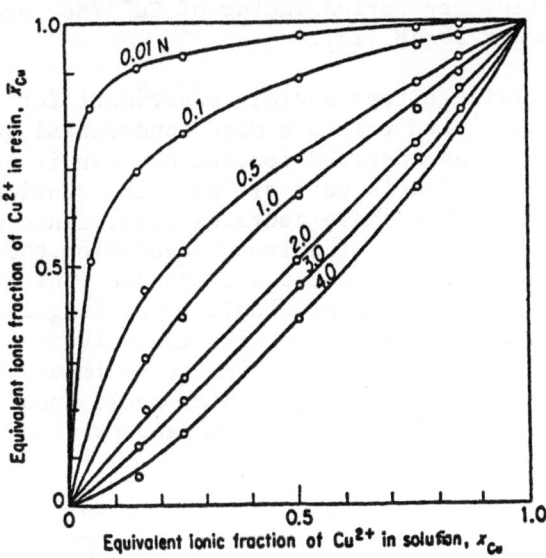

Figure 1- Isotherms for Cu^{2+}/Na^+ exchange on Dowex 50-XB[13] for various total normalities in the bulk solution.

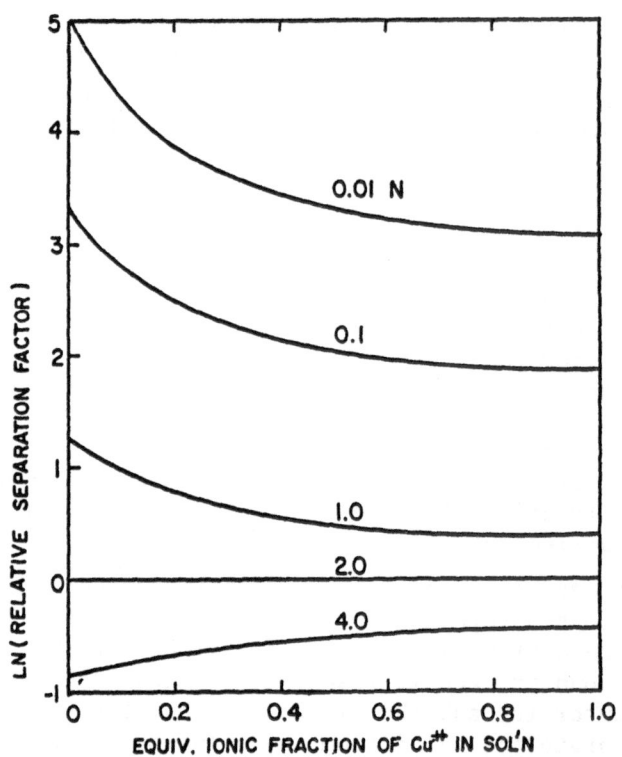

Figure 2- Relative separation factor of Cu^{2+}/Na^+ exchange on Dowex 50-X8 [13].

The problem of devising a reproducible experiment for measuring the adsorbed-phase composition raises a more fundamental question. What is the definition of an adsorbed ion,and how can it be distinguished from another ion which is part of the bulk solution but very close to the surface?The solid phase contains fixed ionic groups and bound counter-ions (A,B).Because of solvent adsorption,the solid phase also contains both solvent molecules and unbound ions or dissolved electrolyte (AX,BX).Not only it is difficult to distinguish bound counter-ions from counter-ions adsorbed with the solvent in pores of the solid,but,from a practical point of view,it is impossible to define a reproducible experimental technique which would consistently separate resin and liquid for different ion exchangers with a wide variety of surfaces and porosities.

This ambiguity about distinguishing the bulk liquid from the adsorbed phase can be resolved by defining the surface excess, a quantity which has been adopted by physical chemists to describe adsorption from liquids on solids but is new in the field of ion exchange.Consider,for example, the exchange of two counter ions A and B.The surface excess of ion A is defined by:

$$n_A^e \equiv n_A^o - (n_A^o + n_B^o) \, x_A \tag{4}$$

where n_i^o refers to the number of ion equivalents of i per gram of solid at the start of the experiment,and x_A is the equivalent ionic fraction of A in the bulk solution at equilibrium.This definition avoids measuring the composition of the solid phase.Since $n_B^e = -n_A^e$ according to Eqn.(4),the sum of surface excesses for both counter ions is zero.

The most straightforward procedure is to contact the solid in B-form with solution containing a single solute AX,where X is the common co-ion.The B-form is prepared by driving the solid to exhaustion with a solution containing concentrated BX electrolyte.Different ratios of solid/solution are contacted and,after equilibrium is established,the composition of the bulk solution is measured.The result is a relationship like that shown on Figure 3,calculated for a solid with a capacity of 1 eq/kg and a relative separation factor of 5. Experiments are conveniently conducted at constant solution normality because ion exchange has no effect on this variable.At equilibrium, the number of ion equivalents of B released from solid to solution (n_B) is equal to $x_B N/R$,where N is the solution normality,R is the mass ratio of solid/solvent (the abcissa of Fig. 3) and $x_B = (1-x_A)$. n_B^o,the number of ion equivalents on the resin in pure B-form, is determined as shown on Figure 4 by extrapolating n_B to "infinite dilution" of the solid. n_A^o at each point of Fig.3 is equal to N/R. Figures 3 and 4,combined with Eqn.(4),yield Figure 5,the surface excess isotherm as a function of ionic fraction in solution at equilibrium.

The establishment of equilibrium can be checked by reversal of the process,which is achieved by contacting the solid in pure A-form with a solution containing solute BX.

The surface excess has more fundamental significance than other possible measurements and is useful because it provides directly the basis for a material balance on the counter ions.The absolute ionic fraction on the surface,while perhaps a more intuitive quantity than surface excess,depends on the procedure used to separate the two phases.

Surface excess is measured in batch experiments without separating the resin from the liquid phase.Ion exchange columns also operate without mechanical separation of solid and liquid.Consider a column with feed of ionic fraction x_A and containing a resin (R) initially in pure B-form which performs the exchange RB $\rightarrow$ RA. Let the total resin capacity for this exchange be n^o. The number of ion equivalents of A exchanged at equilibrium (n_A) is obtained by rearranging Eqn.(4):

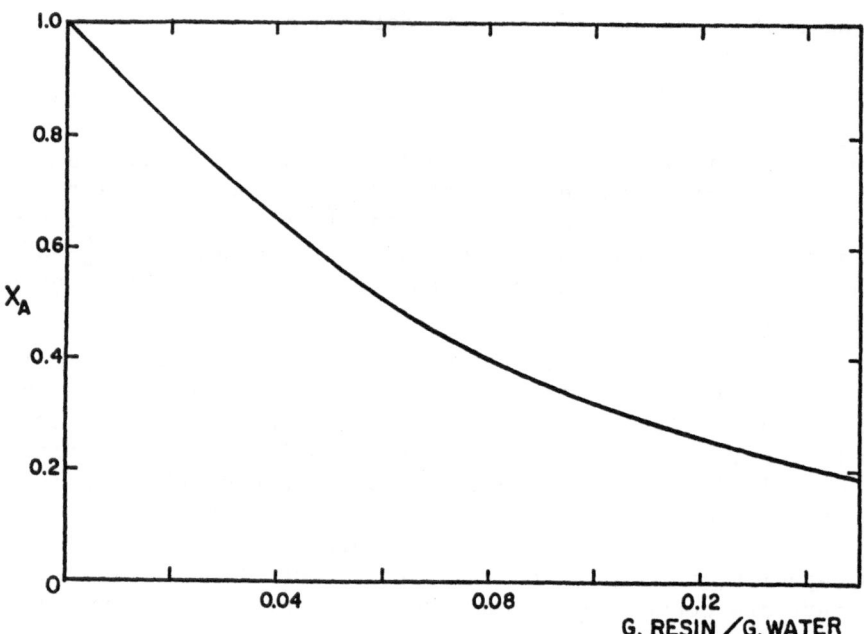

Figure 3- Ionic fraction of ion A in bulk solution at equilibrium as a function of resin/solvent ratio,for a constant relative separation factor $S_{A,B}$=5 and for an exchanger with a capacity of 1 eq/kg.Exchanger initially in B-form is contacted with a solution containing ion A.

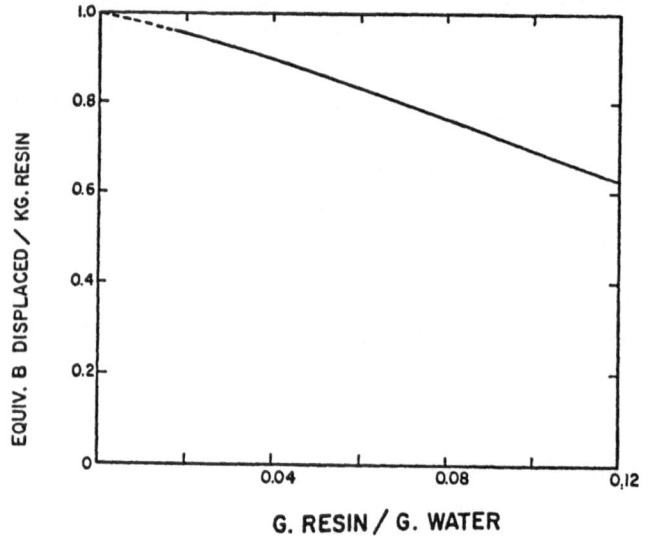

Figure 4- Determination of total capacity of exchanger by extrapolation to "infinite dilution" of resin.System described in Figure 3.

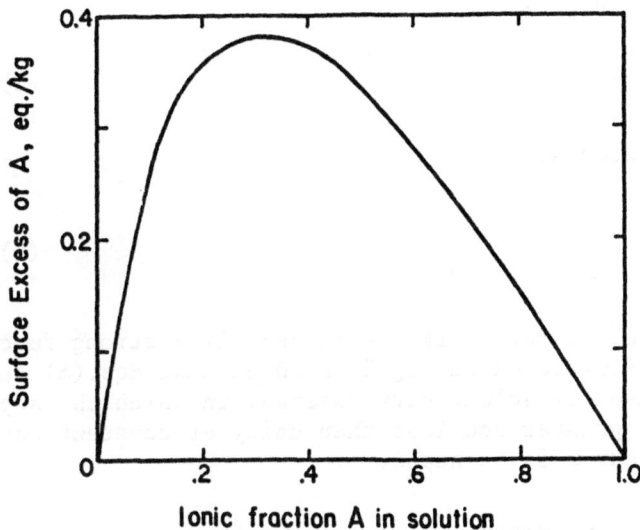

Figure 5- Surface excess isotherm for A/B exchange.
System described in Figure 3.

$$n_A = n_A^e + n^o x_A \qquad (5)$$

Thus, the total amount of A ions exchanged (per unit mass of resin) is the sum of the surface excess and the amount that would be removed from solution ($n^o x_A$) even if the exchanger had no preference for either ion.

Although the absolute surface concentration is not easily determined, an ionic fraction in the solid may be defined in terms of the surface excess. Let n^o be the total capacity of the solid in ion equivalents, measured as described above. The adsorbed-phase ionic fraction is:

$$\bar{x}_A \equiv x_A + n_A^e / n^o \qquad (6)$$

This definition of adsorbed-phase composition relies only on measurements on the bulk solution made before and after the attainment of equilibrium.

The relationship between the relative separation factor and the surface excess follows from Eqs. (3),(4) and (6):

$$n_A^e = \frac{n^o x_A x_B (S_{A,B} - 1)}{S_{A,B} x_A + x_B} \qquad (7)$$

If $n_A^e > 0$ and the relative separation factor ($S_{A,B} = K$) is constant then the maximum in n_A^e is:

$$n_A^e = n^o \frac{(\sqrt{K}-1)^2}{K-1} \tag{8}$$

This maximum is located at:

$$x_A^{max} = \frac{\sqrt{K}-1}{K-1} \tag{9}$$

However, the relative separation factor is usually a strong function of composition as illustrated on Fig.2, in which case Eqs.(8) and (9) do not apply. For cases of selectivity reversal in which the separation factor is both greater and less than unity at constant normality, the surface excess curve is S-shaped.

THERMOTYNAMICS OF ION EXCHANGE

Streat and Brignal [14] have pointed out the "considerable lack of thermodynamic rigor" in the equilibrium models for ion exchange. However, it is possible to derive equations for ion exchange equilibrium which are both rigorous and easy to apply.

Ion exchange isotherms can be reported as $\bar{x}_A$ versus x_A, $S_{A,B}$ versus x_A, or as n_A^e versus x_A. The role of thermodynamics is to provide a network of equations relating these variables. The historical approach is to treat ion exchange as a chemical reaction:

$$\nu_A A(s) + \nu_B B(aq) = \nu_A A(aq) + \nu_B B(s)$$

where the stoichiometric coefficient $\nu_A = |z_B|$ and $\nu_B = |z_A|$. The chemical equilibrium constant is:

$$K = \Pi (a_i)^{\nu_i} \tag{10}$$

a_i in Eq.(10) is the activity of i'th ion. The chemical approach has the advantage that the constant K is independent of normality and composition and varies only with temperature. The difficulty is that the activities in the solid phase are inaccessible by experiment, and this leads to non-trivial problems of defining the standard state for the adsorbed ions.

The chemical equilibrium constant K in Eq.(10) can be written as the product $K_\gamma K_x$, where K_γ contains the effects of nonideality in both phases and, for 1-1 ion exchange, K_x is equal to the relative separation factor of Eq.(3). Experimentally K_x is a strong function of both normality and ionic fraction (see Fig. 2); therefore $K_\gamma = K/K_x$ is also a strong function of the same variables.Given independent measurements of solution-phase activity coefficients,the chemical equilibrium approach allows one to calculate solid-phase activity coefficients from experimental values of K_x. However,these coefficients cannot be measured independently and they do not appear to be correlated with activity coefficients in bulk solution.Consequently the chemical approach to ion exchange,while appealing at first sight,becomes in the final analysis a tautology in which activities are defined in such a way to force K to maintain its constancy.

The solution thermodynamics approach is to specify equality of activities in both phases at equilibrium so that:

$$\bar{a}_i = a_i \qquad (11)$$

The advantage of this method is that the equations of solution thermodynamics may be applied to the equilibria.In addition,most adsorption theories are couched in terms of the variables of phase equilibrium, and they may be applied to ion exchange provided the condition of electroneutrality is introduced to the model.The activity and standard state for the adsorbed phase must still be defined in the formal equations,but these quantities drop out of equations for the surface excess.

For example,an integral thermodynamic consistency test may be derived from the Gibbs adsorption isotherm [8].Consider a system composed of 1-1 counter ions A and B,common co-ion X,solvent S and resin R. The components of the mixture are AX,BX,AR,BR and S.The Gibbs-Duhem equation is written first for the bulk solution and then for the resin phase:

$$n_{AX} \, d\mu_{AX} + n_{BX} \, d\mu_{BX} + n_S \, d\mu_S = 0 \qquad (12)$$

$$\bar{n}_{AR} \, d\bar{\mu}_{AR} + \bar{n}_{BR} \, d\bar{\mu}_{BR} + \bar{n}_{AX} d\bar{\mu}_{AX} + \bar{n}_{BX} d\bar{\mu}_{BX} + \bar{n}_S d\bar{\mu}_S = 0 \qquad (13)$$

Eq.(13) for the solid phase contains terms for adsorbed solvent and for counter ions adsorbed with the solvent in pores of the solid. n refers to number of moles and μ is chemical potential.Isothermal surface and pressure effects are included implicitly in the terms for the solid resin (AR and BR). If these equations are combined and the chemical potentials of AX,BX and S in the solution phase are set equal to the corresponding chemical potentials in the resin phase,and the result is integrated at constant solution normality over three binary pairs of counter-ions formed from three exchangeable counter ions,it can be shown [8] that:

$$\int_{x_A=0}^{1} n_A^e d\ln(a_{AX}/a_{BX}) + \int_{x_B=0}^{1} n_B^e d\ln(a_{BX}/a_{CX}) + \int_{x_C=0}^{1} n_C^e d\ln(a_{CX}/a_{AX}) = 0 \qquad (14)$$

Eq.(14) is a rigorous equation based on experimentally accessible variables: surface excess and bulk solution activity. The three integrals form a triangle rule for three successive binary pairs of counter ions {AB,BC,CA} formed from the set {A,B,C}. If Eq.(14) is not obeyed by experimental data obtained at constant solution normality and temperature,the conclusion is that the data are somehow either erroneous or do not correspond to equilibrium conditions.

Eq.(14) may be simplified under certain conditions.If Harned's rule [8] is obeyed for the activity coefficients of the electrolytes in the bulk solution,then activities may be replaced by ionic fractions.If in addition the total capacity of the resin is the same for all counter-ions (it may not be if smaller ions have access to cavities which cannot be penetrated by larger ions),the result is:

$$\bar{S}_{A,B} \cdot \bar{S}_{B,C} \cdot \bar{S}_{C,A} = 1 \tag{15}$$

where $\bar{S}_{A,B}$ is an average relative separation factor given by:

$$\bar{S}_{A,B} = \exp \left[\int_{x_A=0}^{1} \frac{S_{A,B} - 1}{S_{A,B} x_A + x_B} dx_A \right] \tag{16}$$

with similar equations for the BC and CA pairs of counter ions.In the special case where the relative separation factor is constant $(S_{A,B}=K)$,Eq.(16) reduces to:

$$S_{A,B} = \bar{S}_{A,B} = K \tag{17}$$

Eqs.(15) and (16) also apply,under the same approximations, to counter ions carrying different charges provided their concentration is expressed in equivalent ionic fractions.

Experimental data for pairwise ion exchangers of systems of 3 ions are available for a variety of systems [1,2,3,4,7,9,10,11,12, 16].Some of these were not suitable for analysis,either because of too much scatter in the data or due to an insufficient number of experimental points.The eight systems tabulated in Table 1 were tested for thermodynamic consistency.Consider,for example,the Mg^{2+}/K^+ exchange on KU-2. Eq.(16) was derived for constant normality but should also apply to the experimental conditions of constant ionic strength. The total area for the integral of Eq.(16) was 3.456, so $S=e^{3.456} = =31.69$ .Similar integrations were carried out for the other two ion pairs in the $\{Mg^{2+},Ca^{2+},K^+\}$ system and the product of the three average separation factors was compared to unity according to Eq.(15). The experimental product is $(31.69)(2.246)(0.0160)=1.14$,so these data exhibit a good degree of thermodynamic consistency. A product of 1.33 indicates an error of at least 10 percent in the individual separation factors,which is still acceptable because a 10 percent error in the separation factor corresponds to an average error of 2 or 3 percent in mole fraction.

TABLE 1- Average separation factors from the integral thermodynamic consistency test.

RESIN	BINARY ION PAIR #1	$\bar{S}_{\#1}$	BINARY ION PAIR #2	$\bar{S}_{\#2}$	BINARY ION PAIR #3	$\bar{S}_{\#3}$	PRODUCT
KU-1	$Ca^{2+} - Mg^{2+}$	1.56	$Mg^{2+} - K^+$	7.71	$K^+ - Ca^{2+}$	.095	1.15
KU-2	$Ca^{2+} - Mg^{2+}$	1.57	$Mg^{2+} - K^+$	8.37	$K^+ - Ca^{2+}$	.080	1.05
SULFONATED CARBON	$Ca^{2+} - Mg^{2+}$	1.66	$Mg^{2+} - K^+$	10.12	$K^+ - Ca^{2+}$	.063	1.05
AMBERLITE IRA 400	$NO_3^- - SO_4^{2-}$	4.45	$SO_4^{2-} - Cl^-$	0.77	$Cl^- - NO_3^-$	.270	0.93
DOWEX 50	$NH_4^+ - H^+$	1.75	$H^+ - Li^+$	1.26	$Li^+ - NH_4^+$	.470	1.04
DOWEX 50	$K^+ - NH_4^+$	1.16	$NH_4^+ - H^+$	1.75	$H^+ - K^+$	.480	0.97
DOWEX 50	$Ag^+ - NH_4^+$	2.96	$NH_4^+ - Li^+$	2.13	$Li^+ - Ag^+$	.130	0.82
DOWEX 50 (8% DVB)	$Ag^+ - Na^+$	9.59	$Na^+ - H^+$	2.12	$H^+ - Ag^+$	.066	1.34
DOWEX 50 (16% DVB)	$Ag^+ - Na^+$	2.99	$Na^+ - H^+$	1.47	$H^+ - Ag^+$	.180	0.79

THEORY OF ION EXCHANGE

Ion exchange equilibrium is a very complicated problem, as can be appreciated by studying the complex behavior illustrated on Fig.2. The isothermal separation factor can vary by several orders of magnitude with solution normality. Even at constant normality, very few systems exhibit a constant separation factor. A satisfactory theory must have the capacity to describe the variation of separation factor with solution normality and ionic fraction. The theory should exploit Eq.(15) so that predictions can be made according to the triangle rule. A model capable of predicting the separation factor for A/B exchange from B/C and A/C exchanges can be expected to apply for the ternary {A,B,C} system. Such a theory would allow an N-ion exchange to be predicted from (N-1) constituent pairs.Thus,equilibrium for a quaternary system containing ions {A,B,C,D} could be predicted from experimental data for the three pairs A/B,A/C,A/D.Such a simplification would be extremely useful,because most applications of ion exchange involve more than two counter ions.

Recent theories of physical adsorption of gas and liquid mixtures on solids indicate that energetic heterogeneity of adsorption sites plays an important role.There is no reason why the same heterogeneity effect should not be observed in ion exchange.If the surface were not heterogeneous,one would expect the relative separation factor to be approximately constant.Nonidealities in the bulk and surface phase would produce only minor deviations from this rule. But if the surface contains high and low energy sites,then the separation factor for the preferentially adsorbed species should decrease rapidly with its ionic fraction and with the total solution normality.Figure 2 shows exactly this behavior,and similar results have been observed in many other systems.

Another factor which should be considered in a comprehensive theory of ion exchange is nonideality in the solid phase,which depends on ion-solid and ion-ion interactions,ionic charge,ionic radii, polarizability,and other molecular constants.Also,occlusion effects [1] cause the total solid capacity to depend upon the ionic radii, especially in zeolites.However,a practical theory must strike a compromise and we propose to consider only energetic heterogeneity.Refinements can be added later as required.Specifically we admit at the outset the impossibility of making a priori predictions of ion exchange,even though many useful rules of thumb are available [5]. We adopt instead the approach of using a minimum of experimental measurements within the framework of an approximate theory to make predictions.

A theory of energetic heterogeneity requires the selection of an energy distribution.Many choices are possible,including discrete and continuous distributions.Some candidates from the class of continuous distributions include the normal,gamma and uniform probability functions.The choice is not critical because there is no method available for independent determination of distributions derived from equilibrium measurements.

The moments of a probability distribution are the first (average value),second (variance),third (skewness),etc.The moments are the quantities which have physical significance,and it is unlikely that an equilibrium experiment can distinguish different distributions much beyond their third moments.Here we choose the discrete binomial distribution on the basis of its flexibility and mathematical convenience:

$$P_i = \binom{n}{i} p^i (1-p)^{n-i} \qquad 0 < p < 1 \qquad\qquad (18)$$

p is the skewness parameter, ν is the number of sites, $n=(\nu-1)$ and $i=0,1,2,\ldots,n$. For a symmetric distribution $p=1/2$. For $p=1/2$ the binomial distribution approaches the normal distribution as $\nu \to \infty$. The energy (defined as a positive quantity) of an ion adsorbed on site $\underline{i}$ is:

$$E_i = \bar{E} + \frac{i-np}{\sqrt{np(1-p)}} \sigma \qquad (i=0,1,2,\ldots,n) \qquad (19)$$

$\bar{E}$ is the average energy and σ is the standard deviation of the distribution. The Langmuir equation is chosen to describe the adsorption of ion $\underline{j}$ on site $\underline{i}$:

$$n_{ij} = \frac{n^0 C_{ij} x_j N}{1 + \sum_j C_{ij} x_j N} - \qquad (20)$$

N is the solution normality and x_j is the equivalent ionic fraction of ion $\underline{j}$ in the solution. The neglect of bulk phase activity coefficients in Eq. (20) is justifiable only for experimental data obtained at constant normality. It is assumed that the total ionic capacity of the solid (n^0) is the same for all pairs of counter ions; otherwise the Langmuir equation fails. The constant C_{ij} corresponding to ion $\underline{j}$ on site $\underline{i}$ is given by:

$$C_{ij} = \bar{C}_j \exp\left(\frac{E_{ij} - \bar{E}_j}{RT}\right) \qquad (21)$$

From Eq. (19), the energy for ion $\underline{j}$ on site $\underline{i}$ may be written as:

$$E_{ij} - \bar{E}_j = f_i \sigma_j \qquad (22)$$

where

$$f_i = \frac{i - np}{\sqrt{np(1-p)}} \qquad (i=0,1,2,\ldots,n) \qquad (23)$$

Carrying out the summation over sites $n_j = \sum_i n_{ij} p_i$ using Eqns. (18) and (20) and substituting the result in Eq. (3), it is found that:

$$S_{j,k} = \frac{n_j/x_j}{n_k/x_k} \qquad (24)$$

where

$$n_j/x_j = \sum_{i=0}^{n} \frac{C_{ij} p_i N}{1 + \sum_j C_{ij} x_j N}$$

and

$$n_k/x_k = \sum_{i=0}^{n} \frac{C_{ik} P_i N}{1 + \sum_k C_{ik} x_k N}$$

Eq.(24) is a general equation for the separation factor of counter ion j relative to k, in a solution containing an arbitrary number of counter ions. The parameters are the number of sites (ν), the skewness of the distribution (p), the average energy ($\bar{E}_i$) and standard deviation (σ_i) for each counter ion. Experimental variables are the solution normality (N), the relative ionic fraction in the solution (x_j), and the temperature (T).

In the limit of zero normality the separation factor approaches a value which is independent of both the ionic fraction and other species present. For example, for a symmetric (p=1/2) two-site ($\nu=2$) distribution:

$$\lim_{N \to 0} S_{i,j} = \bar{S}_{i,j} \frac{\cosh(\sigma_i/RT)}{\cosh(\sigma_j/RT)} \tag{25}$$

The usual case is for the component with the higher value of mean energy to possess the larger dispersion ($\sigma \propto \bar{E}$). Therefore the limiting value of the separation factor is normally larger than the mean value ($\sigma_i > \sigma_j$ for $\bar{S}_{i,j} > 1$).

Another interesting limit of Eq.(24) is high normality (N $\to \infty$) for which the ratio n_j/x_j reduces to:

$$n_j/x_j = \sum_{i=0}^{n} \frac{C_{ij} P_i}{\sum_j C_{ij} x_j} \tag{26}$$

with a similar equation for n_k/x_k. For example, for binary exchange of counter ions 1 and 2 on an exchanger with a symmetric (p=1/2) two-site distribution:

$$S_{1,2} = \bar{S} \frac{\eta + x_1(\bar{S} - \eta)}{1 + x_1(\bar{S}\eta - 1)} \tag{27}$$

where:

$$\eta = \cosh[(\sigma_1 - \sigma_2)/RT]$$

and:

$$\bar{S} = \exp[(\bar{E}_1 - \bar{E}_2)/RT]$$

The relative separation factor depends on its mean value and the difference of the standard deviations for the two distributions. When $\sigma_1 = \sigma_2$ the relative separation factor is constant; otherwise its maximum value is at infinite dilution of the more strongly adsorbed ion:

$$\lim_{x_1 \to 0} S_{1,2} = \eta S$$

The minimum value is:

$$\lim_{x_1 \to 1} S_{1,2} = S/\eta$$

Eq.(27) is plotted on Figure 6 for a mean separation factor $\bar{S}=2$ and for various values of $\Delta\sigma/RT$. Selectivity reversal is possible; the curve for $\Delta\sigma/RT=2$ has an azeotrope at $x_1=0.65$.

Constants for a representative ternary system $\{A,B,C\}$ are listed in Table 2. Insertion of the mean selectivities and heterogeneity parameters in Table 3 into Eq.(27) yields the binary exchange curves shown on Figure 7. The ionic fraction in the bulk phase refers to the first ion listed. The B/C exchange exhibits a selectivity reversal at $x_B=0.72$.

TABLE 2 —Energy distributions constants

Counter-ion	$\bar{E}/RT$	σ/RT
A	11.099	4.25
B	10.223	4.00
C	10.000	3.00

TABLE 3 —Constants of Eq.(27) for binary pairs of ions calculated using energy distribution parameters from Table 2

System	$\bar{S}$	$\Delta\sigma/RT$
A,C	3.00	1.25
A,B	2.40	0.25
B,C	1.25	1.00

The thermodynamic consistency of Eq.(27) can be checked by integration according to Eq.(16). Let $x=x_1$; the integrand of Eq.(16) is:

134

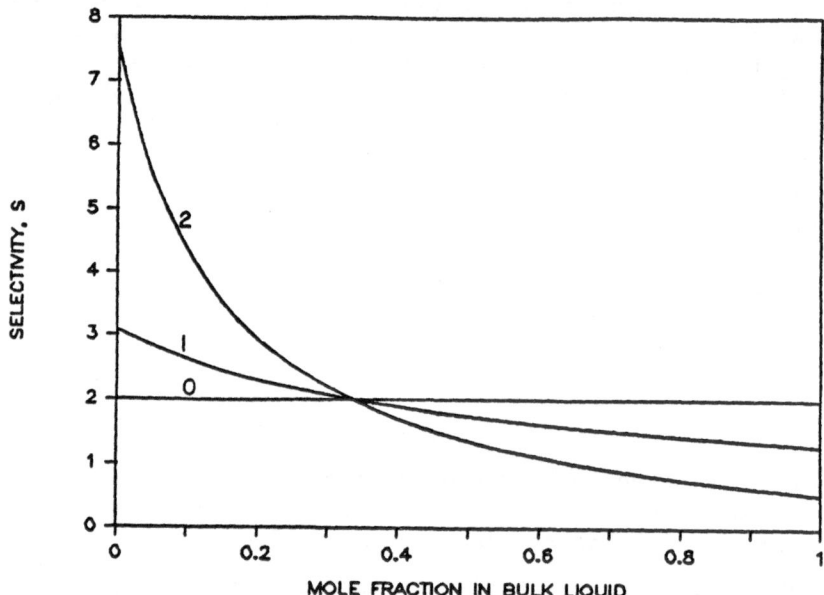

Figure 6 –Relative separation factor of a system with a mean
value of S=2 for various values of the reduced standard deviation ($\Delta\sigma/RT$) of the energy distributions.

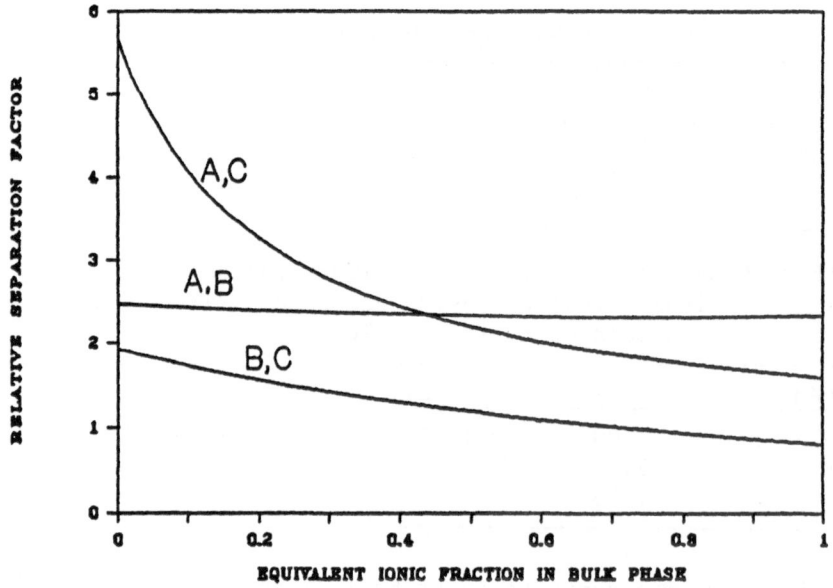

Figure 7 –Relative separation factors for exchange of the three
binary pairs formed from ions {A,B,C} calculated from
Eq.(27) using constants in Table 3. These curves satisfy Eq.(15)

$$\frac{S_{1,2}^{-1}}{S_{1,2}x+(1-x)} = \frac{x\left[\bar{S}^2 -2\bar{S}\eta +1\right] +\bar{S}\eta -1}{x^2\left[\bar{S}^2 - 2\bar{S}\eta+1 \right]+ 2x\left[\bar{S}\eta -1\right]+ 1}$$

This has the form $(1/2)(dU/dx)/U$ and can be integrated analytically. Let I be the definite integral in Eq.(16):

$$I=(1/2)\ \ln\left[\ x^2(\bar{S}^2 -2\bar{S}\eta +1) +2x(\bar{S}\eta -1)+1\right]\ \Big|_0^1$$

$$I=(1/2)\ \ln\left[\ \bar{S}^2\right]=\ln \bar{S}$$

From Eq.(16):

$$\bar{S}_{1,2}=\exp(I)=\exp\left[\ln\bar{S}\right] =\bar{S}$$

Therefore the theory agrees with the triangle rule of Eq.(15) as required by thermodynamics.

The theoretical effect of solution normality is obtained from Eq.(24). Figure 8 shows a plot of $\bar{x}_A$ versus x_A for a binary exchange with the following constants: $p=1/2$, $\nu=2$, $C=200$(reciprocal normality units), $\bar{S}=5$, $\sigma_1/RT=5$, and $\sigma_2/RT=2$. Although the neglect of activity coefficients in the bulk liquid is unjustified, Figure 8 illustrates the strong influence of normality upon the separation factor and agrees qualitatively with the experimental data on Figure 1.

So far only two-site models have been examined, and Figure 9 shows the effect of the shape of the distribution upon the separation factor. The relative separation factor was calculated using Eq.(20), integrated for each distribution at the limit of high normality. All three distributions have the same mean energies ($\bar{S}=2$), standard deviations ($\Delta\sigma/RT=2$), and skewness (they are all symmetric), so that differences in the curves are caused by inequalities in the fourth and higher moments. The two-site model does not differ radically from the others and has the advantage of simplicity (e.g., Eq.27).

However, in comparisons with experiment we found that the two-parameter Eq.(27) lacks the flexibilty needed to fit binary exchange data within experimental error. If the skewness parameter is allowed to assume its full range from zero to unity, then Eq.(26) yields:

$$S_{1,2}=\bar{S}\ \frac{\bar{S}W^{U+V}x_1 + \left[W^U(1-p)+W^V p\right] x_2}{\bar{S}\left[W^V(1-p)+ W^U p\right]x_1 + x_2} \tag{28}$$

where $\bar{S}=\exp\left[(\bar{E}_1-\bar{E}_2)/RT\right]$, $W=\exp\left[(\sigma_1-\sigma_2)/RT\right]$, $U=-p/\left[p(1-p)\right]^{0.5}$ and $V= (1-p)/\left[p(1-p)\right]^{0.5}$.

136

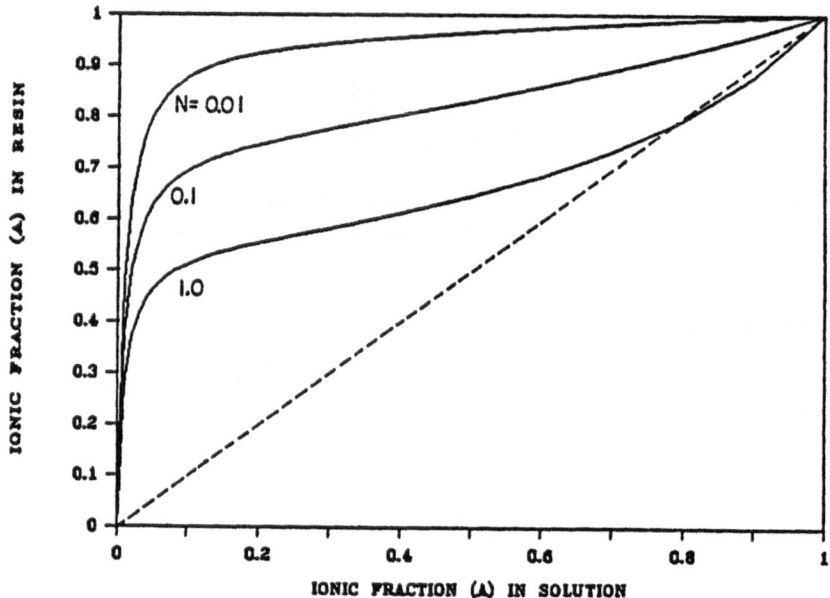

Figure 8- Effect of solution normality upon an ion exchange
with a mean selectivity S=5.Calculated using Eq.(24)
with $\sigma_1/RT=5$ and $\sigma_2/RT=2$.

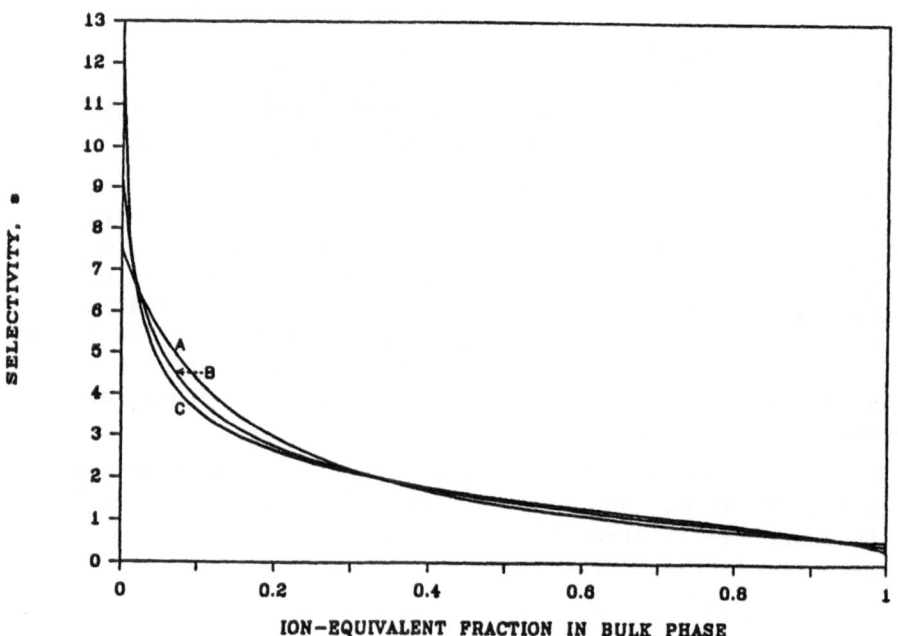

Figure 9- Relative separation factors calculated using various
energy distributions.First three moments of distributions are equal. A: 2-site distribution;B:uniform
distribution; C:Gaussian distribution.

Figure 10 shows the effect of the skewness (p) of the distribution for a system having a mean separation factor $\bar{S}=2$ and difference in heterogeneity $\Delta\sigma/RT=1$. For $p=1/2$, Eq.(28) reduces to Eq.(27).

Eqs.(27) and (28) written for two sites ($\nu=2$) are used here to approximate the heterogeneity of an ion exchanger under the condition of constant normality. The two-site model was first proposed by Barrer and Meier [1] for two kinds of cationic sites in MS-4A. For A/B cationic exchange on solid X:

$$A^+ + BX = B^+ + AX$$

Eq.(10) of Barrer and Meier [1] is:

$$[(1-x_B)K + x_B K_1](K-K_2)(1+\gamma) = K(K_1-K_2)\gamma$$

where K is the relative separation factor:

$$K = \frac{x_B \bar{x}_A}{\bar{x}_B x_A}$$

K_1 and K_2 are mass action equilibrium constants for sites of type 1 and type 2, respectively, present in the ratio $\gamma=N_1/N_2$. The Barrer-Meier equation contains three constants $\{K_1,K_2,\gamma\}$ and the functional relationship between $\bar{x}_B$ and x_B is the same as Eq.(28) .

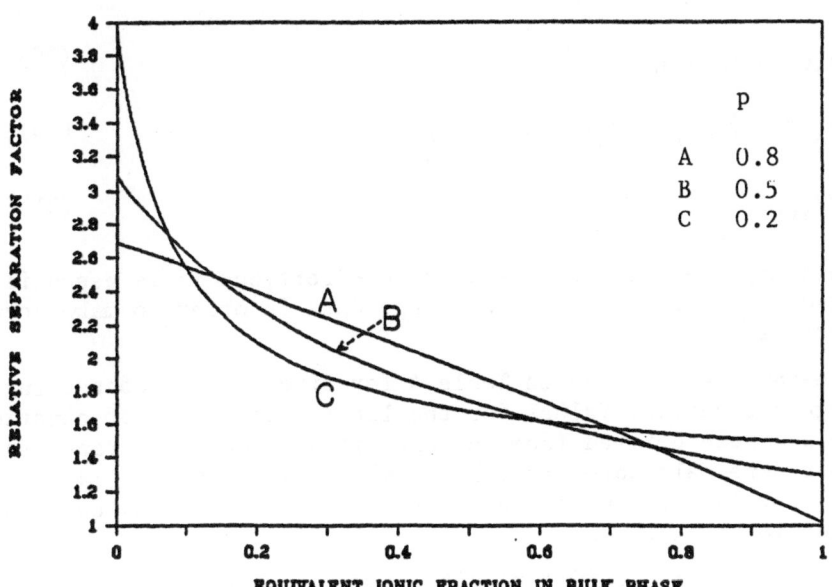

Figure 10- Effect of skewness of distribution upon relative separation factor for binary exchanges having equal differences of mean energies and variances. Calculated from Eq.(28) using various values of p.

COMPARISON OF THEORY WITH EXPERIMENT

After evaluating constants $\{\bar{S}, \Delta\sigma/RT, p\}$ from experimental data for A/B and A/C binary exchanges, the experimental data for the B/C exchange were compared with that predicted by Eq.(28). In all cases, the experimental data were isothermal and at constant normality or constant ionic strength. The effect of normality upon the separation factor was not available.

For each binary exchange, an optimization program was used to derive the set of constants which minimized a residual function equal to the sum of the squares of the differences between calculated (Eq. (28)) and experimental values of ionic fraction in the solid phase. In general, the values of the skewness constant $\underline{p}$ obtained this way were different for each binary exchange. The skewness was forced to be the same for each member of a given system $\{A,B,C\}$ by using an average value. The separation factor shows increasing sensitivity to skewness as the heterogeneity factor $\Delta\sigma/RT$ increases, so instead of an arithmetic average we used:

$$p_{avg} = \frac{(\sigma_A - \sigma_B)p_{AB} + (\sigma_B - \sigma_C)p_{BC}}{(\sigma_A - \sigma_B) + (\sigma_B - \sigma_C)}$$

Then a new set of constants $(\bar{S}, \Delta\sigma/RT)$ was derived for the binary exchanges A/B and A/C. Finally, constants for the third binary exchange B/C were obtained by:

$$\bar{S}_{A,C} = \bar{S}_{A,B}\, \bar{S}_{B,C} \tag{29}$$

$$(\sigma_A - \sigma_C) = (\sigma_A - \sigma_B) + (\sigma_B - \sigma_C) \tag{30}$$

$$p = p_{avg} \tag{31}$$

This may seem like a considerable effort, but it is essential for the theory to fit the binary exchange data in order to make accurate predictions.

Constants are listed in Table 4 for five systems containing three exchangeable cations. Values for the 1st and 2nd pairs of constants in Table 4 were derived from the experimental data; constants for the 3rd pair were calculated from Eqs.(29)-(31). A total of 15 predictions, 3 for each of the 5 systems studied, were made and compared with the experimental points.

The ability of Eq.(28) to fit binary exchange data is illustrated on Figure 11 for Mg^{2+}/K^+ exchange on KU-2. Comparable fits were obtained for the other systems. Figures 12 through 16 show typical predictions of binary exchange data. For a total of 15 binary exchanges, the average of the absolute value of the difference between experimental and theoretical values of resin-phase mole fraction was 0.02.

Differences between experimental and theoretical values of composition can be explained by small thermodynamic consistencies in the experimental data.Consider,for example,Figure 13 for the Mg^{2+}/Ca^{2+} exchange on KU-2. The mean separation factor for the experimental data on Figure 13 is 2.25,but the value predicted from the triangle rule using the Ca^{2+}/K^+ and Mg^{2+}/K^+ exchange data is 62.63/31.69=1.98. Thus,any theory which is thermodynamically consistent will underestimate the resin-phase mole fraction of Ca^{2+},as shown on Fig. 13. Similar results for the other systems lead to the conclusion that Eqs.(29)-(31) provide predictions which are within the accuracy of the experimental data.

MULTICOMPONENT ION EXCHANGE

Equations derived for binary ion exchange can readily be extended to multicomponent (ternary and higher) systems. The separation factor for ion i relative to ion j is defined by:

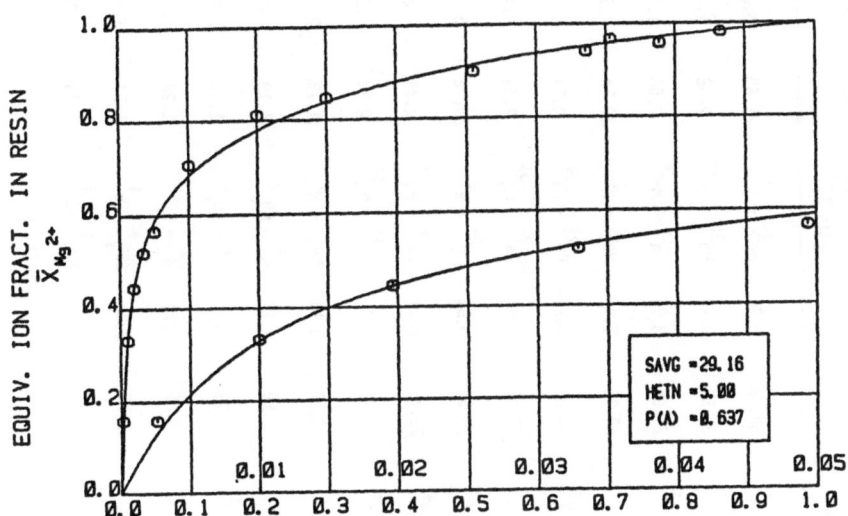

EQUIV. ION FRACT. IN SOLUTION

$X_{Mg^{2+}}$

Figure 11- Equilibrium curve for Mg^{2+}/K^+ exchange on KU-2.
Circles:experimental data [16];solid line:calculated from Eqs.(3) and (28) using constants derived by a best fit of the experimental data

TABLE 4- Predicted binary parameters and component binary parameters

RESIN	BINARY ION PAIR #1	SAVG	HETN	BINARY ION PAIR #2	SAVG	HETN	BINARY ION PAIR #3	SAVG	HETN	P
KU-1	$Ca^{2+} - K^+$	40.83	4.90	$Mg^{2+} - K^+$	21.65	5.06	$Ca^{2+} - Mg^{2+}$	1.89	.97	.719
KU-1	$Ca^{2+} - K^+$	37.95	4.47	$Ca^{2+} - Mg^{2+}$	2.58	1.00	$Mg^{2+} - K^+$	14.71	4.47	.767
KU-I	$Mg^{2+} - K^+$	21.65	5.06	$Ca^{2+} - Mg^{2+}$	2.58	1.00	$Ca^{2+} - K^+$	56.06	5.06	.675
KU-2	$Ca^{2+} - K^+$	60.02	5.47	$Mg^{2+} - K^+$	28.50	4.91	$Ca^{2+} - Mg^{2+}$	2.11	1.11	.709
KU-2	$Ca^{2+} - K^+$	56.83	5.24	$Ca^{2+} - Mg^{2+}$	2.54	1.00	$Mg^{2+} - K^+$	22.37	5.24	.778
KU-2	$Mg^{2+} - K^+$	29.16	5.00	$Ca^{2+} - Mg^{2+}$	2.54	1.00	$Ca^{2+} - K^+$	74.07	5.00	.637
SULFONATED CARBON	$Ca^{2+} - K^+$	70.85	5.89	$Mg^{2+} - K^+$	29.42	5.14	$Ca^{2+} - Mg^{2+}$	2.41	1.15	.644
SULFONATED CARBON	$Ca^{2+} - K^+$	79.09	6.60	$Ca^{2+} - Mg^{2+}$	2.94	.56	$Mg^{2+} - K^+$	26.90	11.87	.574
SULFONATED CARBON	$Mg^{2+} - K^+$	31.30	5.44	$Ca^{2+} - Mg^{2+}$	2.94	.56	$Ca^{2+} - K^+$	92.02	3.02	.578
DOWEX-50	$K^+ - H^+$	2.13	2.18	$NH_4^+ - H^+$	1.73	2.26	$K^+ - NH_4^+$	1.23	0.96	.669
DOWEX-50	$K^+ - H^+$	2.12	2.19	$K^+ - NH_4^+$	1.19	1.00	$NH_4^+ - H^+$	1.78	2.19	.846
DOWEX-50	$NH_4^+ - H^+$	1.73	2.28	$K^+ - NH_4^+$	1.19	1.00	$K^+ - H^+$	2.06	2.28	.500
DOWEX-50	$NH_4^+ - Li^+$	2.15	2.05	$H^+ - Li^+$	1.26	.78	$NH_4^+ - H^+$	1.71	2.63	.600
DOWEX-50	$NH_4^+ - Li^+$	2.16	2.06	$NH_4^+ - H^+$	1.73	2.28	$H^+ - Li^+$	1.25	0.90	.547
DOWEX-50	$H^+ - Li^+$	1.26	1.21	$NH_4^+ - H^+$	1.73	2.28	$NH_4^+ - Li^+$	2.18	1.82	.500

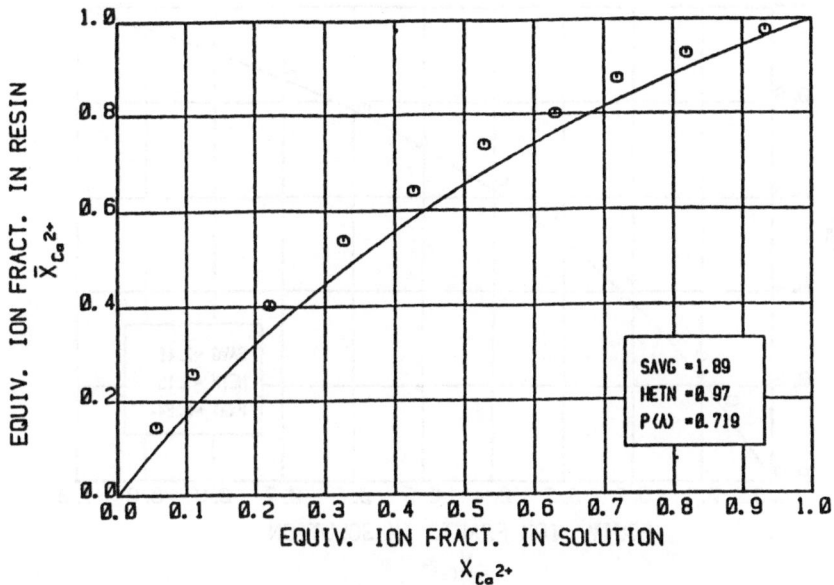

Figure 12-Equilibrium curve for Ca^{2+}/Mg^{2+} exchange on KU-1.
Circles:experimental data [16];solid line:calculated
by Eq.(28) using data for Ca^{2+}/K$^+$ and Mg^{2+}/K$^+$ exchanges.

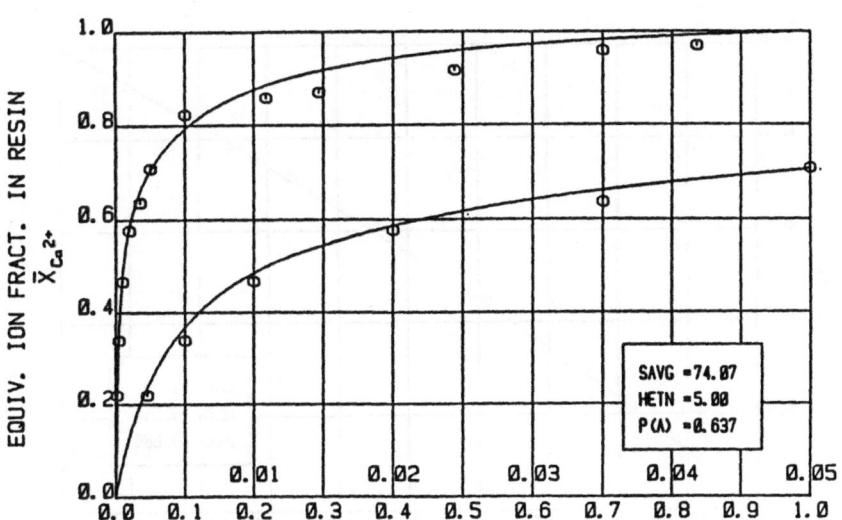

Figure 13-Equilibrium curve for Ca^{2+}/K$^+$ exchange on KU-2.
oo experimental data [16]; --- calculated by Eq.(20)
using data for Mg^{2+}/K$^+$ and Ca^{2+}/Mg^{2+} exchanges.

142

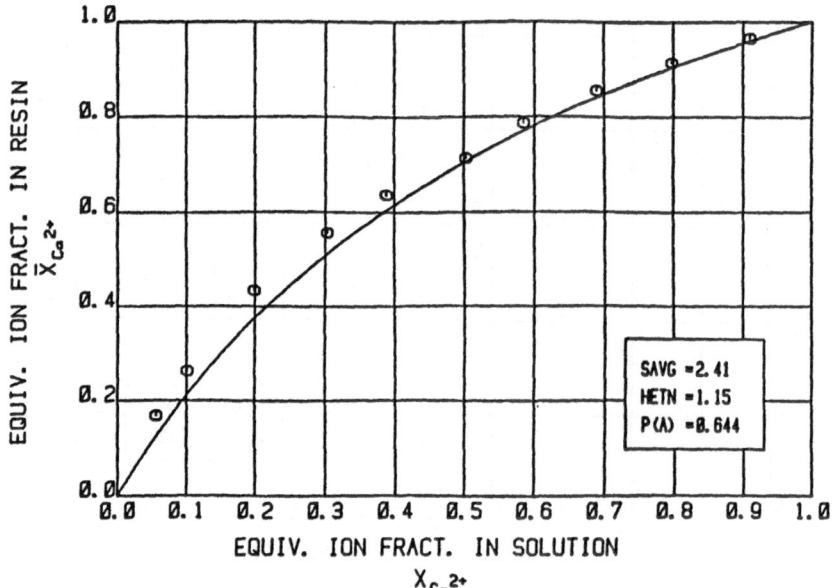

Figure 14- Equilibrium curve for Ca^{2+}/Mg^{2+} exchange on sulfonated carbon. oo experimental data [16] ; —— calculated
by Eq.(28) using data for Ca^{2+}/K^+ and Mg^{2+}/K^+ exchanges

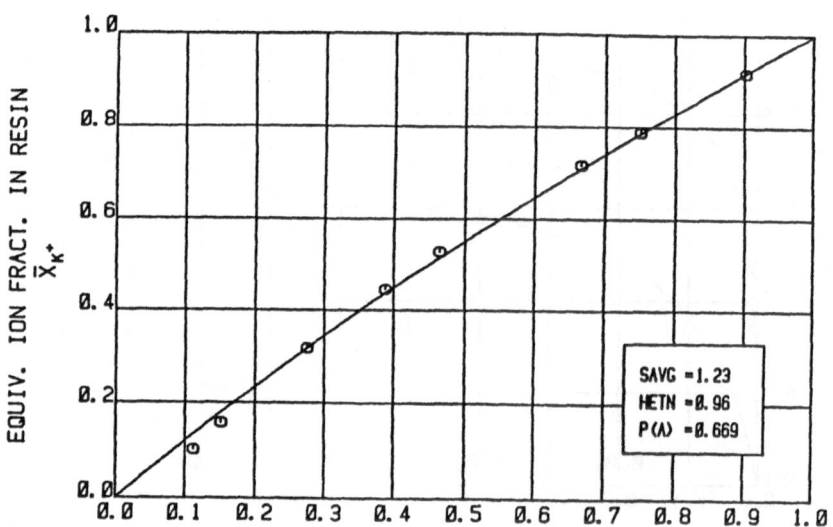

EQUIV. ION FRACT. IN SOLUTION

X_{K^+}

Figure 15-Equilibrium curve for K^+/NH_4^+ exchange on Dowex 50 resin.
oo experimental data [3] ;—— calculated by Eq.(28) using
data for K^+/H^+ and NH_4^+/H^+ exchanges.

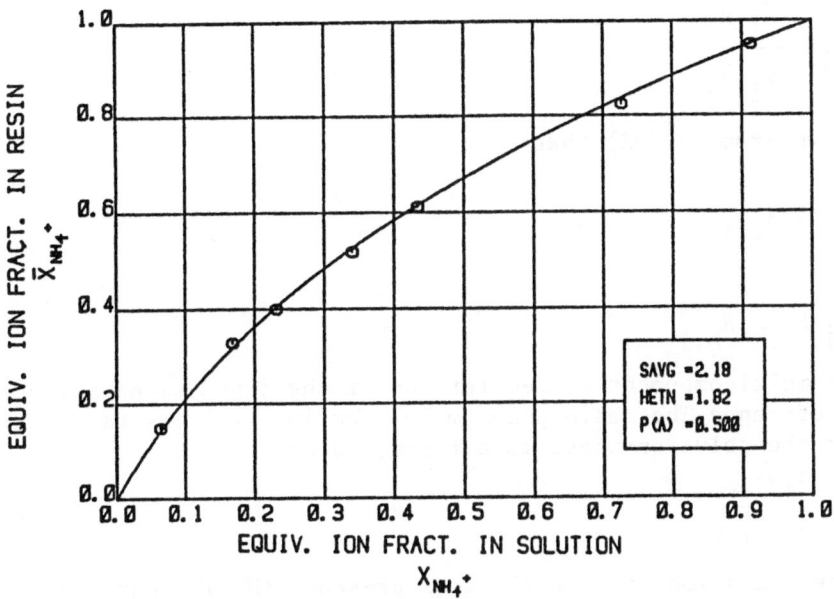

Figure 16- Equilibrium curve for NH_4^+/Li^+ exchange on Dowex 50.
oo experimental data [3];—— calculated by Eq.(28)
using data for H^+/Li^+ and NH_4^+/H^+ exchanges.

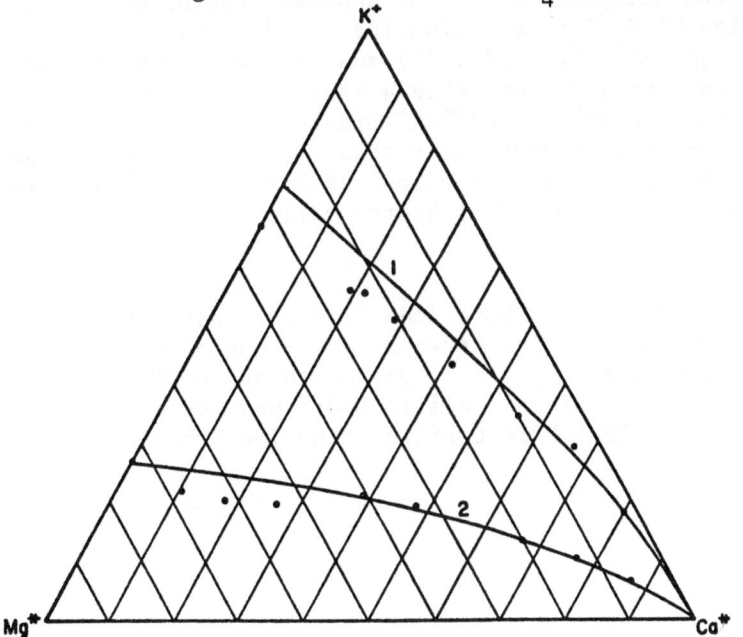

Figure 17- $Ca^{2+}/Mg^{2+}/K^+$ exchange on KU-1.
oo experimental data [16] ;—— calculated by Eq.(28)
using data for Ca^{2+}/Mg^{2+} and Ca^{2+}/K^+ exchanges.Locus
is constant ratio of equivalent ionic fraction K^+/Mg^{2+}
in solution.Curve 1:ratio=65.66;curve 2:ratio=4.

$$S_{i,j} \equiv \frac{\bar{x}_i \, x_j}{x_i \, \bar{x}_j} \tag{32}$$

It follows from Eq.(32) that:

$$S_{j,i} = 1/S_{i,j}$$

$$S_{i,i} = 1$$

$$S_{i,j} = S_{i,k} \, S_{k,j}$$

In a multicomponent system, let one of the ions (k) be selected as the reference. The resin-phase mole fraction ($\bar{x}_i$) may be calculated from the solution-phase mole fractions by:

$$\bar{x}_i = \frac{x_i S_{i,k}}{\Sigma x_i S_{i,k}} \tag{33}$$

where the summation is over all ions present. (N-1) separation factors must be measured for an N-component ion exchange.

Experimental data [16] for the ternary system $Ca^{2+}/Mg^{2+}/K^+$ on KU-1 at 25°C, ionic strength 0.02 and pH 6.0 are shown on Figure 17. The points were measured on loci of constant ratios of equivalent ionic fraction K^+/Mg^{2+} in solution. The solid lines were calculated by means of Eqs.(21),(22),(23),(26) and (33) with the constants tabulated in the second row of Table 4. Thus the predictions are based entirely upon Mg^{2+}/Ca^{2+} and Ca^{2+}/K^+ binary exchanges. Although more studies are needed, this particular result indicates that it may be possible to make reliable predictions of N-component ion exchange from experimental data on (N-1) binary pairs.

ACKNOWLEDGMENT

We wish to thank Professor Linda Wang for instigating this work by pointing out similarities between ion exchange isotherms and isotherms for physical adsorption of gas mixtures on heterogeneous surfaces. Also, we thank Professor Alírio Rodrigues for his invitation to participate in the 1985 NATO Conference on Ion Exchange.

REFERENCES

1. Barrer,R.M. and W.M.Meier. Exchange Equilibria in a Synthetic Crystalline Exchanger.Trans.Faraday Soc.$\underline{55}$,130(1959).

2. Bonner,O.D. and Rhett,V.Equilibrium Studies of the Silver-Sodium-Hydrogen System on Dowex 50.J.Amer.Chem.Soc.$\underline{57}$,254(1953).

3. Bonner,O.D. and W.Payne.Equilibrium Studies of Some Monovalent Ions on Dowex 50. J.Amer.Chem.Soc.$\underline{58}$,183(1954)

4. Dranoff,J. and L.Lapidus.Equilibrium in Ternary Ion Exchange Systems. Ind.Eng.Chem.$\underline{49}$,1297(1957)

5. Helfferich,F.Ion Exchange(Mc Graw-Hill,New York,1962)

6. Kataoka,K. and H.Yoshida.Ion Exchange Equilibria in Ternary Systems. J.of Chem.Engrng. of Japan.$\underline{13}$,328(1980)

7. Manning,M. and S.Melshelmer.Binary and Ternary Ion-Exchange Equilibria with a Perfluorosulfonic Acid Membrane.Ind.Eng.Chem.$\underline{22}$, 311(1983)

8. Novosad,J. and A.L.Myers. Thermodynamics of Ion Exchange as an Adsorption Process. Canadian J.of Chem.Engrng. $\underline{60}$,500(1982)

9. Pal,G.,Chakravarti,A. and M.Sengupta. Studies in Ion Exchange Equilibria II:Some Cation and Anion Exchange Selectivities in Amberlite Resins. Ion Exchange and Membranes $\underline{2}$,21(1974)

10. Pieroni,L. and J.Dranoff.Ion Exchange Equilibrium in a Ternary System.AIChEJournal $\underline{9}$,42(1963)

11. Smith,R.P. and E.T.Woodburn.Prediction of Multicomponent Ion Exchange Equilibria for the Ternary System $SO_4^{2-}/NO_3^-/Cl^-$ from Data on Binary Systems. AIChE Journal $\underline{24}$,577(1978)

12. Soldatov,V.S. and V.A.Bychkova.Ion Exchange in Multicomponent Systems.Calculation of Ion-Exchange Equilibrium in the Ternary System $K^+/NH_4^+/H^+$ from Data for Binary Systems. Russ.J.of Phys.Chem.$\underline{44}$,1297 (1970)

13. Subba Rao,H. and M.M.David. AIChE Journal.$\underline{3}$,187(1957)

14. Streat,M. and W.Brignal. Representation of Ternary Ion Exchange Equilibria. Trans.Inst.Chem.Engrs. $\underline{48}$,T151(1970)

15. Tondeur,D. and G.Klein. Multicomponent Ion Exchange in Fixed Beds. Ind.Eng.Chem.Fund. $\underline{6}$,351(1967)

16. Khoroshko,R.,Kolnenkov,V.,Soldatov,V.,Sudarikova,N. and N.G. Peryshkina. Synthetic Culture Media for the Growth of Plants on the Basis of Ion Exchange Materials. Agrokhimiya .10 (1974)122.

DESIGN METHODS FOR ION-EXCHANGE PROCESSES BASED ON THE "EQUILIBRIUM THEORY"

D.Tondeur and M.Bailly

Laboratoire des Sciences du Génie Chimique,CNRS-ENSIC

1,rue Grandville 54 Nancy - France

1. INTRODUCTION

What is the so-called "equilibrium theory" ?

One is tempted to define it as a model of the mass-transfer in the ion-exchange bed, in which equilibrium between the two phases is assumed at all points and one expects it to give a sort of ideal limiting behaviour. Actually, we shall see that if this assumption is taken strictly, it is impossible to construct a solution which can be considered as a limit of the real behaviour when mass-transfer kinetics become very rapid.

Non-equilibrium regions (constant pattern shocks) must be included in the model to make it physically meaningful. In other words, the "equilibrium" model cannot strictly be an equilibrium model. We shall define it in a less restrictive way as a model which does not account explicitely for mass transfer kinetics.

Do we need an equilibrium theory ?

With high speed computers available to designers and engineers, and capable of handling sophisticated models including various kinetic effects, what is the use of such a primitive model ?

I believe its main merit is to give by simple means (without high speed computer !) the general trends and a good qualitative understanding of the process. And this is essential for conceptual and preliminary design, for feasibility analysis, for trouble shooting, even if sophisticated calculation tools

have to be used in later stages. Even in these stages, it may
help writing efficient computer programs or debugging them. In
addition, in some cases, relatively simple analytical solutions
exist and may give acceptable semi-quantitative estimations.

When is it useful ?

When one does not have enough information to do anything
else. When more information is not needed. When a limiting
behaviour, a reference case is sought. When complex, non-linear,
multicomponent systems are considered, with simple boundary
conditions (step inputs essentially).

When is it useless ?

When dealing with kinetically controlled phenomena. When
accurate quantitative fitting is sought.

What data and tools does one need to use it ?

- At least, the order of selectivity of the components
- Better, reasonably good phase equilibrium data, and
  a fitting of these by equilibrium relations
- Depending on the number of components, a pocket
  calculator or a micro-computer.

2. REPRESENTATION OF COMPOSITION FRONTS

2.1 Effluent histories or breakthrough curves for elution and fixation

Figure 1a shows the experimental concentrations as a
function of effluent volume at the outlet of a column initially
equilibrated with a NaCl solution, and eluted by hydrochloric
acid (data of experiment given in legend). Conversely, the
fixation of Na^+ on the same column under similar conditions is
shown in Fig. 1b. These curves show some common features :

- an initial period during which the outlet composition is
  constant (plateau) and identical to the initial interstitial solution ;

- a final period during which the outlet composition is
  constant (plateau) and identical to the incoming
  solution ;

- between the initial and final plateau, a period of
  varying composition : the exchange front.

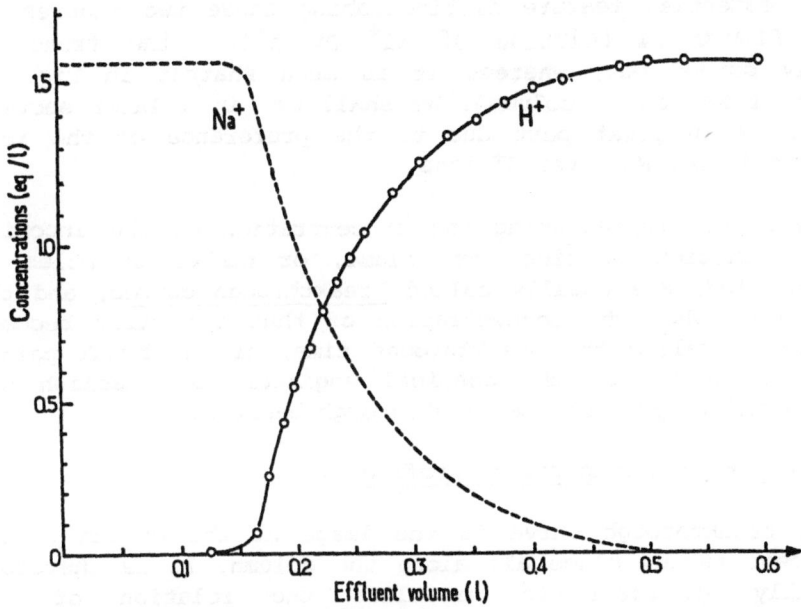

Figure 1a- Experimental effluent history of bed of cation exchan-
            ger initially in Na$^+$ form,receiving as feed HCl 1.49N
            (resin:Duolite C20;bed:0.36m length,0.29 equiv.capaci-
            ty;flowrate:10 ml/min)

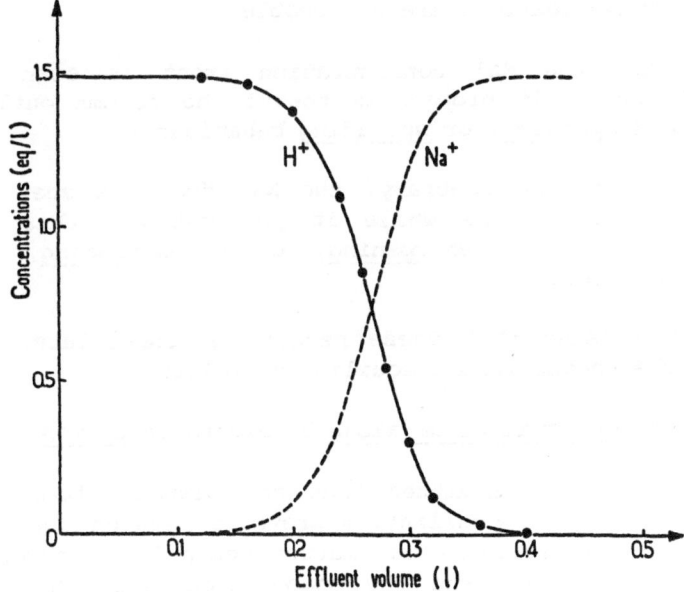

Figure 1b- Experimental effluent history of bed of cation exchan-
            ger in H$^+$ form receiving as feed NaCl 1.55N (same con-
            ditions as in Fig. 1a)

The essential feature distinguishing these two figures is that in Figure 1a (elution of Na^+ by H^+), the front is relatively spread out, whereas it is much sharper in Fig. 1b (fixation of Na^+ on H^+ column). We shall see in a later section that this is in great part due to the preference of the ion-exchange resin for Na^+ over H^+ ions.

The curves representing the concentration of the incoming ion as a function of time, or volume, or number of moles or equivalents fed, are usually called breakthrough curves, and the value of time when the concentration of that ion first becomes noticeable is called the breakthrough time, or the break point. It is an objective of the chemical engineer to establish the properties and to predict the breakthrough behaviour.

2.2 Column internal composition profiles

The breakthrough curve is the image at the column outlet of a process taking place all along the column. It is therefore conceptually important to visualize the relation of the breakthrough curves with the concentration distribution inside the column at any time. Figures 2a and 2b represent schematically the evolution of the internal profiles corresponding respectively to Fig. 1a and 1b. The overall shape of these curves is a sort of mirror image of the breakthrough curves. Two distinctive features are noticeable :

− on Fig. 2a, the Na^+ concentration front is seen to spread more and more as it propagates toward the column outlet : this is called a dispersive, or spreading behaviour ;

− on Fig. 2b, on the contrary, the Na^+ front is seen to assume a sharp constant shape while it propagates ; this is called compressive, or self-sharpening, or non-spreading, or constant-pattern behaviour.

We shall show later that these behaviours are related to the curvature of the thermodynamic equilibrium relation.

2.3 Equilibrium and non-equilibrium along breakthrough curves

It can be reasonably admitted that the plateaus bounding the front are regions where equilibrium prevails between the two phases. If there was no equilibrium, matter would be transfered from one phase to the other and the composition would change. Equilibrium is thus a reasonable assumption inasmuch as in plateaus, the composition is really constant for all species present. This assumption is coherent with physical understanding: the initial plateau is what comes out before the outlet end has felt the effect of the incoming ions, retained in the first column layers ; the final plateau is obtained when the ion-exchanger has taken up all the incoming ions it could, and is thus "saturated" or equilibrated with the incoming feed.

On the other hand, the changing compositions along the front obviously imply mass-transfer, and a driving force for it stemming from a departure from thermodynamic equilibrium. We shall see that this departure from equilibrium is quite different for dispersive and for compressive fronts. The equilibrium model assumes equilibrium is attained in dispersive fronts, but not in compressive ones.

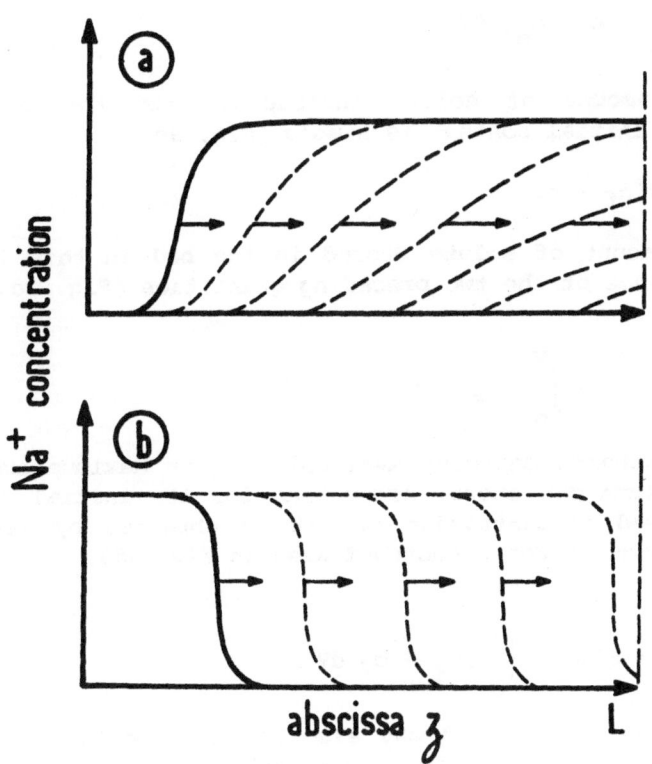

Fig. 2. Internal concentration profiles corresponding to experiments of Figure 1
(schematic)

3. OVERALL RELATIONS GOVERNING BREAKTHROUGH CURVES

3.1 Integrating simple breakthrough curves

Figures 3 show the relationships between the areas determined by a breakthrough curve and the axes, plotted in terms of equivalent concentrations versus effluent volume. These areas represent different elements of an overall material balance on a complete exchange operation and imply no assumption on equilibrium or kinetics.

* The amount of solute that has Leaked out of the column between effluent volumes 0 and V, in excess of the initial interstitial solute, is represented by the hatched area on Fig. 3a, and given by the integral :

$$L(V) = \int_0^V (c - c_o) \, dV \tag{1}$$

* The amount of solute Introduced into the column in excess of the initial content is simply (Fig. 3b)

$$I(V) = V(c_f - c_o) \tag{2}$$

* The amount of solute Stored in the bed in this interval is the difference of the two preceding quantities (Fig. 3c) :

$$S(V) = I - L = \int_0^V (c_f - c) \, dV \tag{3}$$

* The exchange capacity Available is the maximum amount of solute the column can store, accounting for its initial state c_o and of the feed concentration c_f. It is obtained by letting V become very large in Eqn 3 (hatched area in Fig. 3d).

$$A(c_f, c_o) = S(\infty) = \int_0^\infty (c_f - c) \, dV \tag{4}$$

The hatched area in Fig. 3d may also be obtained by integrating "in the other direction", that is by taking c as the integration variable, so that

$$A = \int_{c_o}^{c_f} V \, dc \tag{5}$$

153

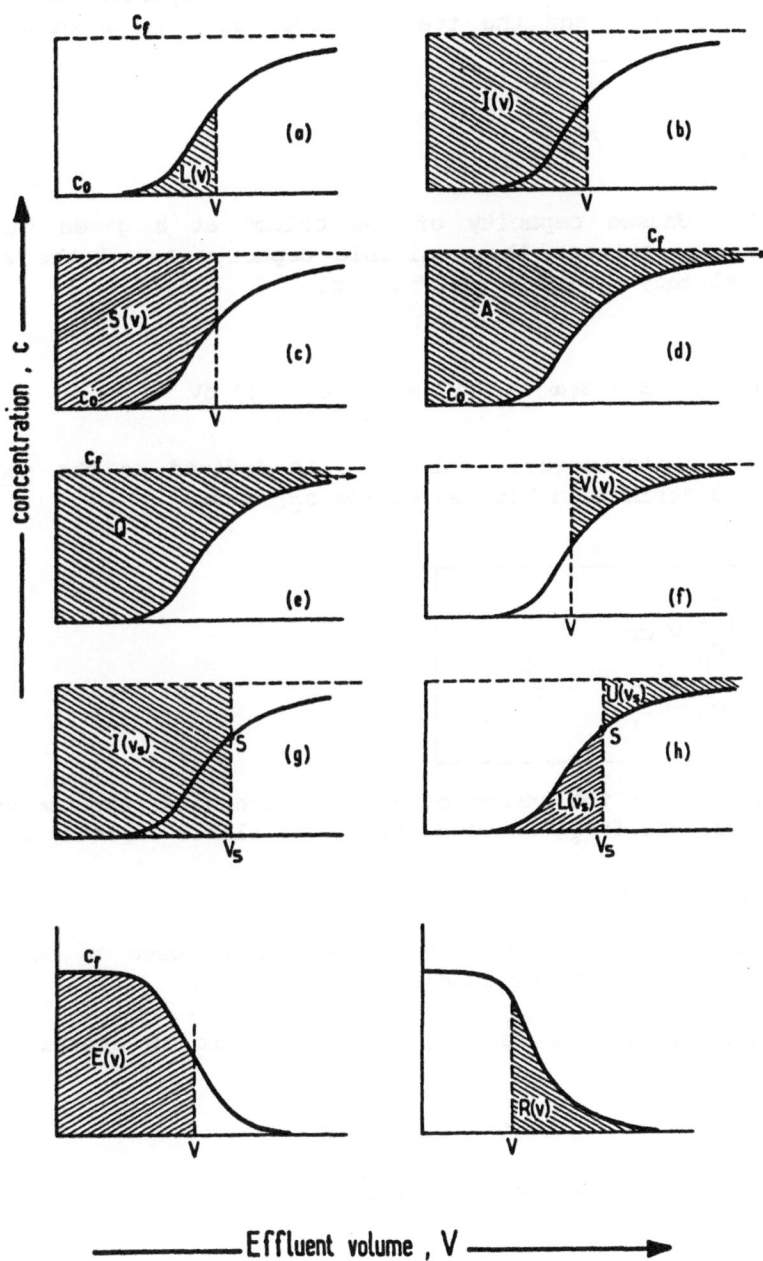

Fig. 3. Areas defined by experimental breakthrough curves (see Section 3.1)

* The total exchange capacity of the column is the value of A when the column does not contain the species considered initially ($c_o = 0$), and the feed consists of that species alone (Fig. 3e)

$$Q = A(c_f, 0) = \int_0^{c_f} V \, dc \qquad (6)$$

* The Unused capacity of the column at a given time is the difference between the available capacity A and the amount of solute already stored $S(V)$ (Fig. 3f)

$$U(V) = A - S = S(\infty) - S(V) = \int_V^{\infty} (c_f - c) \, dV \qquad (7)$$

* The stoichiometric volume V_S is defined as the mean of the volume distribution between c_o and c_f, thus

$$V_s = \frac{\int_{c_o}^{c_f} V \, dc}{c_f - c_o} = \frac{A}{c_f - c_o} \qquad (8)$$

It appears to be the value of volume such that the amount of solute introduced $I(V_S)$ is equal to the available capacity A :

$$I(V_S) = V_S (c_f - c_o) = A \qquad (9)$$

These relations define on the breakthrough curve a so-called stoichiometric point S, of abscissa V_S (Fig. 3g).

Comparing Figures 3a, 3d, 3f and 3g, it is apparent that we have, at $V = V_S$

$$A = I(V_S) - L(V_S) + U(V_S) = S + U \qquad (10)$$

and owing to Eqn. 9, we have necessarily (Fig. 3h)

$$L(V_S) = U(V_S) \qquad (11)$$

Thus, at the stoichiometric volume, the unused capacity is equal to the amount of leakage.

NB 1 : special values of all the foregoing quantities are obtained when the species considered is initially absent from the column ($c_0 = 0$) and present alone in the feed. We have defined here these quantities in the most general case.

NB 2 : we have worked on breakthrough curves of the species fed to the column and saturating the bed. Useful quantities may of course be defined on the effluent history of a species initially present on the bed, but absent (or minor) in the feed. In particular, one may define (Fig. 3 i,j) the amount of that species Eluted E(V) and the amount Remaining on the column R(V) respectively by

$$E(V) = \int_0^V (c - c_f) \, dV \qquad (12)$$

$$R(V) = \int_V^\infty (c - c_f) \, dV \qquad (13)$$

which satisfy

$$E + R = A \qquad (14)$$

A being the available capacity defined in Eqn. 4. For binary exchange, we have clearly

$$E = S \qquad (15)$$

that is the amount of one species stored is equal to the amount of the other species eluted, and

$$R = U \qquad (16)$$

that is the amount of one species remaining to be eluted is the unused capacity for the other species.

3.2 Material balance over a complete exchange operation - Front velocity

The overall material balance can be written easily using the foregoing notions. Let us express the amount of solute stored by a section of column of volume v after going entirely from its initial state c_0 to the final state c_f, in other words the available capacity A of the section. The solute is stored in the solid phase and in the interstitial fluid so that :

$$A = v(1 - \epsilon)(\overline{c_f} - \overline{c_0}) + v \, \epsilon(c_f - c_0) \qquad (17)$$

where v is the volume of the section.

Substituting this expression for A into Eqn. 8, giving V_s:

$$\boxed{\frac{V_s}{v} = \epsilon \left[1 + \frac{1 - \epsilon}{\epsilon} \frac{\Delta \bar{c}}{\Delta c} \right]} \tag{18}$$

If we take v as the total column volume, this expression gives the number of bed volumes (BV) necessary for the stoichiometric point S (the mean of the front) to reach the bed outlet ; or in other words, the bed volumes equivalent to the available capacity of the bed.

Eqn. 18 may be expressed in terms of stoichiometric time t_s:

$$t_s = V_s / F \tag{19}$$

where F is the flowrate ; we define the average front velocity u_s as the average velocity of the stoichiometric point :

$$\boxed{u_s} = \frac{L}{t_s} = \frac{v_{bed}}{\Omega t_s} = \boxed{\frac{u_i}{1 + \frac{1 - \epsilon}{\epsilon} \frac{\Delta \bar{c}}{\Delta c}}} \tag{20}$$

where u_i is the interstitial fluid velocity, defined by

$$u_i = \frac{F}{\Omega \epsilon} = \frac{u}{\epsilon} \tag{21}$$

u being the apparent (empty column) velocity and Ω the column cross-sectionnal area.

3.3. The special case of constant-pattern fronts

Constant-pattern fronts are fronts which move along the column without spreading or deformation. They are obtained in general when a species with strong affinity for the resin displaces a species of weaker affinity. More detailed criteria will be given later.

In order to reach a constant-pattern, a displacement front needs to be fully developed in the bed, meaning that the initial layers of the bed must have reached equilibrium with the inlet solution. The initial formation of a front is thus "changing pattern".

When breakthrough occurs as constant pattern is reached, the relationship between the breakthrough curve and the internal column profiles may be simply visualized on Figure 4. The lower part of this figure represents the breakthrough curve on a time scale, as used earlier. On this curve we have defined the stoichiometric point S, and the corresponding time t_S, and two other special points : the breakthrough point B and the "end" point E. The two latter are somewhat arbitrary, and are defined by the operator for example as a tolerance level, on the outlet concentration, or as a sensitivity level for the detector. They may also be defined as a percentage of the total concentration change in the front, say 5 % for B and 95 % for E.

The upper left part of Figure 4 represents the paths of points B, S and E in a distance vs time diagram. The straight line through the origin, of constant slope u_S represents the path of the mean of the front determined by the material balance (Eqns 18 and 20). This path is that of point S once the constant-pattern is reached. The two other lines represent the paths of points B and E. B is initially faster than the average and E actually appears only when the first layers of the columns are 95 % equilibrated, and is thus delayed. The three paths are parallel when constant pattern is reached. The breakthrough time t_B corresponds to the intersection of the B path with the distance line z = L, and similarly for t_S and t_E. Conversely, at a given time (here chosen as t_B), the positions of B, S and E along the bed are easily obtained from this diagram, and may be used to generate the concentration profile shown on the top right of the figure.

The following quantities are usually defined, to charac- terize the front shape (see Fig. 4).

LOF : Length of Front (sometimes designated by LMTZ, for length of mass transfer zone) ; this is the distance between points B and E at breakthrough time (and thus also in the constant pattern regime).

LUB : Length of Unused Bed : distance between B and S at breakthrough. So-called because if the front were perfectly sharp, it would be entirely located at the abscissa of S, and the upstream part of the bed would be saturated, whereas the downstream part would still be in its initial state (thus unused). The difference L-LUB may be called LEB (length of equilibrated bed).

These quantities may be normalized by L and become thus fractions of the bed length, FLOF, FLUB and FLEB. They are

158

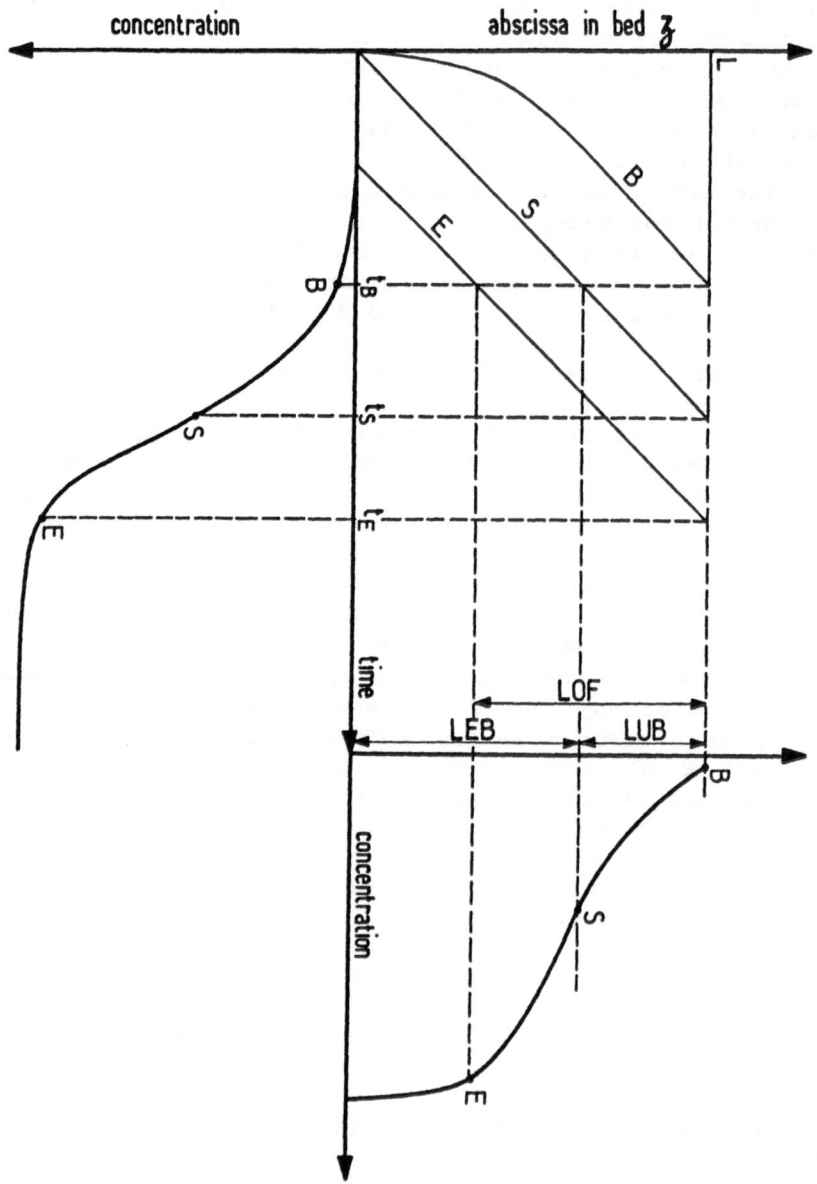

Figure 4- Relations between three representations of breakthrough
curves:distance vs time,concentration vs time,concen-
tration vs distance.

easily determined from the breakthrough curve <u>assuming</u> constant-pattern is reached. To establish this, we write that at time t_B, the stoichiometric point has reached the abscissa

$$z_s = u_s \, t_B \qquad (22)$$

and at $t = t_s$, it has reached the column end, $z = L$

$$L = u_s \, t_s \qquad (23)$$

But LUB is the difference $L - z_s$, so that we have

$$FLUB = \frac{L - z_s}{L} = \frac{t_s - t_B}{t_s} \qquad (24)$$

$$FLEB = 1 - FLUB \qquad (25)$$

A similar approach may be used to determine FLOF : the paths are extrapolated to $t = t_E$ where $z_E = L$ and the fictitious abscissa of B is then

$$z_B = L + LOF = u_s t_E + LUB \qquad (26)$$

wherefrom we obtain, after some calculations, and elimination of u_s

$$FLOF = \frac{LOF}{L} = \frac{t_E - t_B}{t_s} \qquad (27)$$

For constant pattern to be attained, two <u>necessary</u> (but not sufficient) conditions have to be met. The first will be discussed in the next section and relates to the curvature of the equilibrium isotherm. The second requires that the front be comprised between two equilibrium plateaus ; in other words, the front must be comprised entirely in the column, and therefore the length of the front must be smaller than the length of the column : FLOF < 1. It is sometimes assumed that a <u>sufficient</u> condition is that the bed length be 3 or 4 times larger than the front length (FLOF < 0.3).

4. LOCAL RELATIONS GOVERNING BREAKTHROUGH CURVES

After establishing relations describing the properties of fronts and breakthrough curves as a whole, we shall now write the local equations, which may account for the detailed shape of the fronts (De Vault, 1943).

* Material balance on a differential slice of column perpendicular to the main flow direction

$$\epsilon \; \frac{\partial c}{\partial t} + (1 - \epsilon) \; \frac{\partial \bar{c}}{\partial t} + u \; \frac{\partial c}{\partial z} \quad = \quad \frac{\partial}{\partial z} \left[D_a \; \frac{\partial c}{\partial z} \right] \tag{28}$$

$\underbrace{\qquad\qquad\qquad\qquad}$ $\uparrow$ axial diffusion flux

accumulation in   net convective
slice over time dt   transfer flux

The equilibrium model neglects the axial diffusion term and assumes equilibrium between solid phase concentration $\bar{c}$ and liquid phase concentration c, governed by some identified algebraic relation.

* Boundary and initial conditions of the elementary model : step change at the column inlet, and possibly successive steps

Initial : $t = 0$, $z > 0$
             $c = c_0$ = constant in all column   (29)

Boundary : $t > 0$, $z = 0$
             $c = c_f$ = constant for a finite time interval   (30)

* Non-dimensional form of differential equations

The following non-dimensional variables are introduced

- ionic fractions $x_i = c_i/C_{tot}$

$$y_i = \bar{c}_i/\bar{C}_{tot} \tag{31}$$

so that $\Sigma \; x_i = \Sigma \; y_i = 1$ (32)

- capacity ratio :

$$\Lambda = \frac{1 - \epsilon}{\epsilon} \; \frac{\bar{C}_{tot}}{C_{tot}} \tag{33}$$

$$\Lambda = \frac{\text{exchange capacity in a given column section}}{\text{equivalents in solution in same section}}$$

- reduced axial position variable

$$\xi = z/L \tag{34}$$

- reduced and shifted time variable

$$\Theta = \frac{u_i t - z}{L\Lambda} = \frac{1}{\Lambda}\frac{t - t_z}{t_L} \tag{35}$$

with $t_z = z/u_i$, $t_L = L/u_i$

The time Θ is counted positive at z when the inlet fluid gets at abscissa z.

- "throughput ratio" (Vermeulen)

$$\boxed{T = \frac{\Theta}{\xi}} \qquad = \frac{1}{\Lambda}\frac{t - t_z}{t_z} = \frac{u_i t - z}{z\Lambda}$$

$$= \frac{C}{\bar{C}}\frac{ut - z\epsilon}{(1 - \epsilon)z} = \frac{C}{\bar{C}}\frac{V - v\epsilon}{(1 - \epsilon)v} \tag{36}$$

$$= \frac{\text{number of equivalents that have passed abscissa z at time }\Theta}{\text{exchange capacity of column upstream of abscissa z}}$$

The differential material balance then becomes

$$\left[\frac{\partial x}{\partial \xi}\right]_\Theta + \left[\frac{\partial y}{\partial \Theta}\right]_\xi = 0 \tag{37}$$

* **Multicomponent system**

There is then a conservation equation for each species, of which n - 1 are independent, owing to the stoichiometry Eqn. 32. x and y in Eqn 37 may be treated as vectors

$$\underset{\sim}{x} = \left[x_1, x_2, x_4 \cdots, x_n\right]^T ; \underset{\sim}{y} = \left[y_1, y_2 \cdots y_n\right]^T \tag{38}$$

* **Equilibrium relation** : constant relative sorptivity

$$
\alpha_j^i = \frac{y_i}{x_i} \bigg/ \frac{y_j}{x_j} \tag{39}
$$

$$
y_i = \frac{\alpha_n^i x_i}{\sum \alpha_n^j x_j} \qquad\qquad x_i = \frac{\alpha_i^n y_i}{\sum \alpha_j^n y_j} \tag{40}
$$

$$
\alpha_n^1 > \alpha_n^2 > \ldots > \alpha_n^{n-1} > 1 \tag{41}
$$

$$
\alpha_k^i = \alpha_j^i \alpha_k^j \quad ; \quad \alpha_k^i = 1/\alpha_i^k \tag{42}
$$

* **Elimination of solid phase** through equilibrium

In the binary case, $\bar{c}$ is eliminated from Eqn 28 by writing

$$
\frac{\partial \bar{c}}{\partial t} = \frac{d\bar{c}}{dc} \frac{\partial c}{\partial t} \tag{43}
$$

So that the material balance reduces to

$$
\frac{\partial c}{\partial t} + \frac{1 - \epsilon}{\epsilon} \frac{d\bar{c}}{dc} \frac{\partial c}{\partial t} + \frac{u}{\epsilon} \frac{\partial c}{\partial z} = 0 \tag{44}
$$

In the multicomponent case, Eqn 43 becomes

$$
\frac{\partial \bar{c}_i}{\partial t} = \sum_{j=1}^{n} \frac{\partial \bar{c}_i}{\partial c_j} \frac{\partial c_j}{\partial t} \tag{45}
$$

and the system of conservation equations is, in vector-matrix form :

$$
\frac{\partial \underset{\sim}{c}}{\partial t} [\underset{\sim}{I} + \underset{\sim}{J}] + \frac{u}{\epsilon} \frac{\partial \underset{\sim}{c}}{\partial z} = 0 \tag{46}
$$

or in dimensionless form

$$\frac{\partial \underset{\sim}{x}}{\partial \xi} + [\underset{\sim}{J}] \, \frac{\partial \underset{\sim}{x}}{\partial \theta} = 0 \tag{47}$$

where $[\underset{\sim}{J}]$ is the Jacobian matrix, of element $\left[\dfrac{\partial y_i}{\partial x_j} \right]_{x_k}$, $k \neq i,j$

of rank $n - 1$ (De Vault, 1943).

* The wave equation – Concentration velocities

Eqn 44 may be written

$$\frac{\partial c}{\partial t} + u_c \, \frac{\partial c}{\partial z} = 0 \tag{48}$$

with

$$u_c = \frac{u/\epsilon}{1 + \dfrac{1 - \epsilon}{\epsilon} \dfrac{d\bar{c}}{dc}} = \left[\frac{\partial z}{\partial t}\right]_c = - \frac{(\partial c/\partial t)_z}{(\partial c/\partial z)_t} \tag{49}$$

or

$$u_c = \frac{u_i}{1 + \Lambda \dfrac{dy}{dx}} \tag{50}$$

Eqn 48 is a "wave equation", expressing the propagation of any given value c or x of concentration. This type of equation occurs in many areas of engineering and physics, involving the migration of individual waves (Tondeur, 1981) : migration of hydraulic waves, of sound waves, of automobile traffic, sedimentation, multiphase flow, electrophoresis ...

In terms of the dimensionless time and distance variables, we have

$$u'_x = \left[\frac{\partial \xi}{\partial \theta}\right]_x = \frac{1}{dy/dx} \tag{51}$$

u_c or u'_x is the velocity of a value c or x of concentration. It is seen to depend on the slope dy/dx of the y versus x relation. Clearly, if equilibrium is assumed between the phases, dy/dx is the slope of the isotherm.

* Linear relation in constant-pattern fronts

Constant pattern fronts may be treated as "moving steady-states" : an observer that would move at a constant velocity equal to the average front velocity u_s would observe a steady-state countercurrent flow of the two phases. Writing a differential material balance on a slice of column moving at u_s, we have at steady state (Fig. 5)

$$\epsilon(u_i - u_s)(c + dc) + (1-\epsilon) u_s \bar{c} = \epsilon(u_i - u_s)c + (1-\epsilon) u_s(\bar{c} + \bar{dc})$$

flux entering   flux exiting   (52)

which can be arranged into

$$u_s = \frac{u_i}{1 + \frac{1 - \epsilon}{\epsilon} \frac{\bar{dc}}{dc}} \tag{53}$$

expressing that all values of c propagate at the same velocity u_s. Comparing this equation to Eqn. 20, we find that we must have

$$\frac{\bar{dc}}{dc} = \frac{\bar{\Delta c}}{\Delta c} = \text{constant} \tag{54}$$

Eqn. 54 means that $\bar{c}$ and c are <u>linearly related</u> in the front. Actually, we may define an <u>operating line</u> as in counter-curent operations : it is the line relating the concentrations of the two phases which cross each other in the process. This line is thus straight (Sillén, 1950).

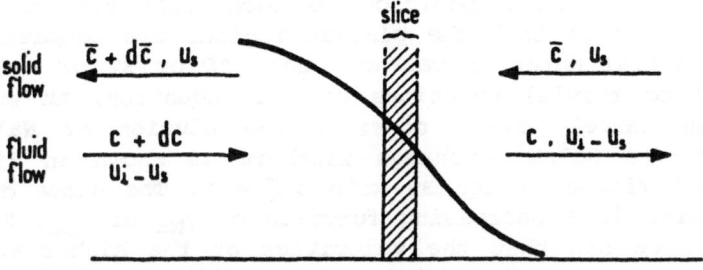

Fig. 5. Fluxes entering and exiting a slice (shaded) moving along the bed at velocity u_s in same direction as fluid.

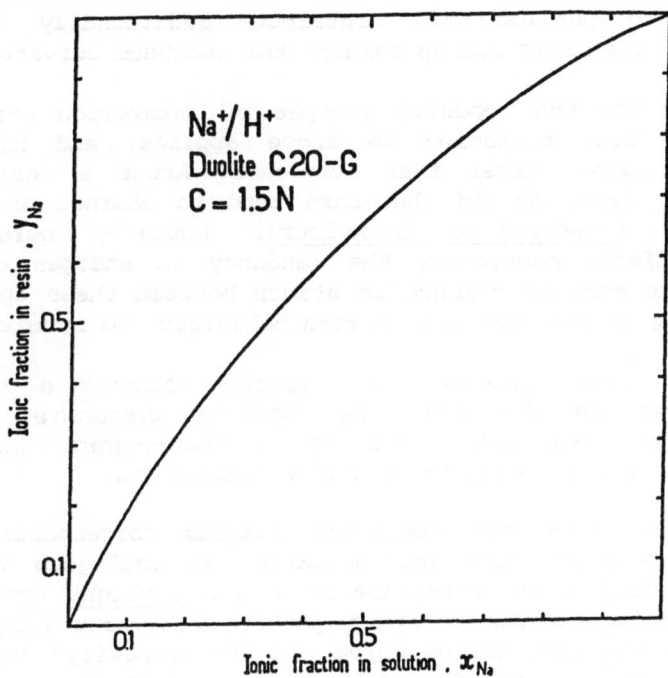

Fig. 6. Equilibrium isotherm for Na^+/H^+ exchange at room temperature.

5. COMPRESSIVE AND DISPERSIVE BEHAVIOUR

We are now in a position to reconsider the examples of Fig. 1. Note first that the plateau regions are compatible with the differential material balance (Eqns 28 or 37 or 44) : they correspond to trivial solutions of this equation, in which all derivatives cancel. Next, consider the elution of Na^+ by H^+ (Fig. 1a). The Na^+/H^+ exchange isotherm is shown on Figure 6, and is well fitted by Eqn 39 with $\alpha_H^{Na} = 2$. The slope dy_{Na}/dx_{Na} of this curve is a decreasing function of y_{Na} or x_{Na}. Referring to Eqn 50, we see that the velocities of the higher values of x_{Na} are higher than that of the lower values. In other words, high Na concentrations travel faster than low values. In Fig. 2a this implies that the front is spreading more and more as it propagates. This <u>dispersive</u> behaviour is thus implied in the variation with composition of the slope dy/dx of the equilibrium isotherm. The existence of diffusional mass-transfer resistances and of axial dispersion will contribute additionally to the spreading, but the basic reason remains the isotherm curvature.

Consider now the opposite example of saturation (Fig. 1b and 2b). The same reasoning as above applies, and high Na concentrations move faster than low ones. After a period of formation, the front should therefore tend to sharpen and even to "overlap" (<u>compressive behaviour</u>). Actually here the diffusional effects counteract the tendency to sharpen of the front, and some sort of balance is struck between these opposite trends, leading to the constant pattern behaviour. To summarize:

When a weakly retained ionic species replaces a strongly retained species on the bed, the front is dispersive. It is compressive and eventually leads to constant-pattern when a strongly retained species replaces a weakly retained one.

Figure 7a shows the operating diagram corresponding to these two behaviours. This representation is analogous to the McCabe-Thiele diagram of distillation : the <u>straight</u> operating line corresponding to the constant pattern behaviour (fig. 1b, 2b) lies below the equilibrium curve (in its concavity) owing to the fact that Na^+ is tranferred from the liquid phase to the solid phase ; thus for a given x_{Na}, the y_{Na} value in the front is <u>lower</u> than the equilibrium value. Conversely, in the case of elution of Na^+ by H^+ (Figs 1a, 2a) Na^+ is transferred from the solid to the liquid, and the y_{Na} value in the elution front must be higher than the equilibrium value. The operating line for elution thus lies <u>above</u> the equilibrium curve. Since this line connects the two extreme plateaus (initial and final concentrations) the elution operating line must obviously be curved, and actually must follow the curvature of the

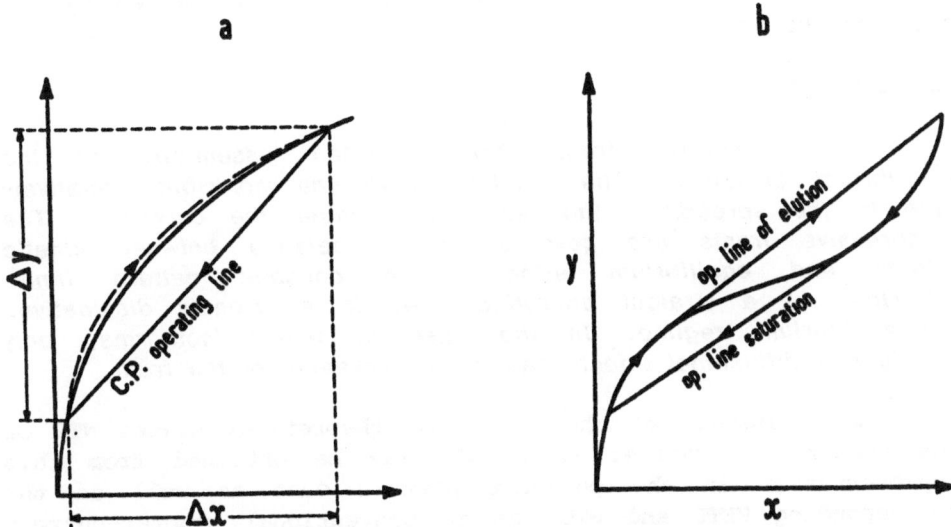

Fig. 7. Operating diagrams for different types of fronts. Straight line of slope $\Delta y/\Delta x$ on 7a corresponds to constant-pattern (C.P.) ; dotted line to dispersive front. The inflected isotherm of Fig. 7b gives rise to mixed fronts.

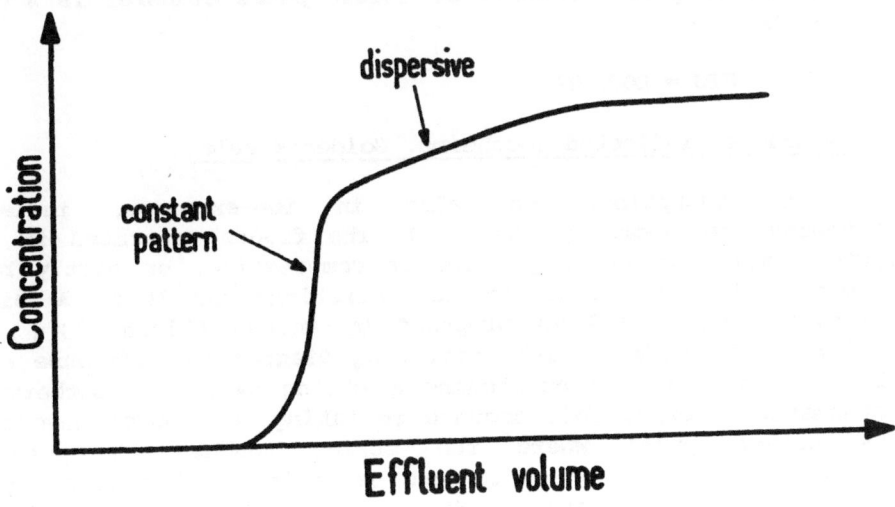

Fig. 8. Schematic breakthrough curve for saturation with an inflected isotherm (see Fig. 7b).

equilibrium curve. This explains that in elution, the slope dy/dx of the real y vs x relation can be approximated by the slope of the isotherm.

To summarize :

The dispersive fronts are governed essentially by the equilibrium curvature. The kinetic phenomena contribute relatively little to the spreading, the less the stronger the curvature. The compressive fronts are governed by a balance between kinetic effects and equilibrium effects. The constant pattern limit, described by a straight operating line, is a strongly dissipative, non-equilibrium regime. In the case of linear isotherms, only kinetic and diffusional effects cause the spreading of the front.

An estimation of the number of theoretical plates NTP or the number of transfer units NTU may be obtained from this representation in the constant pattern case as well as the corresponding HETP and HTU, as in conventional counter-current operations. The NTP is obtained by "drawing steps" between isotherm and operating line from the arbitrary break point B to the end point E defined in Section 3 ; HETP is then simply

$$HETP = LOF/NTP \qquad (55)$$

The NTU is obtained by evaluating the area of the curve $1/(y - y^*)$ versus y_{Na} or $1/(x - x^*)$ versus x_{Na}, according as NTU for solid phase transfer or liquid phase transfer is sought. HTU is then

$$HTU = LOF/NTU \qquad (56)$$

The case of inflected isotherms. Golden's rule

In adsorption, but also in ion-exchange, inflected isotherms are common (Figure 7b). The fronts generated in such systems may thus be dispersive or compressive, or partly both, according to the range of concentrations involved. A simple graphical rule has been proposed by Golden (Klein, 1981a), to construct the McCabe-Thiele operating diagram in such case (Fig. 7b) : it consists in stretching a string over the isotherm (in mathematical terms, this amounts to taking the convex envelope). The convex parts where the string touches the isotherm correspond to zones of continuous variation of the slope dy/dx, thus to dispersive parts ; the concave parts where the string is straight and departs from the isotherm are constant pattern fronts, obeying Eqn. 54. Figure 8 shows an example of the resulting breakthrough curve for the case of saturation of Figure 7b. A sharp front is first observed (saturating species is prefered at low concentration), followed by a dispersive part (displaced species is prefered at low concentration of this species).

6. CONSTRUCTION OF ALGEBRAIC SOLUTIONS FOR CONSTANT RELATIVE SORPTIVITY

6.1 Binary elution curves

Let us first consider binary ion-exchange between species A and B (Na$^+$ and H$^+$ for example) and rewrite the equilibrium relation Eqn 40 as

$$y_A = \frac{\alpha_B^A \, x_A}{\alpha_B^A \, x_A + 1 - x_A} \tag{57}$$

The concentration velocity u'_{xA}, given by Eqn 51, is obtained explicitly as a function of x_A by differentiating Eqn 57 :

$$1 \left/ \frac{dy_A}{dx_A} \right. = \left[1 + x_A (\alpha_B^A - 1) \right]^2 / \alpha_B^A = u'_x \tag{58}$$

and solving for x_A :

$$x_A = \frac{1}{\alpha - 1} \left[\sqrt{\alpha u'_x} - 1 \right] \qquad (\alpha = \alpha_B^A) \tag{59}$$

In order to relate x to operating variables such as effluent volume and abscissa, we have to relate u'_x to these variables, and this requires some knowledge of the structure of the solutions. For that purpose, consider the three-dimensional space (x,z,t) or (x,ξ,θ) of Fig. 9. We have represented in that space an arbitrary initial bed profile x versus ξ at $\theta = 0$, and an arbitrary feed composition x versus θ at $\xi = 0$. For each value of x along the line representing these initial and boundary conditions, we can calculate u'_x from Eqn 58. The straight lines, parallel to the base plane (ξ,θ), and satisfying both Eqns 51 and 58, are called characteristics. They "emanate" from the initial/boundary curve and represent the propagation of the values of x from their initial or boundary values. These characteristics form a surface, which is the integral surface of the differential material balance equations, satisfying equilibrium between phases.

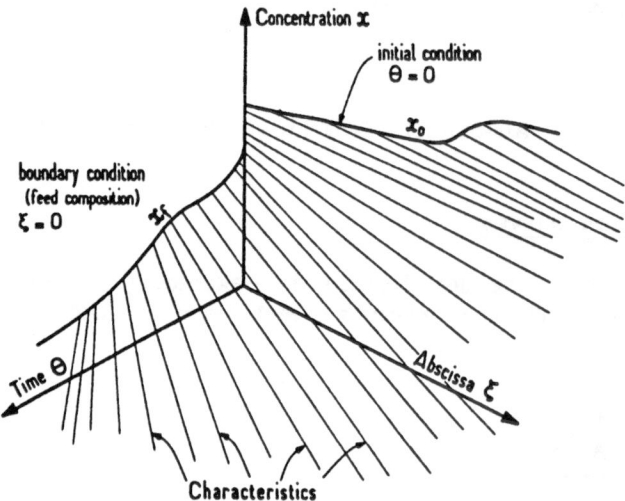

Figure 9- The "space"(concentration,time,distance) and the integral surface for arbitrary but simple initial and boundary conditions and a single independent species(binary exchange)

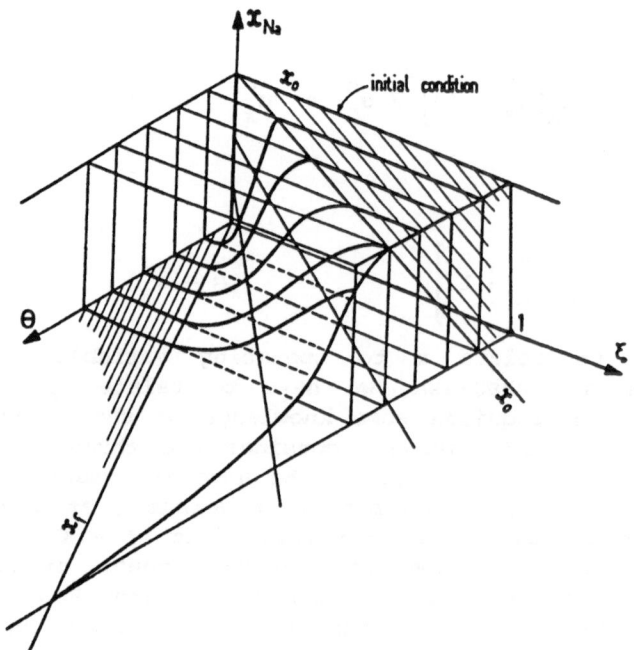

Figure 10-The integral surface in the case of constant initial and boundary conditions(Riemann conditions).Its relation to effluent histories,bed profiles and characteristics diagram.

Figure 10 shows how the more common, two-dimensional bed profiles and breakthrough curves are related to the integral surface : the different bed profiles shown at successive times are obtained by cutting the integral surface by planes parallel to the (ξ, x) plane, that is planes at constant θ ; similarly, the breakthrough curve sketched x versus θ is obtained by cutting the integral surface by a plane at $\xi = 1$. Figure 10 has been constructed for a step change from constant initial conditions to constant boundary conditions ; the integral surface then comprises a plane containing all characteristics emanating from the initial conditions and a plane corresponding to boundary (feed) conditions ; between these two planes, a curved surface containing all characteristics emanating from the x-axis.

Figure 11a shows the projection of the characteristics on the base plane (ξ, θ), for a dispersive front resulting from a step change in feed (elution of Na^+ by H^+). The characteristics corresponding to the front form a divergent fan centered at the origin. Fig. 11b shows the simplified representation of a constant pattern front also centered at the origin. The slope of each characteristic is the velocity u'_x attached to the corresponding concentration x. Since the characteristics go through the origin, we may write (accounting also for Eqn. 36)

$$u'_x = \left[\frac{\partial \xi}{\partial \theta} \right]_x = \frac{\xi}{\theta} = \frac{1}{T} \tag{60}$$

Substituting into Eqn 59, we obtain a relation between x and the throughput ratio T :

$$\boxed{x_A = \frac{1}{\alpha - 1} \left[\sqrt{\alpha/T} - 1 \right]} \tag{61}$$

and from Eqn 57 :

$$\boxed{y_A = \frac{1}{\alpha - 1} \left[\alpha - \sqrt{\alpha T} \right]} \tag{62}$$

These are the general solutions for binary elution fronts with constant α (Walter, 1945).

Figure 12 shows the sketch of the profile y versus 1/T for $\alpha = \alpha^{Na}_H = 2$ (elution of Na^+ by H^+). Clearly, we have $x_A = 1$ for $1/T = \alpha$ and $x_A = 0$ for $T = \alpha$. This figure may be

172

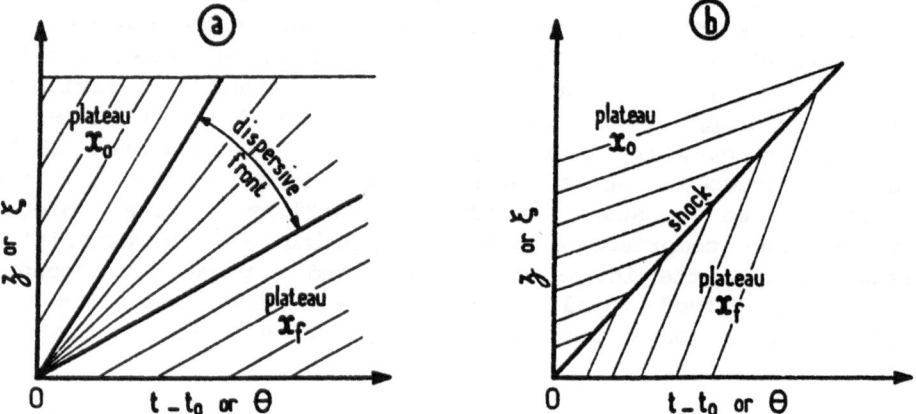

Fig. 11. Characteristic (distance vs time) diagram for dispersive front (11a) and perfectly sharp constant pattern front (shock, 11b).

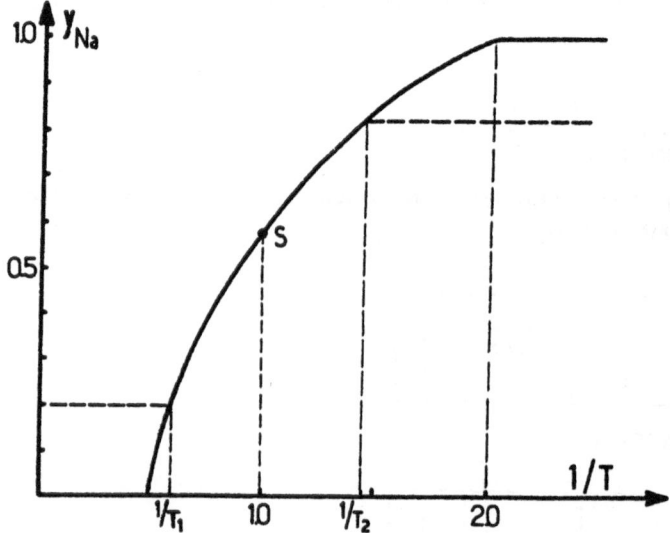

Fig. 12. Dimensionless bed profile for Na^+ elution by H^+ calculated by equilibrium model (Eqn. 62) with $\alpha_H^{Na} = 2$. Dotted horizontal lines indicated the initial and final plateaus in the case of mixed solutions.

considered as a reduced column profile, since $1/T$ is proportionnal to ξ, at constant θ (Eqn 36). Conversely, a plot of x versus T is a generalized breakthrough curve, since at given ξ, T is proportionnal to θ. The "physical" curves may be easily reconstructed from the generalized curves. It is easily verified that the stoechiometric point S is located at $T = 1$, which is the mean value $\overline{T}$ of the distributions $x(T)$ and $y(1/T)$ between $x = y = 0$ and $x = y = 1$.

6.2 Binary saturation curves

The careless application of the foregoing solutions to the reverse case of saturation of bed in H^+ form by a more strongly retained ion Na^+ would lead to "overlapping" profiles (Fig. 13a), which are physically meaningless. The constant-pattern front formed in this situation satisfies Eqn. 53 and 54, which reduce in this case to

$$\left[\frac{\partial \xi}{\partial \theta}\right]_s = u'_s = 1 \bigg/ \frac{\Delta y}{\Delta x} = \frac{1}{T} = 1 \qquad (63)$$

describing the movement of the stoechiometric point. The equilibrium model represents the front by a discontinuity, in y located at $T = 1$ (Fig. 13b).

6.3 Saturation or elution with mixed feed (Fig. 12b, 13b)

If the bed contains initially a mixture of A and B, and the feed is also a mixture, the expressions Eqns 61 and 62 remain valid, except that the bounding plateaus are changed. In other words, the profile is the same as that of Fig. 12 but truncated : the new initial and final plateaus are the dotted lines. The values T_1 and T_2 of T bounding the front are calculated by substituting the initial or feed value of x_A into Eqn 61. The stoechiometric point S can then be shown to obey

$$T_s = \frac{\Delta y}{\Delta x} = \sqrt{T_1 T_2} \qquad (64)$$

In the case of a saturation front, the discontinuity is located at the stoechiometric point S satisfying the same equation, 64 (Fig. 13b)

174

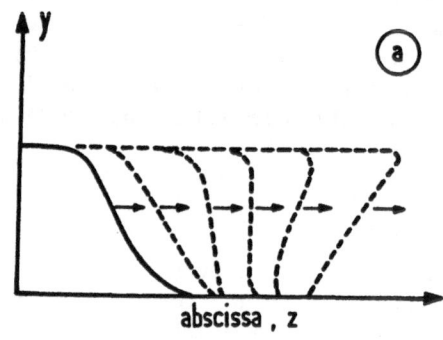

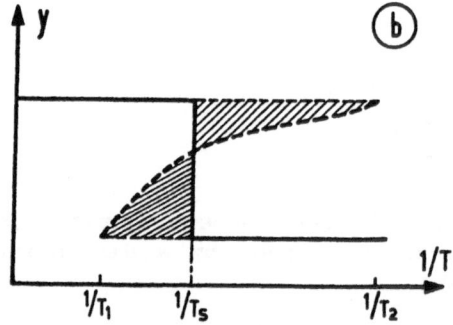

Figure 13 -a) The "compression" effect leading to ficticious overlapping profiles.

b) Relation between stoichiometric value T_S and end point values T_1 and T_2.

6.4 Multicomponent exchange

The foregoing solutions can be modified for multicomponent exchange. The somewhat involved mathematical developments leading to these solutions may be found in their generality in (Helfferich and Klein, 1970 ; Rhee et al., 1970 ; Rhee, 1981) ; in a somewhat simpler but more restricted form in (Glueckauf, 1949 ; Tondeur and Klein, 1967). In the present text, we shall limit ourselves to summarize the main results and the general method, for a single step change in inlet composition.

* Equivalent sorptivity

We define the "equivalent sorptivity" D as the sum of the relative sorptivities α weighted by the respective ion fraction in solution x

$$D = \sum_{1}^{n} \alpha_n^i \, x_i = \frac{1}{\sum y_i \, \alpha_i^n} \tag{65}$$

D is the denominator in the equilibrium law of Eqn 40.

This quantity plays a central role in multicomponent behaviour. It can be viewed as characterizing how rich a mixture is in strongly adsorbed component :

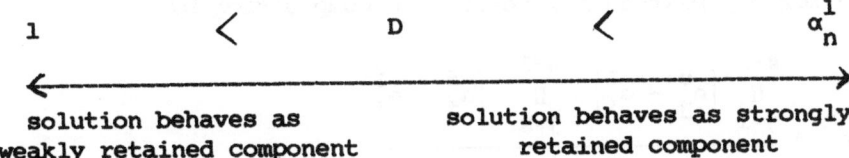

* Characteristic equation

The characteristic equation giving the eigenvalues T of the Jacobian matrix of equilibrium $|J|$ (see Eqn 47) is

$$\det [\underset{\sim}{J} - T\underset{\sim}{I}] = 0 = \sum_{1}^{n} \frac{\alpha_n^i \, x_i}{\alpha_n^i - DT} = H(x,T) \tag{66}$$

Let $h = DT$ (Helfferich and Klein, 1970)   (67a)

$g = Dh = D^2 T$ (67b)

These quantities have invariance properties which are discussed later. Equation 66 has n - 1 real positive distinct roots in T or in h and such that

$$1 < h_{n-1} < \alpha_n^{n-1} < h_{n-2} < \ldots\ldots < h_2 < \alpha_n^2 < h_1 < \alpha_n^1 \quad (68)$$

For a given composition vector $\underset{\sim}{x}$, $H(\underset{\sim}{x},h)$ is a monotonous function of h between two successive $\underset{\sim}{\alpha}$ values, thus the numerical search of the h roots is easy.

* Calculation of plateau compositions

The breakthrough curve resulting from a step change from initial composition $\underset{\sim}{x}^o$ to final composition $\underset{\sim}{x}^f$, will comprise n distinct plateaus, separated by n - 1 fronts. We number these plateaus from 1 to n, going from upstream to downstream ; in other words the plateau Pn is the first observed in the effluent history, and plateau P1 is the last to exit. The plateau compositions are calculated as follows :

- Solve Eqn 66 with $\underset{\sim}{x} = \underset{\sim}{x}^o$, to find the roots h_1^o, h_2^o ... h_{n-1}^o

- Solve Eqn 66 with $\underset{\sim}{x} = \underset{\sim}{x}^f$, to find h_1^f, h_2^f ... h_{n-1}^f

- for any plateau P_k, obtain the composition by

$$x_i^k = \frac{\prod\limits_{j=1}^{k-1} \left[h_j^o - \alpha_i^1\right] \prod\limits_{j=k}^{n-1} \left[h_j^f - \alpha_i^1\right]}{\prod\limits_{j \neq i} \left[\alpha_j^1 - \alpha_i^1\right]} \quad (69)$$

this equation expresses that the plateau P_k is determined by the k - 1 largest h^o and the n-k smallest h^f.

* Construction and properties of fronts. Invariants

The n-1 fronts Γ_1, Γ_2, ... Γ_{n-1} (from upstream to downstream) separating the plateaus P, have the property that along Γ_k, only h_k varies and all h_j with $j \neq k$ are constant. The fronts may then be calculated as follows :

- calculate the values D_k corresponding to each plateau P_k using Eqn 65 and 69.

- determine the dispersive or compressive nature of each front Γ_k with the criteria :

if the ratios $\dfrac{D_{k+1}}{D_k} = \dfrac{h_k^o}{h_k^f}$ Front Γ_k is :

> 1	dispersive (C1)
= 1	non-existent
< 1	compressive

- The value of T at the intersection between a plateau and a front is obtained from Eqn 67a by

$$\text{Plateau } P_k/\text{Front } \Gamma_k \qquad T_{kk} = h_k^o/D_k \qquad \left.\rule{0pt}{36pt}\right\} \quad (70)$$
$$\text{Plateau } P_{k+1}/\text{Front } \Gamma_k \qquad T_{k,k+1} = h_k^f/D_{k+1}$$

- calculate the __invariants__ g_k (g_k does not vary in front Γ_k but all g_i, $i \neq k$ vary) for each front using Eqn 67, written as

$$g_k = D_k \, h_k^o = D_{k+1} \, h_k^f \qquad (71)$$

- the explicit form of the __dispersive__ front Γ_k is then given by (Bailly and Tondeur, 1981)

$$\frac{x_i}{x_i^k} = \frac{D \, \alpha_n^i - g_k}{D_k \, \alpha_n^i - g_k} = \frac{\alpha_n^i \sqrt{g_k/T} - g_k}{D_k \, \alpha_n^i - g_k} \qquad (72)$$

The second expression of Eqn 72 results from replacing D, the current value of D along the front, by its expression as a function of T (Eqn 67b). All the factors in Eqn 72 are known constants, except the variables x_i and T. The linear dependence of x_i on $\sqrt{1/T}$ of the binary case (Eqn 61) is seen to be conserved. g_k can be shown to be the value of $1/T$ for which $x_n = 0$ in front Γ_k.

- the position of the stoichiometric point is given by

$$\bar{T}_{sk} = \frac{g_k}{D_k \, D_{k+1}} \tag{73}$$

which in particular defines the position of the compressive fronts.

7. THE SPECIAL CASE OF TERNARY ION-EXCHANGE

We give some more details on this special case because it presents most of the characteristic features of multicomponent systems, but still allows useful visualisation of the solution presented for constant relative sorptivity. Detailed examples of the use of the foregoing solutions to complex operations (recycle, flow reversal) in the case of ternary ion-exchange are given in (Bailly and Tondeur, 1981 and 1982 ; see also P.C. Wankat in the present volume).

7.1 The Gibbs triangle

Owing to the assumed stoichiometry and the definitions of the ionic fractions (Eqns 32), the ion-exchange process may be represented in the Gibbs equilateral triangle. Each point inside this triangle is characterized by its coordinates as shown on Figure 14a.

In a ternary system, there are two types of fronts, Γ_1 and Γ_2 corresponding to the two roots of the characteristic equation (Eqn 66), which is a quadratic equation. In the Gibbs triangle, the fronts are represented by <u>straight lines</u>. The two families Γ_1 and Γ_2 are easily constructed using the property that the segments intercepted on the sides of the triangle are in a constant ratio (Fig. 14b) :

$$\frac{a}{b} = 1 - \frac{c}{d} = \frac{1 - \alpha_{Na}^{H}}{1 - \alpha_{K}^{H}} \qquad (= 0.63 \text{ here}) \tag{74}$$

The complete network of Γ's is shown on Figure 14c ; through each composition inside the triangle pass one Γ_1 and one Γ_2. The Γ's through the vertices coïncide with the sides of the triangle. A singular point W, called "watershed point", is

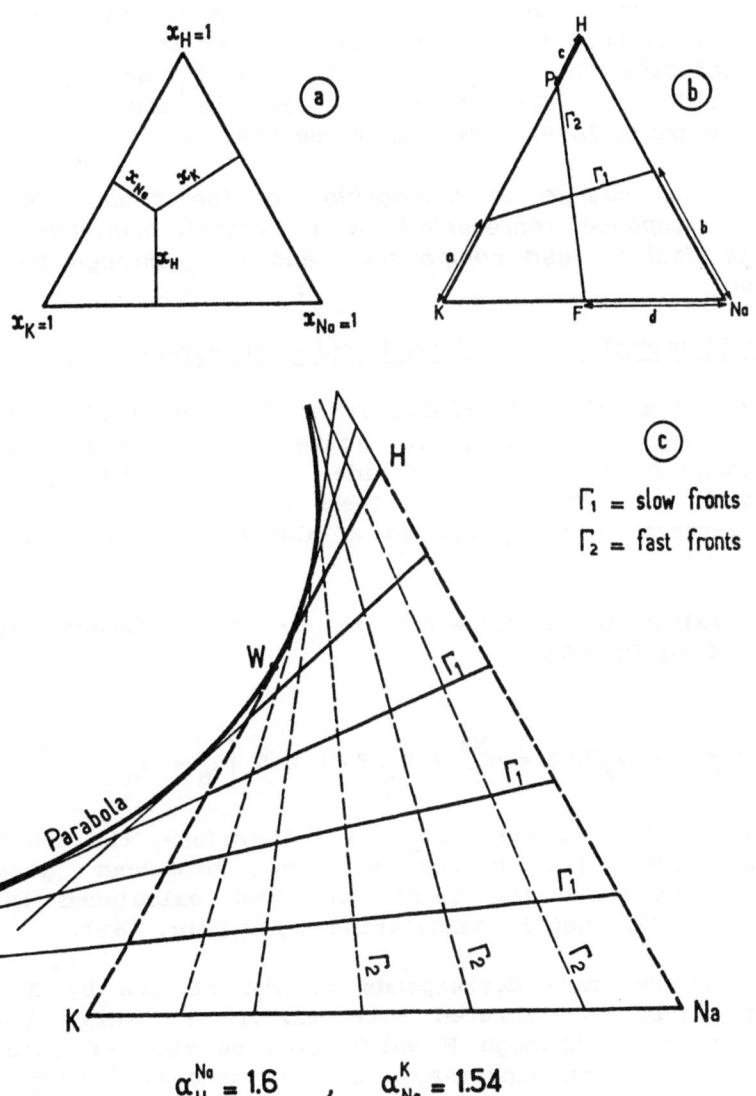

$$\alpha_H^{Na} = 1.6 \quad , \quad \alpha_{Na}^K = 1.54$$

Fig. 14. The Gibbs triangle for ternary ion-exchange
a) The coordinates of a composition (x_K, x_{Na}, x_H)

b) Geometric properties of the straight lines Γ representing the fronts : the intercepts with the sides of the triangle are in a constant ratio (Eqn 74)

c) The complete network of Γ's and their envelope.

located on the side connecting species 1 and n (that is, K^+ and H^+), at a position defined by the ratio of Eqn 74 (that is, $x_{HW} = 0.63$). At this point, the two roots of the characteristic equation coïncide ($h_1 = h_2$, $T_1 = T_2$) and Γ_1 and Γ_2 coïncide with the side K-H. The watershed point is also the contact point of the parabola P, envelope of the Γ's.

Any step change in composition at the column inlet will generate a response represented by a "route" comprising a Γ_1 through the final or feed composition, and a Γ_2 through the initial composition.

7.2 A simple example : binary feed, pure regenerant

Consider a column initially in H^+ form ($x_H = 1$), receiving the feed F ($x_{Na} = x_K = 0.5$). The route comprises the Γ_1 through F which coïncides with the K-Na side, and the Γ_2 through the H-vertex which coïncides with the H-Na side ; their intersection at the Na-vertex corresponds to a plateau of pure Na^+ (fig. 14b).

The values of D corresponding to the different plateaus are (from Eqn 65)

$$D_3(H) = 1 < D_2(Na) = \alpha_H^{Na} < D_1(F) = 0.5 \left[\alpha_H^K + \alpha_H^{Na}\right]$$

and we have $D_3/D_2 < 1$ and $D_2/D_1 < 1$; therefore, the two fronts are compressive. The values of the throughput parameter coresponding to these two fronts are best calculated in this case using Eqn 63, that is calculating $\Delta y/\Delta x$ (Fig. 15a).

The reverse case corresponds to the elution by H^+ of a column initially equilibrated with mixture F. This time the route comprises Γ_1 through H which lies on the H-K side, and the Γ_2 through F which intersects the H-K side at P (Fig. 14b). The construction of the breakthrough curve requires knowledge of the invariants, which in this case can be expressed analytically, and are summarized in the following table.

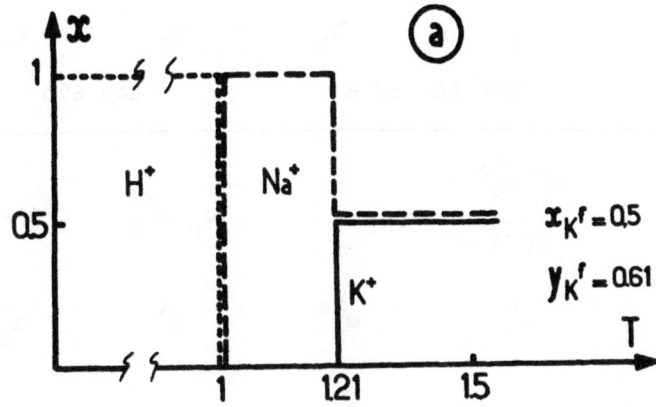

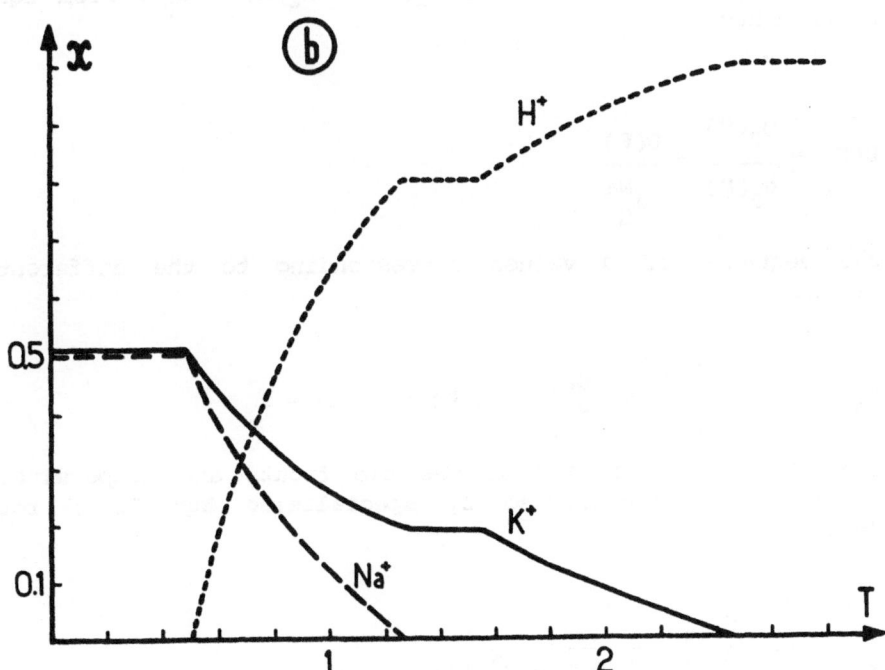

Fig. 15. Calculated dimensionless effluent history for saturation of a H^+ resin with a 50/50 mixture of Na^+ and K^+ (a)
and subsequent elution with H^+ (b).

composition	D	h_1	h_2	g_1	g_2
	Eqn 65	Eqn 66, 67 a		Eqn 67b	
F	$D(F)$	$\dfrac{\alpha_H^K \, \alpha_H^{Na}}{D(F)}$	1	$\alpha_H^K \, \alpha_H^{Na}$	$D(F)$
H	1	α_H^K	α_H^{Na}	α_H^K	α_H^{Na}

To determine the dispersive nature of the two fronts, we need to calculate $D(P)$. This is done by noting that g_2 is constant along Γ_2, and thus $g_2(P) = g_2(F) = D(F)$. Next, h_2 is constant along Γ_1, and therefore $h_2(P) = h_2(H) = \alpha_H^{Na}$. From Eqn 67b, we then have

$$D(P) = \frac{g_2(P)}{h_2(P)} = \frac{D(F)}{\alpha_H^{Na}}$$

and the sequence of D values corresponding to the different plateaus is

$$D_3(F) = 0.5 \left[\alpha_H^K + \alpha_H^{Na}\right] > D_2(P) > D_1(H) = 1$$

Criteria C1 thus indicate that the two fronts are dispersive. The front Γ_2 is constructed by specializing Eqn 72 ; for example :

$$x_K = \frac{x_{KF} \, \alpha_H^K}{\alpha_H^K - 1} \left[\sqrt{1/D_F T} - \alpha_H^K \right]$$

The corresponding reduced breakthrough curve is shown on Fig. 15b.

8. THE QUALITATIVE RULES OF MULTICOMPONENT COMPETITION

8.1 Response of ordered systems to step changes

By ordered system we mean a system in which the different ionic species may be ordered in the sequence of relative sorptivities, and in which this sequence is constant in the process considered. Sorptivity reversals, that is changes in this order, will be discussed only briefly. It can be shown under certain assumptions, that ordered systems present a qualitative pattern of behaviour which is independent of the particular form of equilibrium law, and which is typified by the constant relative sorptivity model (Klein et al. 1967 ; Tondeur and Klein, 1967).

This pattern of behaviour is summarized in the following four properties.

8.2 Basic qualitative properties

* **Property 1** : Number of fronts

The number of fronts generated by a step change in inlet composition is n − 1, that is equal to the number of independent components involved.

* **Property 2** : Relative directions of concentration changes

In any front of order k, Γ_k, the components 1 to k form the high sorptivity group and their concentrations vary in the same direction, opposite to components k + 1 to n forming the low sorptivity group.

NB1 : Fronts are numbered 1 to n−1 from upstream (slow front) to downstream (fast front), and the components are numbered 1 to n in decreasing order of sorptivity.

NB2 : This property gives no information on the absolute direction of change (increasing or decreasing)

NB3 : Any component k switches from the low sorptivity group to the high sorptivity group in front Γ_k, except components 1 (always high) and n (always low).

NB4 : Components k and k+1 are called "key-components" in front Γ_k ; their behaviour is opposite ; they are the components between which competition is strongest, and which will undergo the largest relative changes in concentration.

Property 3 : Appearance and disappearance of components

Components may appear or disappear only in the fronts where they are key components.

Corollaries : in front Γ_k, only components k and k+1 can appear or disappear ; component k can appear or disappear only in fronts k-1 and k, component 1 in front 1, component n in front n-1.

NB : a component is said to "appear" in the effluent curve when it is absent from the initial state, but present in the final state ; it is said to "disappear" in the reverse case ; appearance or disappearance may also be defined with respect to column profiles.

Property 4 : Compressivity and dispersivity

A front is compressive whenever components of the high sorptivity group are taken up by the solid (their concentration decreases going downstream) ; the low selectivity group is then released (desorbed). The front is dispersive in the opposite case.

This set of properties allows the qualitative patterns to be understood, and in many practical cases predicted unambiguously. In the most general cases however, for example when all components are present both in the initial and final state, the properties do not suffice to determine entirely the pattern, and calculations become necessary.

8.3 **Examples**

In the case of ternary systems, properties 1, 2 and 3 are actually expressed in the Gibbs triangle by the geometric properties of the lines Γ_1 and Γ_2 (intercepts with the sides, and slopes). Figures 16 a,b,c,d show four breakthrough patterns as constructed and calculated from the equilibrium model, and corresponding experimental results. These results were indeed obtained under rather careful and controlled conditions (slow flow velocity, relatively narrow granulometry), but they illustrate the semi-quantitative value of the equilibrium approach. Figure 17 shows the Gibbs triangle relative to Figures 16 a,b,c and d.

Figure 18 (Clifford, 1982) shows the effluent history of an anion exchanger, assumed initially in free-base form, receiving a feed containing four components, ranked in the following order of sorptivities

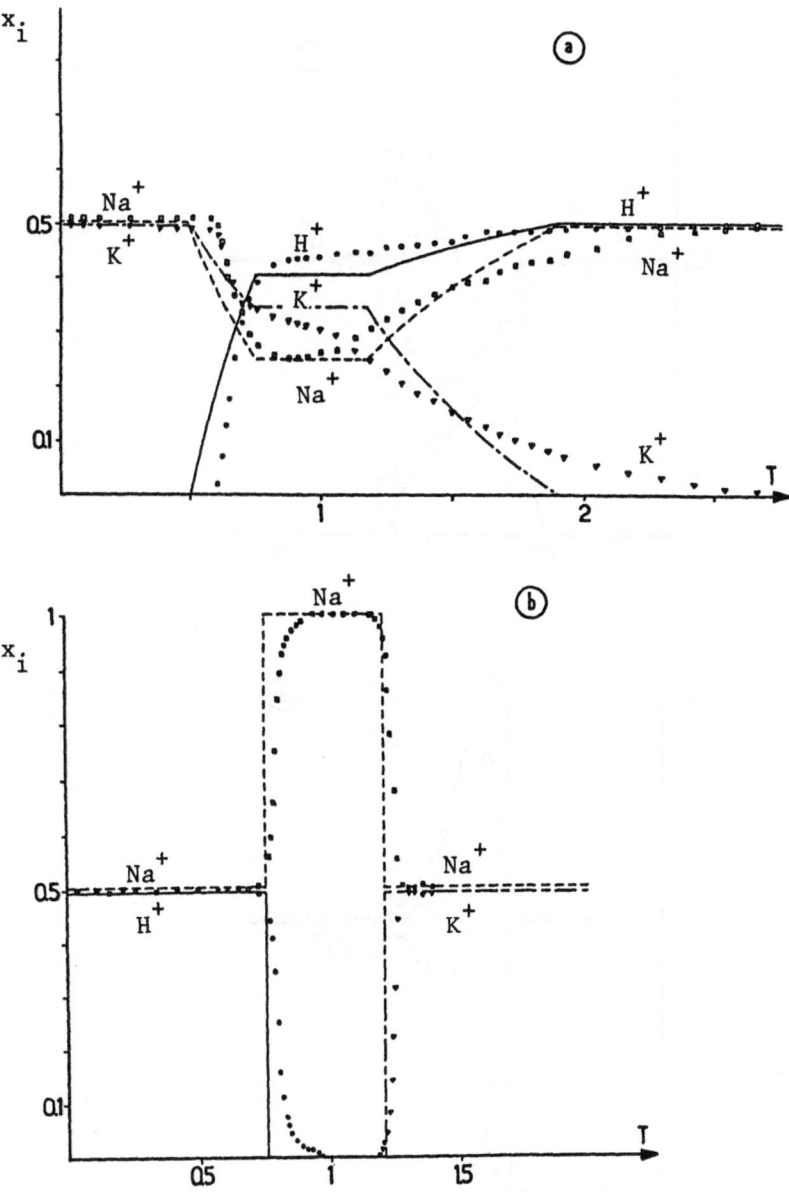

Figures 16. Experimental and calculated effluent histories for successive 50/50 binary feeds. Initial state : $Na^+ + K^+$

a) feed $H^+ + Na^+$

b) feed $Na^+ + K^+$

186

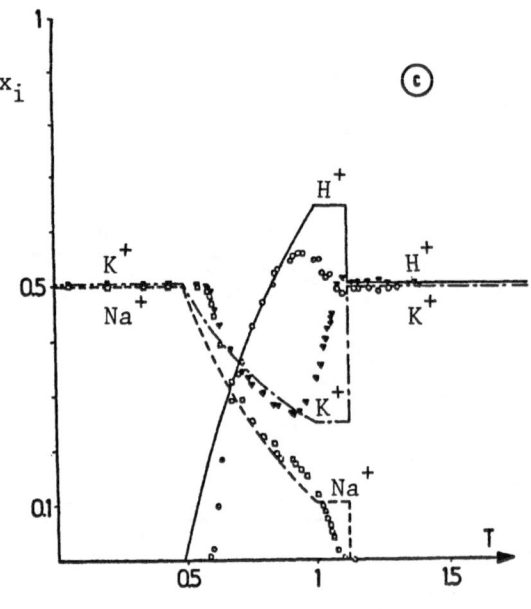

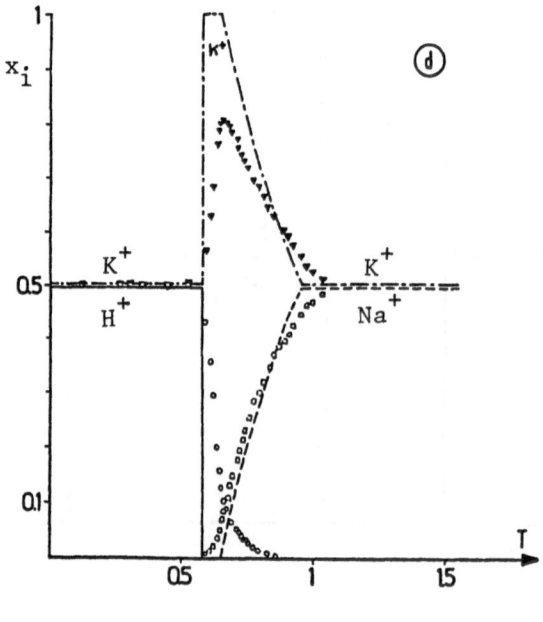

c) feed $H^+ + K^+$

d) feed $Na^+ + K^+$

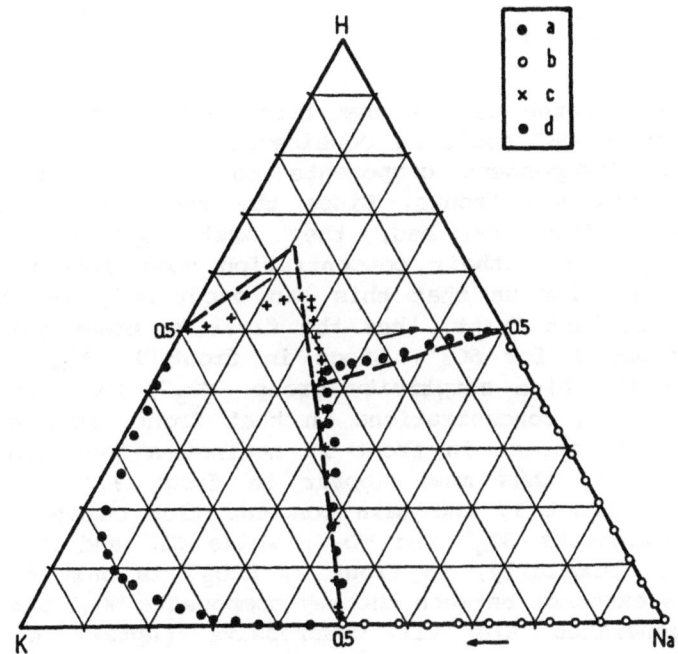

Figure 17- The experiments of Fig.16 represented in the Gibbs triangle.

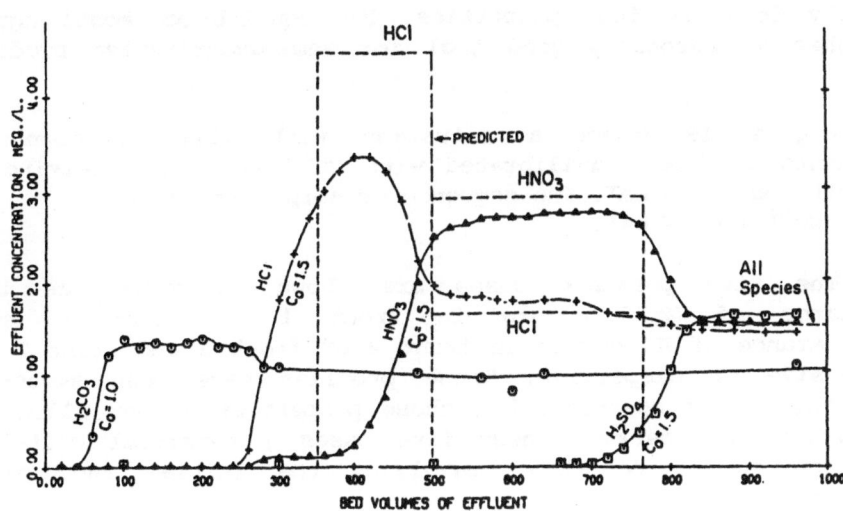

Figure 18- Experimental and calculated effluent history of bed of STY-DVB tertiary amine resin (ES 368;$\alpha_{Cl}^{SO_4}$=2.86; $\alpha_{Cl3}^{NO_3}$=3.87(from Clifford,1982).

$$SO_4^= > NO_3^- > Cl^- > HCO_3^-$$

The resin here behaves as an adsorbent rather than an ion-exchanger (although OH^- could be considered as a component) and there are four independent components to be considered, and therefore four distinct fronts. Since the four components are initially absent from the bed, they must "appear" in the effluent history, i.e., their concentration must increase from zero. Property 3 tells us that this can occur only in front 4 for HCO_3^- (the fastest front, thus the first to break through), and only in front 1 for $SO_4^=$. Since in front 1, $SO_4^=$ is the only member of the high sorptivity group, NO_3^-, Cl^- and HCO_3^- must have decreasing concentrations in that front. This entails that NO_3^- can only appear in front 2, against a decreasing Cl^- concentration ; Cl^- thus must appear in front 3. It can be observed that in front 1, the main concentration changes occurs for the key components $SO_4^=$ and NO_3^-; while Cl^- and HCO_3^- are hardly affected. Similarly, in front 2, HCO_3^- is only slightly affected by the exchange between the key components NO_3^- and Cl^- The fronts generated are all compressive (uptake of high sorptivity component in each front), with some incertitude on the mechanism of uptake of bicarbonate.

In this particular process, which is of practical importance, the qualitative pattern can thus be predicted entirely from the four properties. The equilibrium model again furnishes a reasonably good tool for semi-quantitative predictions.

Figure 19 shows a situation with three independent components : a bed equilibrated with Ca^{++} and K^+, receiving a feed with Mg^{++} and Na^+. The sequence of sorptivities is $Ca^{++} > Mg^{++} > K^+ > Na^+$.

The five possible cases are shown, according as the appearance of Mg^{++} occurs in front 1 or front 2, the disappearance of K^+ occurs in front 2 or front 3 and front 2 is compressive or dispersive. These profiles have been sketched using the four properties, but these properties do not allow us to discriminate between these five cases ; numerical criteria are required to do this, involving the initial and final compositions, and the α's.

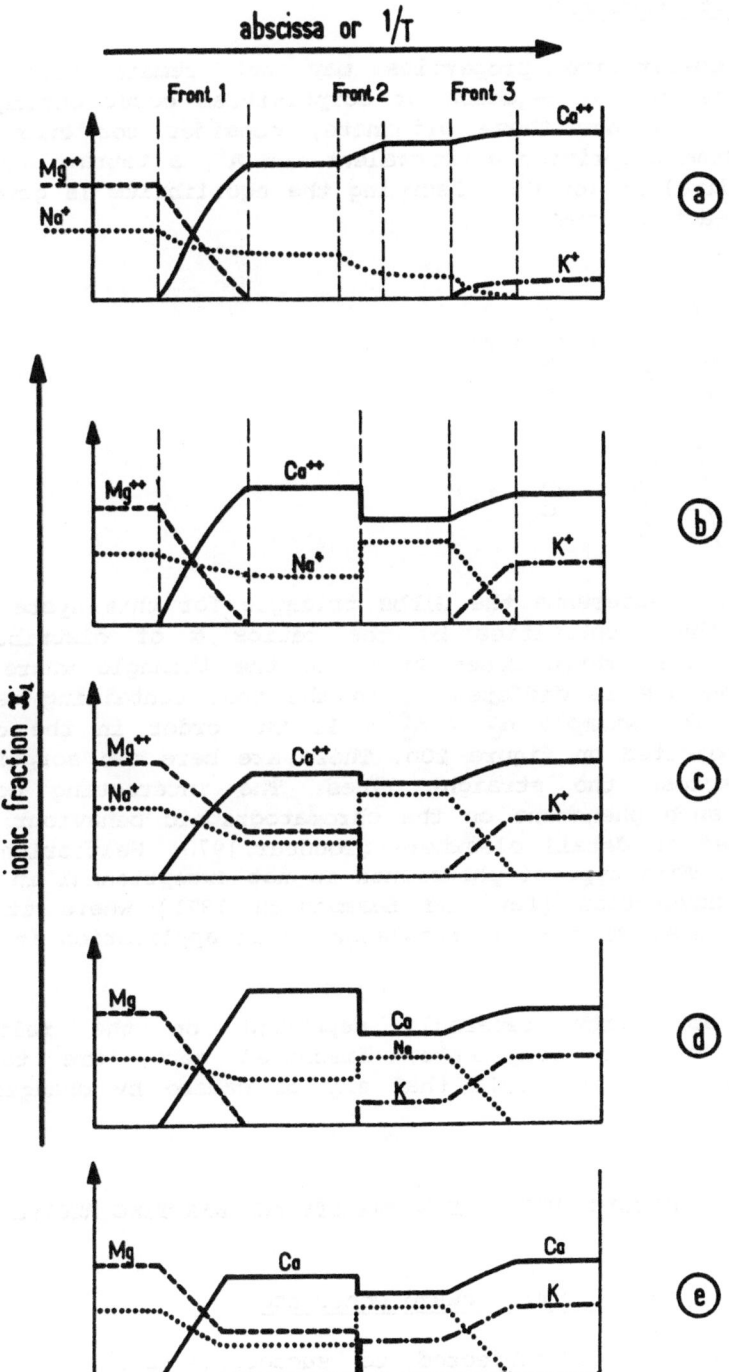

Fig. 19. Schematic composition profiles in four component
ion-exchange, illustrating different possible patterns.

8.4 Sorptivity reversals

The qualitative properties may not remain true when modifications of the sequence of sorptivities occur during the process. To see how these originate, consider for example a ternary system comprising a monovalent ion A^+, a tetravalent ion B^{4+} and a divalent ion C^{2+}. Assuming the equilibrium is governed by the mass-action laws :

$$\left[\frac{y_A}{x_A}\right]^4 \bigg/ \frac{y_B}{x_B} = K_B^A = 2$$

$$\left[\frac{y_A}{x_A}\right]^2 \bigg/ \frac{y_C}{x_C} = K_C^A = 1.7$$

Fig. 20 represents the Gibbs triangle for this system. If we define the sorptivities by the ratios α of distribution ratios y/x, then three zones exist in the triangle where the order of the α's is different ; in the zone containing vertex C, we have for example $\alpha_C^B > \alpha_C^A > 1$. The order in the other zones is indicated on figure 20b. There are here two sorptivity reversals across the straight lines. The interesting consequences of such phenomena on the chromatographic behaviour have been analyzed in detail elsewhere (Tondeur 1970, Helfferich and Klein 1970). This type of phenomenon is not exceptionnal in non-isothermal adsorption (Pan and Basmadjian 1971) where it has been put to use. We have no knowledge of an application in ion-exchange.

N.B. The sorptivity reversals dependent on the relative proportions of the components, discussed here, are to be distinguished from reversals that may be caused by changes in total concentration.

9. NON-STEADY, CYCLIC, VARIABLE NORMALITY AND REACTING REGIMES

9.1 Non-step inputs - Trend toward coherence

If the bed is subjected to successive arbitrary step inputs, or to a continuous change in feed composition, the qualitative patterns are obviously more involved, and the equilibrium model may be handled by the methods described in Helfferich and Klein (1970). The classical cases of elution or

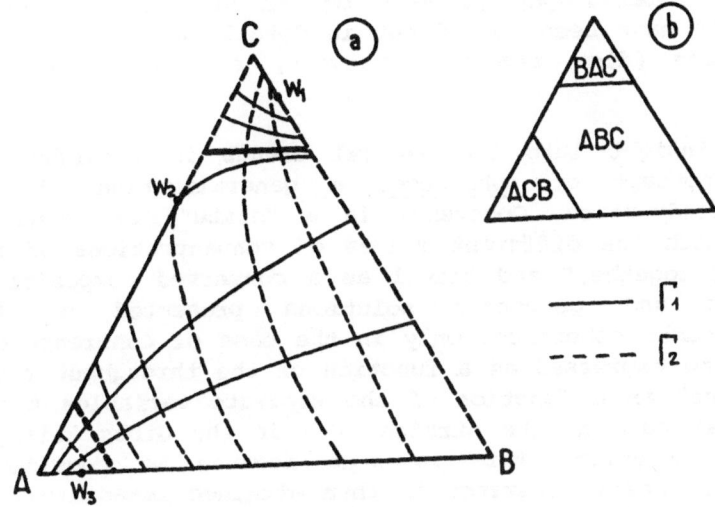

Figure 20- The Gibbs triangle for a ternary system displaying two sorptivity reversals

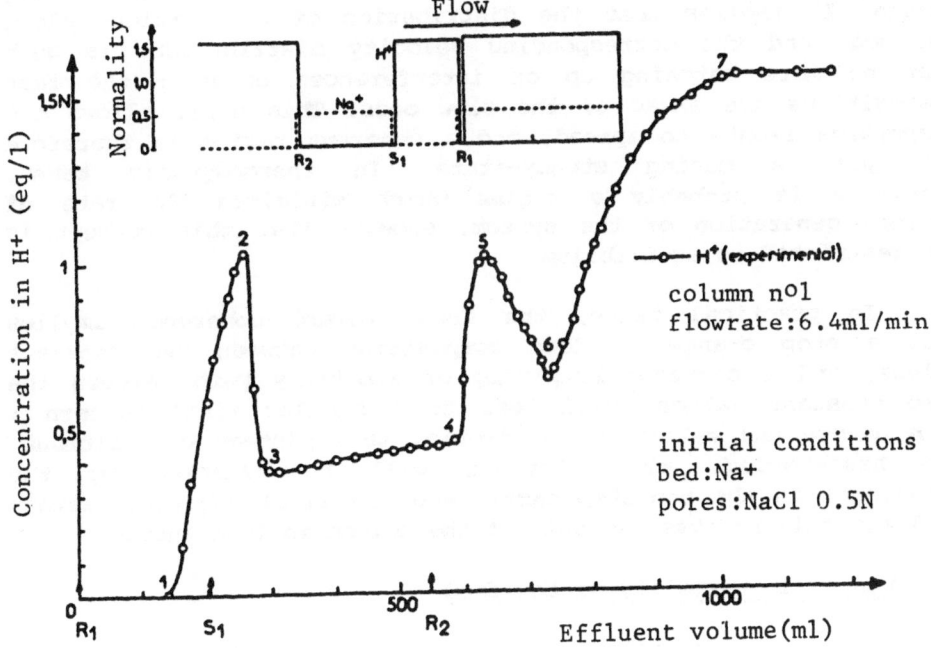

Figure 21- Experimental effluent history illustrating interferences in a column subjected to sucessive changes in composition and normality (the top sketch shows the inlet history) (from Grevillot,10).

displacement chromatography (a step of mixture followed by a step of eluent) have been worked out in detail, as well as some other situations (flow reversal, recycle, see Wankat in the present volume).

A deep insight into the general trends is furnished by Helfferich's concept of <u>coherence</u>, a generalisation of the concept of steady-state. Coherence is a "state", or rather a "regime" in which the different values of concentrations of the species "stick together" and travel as a conserved composition. The algebraic and geometric solutions presented so far implicitly assumed coherence. Only in the case of coherence can the solutions be expressed as a function of the throughput ratio T <u>alone</u>, and not as a function of the separate variables t and z. Only in that case do the straight Γ's in the Gibbs triangle represent the solution. This is true because we considered rigorous <u>step</u> inputs : coherence is then attained immediately in the equilibrium model.

More generally, coherence is a regime toward which the system tends asymptotically when the inlet conditions remain constant after various perturbations and when the bed is long enough. It implies that the distribution of compositions along the bed, and the corresponding velocity distribution, is such that no more catching up or interferences occur : the fast compositions are ahead of the slow ones. This still allows the dispersive fronts to spread, and a coherent regime is therefore not just a moving steady-state. In thermodynamic terms, coherence is probably a regime which minimizes the rate of entropy generation of the system, showing that this concept is not restricted to equilibrium.

In practical terms, the trend toward coherence implies that a step change in feed composition between two constant values, and a corresponding ramp or arbitrary path between the same constant values, will lead to a similar final pattern : same nature and velocities of fronts, same plateau compositions. The breakthrough times however will be affected by the transients in the non-step cases, and the final coherent pattern will actually be observed only if the column is long enough.

9.2 Changes in normality of the feed

In usual ion-exchange practice, the bed may be subjected to important changes in total concentration, accompanying changes in composition. For example in water demineralisation, water of low salt content is treated, then the regeneration is effected with concentrated acid, and this acid is then washed out with water. In other situations, the Donnan electrolyte uptake or release accompanying a change in the nature of the exchanged species may cause a change in total concentration (see "Parametric Pumping" in the present book). This change in normality is propagated through the bed at velocities different from that of the normal exchange fronts, and may interfere with them. In addition, the equilibrium isotherm may undergo a shift with the change in normality, as would be expected for the exchange of ions of different valences.

Such interference phenomena are likely to occur in relatively long columns subject to short cycles, and they may modify radically the qualitative features of the breakthrough curves, and a fortiori its quantitative aspects. These phenomena have been studied in detail by Grevillot and Tondeur (1974) and we give here merely an illustration. Figure 21 shows the effluent history of a bed of sulfonic cation exchanger (0.29 eq. capacity), initially equilibrated with 0.5 N NaCl solution, receiving successively 0.2 liters of HCl 1.55 N (regeneration R1), then 0.34 liters of NaCl 0.5 N (saturation S1), and finally an excess of HCl 1.55 N (regeneration R2). Each of the steps R1, S1, R2 generates two waves or fronts : a wave of total normality change, moving at a velocity close to that of the fluid, and a slower front of ion-exchange. R1 thus generates a normality wave $|C_1|$ which does not affect the H^+ profile, and a dispersive elution front $|de|$ (points 1-2-3-4). This front is split into a part at high normality (1-2) and a part at low normality (3-4) by the decreasing normality wave $|D|$ created by S1, which we observe as a sharp drop in H^+ (2-3). Next, the sharp increase 4-5 is the increasing normality wave $|C_2|$ generated by R2 and is followed by the sharp exchange front $|se|$ (5-6) generated by S1, while $|de|$ (6-7) is the elution front generated by R2.

In this particular example of Na^+/H^+ exchange, the equilibrium isotherm is practically unaffected by the normality changes, implying for example that the pieces of dispersive elution front 1-2, 3-4, 6-7 if properly normalized, fall onto a single curve. This is no longer true say for Ca^{++}/H^+ exchange, where an equilibrium shift is caused by the normality change (Dodds and Tondeur, 1972 II, Grévillot and Tondeur, 1974).

9.3 Cyclic operation

The prediction of the steady cyclic operation of ion-exchangers, especially in multicomponent cases, is a problem in which relatively little conceptual and theoretical progress has been made. In practice, numerical computations or semi-empirical approaches have to be used, and the equilibrium approach does not seem to bring much qualitative insight, except when the successive steps are independent (no interference). An attempt to identify the different types of cyclic regimes attainable in

ternary ion—exchange using the equilibrium model has been made
by Grevillot et al. (1974). The problem of predicting a priori
the cyclic steady—state without calculating intermediate transients is unresolved except in the simplest cases, or with
drastic simplifications (Dodds and Tondeur, 1972, 1974 ; Klein,
1981b).

9.4 Ion—exchange with chemical reaction

See Klein G. in the present volume.

Notations

A	available capacity (Eqn. 4)	equivalents
c	ion concentration in solution	eq/m^3 of solution
$\bar{c}$	ion concentration in solid phase	eq/m^3 of solid
C_{tot}	total concentration in solution	eq/m^3
$\bar{C}_{tot}$	total concentration in solid	eq/m^3
D	equivalent sorptivity (Eqn 65)	N.D.
D_a	axial diffusion or dispersion coefficient (Eqn 28)	$m^2\ s^{-1}$
E	amount of ions eluted (Eqn 12)	eq.
F	flow-rate	$m^3\ s^{-1}$
FLEB, FLOF, FLUB :	fractional length respectively of equilibrated bed, of front, of Unused bed (Eqns 24,27)	N.D.
g,h	invariants (Eqn 67)	N.D.
I	amount of ions introduced in bed (Eqn 2)	eq
$|J|$	Jacobian matrix of equilibrium relation (Eqns 46,47)	—
K	mass action constant (section 8.3)	N.D.
L	amount of ions leaked out of column (Eqn 1)	eq.
L	column length	m
LEB, LOF, LUB :	length of equilibrated bed, of front of unused bed respectively	m
Q	total exchange capacity of column (Eqn 6)	eq
R	amount of ions remaining on bed (Eqn 13)	eq
S	amount of ions stored in the bed (Eqn 13)	eq
t	time	s
T	throughput ratio (Eqn 36)	N.D.
u, u_i, u_s :	velocities respectively in empty tube interstitial, and of the stoichiometric point	$m\ s^{-1}$
v	volume of section of bed of length z	m^3
V	effluent volume	m^3
x_i	ionic fraction of species i in solution	N.D.
$\underset{\sim}{x}$	vector of ionic fractions in solution	—
y_i	ionic fraction of species i in resin	N.D.
$\underset{\sim}{y}$	vector of ionic fractions in resin	—
$\underset{\sim}{z}$	length along column	m

Greek letters

α_j^i	relative adsorptivity of species i with respect to species j (Eqn 39)	N.D.
Γ_i	designates front number i from up to downstream	
ϵ^i	fractional volume of solution in bed (intergrain porosity)	N.D.
Λ	capacity ratio (Eqn 33)	N.D.
ξ	reduced axial position (Eqn 34)	N.D.
θ	reduced and shifted time variable (Eqn 35)	N.D.
Ω	cross-sectional area of bed	m^2

Subscripts and superscripts

i	for interstitial, in u_i
i,j,k	species i,j,k in c,x,y ; front or plateau number
f	final or feed composition
o	initial composition
s	refers to the stoichiometric point

BIBLIOGRAPHY

1. Rodrigues A.E. and Tondeur D. (Eds), 1981 : "Percolation processes : theory and applications", NATO-ASI Series E, n° 33 ; Sijthoff and Nordhoff Intern. Publ., Alphen aan den Rijn, Netherlands

2. Bailly M. and Tondeur D., 1981 : "Two-Way Chromatography", Chem. Eng. Sci., 36, 455-469

3. Bailly M. and Tondeur D., 1982 : "Recycle optimization in non-linear productive chromatography I. Mixing recycle with fresh feed", Chem. Eng. Sci., 37, 1199-1212

4. Clifford D., 1982 : "Multicomponent ion-exchange calculations for selected ion separations", Ind. Chem. Eng. Fundam. 21, 141-153

5. De Vault D., 1943 : "The theory of chromatography", J. Am. Chem. Soc., 65, n° 4, 532-540

6. Dodds J.A. and Tondeur D., 1972, 1974 : "The design of cyclic fixed-bed ion-exchange operations. I. A predictive method applied to a simple softening cycle", Chem. Eng. Sci. 1972, 27, 1267-1281

7. Ibid. "II. The effect of changes in total concentration". Chem. Eng. Sci., 1972, 27, 2291-2398

8. Ibid. "III. Softening solutions containing Na^+, Mg^{++},

Ca^{++}", Chem. Eng. Sci., 1974, <u>29</u>, 611-619

9. Glueckauf E., 1949 : "Theory of chromatography.
    VII. The general theory of two solutes following non-linear
    isotherms". Disc. Faraday Soc. London, <u>7</u>, 12-25

10. Grévillot G. and Tondeur D., 1974 : "Phénomènes
    d'interférences dans un lit fixe d'échangeurs d'ions, sous
    l'effet d'un changement de normalité de l'alimentation",
    The Chem. Eng. J., <u>7</u>, 187-200

11. Grévillot G., Tondeur D. and Dodds J., 1974 :
    "Fonctionnement cyclique d'un lit fixe d'échange d'ions
    avec trois constituants", J. Chromatogr., <u>102</u>, 421-428

12. Helfferich F. and Klein G., 1970 : "Multicomponent
    chromatography", M. Dekker, New York

13. Klein G., Tondeur D. and Vermeulen T., 1967 :
    "Multicomponent ion-exchange in fixed beds. General proper-
    ties of equilibrium systems", Ind. Eng. Chem. Fundam., <u>6</u>,
    339-351

14. Klein G., 1981a : "Ion-exchange and chemical reaction in
    fixed beds", pp. 363-423 in |1|

15. Klein G., 1981b : "Design and development of cyclic
    operations", pp. 427-441 in |1|

16. Pan C.Y. and Basmadjian D., 1971 : "An analysis of
    adiabatic sorption of single solutes in fixed beds :
    equilibrium theory", Chem. Eng. Sci., <u>26</u>, 45-57

17. Rhee H.K., Aris R., and Amundson N.R., 1970 : "On the
    theory of Multicomponent Chromatography", Phil. Trans.
    Royal Soc. London, <u>267</u>, n° A1182, 419-455

18. Rhee H.K., 1981 : "Equilibrium theory of multicomponent
    chromatography", pp. 285-328 in |1|

19. Sillén L.G., 1950 : "On filtration through a sorbent
    layer, V. Final form fronts and broadening fronts", Arkiv
    for Kemi, <u>2</u>, N° 35, 499-512

20. Tondeur D. and Klein G., 1967 : "Multicomponent ion-
    exchange in fixed beds. Constant Separations Factor
    Equilibrium." Ind. Eng. Chem. Fundam., <u>6</u>, n° 3, 351-361

21. Tondeur D., 1970 : "Theory of ion-exchange columns", The
    Chem. Eng. J., <u>1</u>, 337-346

22. Tondeur D., 1981 : "Population migration and wave phenomena in percolation operations", pp. 3-30 in |1|

23. Wankat P.C., 1985 : "Efficient fractionation by ion-exchange", this volume

24. Walter J.E., 1945 : "Multiple adsorption from solutions", J. Chem. Phys., $\underline{13}$, n° 6, 229-234

FIXED-BED ION EXCHANGE WITH FORMATION OR DISSOLUTION OF PRECIPITATE

Gerhard Klein

University of California, Berkeley
Water Thermal and Chemical Technology Center
47th & Hoffman Blvd., Richmond, California 94804

ABSTRACT

The nature, utility, and limitations of fixed-bed ion-exchange systems that involve precipitates are discussed, together with possible applications of the relevant theory in other fields. Local-equilibrium theory is reviewed briefly and extended to exchange accompanied by precipitation or dissolution. The differences between the behavior of beds containing stationary and moving precipitates are discussed qualitatively and with respect to their effect on the continuity relations.

Construction methods for monovariant compound isotherms are developed that consist of a boundary-condition independent saturation branch, and a boundary-condition dependent pure-ion-exchange branch. Based on the local-equilibrium premise and uniform boundary conditions, and with application of Golden's Rule, numerous examples are given of fixed-bed behavior with moving precipitate. For calculating stationary-precipitate cases, a novel graphical method is developed that dispenses with iterative calculations. In this method, integral material-balance curves (if they apply) through the feed- and presaturation-composition points are constructed with the aid of conjugate points from the saturation branch of the overall isotherm. The point on the latter, conjugate to the point of intersection of these curves, is then used in conjunction with Golden's Rule to establish the generalized concentration profiles that determine the ideal fixed-bed behavior for the case of interest. A parallel procedure is presented for constructing differential material-balance curves, and for determining the corresponding fixed-bed behavior.

1 INTRODUCTION

1.1 Applications

The interaction of ion exchange with the formation or dissolution of one or several precipitates can be of interest as a desirable or an undesirable phenomenon, and also because of its analogy to important processes occurring in nature. The presentation of the following illustrative examples does not represent an attempt at being exhaustive.

Precipitation and dissolution have been invoked purposely in a number of process designs, of which only some are being considered actively. However, even in the development of those now temporarily or permanently abandoned, time and effort could in most cases have been saved by application of the principles discussed in the present paper.

During World War II, life rafts were equipped with briquettes of a cation exchanger in the silver-ion form. When put into a small quantity of seawater, the cations in the latter replaced the silver ions of the exchanger, causing precipitation of silver halides, and thus demineralization of the sea water. Formation of sparingly soluble precipitates thus shifted the ion-exchange equilibrium in the desirable direction, while separating a toxic ion species (17). A similar process, but for a column arrangement, has later been proposed for desalination by Glueck (7), but its development has not been pursued.

To reduce the regenerant cost of an ion-exchange process proposed to desulfate brine of the Great Salt Lake, George et al. (6) considered regeneration of an anion exchanger in the sulfate form with barium chloride. Again, precipitation (in this case, of barium sulfate) shifted the equilibrium in the desired direction and separated a toxic ion species.

Utilization of an inexpensive or recycleable regenerant was the objective of the Popper partial desalination process (15). The saline water was passed through a mixed bed of cation and anion exchangers regenerated with a lime slurry. In the exhaustion step, the ions of the solution replace the calcium ion on the cation-, and the hydroxyl ion on the anion-exchanger, producing a lime slurry of lowered solution concentration. Evans and Gomes (5) also used a lime slurry, but to regenerate a weak-base anion exchanger of a mixed-bed demineralizer.

Precipitation as an impediment has been recognized by Applebaum (1), and by Cherney (3), who identified conditions under which a cation-exchanger bed used to soften seawater as a pretreatment for evaporation would become fouled with calcium sulfate precipitate during regeneration with evaporator brine.

A similar problem was envisaged in the early development of a process that bids fair to result in a multimillion-gallon-a-day application, and which is presently being implemented on the demonstration-plant scale by the State of California Department of Water Resources, at Los Banos, California (13,16). Here, hard agricultural drainage water high in sulfate is to be subjected to reverse-osmosis treatment in order to separate it into a concentrated, relatively easily disposable, and a dilute, reusable, stream. The drainage water must be softened in order to prevent precipitation of calcium sulfate in the reverse-osmosis unit. This is done by conventional cation-exchange softening, but, to save the expense of regenerant purchase, and to avoid the disposal of waste chemicals to the environment, regeneration is not carried out with bought sodium chloride, but with the reject reverse-osmosis brine.

During regeneration, calcium sulfate precipitates in the column. To prevent clogging of the exchanger bed, regeneration is done in the upflow mode, at a velocity such that the resin becomes fluidized but does not leave the column through the overflow ports provided, while the smaller precipitate particles do so, being entrained in the effluent.

An analogous scheme has been considered to remove selenate at toxic levels from Kesterson Reservoir in California (12). Because of the chemical similarity of selenate and sulfate ions, it is difficult to prepare economically acceptable anion exchangers that exhibit significant selectivity for selenate as compared to that for sulfate ions. At the present state of technology, it would therefore be necessary to remove all of the sulfate ion in order to remove the selenate ion, present in trace levels. This could be done with a conventional, Type-II anion exchanger, and regeneration with calcium chloride, accompanied by precipitation of calcium sulfate, could reverse the softening equilibrium and yield a solid waste, at a cost that may compare favorably with that of other schemes proposed for the purpose.

In addition to being of potential interest in underground chemical leaching operations, ion exchange accompanied by precipitation can enhance secondary petroleum recovery by flooding with alkaline solutions (2), and studies are in progress on its geochemical role. In a broader scientific context, the theory underlying the phenomena mentioned should be applicable to other fields, notably meteorology, because the "precipitation" of rain, snow, and hail seems analogous to precipitation that accompany ion-exchange phenomena. The intuitive wisdom of the language in recognizing this analogy is remarkable.

1.2 Theoretical Considerations

For fixed-bed contacts at local equilibrium, there is a principal theoretical difference between ion exchange accompanied by the formation or dissolution of precipitate on the one hand; and pure ion exchange, ion exchange accompanied by reaction in solution, and adsorption, on the other. This difference lies in that, for processes in the latter class, the distribution of various chemical or ionic species between the stationary and moving media only needs obey the governing equilibrium and stoichiometric relations, whereas, for the process corresponding to the first of these classes, the amount of precipitate is not an equilibrium property, but can only be derived from the point properties in conjunction with boundary conditions.

In fixed-bed operation, the undesirable effects of precipitation depend on the characteristics of the precipitate, which, in limiting cases, may either remain where it is formed (stationary precipitate), or travel with the velocity of the fluid phase. In other cases, the precipitate may move with a spectrum of velocities of limited or considerable width. The last two types of cases can be referred to as those of "moving precipitate". Precipitate is likely to remain where it is formed if nucleation takes place on or in the sorbent particles, and likely to be of the moving type if nucleation takes place in solution. Finally, precipitates may form that grow as they travel, and then become stationary, possibly to impede further flow of the mobile phases.

The velocity of a precipitate in a fixed exchanger bed is likely to depend on the rate at which the precipitate forms, on flow velocity and particle size, on the type of the exchanger, and on the nature of the precipitate; perhaps also on electro-chemical factors, such as the valences and total concentration of the ions present. For example, at a given flow velocity and particle size, a sparse precipitate might grow into the pores of a macroporous exchange resin but be washed off the surface of the particles of a gel-type resin.

A fully mobile, not-too-voluminous precipitate may flow through a fixed bed without difficulty. Stationary precipitates, on the other hand, will generally exclude the use of standard fixed-bed operation because, even if they form only a thin coating on the exchanger particles, they are likely to interfere with mass transfer; and if they are massive, they will impede flow. In such cases, stirred batch reactors or fluidized beds may be employed to bring about separation of precipitate from the exchanger particles (18).

In view of these considerations, the question arises whether a study of the behavior of fixed beds in which stationary precipitates either form or dissolve warrants the effort. To this, in relation to chemical engineering practice, the answer seems to be affirmative only in certain applications; notably caustic flooding for enhanced oil recovery, where exchanged hardness ions tend to precipitate at the high pH of the solution (2). In other cases, it may be of importance to predict, in the course of preliminary process selection, whether a stationary precipitate could form at all. In addition, there are other reasons that make the pursuit of such studies interesting.

First, investigation of stationary precipitates is tempting not only because it complements the local-equilibrium treatment of the limiting precipitation cases, of which only that of moving precipitate has been analyzed exhaustively, but also because it is greatly more challenging than the analysis of moving-precipitate cases. Second, it furnishes an important and enlightening facet in the discussion and definition of "variance"; and last, but by no means least, availability of a theory for stationary precipitates is likely to provide new approaches to areas of geochemistry and perhaps even meteorology. In geochemistry, extremely low flowrates accompanying deposition and dissolution of solid materials are found in fixed-bed systems that differ from the familiar column systems only by geometry. Although geochemists have begun to concern themselves with the examination of such phenomena with the aid of extensive numerical calculations, it is likely that the more intuitive and general chemical-engineering approach as followed here will contribute to the success of this endeavor.

1.3 Premises and Scope

Local equilibrium, both with respect to ion exchange and the reactions leading to the formation or dissolution of precipitates, and the formation or dissolution of precipitate itself, will be assumed as a simplifying premise, and these phenomena, when the premise applies, may be referred to as "fast" for short. In extension of this terminology, exchanger beds governed by local-equilibrium theory may thus simply be called "fast beds".

The discussion here will be limited to uniformly presaturated beds receiving a feed of constant composition (uniform boundary conditions) that can include the mass of a mobile precipitate per unit volume of mobile medium as a concentration. The volume of precipitate will be neglected. Ad-hoc calculations for rectangular pulses in feed concentrations have been presented by Bunge et al. (2).

For the sake of simplicity, the present discussion addresses itself primarily to binary ion exchange with a single coion species capable of forming a precipitate. Extension to several, nonprecipitating or precipitating coion species is not difficult, and has been dealt with by Golden (8) and reviewed by Klein (10). Inclusion of additional nonprecipitating counterion species increases the variance of the system and leads to eigenvalue problems as formulated generally by Mangelsdorf (14) and not considered here, but which, up to bivariance, are manageable with the aid of the method developed by Vislocky (19).

The formation and dissolution of moving and stationary precipitates is considered. A moving precipitate can be part of the feed stream and later at least partially dissolve in the bed, or it can form in the bed. A stationary precipitate, while meaningless as part of the feed, can, nevertheless, both form and redissolve in the ion-exchange bed.

2 BASIC CONCEPTS

2.1 Introduction

Many of the concepts required for an understanding of fast sorption beds have been summarized previously by the author (10). These are outlined here only to the extent needed for context, together with new concepts required and changes in the symbols used.

2.2 Concentration Variables

Most relations of interest here become simplest if concentrations are expressed in terms of equivalents of an ionic or other chemical species i per unit volume of packed bed. In these terms, concentrations of Species i in the mobile medium will be denoted by C_i, and concentrations of Species i in the stationary medium , by Q_i. Conversion from the concentration units customarily used in ion-exchange equilibria and solubility relations offers no difficulty.

In material-balance equations involving precipitate, shown further below, the amount of precipitate present in unit bed volume is accounted for as a "concentration". For this case, the term "equivalents" as applied to precipitates composed of at least one multivalent ion needs to be defined as the number of equivalents of the species for which the material balance is written.

2.3 Electroneutrality and Equilibrium (Point) Relations

For systems with a single coion species, the species of primary interest here are the (exchangeable) counterions M and N and the (nonexchangeable) coion X. N and X are capable of forming a precipitate. Only precipitates moving with the velocity of the fluid medium, or with zero velocity, will be considered, and their respective "concentrations" will be designated by the symbols C_p and Q_p. Extension of the concepts developed for such simple systems to systems with other precipitating or nonprecipitating coions is readily apparent.

The electroneutrality relation for the mobile medium is

$$C_M + C_N = C^o \tag{1}$$

where C^o is the total solution normality. If X is the only coion species present, Eq. 1 takes the form

$$C_M + C_N = C_X \tag{1a}$$

For the stationary medium, the electroneutrality relation is

$$Q_M + Q_N = Q^o \tag{2}$$

where Q^o is the total exchange capacity, and where bulk electrolyte is assumed to be completely excluded from the exchanger.

The ion-exchange equilibrium relation will be of the general form

$$f_1(C_M,C_N,Q_M,Q_N) = 0 \tag{3}$$

A specific form of this equation frequently used corresponds to a modified mass-action relation with empirical exponents a and b and selectivity coefficient K_{MN}:

$$(Q_M/C_M)^a(C_N/Q_N)^b = K_{MN} \tag{3a}$$

Ideally, a is the absolute value of the valence of Ion Species N, and b that of Ion Species M. It is easy to put Eq. 3a into the form of a two-parameter equation governed by the ratio of a and b and a power of K_{MN}.

Together, Eqs. 1, 2, and 3 define an isothermal equilibrium relation between the variables involved. When C_X is constant, such a relation may take the form of C_M as a function of Q_M.

206

Specifically, for the exchange of two monovalent ion species (a = b = 1 in Eq. 3a) governed by ideal mass-action equilibrium, one obtains

$$C_M = C^o/[1 - K_{MN}(1 - 1/y)] \tag{3b}$$

where the equivalent fraction y of M in the exchanger is given by

$$y \equiv Q_M/Q^o \tag{3c}$$

Finally, the solubility relation will be of the general form

$$f_2(C_M,C_N,C_X) = 0 \tag{4}$$

For monovalent ion species, this may take the specific form of a solubility-product relation:

$$C_N C_X = K \tag{4a}$$

where K is a constant. For the sake of simplicity, the examples presented in this paper are based on this relation, but the changes entailed by other forms of the solubility relation are readily apparent.

Eqs. 4 and 4a only apply at saturation, and then, for $C_X=C^o$, define equilibrium together with Eqs. 1, 2, and 3.

In these equations, which collectively are also called point relations, the dimensions of C_i, Q_i, and Q^o will be taken to be eq dm^{-3}.

2.4 Continuity

The material balances or continuity equations for each species in a fast fixed-sorption bed establish that the appropriately defined relative velocity U of a composition in a simple composition wave, or of a composition shockwave (discontinuity or step) can be obtained respectively as the slope of the isotherm, or of a chord between two points on the isotherm, the isotherm being plotted in terms of C_i vs. Q_i. The criterion for choosing either of these velocities is given further below as Golden's Rule.

U is the velocity of a composition, and U_Δ, that of a composition step, expressed as a fraction of the apparent velocity of the mobile medium as observed from the travelling bed level at which the composition or composition discontinuity exists.

The continuity relations, first derived for simple ion exchange by DeVault (4), may now be applied to the present problems as a differential material balance giving U,

$$d(C_i + C_p)/d(Q_i + Q_p) = U \qquad (5a)$$

or as an integral material balance giving the step velocity U_Δ,

$$\Delta(C_i + C_p)/\Delta(Q_i + Q_p) = U_\Delta \qquad (5b)$$

In these equations, if i is a nonprecipitating species, the precipitate concentrations are zero. If it is a precipitating species, Q_p will be zero for moving, and C_p, for stationary, precipitate, if such is present. The Δ's on the left side of Eq. 5b indicate the difference between the upstream and downstream values of the concentration variables in the parentheses.

2.5 Golden's Rule

For any possible combinations of (constant) boundary conditions and (arbitrary) isotherm configurations, Golden's Rule (8) succinctly indicates the composition ranges over which the respective differential and integral material-balance relations (Eqs. 5a and b) apply.

To understand the basis of the rule, consider a species whose total concentration in the stationary medium, at a given bed level, increases with time. If this concentration changes continuously, then lower concentrations must travel faster than higher ones, or physically impossible multiple concentration values at the same bed level and time would ensue. Since gradual composition changes correspond to the differential material balance, the latter cannot apply here, and the integral material balance is valid.

In the present paper, isotherms, for simplicity, are shown with Q_i as the abscissa and C_i as ordinate, instead of the customary converse. Therefore, the following adaptation of the rule applies: To obtain the generalized concentration profiles (plot of concentrations vs. U or U_Δ), imagine a string stretched around the governing isotherm in the counterclockwise direction from the feed- to the presaturation-composition point. Eq. 5a then gives the velocity of any composition corresponding to a point on a curved portion of the string, and Eq. 5b, the velocity of any concentration step corresponding to a linear portion of the string.

2.6 Generalized Representation of Fixed-Bed Behavior

The composition or step velocities as defined above may be viewed as the rate of change of distance of a composition or a step with contact time, where the latter is defined as the time elapsed since the feed-solution front has reached the bed level of the composition or step under consideration.

For uniform boundary conditions and onedimensional flow, these velocities are constant, so that in a diagram having contact time as abscissa and distance from the bed inlet as ordinate, the trajectories of particular compositions or steps appear as straight lines emanating from the origin. Thus, a plot of concentrations or steps against velocities as defined corresponds to a composition profile at unit contact time, and profiles at any other time are implied by it by proportionality. Similarly, effluent-concentration histories for any column depth can be obtained as crossplots from a contact-time,distance-diagram.

3 MONOVARIANT ION-EXCHANGE ISOTHERMS

3.1 Effect of Total Solution Normality on Simple Ion-Exchange Isotherms

In stoichiometric ion exchange unaccompanied by chemical reaction, and in the absence of swelling and shrinking and of axial dispersion, the total solution normality during fixed-bed operation with uniform boundary conditions will remain constant. When chemical reactions such as precipitation or dissolution, or both, are involved, however, a change in total solution normality will accompany them, so that, as will be seen, it is of interest to view the isotherm as a surface rather than a line. Such a surface may for example be imagined in Q_M, C_M, C^o-space, and represented by a family of contour lines corresponding to constant values of C^o.

Isotherms for ideal homovalent exchange, when plotted in terms of relative normalities, are invariant with respect to the total solution normality, but C_M for a given value of y is proportional to C^o. For heterovalent exchange, electroselectivity causes the isotherm to shift as the total solution normality changes (9), and it can be shown that C_M also increases with C^o, but is less than proportional to it.

For some combinations of meaningful values of a and b in Eq. 3a, closed-form solutions for C_M in terms of Q_M at constant given values of C_X are laborious or nonexistent. Calculation of C_X- or C^o-contours can then be performed numerically with an electronic handcalculator, for which a program has been written (11).

3.2 Saturation Branch

In systems in which ion exchange is accompanied by precipitation, part or all of the C_X-contours (pure-ion-exchange isotherms) may lose their physical meaning, because at saturation, the solubility relation (Eq. 4) takes part in determining the effective equilibrium relation. The portion of the effective overall isotherm generated by a system of equations involving Eq. 4 may be referred to as the saturation branch of the isotherm.

The present section (1) establishes the independence of the configuration (not the extent) of the saturation branch of the boundary conditions, (2) reviews the formal solution for such branches as derived previously (10) for systems for which Eqs. 3b and 4a are valid, (3) presents a generally applicable graphical method of construction, (4) derives limits within which the saturation branch must lie, and (5) discusses the effects of K and K_{MN} on systems governed by Eqs. 3b and 4a.

(1) With the exchange capacity Q^O constant, the four equations 1 through 4 contain the five variables C_M, C_N, C_X, Q_M, and Q_N, so that, if the value of one of them is set, that of the others becomes fixed and, in principle, determinable. The resulting saturation branch of the effective isotherm will thus be unique in the sense that, within the limits of its validity, its shape will be independent of the feed and presaturation compositions.

(2) For the relatively simple systems to which Eqs. 3b and 4a apply, an equivalent of the following analytical solution has been given for the saturation branch (10),

$$1/y = 1 + [(s^{-2} + 1)^{0.5} - 1]/(2K_{MN}) \tag{6}$$

where

$$s \equiv 0.5K^{-0.5}C_M \tag{6a}$$

This definition makes a plot corresponding to Eq. 6 independent of the value of K.

(3) In more complex systems, an analytical solution for the saturation branch may be difficult or impossible to find. In this case, the following simple graphical construction may be employed. First, a set of C_X-contours or pure ion-exchange isotherms is plotted. Next, C_N is eliminated between Eqs. 1 and 4a (if they apply) to yield

$$C_M = C_X - K/C_X \tag{7}$$

210

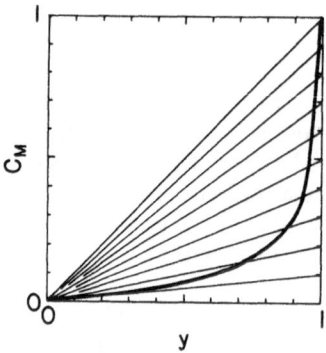

Fig. 1. Construction of saturation branch of isotherm, calculated from Eqs. 3b and 12, with $K = 0.01 eq^2 dm^{-6}$, $K_{MN} = 1$, and $C_X^O = 1N$. Thin lines (linear in this singular case) are pure-ion-exchange isotherms with constant C_X-values. The latter equal the ordinate values of the intersections of these lines with the line $y = 1$.

so that a value of C_M can be determined for every C_X-contour. The curve going through the points so obtained will represent the saturation branch. An example of this procedure is given in Fig. 1, where, for simplicity, the contours were taken to be linear. At $y = 1$, $C_N = 0$, so that $C_M = C_X$. Hence, the C_X-values of the contours can be read off the diagram as the ordinate of their intersection with the line $y = 1$.

(4) For solubility relations of the form of Eq. 4a, the saturation branch will lie between two particular C_X-contours. To show this, we consider that, at $C_M = 0$, with Eq. 1, $C_N = C_X$, and with Eq. 4a, $C_X = K^{0.5}$. At finite values of C_M, C_N, with Eq. 1, will be smaller than C_X and therefore, with Eq. 4a, larger than $K^{0.5}$. The saturation branch of the isotherm must therefore lie entirely above the C_X-contour for which $C_X = K^{0.5}$. As the upper C_X-bound for this branch is the C_X-value C_X^O of any undersaturated region in the bed, it follows that the entire saturation branch must lie between the contours corresponding to this value and the one just derived, i. e., that

$$K^{0.5} < C_X < C_X^O \tag{8}$$

(5) Low solubility of the precipitate will tend to pull counterion Species N out of the exchanger. To maintain electroneutrality, this species must be replaced stoichiometrically by Species M, so that the selectivity for M is enhanced by precipitation. In the isotherm diagrams as plotted here, a high

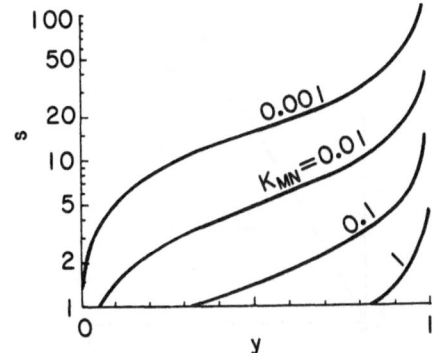

Fig. 2. Effect of selectivity (value of K_{MN}) on saturation branch of isotherm. K_{MN}-contours calculated with Eq. 6.

selectivity for M will be reflected in a low position of the saturation branch. As K approaches zero, the saturation branch will approach the configuration of an isotherm "irreversibly" favorable for Species M.

The effect of the selectivity of the exchanger on the configuration of the saturation branch may be examined through the example of systems governed by Eq. 6, for which representative curves have been plotted in Fig. 2. Here, s on a logarithmic scale is shown as a function of y on a linear scale, for representative values of K_{MN}. Analogous plots for precipitates composed of ions of different valences would show a similar trend. As mentioned, a lower position of a curve indicates increased selectivity for counterion M. It should be kept in mind that, were both coordinate scales linear, all curves would pass through the origin.

3.3 Overall Isotherm

An example of an overall isotherm is shown in Fig. 3, where the respective values of K_{MN}, K and C_X^0 were taken to be 0.25, 0.25 eq^2dm^{-6}, and 1N. The virtual linearity of the saturation branch of the isotherm in this case is a coincidence.

It follows from Eq. 7 that the highest point of the saturation branch will lie at

$$C_M = C_X^0 - K/C_X^0, \text{ or at } s = 0.5K^{-0.5}C_X^0 - 0.5K^{0.5}/C_X^0 \quad (9)$$

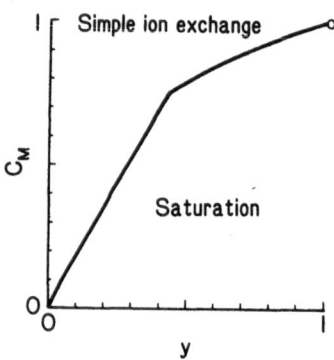

Fig. 3. Overall isotherm for ion exchange and precipitation (K_{MN} = 0.25, K= 0.25eq^{2}dm^{-6}, $C_X{}^O$ = 1N).

and that, for a given value of $C_X{}^O$, this point will lie the higher, the lower the value of K. The pure–ion–exchange branch of the effective isotherm applies between the highest point of the saturation branch and the maximum value of C_M. This implies that the effect of solubility on the overall isotherm closely approaches that, discussed earlier, on the saturation branch. Precipitation can thus be used to strongly modify selectivity.

4 MOVING PRECIPITATE

4.1 Classification of Cases

With moving precipitates, two principal types of boundary conditions can arise, according as precipitate is being formed or dissolved in the bed. In the first of these, the feed is a solution containing the precipitating coion species X and the nonprecipitating counterion species M, and the exchanger initially contains the precipitating counterion species N. (In addition, N may be present in the feed and M initially in the bed.)

As M replaces N on the exchanger, the concentration of N in solution rises, causing precipitate to form if the combination of the initial concentration of N on the exchanger, and the ion-exchange- and solubility-equilibrium relations permit it. For this case, the feed-composition point will lie on the pure-ion-exchange branch, and the presaturation-composition point, on the saturation branch of the effective isotherm. The concentration governing the position of the pure-ion-exchange branch will be the total normality of the feed solution.

In the second case, the feed is a suspension of precipitate in equilibrium with a solution containing Ions N and X (and possibly M), and initially, the bed contains M (and possibly N). Precipitate dissolves as it passes through the bed. In this case, the feed-composition point must lie on the saturation branch, and the presaturation-composition point, on the pure-ion-exchange branch of the effective isotherm. The position of the latter will be governed by the sum of C_p and C_X, which remains constant throughout the process.

4.2 Prediction of Composition Profiles

As discussed earlier, given boundary conditions and given ion-exchange- and solubility-equilibrium relations imply a unique overall isotherm in terms of the concentrations of the non-precipitating counterion species M, so that Golden's Rule, in conjunction with Eqs. 5a and b, will lead to generalized C_M- and Q_M-profiles. Eqs. 1 through 4 will then permit the associated values of C_N, Q_N and C_X to be obtained. The precipitate concentration C_p is readily obtained from the consideration that X remains completely in the mobile medium, so that

$$C_p = C_X{}^o - C_X \tag{10}$$

As an example, consider a bed initially containing exchanger in the N-form and receiving an MX solution as feed. Fig. 4a shows the effective overall isotherm in terms of y and x, calculated with Eqs. 3b and 6, for $K = 10^{-10} eq^3 dm^{-6}$, $K_{MN} = 0.05$, and $C_X{}^o = 10^{-4}N$. x is defined by

$$x \equiv C_M/C^o \tag{11}$$

Also shown is the application of Golden's Rule, the imaginary string hugging the effective isotherm near its ends, but becoming a chord near its middle portion. Fig. 4b represents the corresponding generalized mobile-medium-concentration profile, in terms of x vs. U.

The profile is seen to consist of (1) a short plateau zone, corresponding to the feed-composition point on the isotherm, in which the composition of the moving medium is that of the feed, and the composition of the stationary medium, that in equilibrium with the feed, (2) a short segment of a simple wave, corresponding to the effective segment of the pure-ion-exchange isotherm, and so nearly vertical as to resemble a shockwave, (3) a relatively long middle-plateau zone corresponding to the intersection

214

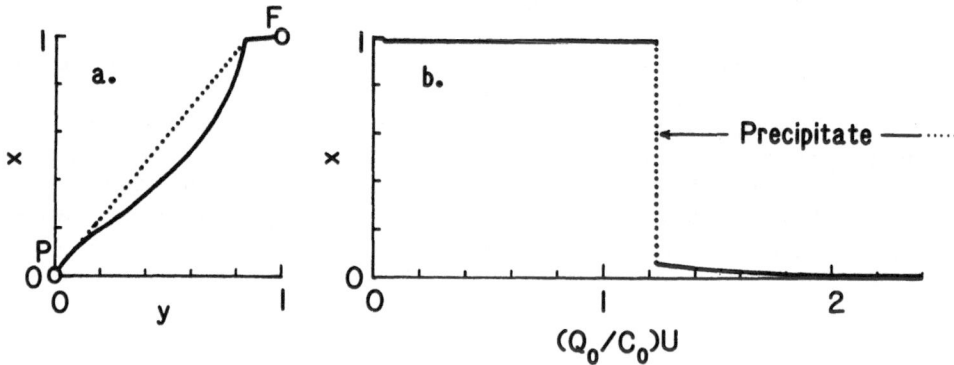

Fig. 4. Example of system with moving precipitate. a. Golden's Rule applied to effective overall isotherm calculated with Eqs. 3b and 6 (for K_{MN} = 0.05, K = $10^{-10}eq^2dm^{-6}$, and C^O = $10^{-4}N$). Feed-composition point at F; presaturation-composition point at P. b. x-profile. Shockwave shown as dashed line.

of the effective portion of the pure-ion-exchange isotherm and the chord to the saturation branch, (4) a segment of a simple wave, corresponding to the saturation branch of the isotherm below the point of tangency of the chord, and finally (5) a plateau zone, extending to infinity in the U-direction, corresponding to the presaturation-composition point. In the interplateau portions of the overall profile, x, and therefore, C_M, is seen to decrease monotonically with U. Precipitate is present downstream of the shockwave (and in the effluent), since this part of the profile corresponds to the saturation branch of the isotherm.

The phenomena expressed in Fig. 4b are a reduction of C_M along the pure-ion-exchange isotherm in the direction away from the feed-composition point, i. e., in the downstream direction. As x and C_M decrease along the first, short segment of the first simple upstream wave of the x-profile, C_N increases and C_X remains constant. Where the pure-ion-exchange and saturation branches meet, the solubility limit has been reached. However, since, at this point, C_p, with Eq. 10, is zero, precipitate does not yet form. It does form in the shockwave downstream of the middle plateau zone. Then, x and C_M decrease in the simple wave, down to the third-plateau values (zero). Any precipitate is being swept downstream with the velocity of the moving medium.

In the shockwave, a sharp drop in C_M is accompanied in the mobile medium by a sharp rise in the supply of N, most of which

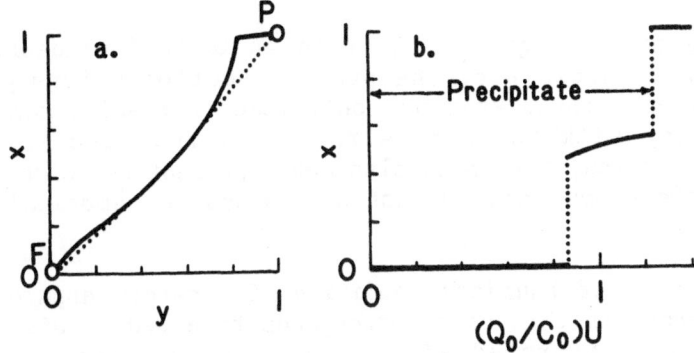

Fig. 5. Example of system with moving precipitate. Constants as in Fig. 4, but feed- and presaturation-composition points interchanged.

precipitates, also sharply reducing the value of C_X. A slight, gradual increase in C_p occurs in the downstream simple wave.

For the same value of K, variants of this example correspond to a saturation branch without inflection point (e.g., for $K_{MN} = 01$), which causes disappearance of the downstream simple wave, and values of K_{MN} greater than unity, which lead to the disappearance of both simple waves.

If the feed- and presaturation-composition points in Fig. 4a are interchanged, application of Golden's Rule will lead to an upstream plateau zone containing precipitate, but free of M, followed by a simple wave within which precipitate is dissolving progressively, and at the downstream end of which it is fully dissolved. This simple wave is followed immediately by another shockwave, that leads to the composition corresponding to the feed-composition point. The x-profile is again monotonic, but rises instead of falling. The corresponding diagrams are shown in Fig. 5.

5 STATIONARY PRECIPITATE

5.1 Characteristics of Computations

The principal difference between problems involving moving and stationary precipitates is that, for the former, the precipitate concentrations can be obtained readily with the aid of Eq. 10, while, for stationary precipitate, material balances involving the feed- and presaturation-compositions must be solved simultaneously. This makes the stationary-precipitate case considerably more difficult.

As a purely algebraic approach would be formidably laborious, if not impossible, the present section offers graphical construction methods that not only lead to results commensurate in accuracy with the errors resulting from the simplifying premises adopted, but which also make apparent the classification of possible cases, and suggest approaches to numerical solution procedures.

The type of boundary condition of greatest apparent potential interest in this case corresponds to a bed receiving a feed solution not saturated with respect to precipitate, throughout which precipitate initially is distributed uniformly. This means that the presaturation-composition point for Species M lies on the saturation branch of the isotherm for this species, and the feed-composition point, on the pure-ion-exchange branch.

The key to the construction methods employed lies in plotting the isotherm for M and "conjugate" material-balance curves for X on the same graph. One of the latter is constructed through the feed-composition point and another through the presaturation-composition point, for X. The abscissa of the intersection of these curves then corresponds to the concentration of precipitate.

An example of plotting the feed- and presaturation-composition points on such a graph is shown in the diagram of Fig. 6, where the horizontal axis is calibrated in units of Q_M or

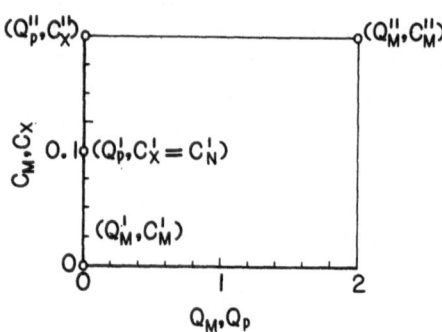

Fig. 6. Example of plot of feed-composition points (Q_i'', C_i'') and presaturation points (Q_i', C_i'). (K = 0.0064eq^{2}dm^{-6}).

Q_p, and the vertical axis, in units of C_M or C_X. The maximum possible value of Q_M is the exchange capacity Q^o, here taken to be 2.00 eq dm^{-3}. On such a diagram, the feed composition is represented by <u>two</u> points, here (Q_M''=2.00,C_M''=0.16) and (Q_p''=0,C_X''=0.16). The presaturation composition is also represented by two points, here (Q_M'=0,C_M'=0) and (Q_p'=0,C_X'=0.08). This, with C_N' = 0.08, corresponds to a value for K of 0.0064. The units of all these quantities are the same as before.

5.2 Construction for Shockwaves

The procedure for constructing the integral-material-balance line for Species X, emanating from the presaturation point for this species is indicated in Fig. 7, where K and the feed- and presaturation points are the same as in Fig. 6, and K_{MN} = 1. The overall isotherm is constructed as indicated before, and the saturation branch has been extended (dashed-and-dotted line) to beyond the point of intersection with the pure-ion-exchange isotherm, because this part of it will be seen to be necessary in subsequent steps of the the construction.

Points on the curve sought are found by (1) drawing an arbitrary chord to the saturation branch of the isotherm through the presaturation point for Species M (amounting to setting a step velocity); (2) reading the C_M-ordinate of the intersection A_M of this chord with the saturation isotherm; (3) calculating

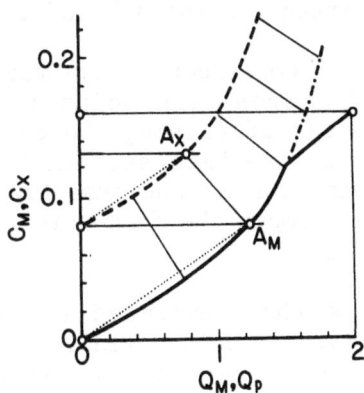

Fig. 7. Stationary precipitate. Construction of integral material-balance line (dashed) through feed-composition point for precipitable coion X (K and boundary conditions as in Fig. 6; K_{MN} = 1).

the equilibrium C_X-value corresponding to this C_M-value with the aid of Eq. 9, in the form

$$C_X = [C_M + (C_M^2 + 4K)^{0.5}]/2 \qquad (12)$$

(4) drawing a horizontal line corresponding to this C_X-value; and (5) finding the intersection A_X of this line with a line going through the presaturation point for X and parallel to the chord to the saturation branch of the isotherm. This intersection is the point sought.

The rationale of this construction is that the value of C_X represented by the ordinate of the horizontal line reflects equilibrium with C_M at Point A_M, and that, because of equilibrium, the velocity of the step in the X-profile must equal that of the step in the M-profile. The velocity of the latter is given by the slope of the chord to the M-isotherm (with Eq. 5b, written for M); and that of the former, by the slope of the chord to the integral material-balance line for X (given by the same equation, written for X). The sought point on the latter, which must satisfy both the applicable point relations and the material balance, is therefore constrained to be the intersection of the horizontal line with the chord to the material-balance line for X. Points A_M and A_X may be regarded as conjugate points and connected by a tieline, as shown.

The figure shows the integral-material-balance line for X as a dashed, smooth curve drawn through several points constructed in this manner. The details of the construction of the other points have been omitted for the sake of clarity.

The procedure for constructing the material-balance line for X through the feed-composition point is analogous, with the exception that, instead of drawing the chords to the M-isotherm through the origin, a line intersecting that isotherm is drawn through the feed point for M, and that the line parallel to this first line is drawn through the feed-point, instead of the presaturation point, for X.

Fig. 8a shows the completed construction, with both material-balance curves as dashed lines. The coordinates of their intersection B_X are the precipitate concentration Q_p, and the fluid-phase concentration C_X in the precipitate zone. The potentially useful range of the material-balance curve for X is limited to the shaded area.

The concentrations of M in the precipitate-plateau zone are the coordinates of the point B_M on the saturation branch of the isotherm, conjugate with B_X.

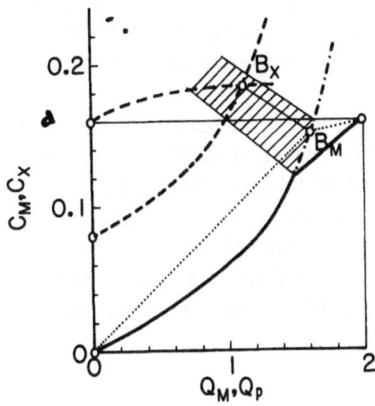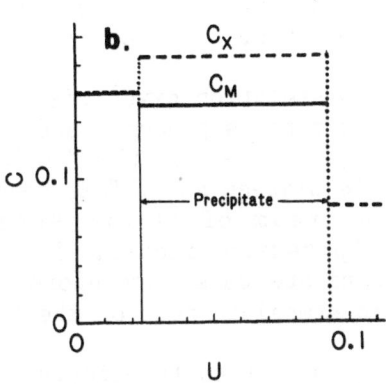

Fig. 8. Stationary precipitate. Determination of composition in precipitate zone, and application of Golden's Rule to example of Fig. 7.

Golden's Rule may now be applied to the M-isotherm. In stretching the imaginary string from the feed-composition point for M in the counterclockwise direction, one must include Point B_M, so that the material-balance relations with respect to feed- and presaturation-compositions for X are obeyed. The string then continues linearly to the presaturation point for M at the origin. As indicated by the dotted lines, there result two shock-waves, upstream and downstream of a plateau zone corresponding to the composition indicated by the coordinates of Points B_M and B_X. Since B_M is on the extended saturation branch of the isotherm, the solubility relation is obeyed, and precipitate is present in this zone. The precipitate concentration Q_p is the abscissa-value for the conjugate point B_X. The slopes of the two straight-line segments of the imaginary string give the velocities of the shockwaves in terms of U_Δ. The profiles corresponding to Fig. 8a are shown in Fig. 8b. Analogous, iterative numerical calculations confirm these results.

The response of the fixed bed to the feed composition of this example is that to be expected qualitatively from physical considerations. In principle, there will be an upstream plateau zone in which the exchanger is in equilibrium with the feed, and from which precipitate is absent because the feed does not contain the precipitable counterion species N.

Since, when feed is beginning to enter the bed, all zones in the latter are vanishingly short, there actually will be precipitate formed right at the inlet end, but it will dissolve progressively as more feed enters the bed. In the limiting case of a completely insoluble precipitate, this first plateau zone would be absent, and, because of the small solubility of many precipitates, this condition often will be approached closely.

Dissolution gives rise to a shockwave, downstream of which precipitate is present, and in which the total solution normality is higher than in the feed. These conditions prevail in the middle plateau zone. This zone and the presaturation plateau zone downstream of it are separated by a shockwave in which precipitation occurs. In contradistinction to the moving-precipitate case, therefore, the precipitate is present only in an intermediate zone of the bed.

5.3 Construction for-Simple Waves

In the usual case that C_X is not a linear function of C_M, it can be shown that, for given boundary conditions, the integral and differential material-balance lines for X are not identical.

For an adaptation of the construction method for shockwaves to simple waves, consider Fig. 9, which shows the part of the saturation branch of a hypothetical isotherm for Species M near the origin, which corresponds to the presaturation composition. The curve is convex upward, so that, with Golden's Rule, a simple wave is likely to result. The presaturation point for Species X and the value of K are the same as in Fig. 6. Also, the relation between C_X and C_M is again considered to be given by Eq. 12. The Calibration of the Q_M, Q_p-axis has been omitted as irrelevant.

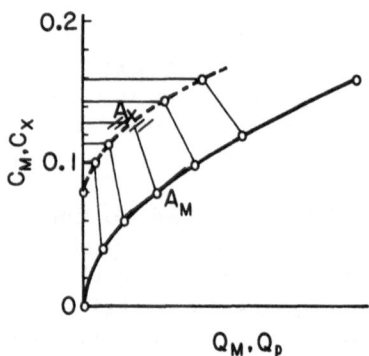

Fig. 9. Stationary precipitate. Construction of differential material-balance line (dashed) through presaturation-composition point for precipitable coion X.

To proceed with the construction, (1) pick a point A_M on the isotherm, (2) read the C_M-value of its ordinate, (3) calculate the corresponding C_X-value with Eq. 12 and draw a horizontal line representing this value, (4) in the vicinity of where the differential-material-balance line for X (the dashed curve in the figure) is expected to fall, draw a few short line segments parallel to the isotherm at Point A_M, so that they intersect the horizontal line just constructed, (5) repeat this procedure for other selected points. The line segments obtained now constitute a direction field through which, as well as through the presaturation-composition point for X, a curve (shown as dashed line here) may be drawn so that it is parallel to the line segments in its vicinity. This is the differential-material-balance curve for X. (6) Connect the points established on this line with tielines to their conjugate points on the isotherm, as shown.

In practical constructions, it is most convenient to start at the boundary-condition point (here, the origin), so that the area in which the direction field is of interest becomes apparent as the material-balance line becomes established progressively.

In contradistinction to the construction based on integral material balances (Eq. 5b), which is required only in the vicinity of the intersection of the two material-balance lines for Species X, here, it is necessary to include the entire part of the curve from the boundary-condition point to the point of interest, since application of the direction-field method amounts to an integration.

The question next arises how such constructions can be used to provide a basis for the application of Golden's Rule. In looking for an appropriate example, one may be tempted to simply exchange the feed- and presaturation compositions of systems of the type of that of Fig. 7. However, with stationary precipitate, the feed cannot contain the latter, and the feed-composition plateau zone will be saturated with respect to precipitate but not contain the latter, and precipitate will not be present anywhere in the bed. Instead, we use as an example the same boundary conditions and K-value as in Fig. 6, but a hypothetical overall isotherm with an inflection point, and with a convex-upward part near the origin, as in Fig. 9. The tentative assumption that the differential material balance (Eq. 5a) is valid in this region does not involve risking an error, as nonvalidity would become apparent automatically in the course of the construction.

The method, illustrated in Fig. 10, involves the following steps: (1) Construct the overall isotherm, with the saturation branch extended beyond its intersection with the simple ion-exchange isotherm, as in Fig. 7; (2) Construct the differential-

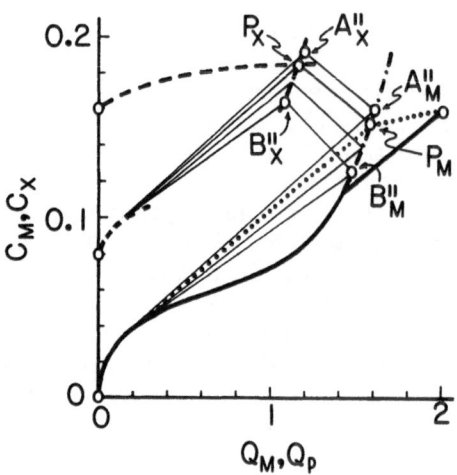

Fig. 10. Stationary precipitate. Construction of differential-material-balance line through feed-composition point for precipitable coion X (hypothetical isotherm; K-value and boundary conditions as in Fig. 8).

material-balance line for Species X, as in Fig. 9; (3) Assume representative "jump-off" points A_M', B_M', etc. (not shown to avoid confusion), on the saturation branch of the isotherm for M, i. e., points at which the integral material balance begins to apply. Draw tangents to the curve at these points and mark their respective intersections with the extended isotherm A_M'', B_M'', etc; (4) Draw parallel tangents to the X-curve; (5) Use Eq. 12, or an analogous, equivalent relation to determine the points A_X'', B_X'', etc. on these tangents that are conjugate to Points A_M'', B_M'', etc., respectively; (6) Draw a curve through Points A_X'', B_X'', etc., as the locus of compositions corresponding to the downstream end of the saturation zone; (7) Construct the integral-material-balance line for X through the feed-composition point for X, as in Fig. 8a. (Whether cases can exit in which a differential material-balance line through this point is valid, and problems arising as a result of applying this material balance have not been fully resolved, and these questions are not further pursued here.) (8) The intersection P_X of this curve with the one constructed before is analogous to Point B_X in Fig. 8; (9) Using Eq. 9, find the conjugate point, P_M. (10) Apply Golden's Rule, as indicated by the dotted lines.

Fig. 11 shows the generalized concentration profiles derived from Fig. 10. They are qualitatively similar to those of Fig. 8b, except that the downstream shockwave has been replaced by a

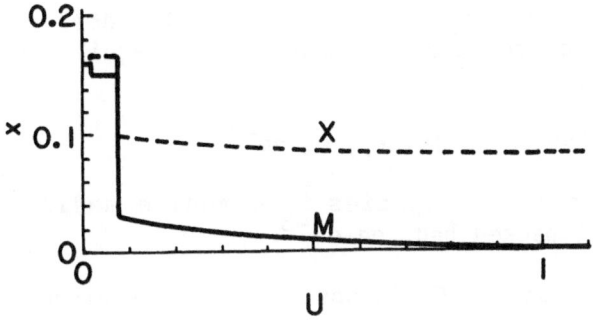

Fig. 11. Generalized concentration profiles corresponding to Fig. 10.

combination of a shockwave and a simple wave downstream of it. Because of the length of the latter, the U-scale had to be changed. Precipitate again occurs between the upstream and downstream shockwaves, but also in the simple wave. Its concentrations can be read as the abscissae of Point P_X (1.17eq dm^{-3}) and of points on the curve corresponding to the simple wave in Fig. 10. Precipitate forms in the simple,shock-wave-combination and redissolves in the upstream shockwave.

Although these profiles are based on the hypothetical data underlying Fig. 10, they appear to indicate that, with stationary precipitate, simple waves are possible.

ACKNOWLEDGMENTS

The author wishes to express his gratitude for the stimulating discussions on this subject he has had over the years with the late Professor Vermeulen and a number of his students; to Professor and Mrs. Vermeulen, for the legacy and gift of a word processor; to Dr. Cal Herrmann for invaluable help in its use; to the University of California Thermal and Chemical Technology Center, and its director, Professor Laird, for providing office space and other facilities; and to Judy Sindicic and Rebecca Eaton for their kind assistance with the production of the graphs and the camera-ready version of the text.

SYMBOLS

a,b   Valences, or empirical exponents, in ideal or modified mass-action relation for ion-exchange equilibrium, dimensionless.

C^o Total solution normality, eq dm^{-3}.

C_i Concentration of Species i in mobile medium, per unit volume of packed bed, eq dm^{-3}.

C_X^o Constant value of C_X in undersaturated regions.

f_1, f_2 Functions

K   Solubility product for monovalent ions, $eq^2 eq^{-3}$.

K_{MN} Selectivity coefficient for ion-exchange equilibrium, dimensionless.

M,N,X   Nonprecipitable and precipitable counterion; precipitable coion.

Q^o Total exchange capacity, eq dm^{-3} (of packed bed).

Q_i Concentration of Species i in stationary medium, per unit volume of packed bed, eq dm^{-3}.

s   Normalized concentration of Species M in mobile medium (cf. Eq. 6a), dimensionless.

U   Composition velocity as fraction of apparent velocity of mobile medium, dimensionless.

U_Δ Velocity of a shockwave as fraction of apparent velocity of mobile medium, dimensionless.

x   Equivalent fraction of Species M in solution, dimensionless.

y   Equivalent fraction of Ion Species M in exchanger, dimensionless.

Sub- and Superscripts

p   Precipitate

'   Downstream, or presaturation, composition.

''   Upstream, or feed, composition.

REFERENCES

1. Applebaum, S.B. Demineralization by Ion Exchange. (New York, Academic Press, 1968).
2. Bunge, A.L., G. Klein and C.J. Radke. Divalent Ion Exchange with Alkali. Society of Petroleum Engineers Symposium on Oilfield and Geothermal Chemistry, Stanford (1980).
3. Cherney, S. Scale-Prevention Studies and Feasibility Analysis of Seawater Softening by Fixed-Bed Ion Exchange. M.S. thesis in Chemical Engineering, University of California, Berkeley   (1966).
4. DeVault, D. The Theory of Chromatography, J.Am.Chem.Soc. 65 (1943) 532-540.
5. Evans, S. and R.M. Gomes. Desalination of Rhine River Water by Ion Exchange. Part 1, Pilot Plant Studies of Lime Slurry Regeneration. Desalination 19 (1976) 433-438.
6. George, D.R., J.M. Riley and L. Crocker. Preliminary Process Development Studies for Desulfating Great Salt Lake Brines and Sea Water. Bureau of Mines Report of Investigation 6928, U.S. Department of the Interior, Washington, D.C. (1967).
7. Glueck, A.R. Desalination by an Ion-Exchange-Precipitation-Complex Process. Desalination 4 (1973) 32-37.
8. Golden, F.M. Theory of Fixed-Bed Performance for Ion Exchange Accompanied by Chemical Reaction. Doctoral dissertation in Chemical Engineering, University of California, Berkeley (1972).
9. Helfferich, F. Ion Exchange (New York, McGraw-Hill, 1962).
10. Klein, G. Ion Exchange and Chemical Reaction in Fixed Beds. In "Percolation Processes: Theory and Applications", A.E. Rodrigues and D. Tondeur, editors. (Alphen aan den Rijn, Sijthoff & Noordhoff, 1981).
11. Klein, G. Calculation of Ideal or Empirically Modified Mass-Action Equilibria in Heterovalent, Multicomponent Ion Exchange. Computers and Chemical Engineering 8 No. 3/4 (1984) 171-178.
12. Klein, G. Selenium Removal Using Ion Exchange - Application to Kesterson Reservoir. Paper presented at Symposium on Selenium in the Environment, California State University, Fresno, Division of Extended Education (1985).
13. Klein, G., T.J. Jarvis and T. Vermeulen. Fluidized-Bed Ion Exchange with Precipitation - Principles and Bench-Scale Development. In "Recent Developments in Separation Science", V, Norman N. Li, editor. (West Palm Beach, Florida, CRC Press, 1979).
14. Mangelsdorf, P.C. Difference Chromatography. J.Anal.Chemistry 38 (1966) 1540-1546.
15. Popper, K., R.J. Bouthilet and V. Slamecka. Ion-Exchange Removal of Sodium Chloride with Calcium Hydroxide as Recoverable Regenerant. Science 141 (1963) 1083-1089.

16. State of California Department of Water Resources. Agricultural Waste Water for Power Plant Cooling. Development and Testing of Treatment Process (1978).
17. Tiger, H.L., S. Sussman, M. Lane and V.J. Calise. Desalting Sea Water. Ind. Eng. Chemistry 38, No. 11 (1946) 1130–1133.
18. Van der Meer, Adries Piet. On Countercurrent Fluidized Ion-Exchange Columns. Doctoral dissertation, Technische Hogeschool, Delft (1985).
19. Vislocky, J.M. Local Equilibrium Theory for Bivariant Fixed-Bed Sorption Systems. Doctoral dissertation in Chemical Engineering, University of California, Berkeley (1982).

NUMERICAL METHODS

C. Costa,A. Rodrigues and J. Loureiro

Department of Chemical Engineering
University of Porto,Porto-Portugal

INTRODUCTION

Model equations of ion exchange processes (usually PDEs) have analytical solutions only in few cases.Thus in most situations the use of numerical methods is required.

Even in the case of linear unsteady state models where Laplace transform methods can be applied a complex transfer function is often obtained.Its numerical inversion is performed by Fast Fourier Transform methods [1] or numerical integration of the convolution integral by quadrature [2];this topic will not be covered here.

For multicomponent stoichiometric ion exchange equilibrium models a convenient mathematical tool has been developed in two versions: Helfferich and Klein's h-transformation [3] and the method of characteristics [4,5].This method leads to analytical or quasi-analytical solutions (solution of high degree algebraic equations);this topic will not be covered in this paper as well.

Models for ion exchange processes can be grouped in view of two main features:time and spacial dependency of the dependent variables (generally species concentrations).Table I presents a model classification showing the type of equations,some related numerical methods and some examples of ion exchange processes.

Complex models (unsteady and distributed) are described by PDEs. The strategy for the numerical solution of these equations implies the reduction in one or more independent variables which leads to the solution of a system of ODEs or algebraic equations.This strategy is behind all the numerical methods to be discussed.The basic idea is to generate approximations that,at the end,result in an algebraic system which has to be solved.The numerical methods for the solution of these systems will not be discussed also and the interested reader is addressed to two reviews on the topic 6,7 .

Table I - Some models, equations and related numerical methods in ion exchange.

Models	Equations	Some Numerical Methods of Solution	Examples in Ion Exchange
Unsteady state, lumped	ODEs - initial value problems	Explicit and semiimplicit Runge-Kutta and linear multistep methods	Slurry processes and cells-in-series models using a lumped representation for mass transfer
Steady state, distributed	ODEs - boundary value problems Elliptic PDEs	Discretization to algebraic equations using finite differences or weighted residuals methods	Steady state counter-current processes
Unsteady state, distributed	Hyperbolic and parabolic PDEs	Discretization to either algebraic equations or to initial value problems using finite differences or weighted residuals Combination of independent variables by the method of characteristics (hyperbolic PDEs) and solution of the remaining initial value problem	Fixed beds Parametric pumping

ORDINARY DIFFERENTIAL EQUATIONS

Going back to Table I we can see that steady state processes with spatially distributed parameters and unsteady state processes with lumped parameters can be modeled by using ODEs

$$F(x,y',\ldots,y^{(n)})=0 \tag{1}$$

with n conditions. If these conditions are specified at a unique point the problem is called an initial value one; if not it is a boundary value problem.

Initial Value Problems

Any n-order ODE can be transformed in a system of n first order ODEs. Thus only methods for the solution of first order equations will be discussed. Let us consider a system of 1st order ODEs represented by

$$\frac{d\bar{y}}{dt} = f(\bar{y},t) \tag{2}$$

with the initial condition

$$\bar{y}(0) = \bar{y}_o \tag{3}$$

Algorithms for the numerical solution of these equations start at the initial condition and produce approximations to the dependent variables at discret values (t_i) of the independent variable using constant or variable step $(h_i=t_{i+1}-t_i)$.

The Euler method is based on a Taylor series expansion where the 2nd and higher order terms are neglected

$$y_{i+1} = y_i+hf(t_i,y_i) \tag{4}$$

In the absence of round-off errors the local truncation error is proportional to h^2 : $O(h^2)$ and the global truncation error (propagated + local) is of $O(h)$. In general for a method of order r the local truncation error is of order r+1 and the global truncation error is of order r.

Since the local error is proportional to the step used this method is convergent to the solution and more accurate as h goes to zero.

Even in the case of negligible round-off and truncation errors numerical methods may show instabilities which means that the error may became unbounded as the number of steps increases. These instabilities are due to the substitution of the original ODEs by the numerical approximations which are difference equations. Solutions of ODEs are generally of the form Ce^{pt} where p is the root of the characteristic equation; these difference equations have solutions of the form Cp^i where i is the number of steps. Thus instabilities may arise due to the different type of the two solutions.

Let us use the Euler method for the simple linear ODE

$$\frac{dy}{dt} = \lambda y \tag{5}$$

with $\lambda < 0$ and $y(0) = 1$ (finite).The analytical solution is then

$$y = e^{\lambda t} \tag{6}$$

and $\lim_{t \to \infty} y = 0$.

Using now the Euler method for the same equation and assuming a solution of the form Cp^i we obtain

$$y_i = Cp^i = C(1+\lambda h)^i \tag{7}$$

For ensuring stability of the numerical method we need that $\lim_{i \to \infty} y_i = 0$.This condition leads to $|p(h\lambda)| < 1$ or to $|h\lambda| < 2$ which means that depending on the value of $\lambda < 0$ the maximum value allowed for h is bounded - the method is said to be conditionally stable. Note that oscillations will be present if $|h\lambda| > 1$.

Let us take now a modified version of the Euler method - the trapezoidal rule

$$y_{i+1} = y_i + \frac{h}{2} (f(y_i) + f(y_{i+1})) + O(h^3) \tag{8}$$

Similar calculations lead to

$$y_i = C(\frac{1+h\lambda/2}{1-h\lambda/2})^i \tag{9}$$

In this case we have always $|p(h\lambda)| < 1$ provided $\lambda < 0$; the method is called A-stable.Nevertheless as $|h\lambda| \to \infty$, p goes to -1 and we get oscillatory behaviour.To avoid this we need to use $|h\lambda| < 2$.

If we take now the implicit Euler method

$$y_{i+1} = y_i + hf(y_{i+1}) + O(h^2) \tag{10}$$

we obtain

$$y_i = C(\frac{1}{1-h\lambda})^i \tag{11}$$

Being A-stable this method also verifies the condition $\lim_{h\lambda \to \infty} |p(h\lambda)| = 0$. It is then called strongly A-stable (no oscillatory behaviour).

Up to now we discussed the stability of one linear ODE.The extension to systems of ODEs may be done just by using for λ the absolute value of the largest eigenvalue of the Jacobian of the system.This is at least valid in the vicinity of t_i.In the case of nonlinear systems this eigenvalue may change with time and the maximum step size should be set in order to guarantee stability in the whole time interval.A more detailed discussion on stability can be found in $|8|$.

The importance of these stability considerations can be shown by means of a simple example

$$\frac{d\bar{y}}{dt} = A\bar{y} \qquad \text{with} \quad A \equiv \begin{vmatrix} -1 & 0 \\ 1 & \beta \end{vmatrix} \; ; \beta < 0 \qquad (12)$$

The solution is of the form

$$y_1 = C_0 e^{-t} \qquad ; \; y_2 = C_1 e^{-t} + C_2 e^{-\beta t} \qquad (13)$$

The eigenvalues of the Jacobian are $\lambda_1 = -1$ and $\lambda_2 = \beta$. If $|\beta|$ is large say 10^4 and if we want to use the explicit Euler method to obtain the numerical solution we are restricted to a maximum step size of $2/10^4$ due to stability considerations.

Figure 1 sketches the solution of Eq.(12).As it can be seen we have a fast transient corresponding to $\exp(-\beta t)$ and a slow changing solution corresponding to $\exp(-t)$.So we have to use a small step initially but it may seem reasonable to increase the step size beyond $2/10^4$ after the fast transient vanishes.This is not possible because the unstable propagation of error due to the fast transient will lead to an unstable solution.These systems are called stiff; A-stable or strongly A-stable methods are required for their numerical integration.

Stiffness is usually characterized by the ratio between the absolute values of the greatest and the smallest eigenvalues of the Jacobian.

$$SR = \frac{\max \; |Re(\lambda_i)|}{\min \; |Re(\lambda_i)|} \qquad (14)$$

For practical purposes and using SR (stiffness ratio) as a measure of the system stiffness we can say that

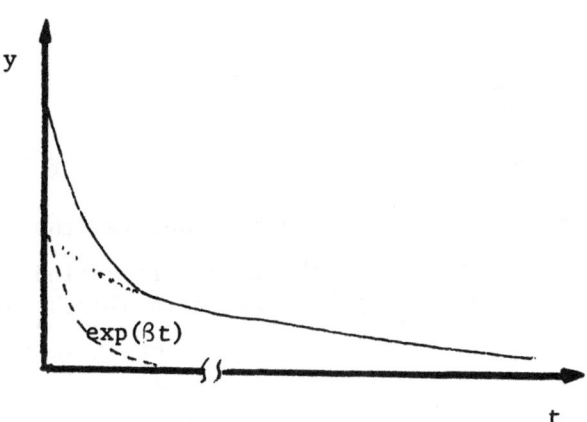

Figure 1 - Sketch of a possible solution of the system of ODEs in
Eq. (12)

SR < 100   – nonstiff
100 < SR <1000   – midly stiff
SR > 1000   – stiff

Generally speaking as a physical system stabilizes stiffness increases.For example as a steep front dissipates,the system stabilizes and the ODEs stiffens.

Runge – Kutta methods. Up to now integration formulae were based in the truncation of 2nd and higher order terms in the Taylor series expansion,thus leading to low order and low accuracy methods.Accuracy can be improved by keeping more terms in that expansion.The problem will be then the cumbersome calculation of high order total derivatives.To avoid this drawback a class of methods was developed by Runge and Kutta which keep the form of a series expansion where only f(t,y) has to be evaluated at different points in the interval. These function values are conveniently weighted in order to get a match with the corresponding Taylor series expansion.

A general r order representation for this class of methods is

$$\overline{y}_{i+1} = \overline{y}_i + \sum_{j=1}^{v} \omega_j \overline{k}_j \tag{15}$$

$$\overline{k}_j = h \overline{f} (t_i + b_j h, \overline{y}_i + \sum_{\ell=1}^{v} a_{j\ell} \overline{k}_\ell) \tag{16}$$

where ω_j, b_j and $a_{j\ell}$ are adjustable parameters and v is the number of intermediate values of $\overline{f}(t,\overline{y})$ that must be evaluated in each time step.An useful representation of these coefficients is the so called "Butcher block"

$$
\begin{array}{c|ccc}
b_1 & a_{11} & \cdots & a_{1v} \\
\cdot & \cdot & \cdots & \cdot \\
\cdot & \cdot & \cdots & \cdot \\
\cdot & \cdot & \cdots & \cdot \\
b_v & a_{v1} & \cdots & a_{vv} \\
\hline
& \omega_1 & \cdots & \omega_v
\end{array}
\qquad
\begin{array}{c|c}
\overline{C} & \overline{\overline{A}} \\
\hline
& \overline{W}
\end{array}
\tag{17}
$$

From the form of the $\overline{\overline{A}}$ matrix three main classes of Runge-Kutta methods may be devised:

- $a_{j\ell}=0$ for $\ell>=j$ the $\overline{\overline{A}}$ is lower triangular,the $\overline{k}_j$ can be calculated sequencially and the method is called explicit;
- $a_{j\ell}=0$ for $\ell>j$ but $a_{jj} \neq 0$, $\overline{\overline{A}}$ is triangular,$\overline{k}_j$ appear in both sides of Eq.(16) which implies an iteractive solution for $\overline{k}_j$ in each step – the method is said semi-implicit;
- $a_{j\ell}\neq 0$ for some $\ell>j$ then the $\overline{k}_j$ have to be calculated by the solution of nonlinear algebraic equations and the method is called implicit.

One of the most popular explicit methods is the 4th order Runge-

-Kutta-Gill

$$
\begin{array}{c|ccccc}
0 & 0 & 0 & 0 & 0 \\
1/2 & 1/2 & 0 & 0 & 0 \\
1/2 & -1/2+1/\sqrt{2} & 1-1/\sqrt{2} & 0 & 0 \\
1 & 0 & -1/\sqrt{2} & 1+1/\sqrt{2} & 0 \\
\hline
& 1/6 & (1-1/\sqrt{2})/3 & (1+1/\sqrt{2})/3 & 1/6
\end{array}
\tag{18}
$$

This method involves four function evaluations per step and is $O(h^5)$ locally accurate. However it is only conditionally stable, as all the explicit Runge-Kutta methods (see Table II). Using a standard 4th order Runge-Kutta, Kolcini |9| solved several multistaged and deep fluidized bed ion exchange models.

On the contrary semi-implicit and implicit Runge-Kutta methods are A-stable or strongly A-stable but involve the solution of non--linear algebraic equations at each step which increases computations.

To avoid this problem Rosenbrock |10| developed a special class of semi-implicit methods by performing a suitable linearization of Eq.(16) which does not affect the order of accuracy and stability. However this implies Jacobian evaluations at each step. This original method was 3rd order involving two derivative evaluations and two Jacobian evaluations and inversions per step; the method is A-stable with an oscillation limit of $|p| = 0.8$. Calahan |11| further improved the method by making only one Jacobian evaluation. Caillaud and Padmanahan |12| developed a strongly A-stable 3rd order method which needs only two derivative evaluations, one Jacobian evaluation and a matrix multiplication per step. This method was further improved by Michelsen |13| by eliminating the matrix multiplication

$$
\overline{y}_{i+1} = \overline{y}_i + \sum_{j=1}^{3} \omega_j \overline{k}_j
\tag{19}
$$

$$
\overline{k}_1 = h(\overline{I} - ha\overline{J}_i)^{-1} \overline{f}(y_i)
\tag{20}
$$

$$
\overline{k}_2 = h(\overline{I} - ha\overline{J}_i)^{-1} \overline{f}(\overline{y}_i + b_2\overline{k}_1)
\tag{21}
$$

$$
\overline{k}_3 = (\overline{I} - ha\overline{J}_i)^{-1} (b_{31}\overline{k}_1 + b_{32}\overline{k}_2)
\tag{22}
$$

where $\overline{I}$ is the identity matrix, $(J_{\ell j})_i = (\frac{\partial f_\ell}{\partial y_j})|_{y_i}$ and

$a^3 - 3a^2 + 3a/2 - 1/6 = 0$, $a = 0.4358...$

$b_2 = 0.75$ $\omega_1 = 11/27 - b_{31}$

$b_{31} = -(8a^2 - 2a + 1)/(6a)$ $\omega_2 = 16/27 - b_{32}$

$b_{32} = 2(6a^2 - 6a + 1)/(9a)$ $\omega_3 = 1$

This method has been found to be one of the most efficient for solving systems of ODEs.

The principal advantages of the Runge-Kutta methods can be summarized as follows:

- they are self-starting
- the step length can be easily changed because each new step can be viewed as a new initial value problem
- they are accurate and can be made A-stable or strongly A-stable
- they do not require iterative procedures
- no special storage requirements are necessary since only the information of the last step is needed.

However they present some drawbacks,namelly:

- the requirement of many function evaluations per step
- the difficulty to estimate the local truncation error.

Since simple expressions for the truncation error are not known it can be approximated by a one-step two-step strategy based on the so-called Richardson extrapolation technique |13| which also improves accuracy.Then a procedure can be developed to decide when to increase decrease or mantain the step size.

Another strategy is the simultaneous use of two formula of orders r and r+1 to calculate an approximate value of the local truncation error.Semi-implicit 3rd and 2nd order methods can be imbedded in this way.The extra computational cost can be mantained at a minimum by using the same constants for the two methods.Recently Cash |14| developed strongly A-stable 3rd order imbedded methods and by tracking the stiff component Prokopakis and Seider |15| developed a 3rd order adaptative A-stable method.

Table II summarizes stability data for some Runge-Kutta methods.

Table II - Comparison among some Runge-Kutta methods

Method	Order of accuracy	Stability
Explicit Euler	h	$CS;2.0^a$
Runge-Kutta-Gill	h^4	$CS;2.8^a$
Modified Euler	h^2	$AS;2.0^b$
Implicit Euler	h	$SAS;\infty$
Michelsen	h^3	$SAS;\infty$

CS-conditionally stable;AS-A-stable;SAS-strongly A-stable;a-stability limit on $|h\lambda|$;b-oscillation limit on $|h\lambda|$

Linear multistep methods.A second important class of methods for the solution of ODEs are the linear multistep methods (LMS) of the form

$$\overline{y}_{i+1} = \sum_{j=1}^{v}\alpha_j\overline{y}_{i-j} + h\sum_{j=-1}^{v}\beta_j\overline{f}(t_{i-j},\overline{y}_{i-j}) \qquad (23)$$

where α_j and β_j are constants.Note that as these methods require the knowledge of up to v+1 prior values of $\overline{y}$ and $\overline{f}$ they are not self starting.If $\beta_{-1}=0$ then the method is explicit in $\overline{y}_{i+1}$ and is called a predictor method.If $\beta_{-1} \neq 0$ then we get an implicit method-corrector. The referred constants (say r+1 nonzero) can be calculated by stating that the formulae is exact for a polynomial of degree r or less.The local truncation error can be found by assuming the form $Ch^{n+1}f^{(r)}$.

The most used predictor multistep methods are probably those of Adams-Bashforth:

$$\bar{y}_{i+1} = \bar{y}_i + h\sum_{j=0}^{v} \beta_j \bar{f}(t_{i-j}, \bar{y}_{i-j}) \tag{24}$$

Table III shows the coefficients β_j and the constant C of the error formulae for various orders.

Table III - Coefficients for the Adams-Bashforth predictors

order	β_0	β_1	β_2	β_3	β_4	β_5	C
1	1						1/2
2	3/2	-1/2					5/12
3	23/12	-16/12	5/12				9/24
4	55/24	-59/24	37/24	-9/24			251/720
5	1901/720	-2774/720	2616/720	-1274/720	251/720		475/1440
6	4277/1440	-7923/1440	9982/1440	-7298/1440	28771/1440	-475/1440	19087/60480

Adams-Moulton derived equivalent corrector formulae:

$$\bar{y}_{i+1} = \bar{y}_i + h\sum_{j=-1}^{v} \beta_j \bar{f}(t_{i-j}, \bar{y}_{i-j}) \tag{25}$$

The corresponding coefficients are given in Table IV.

Table IV - Coefficients for the Adams-Moulton correctors

order	β_{-1}	β_0	β_1	β_2	β_3	β_4	C
2	1/2	1/2					-1/12
3	5/12	8/12	-1/12				-1/24
4	9/24	19/24	-5/24	1/24			-19/720
5	251/720	646/720	-264/720	106/720	-19/720		-27/1440
6	475/1440	1427/1440	-798/1440	482/1440	-173/1440	27/1440	-863/60480

For predictors and correctors of the same order,the corrector formula is more accurate (see the C values in Tables III and IV).

The predictor-corrector methods use a predictor to get a first estimate of $\bar{y}_{i+1}$ and then the corrector formulae (more accurate) is used in an iterative way (usually no more than one iteration if h is properly chosen) to calculate a more accurate value for $\bar{y}_{i+1}$.

As these methods are not self-starting this usually may be done in two different ways:
- using an equal order single step method (e.g. Runge-Kutta) for various steps in order to generate the values of f needed
- using one step formula to provide starting values for the two step formula and so on.

Since the local truncation errors are known they can be used to adjust the step size.Step change in LMS methods implies restarting which is usually overcomed by defining coefficients that depend upon past step sizes.

LMS methods are characterized by high order difference equations with multiple roots whereas the ODE characteristic equation only has one root.Those extra roots may lead to instabilities.The general solution of such difference equations is

$$y_i = C_1 p_1^i + C_2 p_2^i + \ldots + C_k p_k^i \qquad (26)$$

One principal root,say p_1,corresponds to the solution of the ODE and the other k-1 roots are extraneous.If $|p_i| < 1$ then $\lim_{i \to \infty} y_i = 0$.So,a LMS method is called A-stable or absolutely stable if $|p_i| < 1$ for $i=1,\ldots,k$.For the case of the Adams 4th order predictor-corrector the stability limit is $|p| = 1.3$.

Predictor-corrector methods present some advantages when compared with Runge-Kutta methods,namely:
- direct knowledge of the local truncation errors
- the number of function evaluations per step is smaller
and some disadvantages including:
- non self-starting
- step-size changes are not easy
- more storage requirements for past information.

One important development in these methods has been achieved by Gear |16| by using the so-called backward difference methods.The general expression for this family is

$$\bar{y}_{i+1} = \sum_{j=0}^{v-1} \alpha_j \bar{y}_{i-j} + h\beta_{-1}\bar{f}(t_{i+1},\bar{y}_{i+1}) \qquad (27)$$

Table V presents the values for the coefficients of Eq.(27).

Near A-stability (called by Gear "stiff stability") is achieved by adjusting order and step.It was stated that stability was not necessary for values of $h\lambda$ close to the imaginary axis but sufficiently far from the origin.A method is called "stiffly stable" if it is absolutely stable for all values of $Re(h\lambda)$ to the left of some negative constant A,and accurate close to the origin.The multistep methods developed by Gear are stiffly stable for orders up to 6.The upper bound for the stable region,A, goes from -6.1 to 0 as the order goes

Table V - Coefficients for the equation (27)

order	α_0	α_1	α_2	α_3	α_4	α_5	β_{-1}
1	1						1
2	4/3	-1/3					2/3
3	18/11	-9/11	2/11				6/11
4	48/25	-36/25	16/25	-3/25			12/25
5	300/137	-300/137	200/137	-75/137	12/137		60/137
6	360/147	-450/147	400/147	-225/147	72/147	-10/147	60/147

from 6 to 2.

In conclusion we can say that if the problem to solve is stiff then the choice is essentially between 3rd order semi-implicit Runge--Kutta methods and the Gear's backward difference formula. For non--stiff problems explicit Runge-Kutta and LMS predictor-corrector methods may be used.

Differential-Algebraic systems. When a discretization technique is applied to the spatial coordinates of PDEs, algebraic equations mixed with ODEs can appear. This is the case when a finite element colloca-tion technique with Jacobi polynomials is used.

Michelsen [17] extended the 3rd order semi-implicit Runge-Kutta method to the case of the solution of a coupled differential-algebraic system of the form

$$\frac{d\bar{y}}{dt} = \bar{f}(\bar{y}, \bar{z}) \qquad (28)$$

$$\bar{0} = \bar{g}(\bar{y}, \bar{z}) \qquad (29)$$

This is achieved by transforming first Eq.(29) in a high stiff ODE

$$\varepsilon\frac{d\bar{z}}{dt} = \bar{g}(\bar{y}, \bar{z}) \qquad (30)$$

with $\varepsilon \ll 1$ and then applying Eqs.(19) to (22) and taking the limit when ε goes to zero.

This method was further generalized by Holland [18,19] to the case where the ODEs are implicit.

For the backward difference formulas, Gear [20,21] also developed methods for the solution of simultaneous algebraic-differential sys-tems in explicit or implicit form.

Software. Common computer libraries like IMSL and NAG contain usually an extensive collection of routines generally including Runge-Kutta explicit,LMS predictor-corrector and some version of Gear's method. Table VI shows some of the software on initial value problems that usually is not included in these general libraries.

Generally the user of these codes has to provide a main program where some flags,initial conditions and other requested or necessary information is given.The equations are usually described in a subroutine as well as the Jacobian of the system.Some of those codes like the LSOLVERs and EPISODEs have a built in capability of numerically generating the Jacobian.

In the case of differential-algebraic systems the matrix that multiplies the derivatives vector is usually provided also in a separate subroutine.

Example. Consider a CSTR where a stoichiometric ion-exchange process is taking place.Assume an idealized situation where only film mass transfer resistance is present with a constant film mass transfer coefficient (k_f).The ion-exchange equilibrium is represented by a mass action type law.The model equations may be stated as $|30,31|$:

$$\frac{dX}{dt} = (1+\xi)\{1-(1+N_f)X+ \frac{N_fY}{Y+k(1-Y)}\} \tag{31}$$

$$\frac{dY}{dt} = \frac{(1+\xi)N_f}{\xi} \{X- \frac{Y}{Y+k(Y-1)} \} \tag{32}$$

$$X(0)=Y(0)=0$$

where $\xi=(1-\varepsilon)Q/(\varepsilon c_o)$ is the mass capacity factor and $N_f=(1-\varepsilon)k_fa_p\tau/\varepsilon$ is the number of film mass transfer units, $\tau=V\varepsilon/U$ is the space time, ε the porosity,V the CSTR volume,U the flowrate,a_p the specific area, Q the ion-exchange capacity,c_o the input concentration,X and Y are normalized concentrations respectively by c_o and Q,k is the equilibrium constant and t is time.

Now the Jacobian of this system is

$$J = \begin{vmatrix} -(1+\xi)(1+N_f) & (1+\xi)N_f \dfrac{k}{\{Y+k(1-Y)\}^2} \\ N_f(1+\xi)/\xi & -N_f\dfrac{1+\xi}{\xi} \dfrac{k}{\{Y+k(1-Y)\}^2} \end{vmatrix} \tag{33}$$

The eigenvalues (λ) of the Jacobian matrix are given by

$$\{Y+k(1-Y)\}^2 \lambda^2 + \{N_f\frac{1+\xi}{\xi}k + (1+\xi)(1+N_f)\{Y+k(1-Y)\}^2\}\lambda +$$

$$+ \frac{(1+\xi)^2}{\xi} N_fk = 0 \tag{34}$$

As it can be seen in Table VII even this simple system shows stiffness for some values of the parameters and its stiffness ratio

Table VI - Software for the solution of initial value problems

Name	Method	Comments	Referenc.
STRIDE	Implicit Runge-Kutta	Stiff ODEs	|22|
STIFF3	3rd order semi-implicit Runge-Kutta	Stiff ODEs	|23|
LSODE	Adams and Gear formula	Non-stiff and stiff ODEs.Full or banded Jacobian.	|24|
LSODA	"	Same as LSODE but with automatic switching between stiff and non-stiff formula	
LSODES	"	Sparce Jacobian	|25|
EPISODE	"	Reocurring stiffness,full Jacobian	|25|
EPISODEB	"	Reocurring stiffness,banded Jacobian	|26|
LSODI	"	Implicit ODEs and differential-algebraic	|26|
DASSL	"	"	|24|
DFASUB	"	"	|27|
EPISODEIB		Same as EPISODE but for implicit ODEs and differential-algebraic with banded Jacobian	|28|
DASP3	4th order Runge-Kutta 3rd order backward difference formula	Non-automatic partioning and differential-algebraic	|26|
			|29|

varies with time.Table VIII compares computing time (for a precision of 10^{-3}) for the solution of this problem using four algorithms: IBM 4th order Runge-Kutta-Gill (RKG),STIFF3 and both Adams(ADM) and backward difference formulae (BDF) of GEARB (an early version of LSODE).

Table VII – Evolution of the stiffness ratio as a function of N_f for fixed values of $\xi=100$ and $k=10$.

Case	N_f	Stiffness ratio	
		Y=0	Y=0.5
1	0.1	1.2×10^4	1.9
2	0	20.8	1.8
3	5	1.4	4.1
4	10	1.2	2.0

Table VIII – Comparison among various codes for the integration of the system of equations (32) and (33)

Case	Computing time				number of steps			
	RKG	STI	ADM	BDF	RKG	STI	ADM	BDF
1	40	2	12	12	2002	17	277	259
4	40	6	66	65	2002	33	2792	2717

As we can see STIFF3 performs better here but if we consider for example a series of 20 CSTR then this is no longer valid.In this case BDF formula perform better specially if a higher precision is required |30| .

Boundary Value Problems

Methods for the solution of boundary value (BV) problems are not so general as those for initial value problems . The BV problems can be transformed in a system of initial value ODEs but then some of the initial values are not known.To avoid this difficulty a class of methods was developed – the shooting methods |32,33|.

The solution of BV problems is usually carried out by using finite differences and the method of weighted residuals (MWR).

Finite differences. The idea is to divide the domain in a set of grid points and produce approximations for the derivatives at those points; the BV problem is transformed into a set of algebraic equations.

The derivative approximations may be calculated from convenient Taylor series expansions:

$$\frac{dy}{dx}\Big|_i = (y_{i+1} - y_i)/h + O(h) \qquad \text{forward difference} \qquad (35)$$

$$\frac{dy}{dx}\Big|_i = (y_i - y_{i-1})/h + O(h) \qquad \text{backward difference} \qquad (36)$$

$$\frac{d^2y}{dx^2}\Big|_i = (y_{i+1} - 2y_i + y_{i-1})/h^2 + O(h^2) \qquad (37)$$

$$\frac{dy}{dx}\Big|_i = (y_{i+1} - y_{i-1})/(2h) + O(h^2) \qquad \text{central difference} \qquad (38)$$

Higher order approximations can be derived by keeping more terms in the Taylor series expansion and taking more grid points. We will come back to this topic when discussing methods for the numerical solution of PDEs.

Method of weighted residuals. The idea here is to represent a trial solution as a linear combination of basis functions which are required to be piecewise continuous and linearly independents; the trial solution also obeys to the boundary conditions

$$y \simeq y_N = \sum_1^N \alpha_i \, \phi_i(x) \qquad (39)$$

If this trial solution is substituted in the original ODE then a residual $R_N(\overline{\alpha}, x)$ is obtained which is to be minimized in some way. In the MWRs it is an average value of that residual multiplied by a suitable weighting function that is minimized

$$\int_0^1 W(x) \, R_N(\overline{\alpha}, x) \, dx = 0 \qquad (40)$$

The choice of the weighting functions determined the development of several MWRs. In the Galerkin method those functions are the trial functions. The integral in Eq.(40) may be evaluated with a quadrature formulae

$$\int_0^1 W(x) \, R_N(\overline{\alpha}, \vec{x}) \, dx = \int_0^1 \phi(x) \, R_N(\overline{\alpha}, x) \, dx \simeq \sum_{i=1}^N w_i R_N(\overline{\alpha}, x_i) \phi_i(x_i) = 0 \qquad (41)$$

If we choose the x_i as the zeros of a shifted Jacobi polynomial of degree N , $P_N^{(\alpha, \beta)}(x)$, then the quadrature is exact. Moreover $\overline{\alpha}$ can be readilly calculated by solving the homogeneous linear system in Eq.(41)

$$R_N(\overline{\alpha}, \overline{x}) = 0 \qquad (42)$$

The shifted Jacobi polynomials are an orthogonal set with respect to the weighting function $(1-x)^\alpha x^\beta$ where $\alpha, \beta > -1$

$$\int_o^1 (1-x)^\alpha x^\beta P_N^{(\alpha, \beta)}(x) \, \Gamma_M^{(\alpha, \beta)}(x) \, dx = 0 \qquad N \neq M$$

$$\int_0^1 (1-x)^\alpha x^\beta \{P_N^{(\alpha,\beta)}\}^2 \, dx > 0 \qquad\qquad N=M \qquad\qquad (43)$$

In the orthogonal collocation (OC) method we use orthogonal polynomials as trial functions and the respective zeros (collocation points) as the points where the residual is zeroed. This method was first proposed by Stewart and Villadsen |34| for the solution of BV problems.

In order to avoid problems in the calculation of $\bar\alpha$ it was proposed to restate R_N in terms of the collocation points ordinates |8|. This procedure involves a matrix inversion which can be avoided by using Lagrange polynomials as trial functions |23| defined as:

$$\ell_i(x) = \frac{P_{N+1}^{(\alpha,\beta)}(x)}{(x-x_i)P_{N+1}^{(\alpha,\beta)}{}^{(1)}(x)} \qquad\qquad (44)$$

Since $\ell_i(x) = \begin{cases} 0 & i \neq k \\ 1 & i=k \end{cases}$ and taking into account Eq.(39), $y_N(x_k)=\alpha_k$; the expansion coefficients are the ordinates of the approximating functions at the collocation points.

The derivatives of the trial functions are easily calculated

$$\frac{d^k y_N}{dx^k}\bigg|_{x_j} = \sum_1^N y_N(x_i)\left\{ \frac{d^k}{dx^k} \ell_i(x)\right\}\bigg|_{x_j} \qquad\qquad j=1,\ldots,N \qquad (45)$$

For the case of 1st and 2nd derivatives

$$\frac{d\bar{y}_N}{dx} = \bar{A}\,\bar{y}* \qquad \text{and} \qquad \frac{d^2\bar{y}_N}{dx^2} = \bar{B}\,\bar{y}* \qquad\qquad (46)$$

Software has been developed for the calculation of the zeros of the Jacobi polynomials (JCOBI), the weights of the quadrature formula (RADAU) and matrices $\bar{A}$ and $\bar{B}$ (DFOPR) |8,23| .

The error in these approximations may be obtained by considering that if $N=\infty$ in Eq.(39) then the residual would be zero in all the points of the domain and thus the infinite series represents the exact solution. If we only use a finite number of terms in Eq.(39) the error

$$E(x) = y(x)-y_N(x) = \alpha_{N+1}\,\phi_{N+1} + \cdots \qquad\qquad (47)$$

would be of the order of the 1st term neglected assuming that

$$\sum_{i=1}^\infty \alpha_{N+i}\,\phi_{N+i} \text{ is rapidely convergent } |35| .$$

When the BV solution presents steep profiles or even discontinuities its representation is poor or requires a lot of collocation points which means the solution of a large system of algebraic equations (which are nonlinear if the original BV problem is nonlinear).In order to deal with this situation Carey and Finlayson |36| proposed the so-called collocation in finite elements.In this method the domain is divided into M subintervals,not necessarilly of equal length,and OC is applied within each of them;continuity of the solution and of its 1st derivative is usually required at each knot.The coefficient matrix of the resulting algebraic system is of block diagonal form.For further details see |8| .

Another set of orthogonal polynomials are the Hermite polynomials which are defined with respect to exp($-x^2$) as weighting function.The 3rd order Hermite polynomials can be constructed in order to ensure authomatically the continuity of the solution and of its 1st derivative in the knots,thus avoiding some extra equations |8| .A more general class of interpolating functions with these properties are the so-called B-splines developed by deBoor |37,38| .

The efficiency of the orthogonal collocation methods in finite elements heavilly depend on the subintervals localization.Some methods were proposed to optimize that localization in order to minimize errors |36,39| .

Recently developed software,COLSYS |40|,uses collocation in finite elements with B-splines incorporating a mesh refinement algorithm.

Stability of collocation methods is not yet well known.Some new developments have been published recently by Arnold and Saraven |41|.

Some examples of application of OC methods will be given in the discussion of numerical methods for PDEs.

PARTIAL DIFFERENTIAL EQUATIONS

Most of the mathematical models for ion-exchange processes involve PDEs.Let us consider a general PDE in one independent spatial variable

$$a \frac{\partial^2 y}{\partial z^2} + 2b \frac{\partial^2 y}{\partial z \partial t} + c \frac{\partial^2 y}{\partial t^2} + d \frac{\partial y}{\partial z} + e \frac{\partial y}{\partial t} + fy = g \qquad (48)$$

As many of the solution properties are determined by the value of b^2-ac,PDEs are usually classified into three groups

$$b^2-ac \begin{cases} > 0 & \text{hyperbolic} \\ = 0 & \text{parabolic} \\ < 0 & \text{elliptic} \end{cases}$$

When a=b=c=0 then a 1st order PDE results which in many instances shows properties that are analogous to the hyperbolic group.

The integrated form of those equations is dependent on some unknown functions which are further specified by the initial and boundary conditions.These are usually classified in three groups. Consider a general form for the boundary conditions of a PDE

$$\alpha(z*,y)y + \beta(z*,y) \left.\frac{dy}{dz}\right|_{z*} = \gamma(z*,y) \tag{49}$$

where z* is at the boundary;then the boundary conditions are called
      Dirichlet conditions   - open to diffusion   ($\beta=0$)
      Neumann conditions   - closed to diffusion at outlet ($\alpha=0$)
      Robin (mixed) conditions - closed to diffusion at inlet ($\alpha,\beta\neq0$)
If $\gamma=0$ then these conditions are said homogeneous.

Methods for the numerical solution of PDEs are reported in Table I : .global methods   - the PDEs are transformed
         directly into a system of algebraic equations
    .numerical method of characteristics - the independent variables
         are combined and the IV ODEs obtained are then solved
    .method of lines   - the PDEs are transformed
         into a system of IV ODEs which is then solved with suitable
         methods.

We will not discuss the case of the elliptic PDEs which,in the case of regular boundary conditions can be solved in a straightforward manner by using orthogonal collocation techniques in two dimensions. For a discussion on methods to solve this type of problems refer to Lapidus and Pinder |42|.

Global Methods

Finite differences are well established methods for the solution of parabolic and hyperbolic PDEs.

Referring first to the parabolic case forward,backward and central difference formula (see Eqs.(35),(36) and (38)) can be applied.The method is explicit or implicit as spatial derivatives are evaluated at time j or j+1,respectively.Explicit methods have stability bounds whilst the implicit ones are unconditionally stable; the cost for stability is the solution of a tridiagonal algebraic system.

Boundary conditions are usually treated in two ways:the so-called false boundary and the integral methods |32|.

One important implicit method which is often used is the Crank--Nicholson method |43|.The spatial derivatives are averaged at old and new times.For example the 2nd derivative is approximated at the new time level by

$$\frac{\partial^2 y}{\partial z^2} \simeq \frac{1}{2h^2} \{(y_{i-1,j}-2y_{ij}+y_{i+1,j})+(y_{i-1,j+1}-2y_{i,j+1}+y_{i+1,j+1})\}$$

$$(50)$$

and the time derivative

$$\frac{\partial y}{\partial t} \simeq \frac{1}{k} (y_{i,j+1}-y_{i,j})$$

$$(51)$$

Some explicit methods are stable and as accurate as the Crank-Nicholson formula which are $O(h^2+k^2)$. One of these methods is the Saul'yev procedure |44| which is a two step method:
- first advance to time level j+1 by proceeding in the z direction with

$$\frac{\partial^2 y}{\partial z^2} \simeq (y_{i-1,j+1}-y_{i,j+1}+y_{i+1,j})/h^2$$

$$(52)$$

$$\frac{\partial y}{\partial t} \simeq (y_{i,j+1}-y_{i,j})/k$$

$$(53)$$

- then advance to time level j+2 by proceeding in the negative z direction with

$$\frac{\partial^2 y}{\partial z^2} \simeq (y_{i-1,j+1}-y_{i,j+1}-y_{i,j+2}+y_{i+1,j+2})/h^2$$

$$(54)$$

$$\frac{\partial y}{\partial t} \simeq (y_{i,j+2}-y_{i,j+1})/k$$

$$(55)$$

Many applications of finite differences methods may be found in ion-exchange and related fields. Marra and Cooney |45| studied swelling and shrinking using explicit finite differences. Crittenden |46| studied multicomponent adsorption using backward differences in time and Crank-Nicholson for the particles. Svedberg |47| also used Crank-Nicholson for periodic countercurrent processes. Rolke and Wilhelm |48| who studied parametric pumping considering axial dispersion and diffusion into the particles used explicit 1st order differences for time and axial distance and the Crank-Nicholson method for the particles. Hashimoto et al |49| used explicit differences for the solution of model equations for a continuous moving bed for separating frutose-glucose mixtures.

The use of orthogonal collocation for solving parabolic PDEs was proposed by Villadsen and Sorensen |50| . The idea was to apply simultaneously OC in space and finite element collocation in time. As the right boundary condition is unknown it was calculated by extrapolation. A similar idea was latter proposed by Birnbaum and Lapidus |51| using Hermite or Laguerre polynomials for the time variable. Since these polynomials are defined in an infinite domain a global collocation technique could be used. Both methods present drawbacks as in the first there is an error propagation due to extrapolation and in the second although eliminating this problem performs a poor representation in time due the dispersion of collocation points in the interval $(0,\infty)$.

A more detailed description of finite difference and weighted residual methods for the solution of parabolic PDEs can be found in |42,52,53|.

Hyperbolic PDEs are generally considered a very difficult problem as they can propagate and create discontinuities (weak solutions).

Weighted residuals methods are not yet competitive for the solution of these equations.

We will restrict ourselves to finite differences methods which may be made conservative.These methods are extensions of the pioneer work of Lax and Wendroff |54,55|.The errors introduced by these methods are of two types:

in the vicinity of steep fronts the numerical solution is characterized by oscillations and overshoot or dumping.The causes for these phenomena are due to the wave propagation patterns of the numerical methods.If a phase lag exists between waves we have oscillations (numerical dispersion).When the waves are numerically dumped we have dumping (numerical dissipation)

The application of the method of Lax-Wendroff can lead to stability problems;then a flux corrected version may be used |56,57,58,59|.

Examples of application of these methods to the solution of non--linear coupled adsorption and chemical reaction are extensivelly discussed in |60|.

For a detailed review and discussion on these methods refer to |42,52,61,62,63|.

Numerical Method of Characteristics

It was Acrivos |64| who introduced in the chemical engineering field the numerical method of characteristics.This method generally enables the solution of a nonreducible system of 1st order hyperbolic PDEs by reducing it to an equally sized system of IV ODEs.

Let us consider the model equations for fixed bed dilute ion--exchange when plug flow is assumed,

$$u\frac{\partial c}{\partial z} + \frac{\partial c}{\partial t} = f_1(c,q) \tag{56}$$

$$\frac{\partial q}{\partial t} = f_2(c,q) \tag{57}$$

The characteristic directions are:

$$\frac{dz}{dt} = u \text{ (direction I) and } dz = 0 \text{ (direction II)}.$$

Along characteristic directions we have

$$\left(\frac{dc}{dz}\right)_I = f_1(c,q) \tag{58}$$

$$\left(\frac{dq}{dt}\right)_{II} = f_2(c,q) \tag{59}$$

which is an initial value problem. The initial conditions can be set according to the initial conditions of the original hyperbolic equations. For example for a clean bed and a step concentration input

$$t=0 \;\rightarrow\; q=0 \;\text{and}\; z=0 \;\rightarrow\; c=c_o$$

we can calculate from Eq.(58)

$$\frac{dc(0,z)}{dz} = f_1(c,0) \tag{60}$$

$$\frac{dq(t,0)}{dt} = f_2(c_o,q) \tag{61}$$

The initial value problem can then be solved by using one of the explicit methods already described. Usually high order methods are not used since LMS methods need a lot of storage of historical data and it is generally accepted that it is better to use more steps of a low order method. The modified Euler method is often the selected one. It is used as an implicit method or in a predictor-corrector scheme using the proper predictor as the explicit Euler method.

Many authors used this method in ion-exchange and related fields like Omatete et al |65| for ternary ion-exchange and Vaillard et al |66| for the staged fluidized bed. Recently Tan and Spinner |67| established algorithms for countercurrent ion-exchange where perturbations in one or both ends of the bed can be dealt with.

The method is not directly applicable to a system of hyperbolic and parabolic equations that arises in the case of the existence of an internal diffusion mechanism described by a Nernst-Planck or Ficks laws. This problem can however be circumvented by solving the boundary value problem in the parabolic PDE by using finite differences |68| or collocation |69|.

The meaning of the characteristic directions hereto derived is that the fluid moves at a certain velocity (dz/dt=u) whereas the solid phase stays stationary. Based on this idea Sweed and Wilhelm |70| developed the STOP-GO algorithm. Here the bed height is divided into cells. Then for each time step the fluid moves to the next cell without mass transfer (GO) and then STOPs and interchange is allowed. A new time step is then taken and all the sequence reworked. They applied this algorithm to a parametric pumping model where mass transfer is represented by a linear driving force law.

Method of lines

When dealing with nonlinear problems the application of global methods leads to the solution of systems of nonlinear algebraic equations which are usually difficult to solve numerically.

Another way of dealing with PDEs is to split the problem into two parts :
- first solve the boundary value problem by transforming the PDE into a system of IV ODEs
- then solve the initial value problem with an appropriated integrator.

This is what is usually called the numerical method of lines |71|.

The approximation for the spatial derivatives may be obtained by using finite differences or weighted residuals methods.

The remaining IV ODE system must then be solved.Normally the stiffness of this system increases with the number of discretization points (at least for the diffusion equation when using finite differences this has been demonstrated by several authors,e.g. |72|) and often a differential-algebraic system is obtained.Also in many applications the type of solution that makes the problem stiff moves as a wave in space and so there is no interest in using methods that are good just because they are able to take larger time steps in the integration interval.Another important consideration is that if we use a poor spatial discretization it will generally cause oscillations that will be carefully followed by the integrator thus preventing the time step to increase.

Usually errors due to spatial discretization are not small;so there is no need for demanding too much precision for the ODE solver.

This method of lines generally works well with parabolic PDEs. There are many examples in the literature of the application of this method like in multicomponent adsorption in batch and fixed beds considering intraparticle diffusion |73,74|,periodic counter-current operation |75|,fixed bed and batch adsorption with film and pore diffusion |76,77| and regeneration of polymeric adsorbents |78,79|.Most of those studies used collocation (global and finite elements) for the discretization of spatial coordinates.

The method of lines may be applied also to the solution of elliptic PDEs by transforming them into the corresponding parabolic PDEs and then integrating these up to the stationary state.

Systems of 1st order hyperbolic PDEs may also be integrated with this method if some sufficiently dispersive phenomena is present. In these cases and even for parabolic equations with a quasi-hyperbolic behaviour we have usually steep profiles that move along the space.The representation of the solution needs many grid or collocation points giving rise to a large system of IV ODEs.Some new approaches have been recently proposed.The idea is to develop a

method in which grid points concentrate in the vicinity of steep
fronts and travel with them.

Jensen and Finlayson |80| proposed an integral change of varia-
bles

$$z* = z - \int_0^t \lambda(t')dt' \qquad (62)$$

which allows to mantain the profile stationary.The drawback of this
method is that the transformed system have moving boundary conditions
with integrals to be solved at these boundaries unless the value of
λ (velocity of the front) is known (which is not the case in a non-
linear system).Then the new problem can be solved by using the
method of lines with collocation.

Ramachandran and Dudukovic |81| also proposed another change of
variable

$$z* = z/\lambda \qquad (63)$$

where the position λ is fixed by a zero flux condition.They proposed
the use of OC to discretize the new equations.With this change of
variable they tried to concentrate the collocation points in a narrow
zone by assuming that initially it is there where the profile is
steeper.

Hu and Schiesser |82| proposed a moving grid method.This scheme
concentrates collocation points near the steep fronts and is able to
add or remove collocation points;this is done by 2nd derivative
analysis.This was applied to multicomponent chromatography problems
and compared to the solutions obtained by the method of characteristics.

The moving finite element method was developed by Gelinas and
Doss |83,84|.This method is based in theoretical grounds and is also
able to follow steep fronts,concentrating there the finite elements,
the number of which is mantained constant.Recently Sereno |85| was
able to couple this method with collocation in order to solve fixed
bed adsorption problems with intraparticle diffusion.

In conclusion we may say that the method of lines is convenient
to solve parabolic PDEs as it is simple to handle and give good
results.Usually orthogonal collocation is used for the discretization
of spatial coordinates but finite differences may also be used.The
initial value problem can be solved by using explicit or stiff methods
depending on the properties of the system.

In the case of systems of hyperbolic PDEs the numerical method
of characteristics is good and generally easy to implement.If due to
the complexity of the equations it is difficult to know the characteristic directions then a recently appeared method may be used.This
method developed by Carver |86| and called pseudo-characteristics
method of lines uses biased finite differences for the spatial
coordinates and authomatically searches for the characteristic

directions thus putting the bias in the proper positions avoiding numerical dispersion.

Since the usual objective of people working in ion-exchange is to get their problems solved,little attention had been paid to the comparison of various numerical methods in the solution of ion--exchange problems.One of those few studies was done by Carra and coworkers |87| for a multicomponent nonlinear fixed bed adsorption model assuming no axial dispersion and intraparticle diffusion represented by a linear approximation.They used a modified version of Lax-Wendroff method,the numerical method of characteristics,finite differences and collocation methods.These methods were compared in terms of integration time step,accuracy and computing time.

Software

In the case of PDEs general computer libraries present some software.This software is not so general as that for IV ODEs.

Table IX presents some of the available software that can be suitable for ion-exchange applications.It should be noted that this software applies usually to parabolic PDEs except DYLA which uses moving finite elements.All the others had been tested with the Burger's equation for the ability to treat hyperbolic equations.They seem good but the direct application to the solution of ion-exchange models may cost a lot of storage and CPU time |32,94|.

Table IX - Some software for the solution of PDEs

Name	Spatial discretization	Space variables	Boundary conditions	IV solver	References
MOL1D	FD	1		Gear	|88|
PDECOL	CO(B-splines)	1		Gear	|89|
DSS/2	FD	2 or 3	rectangular	Gear+RK	|90|
FORSIM VI	FD	2 or 3	"	Gear+RK	|91|
PDETWO	FD	2	"	Gear	|92|
DISPL	Galerkin (B-splines)	2	"	Gear	|93|
DYLA	Galerkin (linear elements)	1	"	Gear	|83|

FD-finite differences;CO-collocation;RK-Runge-Kutta

REFERENCES

1. Higgins,R.J..Amer. J. Phys. 44(1976)766.
2. Provencher,S.W.CONTIN(version 2) Users Manual.Tehcnical Report EMBL-DA05 (Europeen Molecular Biology Laboratory,1982).
3. Helfferich,F. and G. Klein.Multicomponent Chromatography (New York,Marcel Dekker,1970).
4. Aris,R. and N.R. Amundson.Mathematical Methods in Chemical Engineering.vol 2 (New Jersey,Prentice-Hall,1973).
5. Rhee,H.K.Equilibrium Theory of Multicomponent Chromatography in Percolation Processes:Theory and Applications edited by A. Rodrigues and D. Tondeur (The Hague,Sijthoff and Noordoff,1981).
6. Sargent,R.W.A Review of Methods for Solving Nonlinear Algebraic Equations in Foundations of Computer-Aided Chemical Process Design edited by R.S. Mah and W.D. Seider.vol 1 (New York,Engineering Foundation Conferences,1981).
7. Gustavson,F.G.Some Aspects of Computation With Sparce Matrices in Foundations of Computer-Aided Chemical Process Design edited by R.S. Mah and W.D. Seider.vol 1 (New York,Engineering Foundation Conferences,1981).
8. Finlayson,B.A.Nonlinear Analysis in Chemical Engineering (New York,McGraw-Hill Int. Book Comp.,1980).
9. Koloini,T. and M. Zunier.Can. J. Chem. Eng. 57(1979)770.
10. Rosenbrook,H.H.Comput. J. 5 (1963)329
11. Calahan,D.A.Proc.IEEE 56 (1968)744.
12. Caillaud,J. and L. Padmanabhan.The Chem. Eng. J. 2 (1971)227.
13. Michelsen,M.L.AIChE J. 22 (1976)594.
14. Cash,J.R.The Chem. Eng. J. 20(1980)219.
15. Prokopakis,G.J. and W.D. Seider.Ind. Eng. Chem. Fund. 20 (1981)255.
16. Gear,C.W.Numerical Initial Value Problems in Ordinary Differential Equations (New Jersey,Prentice-Hall,1971).
17. Michelsen,M.L. and J. Villadsen.Polynomial Solution of Differential Equations in Foundations of Computer-Aided Chemical Process Design edited by R.S. Mah and W.D. Seider.vol 1 (New York,Engineering Foundation Conferences,1981).
18. Feng,A,C.D. Holland and S.E. Gallun.Comp. Chem. Eng. 8 (1984)51.
19. Holland,C.D. and A.I. Liapis.Computer Methods for Solving Dynamic Separation Problems (New York,McGraw-Hill Book Comp.,1983).
20. Gear,C.W.IEEE Trans. Circuit Theory CT18 (1971)89.
21. Gear,C.W. and L.R. Petzold.SIAM J. Num. Anal. 21 (1984)716.
22. Butcher,J.C.,K. Burrage and F.H. Chipman.STRIDE-Stable Runge-Kutta Integrator for Differential Equations.Report nº 150 (New Zealand, Dept. of Mathematics-University of Auckland,1979).
23. Villadsen,J. and M.L. Michelsen.Solution of Differential Equation Models by Polynomial Approximation (Prentice-Hall Inc.,1978).
24. Hindmarsh,A.C.ACM-SIGNUM Newsletter 15 (1980)10.
25. Hindmarsh,A.C.Stiff System Problems and Solution at LLNL in Stiff Compututation edited by R. Aiken (Oxford Univ. Press,1985).
26. Byrne,G.D. and A.C. Hindmarsh.ACM Trans. Math. Soft. 1 (1975)71.
27. Petzold,L.R.A Description of DASSL:A Differential/Algebraic

System Solver in Scientific Computation edited by R. Stepleman et al (North-Holland Pub. Comp.,1983).

28. Brown,R.L. and C.W. Gear.Documentation for DFASUB-A Program for the Solution of Simultaneous Implicit Differential and Nonlinear Equations.Report UIUCD-R-73-575 (Urbana,Dept. of Computer Science University of Illinois,1973).

29. Soderling,G.DASP3-A Program for the Numerical Integration of Partitioned Stiff ODEs and Differential-Algebraic Systems.Report TRITA-NA-8008 (Stockolm,Dept. of Numerical Analysis and Computer Science,The Royal Institute of Technology,1980).

30. Rodrigues,A. et al.Computational Aspects of the Dynamics of Sorption Operations (12th Symposium on Computer Applications in Chemical Engineering,Montreux,1979).

31. Rodrigues,A. et al.Dynamics of Fixed Bed Adsorbers in Stiff Computation edited by R. Aiken (Oxford Univ. Press,1985).

32. Davis,M.E.Numerical Methods and Modeling for Chemical Engineers (John Wiley & Sons,1984).

33. Kubicek,M. and V. Hlavacek.Numerical Solution of Nonlinear Boundary Value Problems With Applications (New Jersey,Prentice--Hall,1983).

34. Stewart,W.E. and J. Villadsen.Chem. Eng. Sci. 22 (1967)1483.

35. Birnbaum,I. and Lapidus,L.Chem. Eng. Sci. 33 (1978)427.

36. Carey,G.F. and B.A. Finlayson.Chem. Eng. Sci. 30 (1975)587.

37. de Boor,C.SIAM J. Num. Anal. 14 (1977)441.

38. de Boor,C.Practical Guide to Splines (New York,Springer-Verlag, 1978).

39. Birnbaum,I. and L. Lapidus.Chem. Eng. Sci. 33 (1978)443.

40. Ascher,U.,J. Christiansen and R.D. Russell.ACM Trans. Math. Soft. June (1981).

41. Arnold,D.N. and J. Saraven.On the Assymptotic Convergence of Spline Collocation Method for PDEs (Finland,Dept. of Mathematics, University of Oulu,1983).

42. Lapidus,L. and G. Pinder.Numerical Solution of Partial Differential Equations in Science and Engineering (John Wiley & Sons, 1982).

43. Eigenberger,G. and J.B. Butt.Chem. Eng. Sci. 31 (1976)681.

44. Towler,B.F. and R.K. Yang.The Chem. Eng. J. 12 (1976) 81.

45. Marra,R.A. and D.O. Cooney.AIChE Symp. Ser. 71 (1976)148.

46. Crittenden,J.C. et al.J. Environm. Eng. Div. EE6 (1978)1175.

47. Svedberg,U.G.Chem. Eng. Sci. 31 (1976)345.

48. Rolke,R.W. and R.H. Wilhelm.Ind. Eng. Chem. Fund. 8 (1969)235.

49. Hashimoto,K. et al.Biothecn. & Bioeng. 25 (1983)2371.

50. Villadsen,J. and J.P. Sorensen.Chem. Eng. Sci. 24 (1969)1337.

51. Birnbaum,I. and L. Lapidus.Chem. Eng. Sci. 33 (1978)455.

52. Ames,W.F.Numerical Methods for Partial Differential Equations (New York,Academic Press,1977).

53. Mitchell,A.R. and D.F. Griffiths.The Finite Difference Method in Partial Differential Equations (Chichester,John Wiley & Sons, 1980).

54. Lax,P. and B. Wendroff.Comm. Pure Appl. Math. 13 (1960)217.
55. Lax,P. and B. Wendroff.Comm. Pure Appl. Math. 17 (1964)381.
56. Boris,J.P. and D.L. Book.J. Comput. Phys. 11 (1973)38.
57. Book,D.L. et al.J. Comput. Phys. 18 (1975)248.
58. Boris,J.P. and D.L. Book.J. Comput. Phys. 20 (1976)397.
59. Zalesak,S.T.J. Comput. Phys. 31 (1979)335.
60. Loureiro,J.L.PhD Thesis (University of Porto,1985).
61. Sod,G.A.J. Comput. Phys. 27 (1978)1.
62. Woodward,P. and P. Corella.J. Comput. Phys. 54 (1984)115.
63. Carver,M.B. and W.E. Schiesser.Biased Upwind Difference Approximations for 1st Order Hyperbolic (convective) PDEs (73rd Annual Meeting of the AIChEngrs.,1980).
64. Acrivos,A.Ind. Eng. Chem. 48 (1956)703.
65. Omatete,O.O.,R.N. Clazie and T. Vermeulen.The Chem. Eng. J. 23 (1970)241.
66. Vaillard,R.G.,L.S. Kershenbaum and M. Streat.Chem. Eng. Sci. 36 (1981)307.
67. Tan,K.S. and I.H. Spinner.AIChE J. 30 (1984)770.
68. Klaus,R.,R. Aiken and D.W.T. Rippin.AIChE J. 23 (1977)579.
69. Barba,D.,G. Re and P.U. Foscolo.The Chem. Eng. J. 26 (1983)33.
70. Sweed,N.H. and R.H. Wilhelm.Ind. Eng. Chem. Fund. 8 (1969)221.
71. Carver,M.B.Method of Lines Solution of Differential Equations-Fundamental Principles and Recent Extensions in Foundations of Computer-Aided Chemical Process Design edited by R. Mah and W. Seider.vol 1 (New York,Engineering Foundation Conferences,1981).
72. Finlayson,B.A.Solution of Stiff Equations Resulting from PDEs in Stiff Computation edited by R. Aiken (Oxford Univ. Press,1985).
73. Liapis,A.I. and D.W. Rippin.Chem. Eng. Sci. 32 (1977)619.
74. Raghavan,N.S. and D.M. Ruthven.Chem. Eng. Sci. 39 (1984)1201.
75. Liapis,A.I. and D.W. Rippin.AIChE J. 25 (1979)455.
76. Costa,C. and A. Rodrigues.Chem. Eng. Sci. 40 (1985)983.
77. Costa,C. and A. Rodrigues.AIChE J.,in press.
78. Costa,C. and A. Rodrigues.Chem. Eng. Sci. 40 (1985)707.
79. Costa,C. and A. Rodrigues.AIChE J.,in press.
80. Jensen,O.K. and B.A. Finlayson.Adv. Water Resources 3 (1980)9.
81. Ramachandran,P.A. and M.P. Dudukovic.Chem. Eng. Sci. 39 (1984) 1321.
82. Hu,S.S. and W.E. Schiesser.An Adaptative Grid Method in the Numerical Method of Lines in Advances in Computer Methods for PDEs edited by R. Vichnevetsky and R.S. Steplemen (IMACS,1981).
83. Gelinas,R.J. and S.K. Doss.J. Comput. Phys. 40 (1981)202.
84. Miller,K. and R.N. Miller.SIAM J. Num. Anal. 18 (1981)1019.
85. Sereno,C. Simulation of an adsorption fixed bed by using orthogonal collocation and moving finite elements.Report DTH (1983).
86. Carver,M.B.J. Comput. Phys. 35 (1980)57.
87. Morbidelli,M. et al.Ing. Chim. Ital. 19 (1983)47.
88. Hyman,M.Report LA-7595-M (Los Alamos Scientific Laboratory,1979).
89. Madsen,N.K. and R.F. Sincovec.ACM Trans. Math. Soft. 5 (1979)326.
90. Schiesser,W.DSS/2-An Introduction to the Numerical Method of

Lines Integration of PDEs (Bethlehem,Lehigh University,1976).

91. Carver,M. et al.Report AECL-5821 (Ontario,Chalk River Nuclear Laboratories,1978).

92. Melgaard,D. and R. Sincovec.ACM Trans. Math. Soft. 7 (1981)106.

93. Leaf,G.K. et al.DISPL:A Software Package for One and Two Spatially Dimensioned Kinetic-Diffusion Problems.Report ANL-77-12 (Illinois, Argonne National Laboratory,1977).

94. Machura,M. and R.A. Sweet,ACM Trans. Math. Soft. 6 (1980)461.

MODELLING OF MULTICOMPONENT FIXED BED ION EXCHANGE OPERATIONS

F.Evangelista and F.C.Di Berardino

Dipartimento di Chimica,Ingegneria Chimica e Materiali
Università degli Studi dell'Aquila,67100 L'Aquila,Italy

INTRODUCTION

In industrial ion exchange operations,multicomponent systems are often encountered.The design of fixed beds and the prediction of their performance in such cases are a challenge for design engineers when accurate knowledge of breakthrough curves for all components is required.Their location,breadth and shape are determined by the combined effects of equilibrium and of mass transfer rates.These in turn are affected by the physicochemical properties of the resin and its initial composition,solution concentration and composition and fluid dynamic conditions.Moreover,interferences between waves may occur.

Several mathematical models,with different descriptive capabilities,have been developed in the past years.The simplest model for multicomponent exchange is based on the equilibrium theory |1|.This is approximately valid in many cases for constant or proportionate pattern and for constant or variable separation factors.Also,an analytical method,valid for any number of components but only for constant separation factors,has been developed |2,3|.Recently,a computerized version of the later,able to handle variable initial and feed composition,has been implemented |4|.Other conceptual models useful in predicting column behaviour in multicomponent ion exchange have also been worked out |5|.

Starting with the transfer unit concept,equilibrium and kinetic effects have been combined for constant separation factor isotherms and for constant-pattern behaviour with arbitrary isotherms |6|.

All these methods are simple but they are characterized by low accuracy for real cases and have been successful only in a limited number of simple situations.However,they are useful for quick initial ootimates and to evaluate the limiting performance of a process.

Less detailed attention has been paid to mass transfer rates because this requires qualitative and quantitative accounting of all steps involved in ion exchange process,as reflected in the solution of differential mass balance equations.

Two main resistances have been identified and quantified by mass transfer coefficients in the solution and in the resin phase,respectively |7|.These coefficients would account for the effective motion of ions in multicomponent systems under concentration and electric field gradients.The treatment has been then simplified by trying to establish a priori which of them is the controlling step.However, this process is not so straightforward and there is still much uncertainty left.

In this paper a more generalized approach to multicomponent systems is described accounting for liquid-phase resistance by the film model and for solid-phase resistance by the effective concentration profile inside the resin bead.

The method of lines |8| has been used to solve the partial differential equations resulting from differential mass balances.The resulting system of ordinary differential equations can be easily integrated with robust variable-step integrators |9|.The method has been improved by incorporating in the integration scheme a procedure to self-adjust the spatial grid to wave dynamics.This will often lead to considerably shortening the computation time.

EQUILIBRIUM AND MASS TRANSFER RATES

Equilibrium evaluation is the first step to address the modelling of ion exchange operations.The most used equations for this purpose are those derived from theories of mass action law type.In general, for more than two heterovalent counterions,non-linear equations must be solved by a Newton-Raphson technique |10| .

Mass transfer in ion exchange processes is due to differences in electrochemical potentials and is controlled by several diffusional resistances.Diffusion through the external boundary layer and diffusion inside the exchanger are by far the most relevant steps in either directions.Simpler cases are also encountered in practice in which one of them is thought to be the controlling step.However,the great variety of operating conditions and concentrations does not always allow to clearly identify it a priori.In this paper both resistances are taken into account and equilibrium conditions are assumed to exist at the interface.

The Nernst-Planck equation for counter-ion $\underline{i}$ and for the resin phase can be written as follows:

$$J_i = -\frac{\rho Q}{1-\varepsilon}\bar{D}_i(\nabla Y_i + z_i Y_i \frac{F}{RT}\nabla\phi) \qquad i = 1,2,\ldots,n \qquad (1)$$

This is strictly valid if we assume quasi-homogeneity of the exchanger material,negligible interactions between $\underline{i}$ and the other counter-ions and negligible convective transport flux.

Moreover, the zero current and the electroneutrality conditions require that:

$$\sum_{j=1}^{n} J_j = 0 \qquad (2)$$

After combining Eqs.(1) and (2) and using the electroneutrality condition for the resin phase, the electric gradient can be eliminated from Eq.(1); the result is:

$$J_i = \frac{\rho Q}{1 - \varepsilon} \sum_{j=1}^{n-1} \bar{D}_{ij} \nabla Y_j \qquad i = 1, 2, \ldots, n-1 \qquad (3)$$

with

$$\bar{D}_{ij} = \bar{D}_i \frac{z_i Y_i (\bar{D}_j - \bar{D}_n)}{z_n \bar{D}_n + \sum_{k=1}^{n-1} (z_k \bar{D}_k - z_n \bar{D}_n) Y_k} \qquad (4)$$

$$\bar{D}_{ii} = \bar{D}_i \frac{z_i Y_i (\bar{D}_i - \bar{D}_n)}{z_n \bar{D}_n + \sum_{k=1}^{n-1} (z_k \bar{D}_k - z_n \bar{D}_n) Y_k} - \bar{D}_i \qquad (5)$$

All quantities have been worked out as for n-1 components, since all quantities referring to the n-th component can be obtained knowing those of the other n-1 components. This procedure reduces the number of equations to be solved.

For spherical symmetry, Eq.(3) becomes:

$$J_i = \frac{\rho Q}{1 - \varepsilon} \sum_{j=1}^{n-1} \bar{D}_{ij} \frac{\partial Y_j}{\partial r} \qquad i = 1, 2, \ldots, n-1 \qquad (6)$$

Following the same procedure for the liquid phase, a similar expression is obtained. It can be derived from Eq.(3) by substituting the liquid equivalent fraction X_i for the solid equivalent fraction Y_i, the total liquid concentration C for the total solid concentration $\rho Q / (1-\varepsilon)$ and D_{ij} for $\bar{D}_{ij}$. The result is:

$$J_i = C \sum_{j=1}^{n-1} D_{ij} \nabla X_j \qquad i = 1, 2, \ldots, n-1 \qquad (7)$$

Applying the film theory to the boundary layer, Eq.(7) becomes:

$$J_i = C \sum_{j=1}^{n-1} k_{ij} (X_j - X_j^*) \qquad i = 1, 2, \ldots, n-1 \qquad (8)$$

where k_{ij} is the multicomponent mass transfer coefficient and can be calculated from mass transfer coefficients for a single ion by means of relationships of the form (4) and (5). The k_i for the individual ion can be calculated from dimensionless relationships |11,12| knowing the diffusion coefficients and fluid dynamic conditions.

MASS BALANCE EQUATIONS

The following basic equations are based on the usual assumptions for fixed bed ion exchange operations, i.e., plug flow, spherical particles and constant geometrical and physical characteristics for the bed as well as for the resin beads. Thus, by using Eq. (8), a differential mass balance for the liquid phase leads to:

$$\frac{\partial X_i}{\partial t} = -v\frac{\partial X_i}{\partial z} + \frac{a_p}{\varepsilon}\sum_{j=1}^{n-1} k_{ij}(X_j - X_j^*) \qquad i = 1,2,\ldots,n-1 \qquad (9)$$

with initial conditions:

$$X_i = X_i^o \quad \text{for } t = 0 \qquad\qquad i = 1,2,\ldots,n-1 \qquad (10)$$
$$0 \leq z \leq Z$$

and boundary conditions:

$$X_i = X_i^F \quad \text{for } t > 0 \qquad\qquad i = 1,2,\ldots,n-1 \qquad (11)$$
$$z = 0$$

A differential mass balance for the solid phase can also be written by using Eq. (6); the result is:

$$\frac{\partial Y_i}{\partial t} = -\frac{1}{r^2}\sum_{j=1}^{n-1}\frac{\partial}{\partial r}\left(r^2 \bar{D}_{ij}\frac{\partial Y_j}{\partial r}\right) \qquad i = 1,2,\ldots,n-1 \qquad (12)$$

with initial conditions:

$$Y_i = Y_i^o \quad \text{for } t = 0 \quad 0 \leq r \leq R \quad i = 1,2,\ldots,n-1 \qquad (13)$$
$$0 \leq z \leq Z$$

and boundary conditions:

$$\frac{\partial Y_i}{\partial r} = 0 \qquad \begin{array}{l} \text{for } t > 0 \quad r = 0 \\ 0 \leq z \leq Z \end{array} \quad i = 1,2,\ldots,n-1 \qquad (14)$$

$$C\sum_{j=1}^{n-1} k_{ij}(X_j - X_j^*) = \frac{\rho Q}{1-\varepsilon}\sum_{j=1}^{n-1}\bar{D}_{ij}\frac{\partial Y_j}{\partial r} \qquad \begin{array}{l} \text{for } t > 0 \\ r = R \\ 0 \leq z \leq Z \end{array} \qquad (15)$$
$$i = 1,2,\ldots,n-1$$

Eqs. (14) and (15) follow from the zero flux condition at the center of the particle and the equality of fluxes at the interface, respectively. Constant feed composition and uniform initial bed composition have been assumed. Other situations, that is variable feed composition and variable initial bed composition, can be easily handled by changing initial and boundary conditions.

METHOD OF SOLUTION

Many methods of solution for PDEs equations have been developed; however, most of them deal with particular cases. One general method of solution is the method of lines |8|. It consists of discretizing only the spatial derivatives of time dependent PDEs. The resulting approximate system of ODEs can be then integrated with robust integrators. Software for systems of parabolic PDEs in one space |13| and two space dimensions |14| have been developed. Also systems of hyperbolic PDEs have been solved by this method |15|. Up to now no attempt has been made to solve nonlinear systems of coupled hyperbolic and parabolic PDEs in two space dimensions as in the present study.

The method of lines, in this case, consists on the discretization of the particle radius by n_r points and the bed height by n_z points. Using a three point difference scheme for all spatial derivatives, Eq.(9) becomes:

$$\frac{dX_{i,1}}{dt} = 0 \qquad\qquad i = 1,2,\ldots,n-1 \qquad (16)$$

$$\frac{dX_{i,k}}{dt} = -\frac{v}{A+B}\left[A\frac{X_{i,k+1}-X_{i,k}}{z_{k+1}-z_k} + B\frac{X_{i,k}-X_{i,k-1}}{z_k-z_{k-1}}\right] +$$

$$+ \frac{a_p}{\varepsilon}\sum_{j=1}^{n-1} k_{ij,k}\left(X_{j,k} - X^*_{j,k}\right) \qquad \begin{array}{l} i = 1,2,\ldots,n-1 \\ k = 2,3,\ldots,n_z-1 \end{array} \qquad (17)$$

$$\frac{dX_{i,n_z}}{dt} = -v\frac{X_{i,n_z} - X_{i,n_z-1}}{z_{n_z} - z_{n_z-1}} + \frac{a_p}{\varepsilon}\sum_{j=1}^{n-1} k_{ij,n_z}\left(X_{j,n_z} - X^*_{j,n_z}\right)$$

$$i = 1,2,\ldots,n-1 \qquad (18)$$

Equation (16) is derived from boundary conditions (11) at the inlet of the bed. Eqs.(17) are valid for all internal points where the spatial derivative has been approximated by a weighted average of A forward differences and B backward differences. This scheme has been proved |15| to avoid instability, usually present in discretizing hyperbolic PDEs. In Eq.(18), that is at the exit of the bed, the backward difference only has been used.

Discretization is now applied to Eq.(12) which is thus transformed in:

$$\frac{dY_{i,1,k}}{dt} = \frac{C(1-\varepsilon)}{\rho Q} \frac{2(2r_1 - r_2)}{r_1(r_1 - r_2)} \sum_{j=1}^{n-1} k_{ij,k}\left(X_{j,k}^* - X_{j,k}\right)$$

$$+ \frac{2}{(r_1 - r_2)^2} \sum_{j=1}^{n-1} \bar{D}_{ij,1+\frac{1}{2},k}\left(Y_{j,1,k} - Y_{j,2,k}\right)$$

$$i = 1,2,\ldots,n-1$$
$$k = 1,2,\ldots,n_z \tag{19}$$

$$\frac{dY_{i,\ell,k}}{dt} = -\frac{3}{r_{\ell+\frac{1}{2}}^3 - r_{\ell-\frac{1}{2}}^3} \left[\frac{r_{\ell+\frac{1}{2}}^2}{r_{\ell+1} - r_\ell} \sum_{j=1}^{n-1} \bar{D}_{ij,\ell+\frac{1}{2},k}\left(Y_{j,\ell+1,k} - Y_{j,\ell,k}\right) \right.$$

$$\left. - \frac{r_{\ell-\frac{1}{2}}^2}{r_\ell - r_{\ell-1}} \sum_{j=1}^{n-1} \bar{D}_{ij,\ell-\frac{1}{2},k}\left(Y_{j,\ell,k} - Y_{j,\ell-1,k}\right) \right]$$

$$i = 1,2,\ldots,n-1 \qquad k = 1,2,\ldots,n_z$$
$$\ell = 2,3,\ldots,n_r-2 \tag{20}$$

$$\frac{dY_{i,n_r-1,k}}{dt} = \frac{3r_{n_r-1-\frac{1}{2}}^2}{\left(r_{n_r-1} - r_{n_r-2}\right)\left(r_{n_r-\frac{1}{2}}^3 - r_{n_r-1-\frac{1}{2}}^3\right)} \cdot$$

$$\sum_{j=1}^{n-1} \bar{D}_{ij,n_r-1-\frac{1}{2},k}\left(Y_{j,n_r-1,k} - Y_{j,n_r-2,k}\right) \quad \begin{array}{l} i = 1,2,\ldots,n-1 \\ k = 1,2,\ldots,n_z \end{array} \tag{21}$$

with

$$r_{\ell\pm\frac{1}{2}} = \frac{r_{\ell\pm1} + r_\ell}{2} \tag{22}$$

$$\bar{D}_{ij,\ell\pm\frac{1}{2},k} = \bar{D}_{ij}\left(t, r_{\ell\pm\frac{1}{2}}, \underline{Y}_{\ell\pm\frac{1}{2}}\right) \tag{23}$$

and

$$Y_{\ell \pm \frac{1}{2}} = \left(\frac{Y_{1,\ell \pm 1,k} + Y_{1,\ell,k}}{2} , \ldots , \frac{Y_{n-1,\ell \pm 1,k} + Y_{n-1,\ell,k}}{2} \right) \quad (24)$$

Eq.(19) results from boundary condition (15) at the interface, Eqs.(20) are for the internal points and Eq.(21) for the n_r-1 th point. The Y_{i,n_r} equivalent fraction is equal to Y_{i,n_r-1} due to the boundary condition (14).

Due to the higher concentration gradient on the particle surface, points have been spaced in this region by arbitrarily assuming an exponential variation of the grid intervals.

The $2 \times (n-1)$ PDEs, Eqs.(9) and (12) are thus transformed into $n_e = (n-1) \cdot n_r \cdot n_z$ ODEs, Eqs.(16) to (21). These can be rewritten in compact form as:

$$\frac{du_m}{dt} = f_m (t, u_1, u_2, \ldots, u_{n_e}) \qquad m = 1,2,\ldots n_e \qquad (25)$$

which is a system of ODEs with initial conditions (10) and (13). Then Eqs.(25) can be integrated with robust integrators [9].

The $X_{i,k}$ and $Y_{i,\ell,k}$ variables are put into the vector u_m. Correspondence between m and i,ℓ,k is provided by the following relationship:

$$m = (k-1)(n-1)n_r + (\ell-1)(n-1) + i \qquad \begin{array}{l} i = 1,2,\ldots,n-1 \\ \ell = 1,2,\ldots,n_r \\ k = 1,2,\ldots,n_z \end{array} \qquad (26)$$

Due to the generally large number of ODEs to be solved the system tends to be stiff. Therefore, implicit integration methods which use Jacobian matrices are often required. In this work a modified version of the GEARBI package [9] has been used, taking particular advantage of the favorable structure of the Jacobian matrix. Letting the indices i,ℓ,k vary in the given order, the matrix has a block-banded structure. The associated linear system is solved by a block iterative treatment of the whole Jacobian. All non-zero entries of the matrix are evaluated by analytical expressions which, for the sake of simplicity, are not reported here.

The GEARBI package is a variable-step, variable-order integrator, so that it is useful to adjust the step size and the order dynamically as integration proceeds. However, these capabilities are not enough in this case and computation time tends to be high also. A big improvement can be made if we examine the ion exchange process thoroughly. For n exchangeable ions n-1 transition zones and n plateau zones are generated inside the bed [1]. The exchange process takes place only in the transition zones.This means that all components of the vector $\underline{f}$ of Eq.(25) related to the plateau zones are zero and that the corresponding $\underline{u}$ components remain unchanged,at least for some time.Therefore,points placed in the plateau zones can be removed,thus lowering the total number of equations to be solved.However,front waves and plateau zones move along the bed and the spatial grid should be able to dynamically follow these changes as time goes on.Again,we are in the favorable situation that all $Y_{i,\ell,k}$ are constant if the corresponding $X_{i,k}$ are constant.So,fixing the point corresponding to the inlet of the bed,an appropriate fine mesh of constant intervals is first established in the inlet zone of the bed and integration started.After a certain time interval,integration is stopped and the $\dot{X}_{i,k}$ are evaluated.The new spatial grid is then set up by taking points k where the $\dot{X}_{i,k}$ are greater than a given minimum.Few more points are also added to allow the movement of the exchange zones in that time interval.Checks are also made to ensure that the front waves have not spread out of their corresponding fine-mesh grids.If this occurs,the solution is discarded,the spatial grid enlarged and the integration restarted from the previously accepted solution.As it can be easily worked out,this method is most effective if the exchange zone is contained in a small region,that is for constant--pattern behaviour.Systems with proportionate pattern can also be handled,but with lower effectiveness.The later is reduced further if the number of components increases and/or diffusional resistances make waves to spread out over the total bed height.

Figure 1a shows the spatial grid with the fine mesh for one transition zone only.More fine meshes would have been generated for multicomponent systems.Figure 1b shows the block diagram of the corresponding computer program.

EXPERIMENTAL

A non-commercial strong cationic resin was used in all experiments.Selected particles were obtained from a sample by wet screening followed by a water elutriation technique.The average,minimum and maximum diameters of the beads were determined by a microscope over a sample of one hundred particles.Then a jacketed glass column of 1 cm ID was filled with them.Particular attention was paid to the packing procedure to ensure voids uniformity.Particles were let to settle gently on a concentrated sodium chloride solution.To ensure uniform flow distribution the column ends were filled with glass beads and glass wool.The physicochemical properties of the resin and bed characteristics are reported in Table I.

All experiments were carried out at a temperature of 40 C held constant by water flowing through the column jacket.The solutions were preheated to 40 C by a double pipe heat exchanger to ensure temperature uniformity and,at the same time,were degasified to prevent bubble formation inside the bed.Synthetic solutions prepared from reagent grade chemicals and distilled water were fed to the

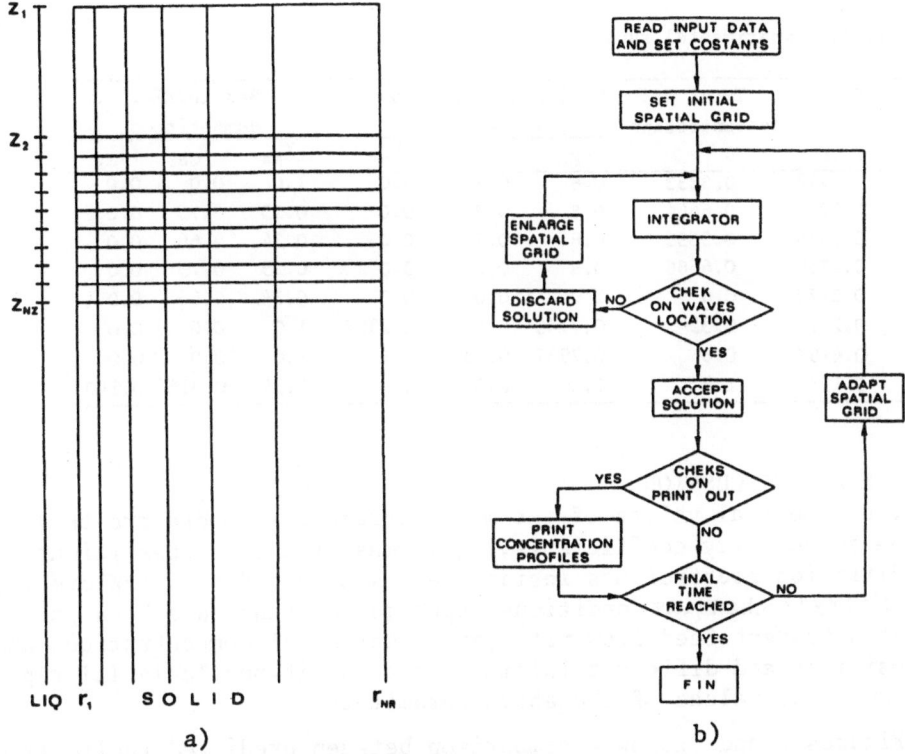

Figure 1- Spatial grid (a) and the block diagram (b) of the computer program.

column at the desired velocity by a three piston pump.Fractions of 30 cc were collected by an automatic collector.The operating conditions of the experimental runs,carried out on binary and ternary solutions of Ca^{2+},$Mg^{2+}$and Na^+,are reported in Table II.Complexometric methods were used to determine Ca and Mg contents of the eluted fractions.Na^+ was usually obtained as a difference.For checking purposes its concentration was occasionally determined by an atomic absorption technique.

Table I. Physicochemical properties of the resin and bed characteristics.

Resin bead radius	0.0325±0.003	cm
Bed diameter	1.0	cm
Bed height	91.0	cm
Void fraction	0.45	
Bed density	0.585	g of dry resin/cm^3
Specific surface area	50.77	cm^2/cm^3
Exchange capacity	4.5	meq/g of dry resin

Table II. Experimental runs carried out.

Run No	Feed flow rate v	Total concentration C	Equivalent fractions in feed			Bed initial composition		
			X_{Na}	X_{Ca}	X_{Mg}	Y_{Na}	Y_{Ca}	Y_{Mg}
1	0.2778	0.3333	0.8	0.2	0.0	1.0	0.0	0.0
2	0.2778	0.6666	0.8	0.2	0.0	0.25	0.75	0.0
3	0.5556	0.3333	0.8	0.2	0.0	0.34	0.66	0.0
4	0.5556	0.6666	0.8	0.2	0.0	0.25	0.75	0.0
5	0.2778	0.6	1.0	0.0	0.0	0.34	0.66	0.0
6	0.2778	0.35	0.7143	0.0	0.2857	1.0	0.0	0.0
7	0.4167	0.6	0.7957	0.0308	0.1735	1.0	0.0	0.0
8	0.1389	1.8	1.0	0.0	0.0	0.545	0.315	0.140

RESULTS AND DISCUSSION

The numerical values of the model parameters needed are those of the selectivity coefficients,liquid mass transfer coefficients and diffusion coefficients inside the resin.In order to explore their sensitivity to changing conditions experimental runs have been performed at different feed flow rates,different total concentrations and compositions and different initial bed compositions.Table III reports the numerical values of the above parameters.

Figures 2 and 3 show a comparison between predicted (solid lines) and experimental breakthrough curves for each run.The selectivity coefficients have been calculated from experimental breakthrough curves.They show a strong dependence on total concentration and depend less on composition.Knowing the diffusion coefficients at infinite dilution single mass transfer coefficients have been calculated by the following relationship |11|:

$$k_i = 1.85 \cdot v \cdot (Re \cdot Sc)^{-2/3} \cdot (\varepsilon/(1 - \varepsilon))^{-1/3} \qquad (27)$$

which is very suitable for Re< 30.No attempt has been made to account for its variation with total concentration and composition.In spite of the high solution concentration diffusion through the liquid boundary layer seems to affect the overall exchange rate in both favorable (runs 1 and 3) and unfavorable (runs 2 and 4) equilibria.

A different procedure has been followed to evaluate the ionic diffusion coefficients inside the resin.First,the diffusion coefficient of the Na^+-ion has been fixed to a value found in the literature |16,17| and the diffusion coefficients of Ca^{2+} and Mg^{2+} ions have been evaluated by fitting binary experimental breakthrough curves (Figure 2).Then we tried to adjust the diffusion coefficient of Na^+.However,no improvement was found since the less diffusing species (Ca^{2+} and Mg^{2+}) control the exchange rate and determine the shape of the breakthrough curves.

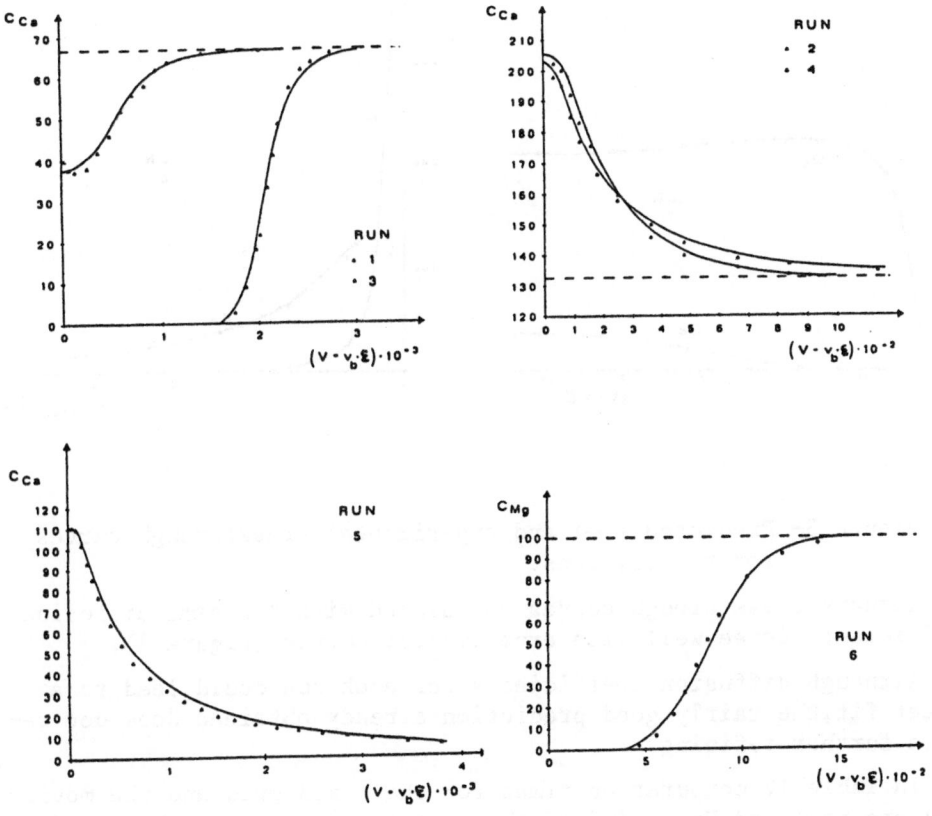

Figure 2- Predicted (——) and experimental breakthrough curves.

Table III. Numerical values of model parameters.

Run No	$K_{Ca/Na}$	$K_{Mg/Na}$	$k_{Na} \cdot 10^3$	$k_{Ca} \cdot 10^3$	$k_{Mg} \cdot 10^3$	$\bar{D}_{Na} \cdot 10^6$	$\bar{D}_{Ca} \cdot 10^8$	$\bar{D}_{Mg} \cdot 10^7$
1	39.00	–	8.158	5.760	–	1.6	6.0	–
2	18.50	–	8.158	5.760	–	1.6	6.0	–
3	39.00	–	10.276	7.255	–	1.6	6.0	–
4	18.50	–	10.276	7.255	–	1.6	6.0	–
5	20.00	–	8.158	5.760	–	1.6	6.0	–
6	–	2.35	8.158	–	5.336	1.6	–	2.5
7	21.65	1.72	9.342	6.595	6.110	1.6	6.0	2.5
8	6.90	0.54	6.477	4.572	4.236	1.6	6.0	2.5

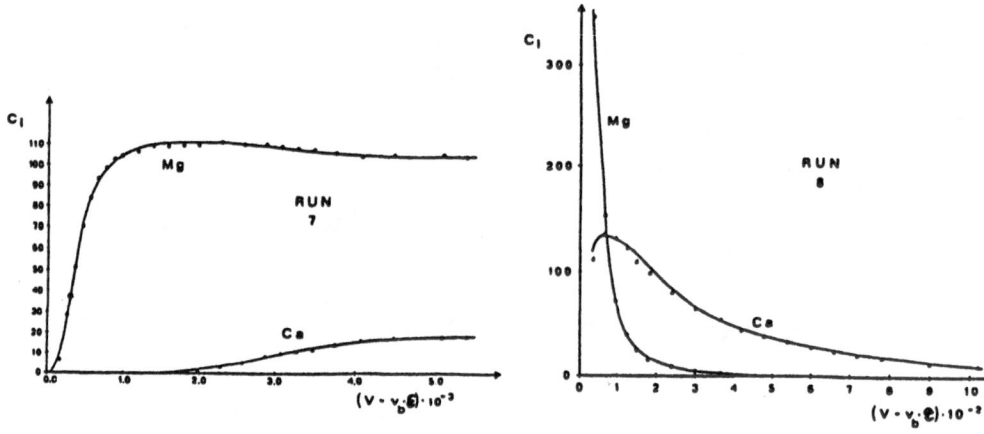

Figure 3- Predicted (——) and experimental breakthrough curves
for ternary runs.

Ternary breakthrough curves calculated with the same diffusion
coefficients agree well with experimental curves (Figure 3).

Although diffusion coefficients for each run could lead to a
better fit,the fairly good prediction already obtained does not require further refining.

In Table IV computation times for the fixed grid and the moving
grid are reported.Up to 50% of the computation time can be saved by
the moving technique for binary systems with constant pattern behaviour (runs 1,3 and 6).Less effectiveness is obtained for binary
systems with proportionate pattern (runs 2,4 and 5) and ternary systems (runs 7 and 8).However,this behaviour should be indicative only
since computation time and savings depend on operating conditions,
mainly feed concentration and composition and initial bed composition (see Eq.(19)).Less computation time must be expected for more
dilute solutions which are more often encountered in ion exchange
operations.

TABLE IV- Computation times(minutes) UNIVAC 1100/60

				Run				
	1	2	3	4	5	6	7	8
Moving grid	5.2	6.9	4.7	6.5	4.8	2.4	27.0	20.3
Fixed grid	10.8	8.1	6.9	7.9	5.2	4.1	32.8	24.1

CONCLUSION

A numerical algorithm for multicomponent ion exchange operations on fixed beds has been worked out. Starting with the Nernst-Planck equation, both liquid and resin phase resistances have been taken into account using the liquid film model and the effective concentration profile inside the bead. Single mass tranfer coefficients and single ionic diffusion coefficients have been found to be quite satisfactory in predicting effluent concentration histories in binary and ternary systems.

The method of lines, coupled with the moving grid technique, has been proved to be effective in solving systems of hyperbolic and parabolic PDEs.

ACKNOWLEDGEMENT

The authors wish to express their sincere thanks to G.Klein and A.Bunge for useful comments and suggestions.

NOTATION

A = weight for forward differences, here $A = 1$.

a_p = specific interfaccial area, cm^2/cm^3

B = weight for backward differences, here $B = 2$.

C = total solution concentration, meq/cm^3

C_i = solution concentration of ion i, meq/cm^3

D_i = individual diffusion coefficient of ion i, cm^2/s

D_{ij} = multicomponent diffusion coefficient of ion i, cm^2/s

F = Faraday constant, Coulomb/eq

f = defined in Eq (25)

J_i = flux of ion i, $meq/cm^2 \cdot s$

K_{ij} = dimensionless selectivity coefficient

k_i = individual mass transfer coefficient of ion i, cm/s

k_{ij} = multicomponent mass transfer coefficient of ion i, cm/s

n = number of counter-ions

n_e = total number of ordinary differential equations

n_r = number of points inside the resin

n_z = number of points along the bed height

Q = exchange capacity of the resin, meq/g of dry resin

R = universal gas constant, joule/°K·mol

R = radius of the resin bead, cm

r = radial coordinate, cm

T = absolute temperature, °K

t = time, s

u = defined in Eq (25)

v = liquid velocity through the bed, cm/s

v_b = bed volume, cm^3
V = volume of eluted solution, cm^3
X_i = liquid equivalent fraction of ion i
Y_i = solid equivalent fraction of ion i
Z = bed height, cm
z = axial coordinate, cm
z_i = valence of ion i

Greek letters

ε = void fraction of the bed
ρ = density of the bed, g/cm^3
ϕ = electric potential, volt
∇ = gradient operator, 1/cm

Superscripts

$-$ = refers to the resin phase
$*$ = denotes equilibrium quantities
o = denotes initial conditions
F = refers to the feed
$\cdot$ = denotes time derivatives

Subscripts

$-$ = indicates vector quantities
i = refers to counter-ions
j = refers to counter-ions
k = refers to counter-ions or to discretizing points
ℓ = refers to discretizing points
m = refers to components of vector u

REFERENCES
1. Klein,G.,D.Tondeur and T.Vermeulen.Ind.Eng.Chem.Fundamentals.6 (1967)339
2. Helfferich,F.Ind.Eng.Chem.Fundamentals.6(1967)367.
3. Helfferich,F. and G.Klein.Multicomponent Chromatography.(Marcel Dekker,New York,1970)
4. Klein,G.,M.Nassiri and J.Vislocky.AIChE Symp.Series 80(1984)14
5. Helfferich,F.AIChE Symp.Series 80(1984)1
6. Basmadjan,D. and C.Karayannapoulos.I&EC Proc.Des.Dev.24(1985)140
7. Clazie,R.,G.Klein andT.Vermeulen.OSW Res.Dev.Prog.Rep.No 326,US Dept.of Interior(1968)
8. Madsen,K. and R.Sincovec.Computational Methods in Non Linear Analysis.(Austin,Texas Institute for Computational Mechanics,ed.J. Oden et al.,1974)

9. Hindmarsh,A.UCID-30149,Lawrence Livermore Laboratory Report(1976)
10.Klein,G.Comp.Chem.Eng.8(1984)171
11.Kataoka,T.,H.Yoshida and T.Yamada.J.Chem.Eng.Japan.6(1973)172
12.Carberry,J.AIChEJ.4(1960)460
13,Sincovec,R. and N.Madsen.ACM Trans.Math.Software.1(1975)261
14.Melgaard,D. and R.Sincovec.ACM Trans.Math.Software.7(1981)126
15.Heidweiller,J. and R.Sincovec.J.Comp.Phys.22(1976)377
16.Helfferich,F.Ion Exchange.(New York,McGraw Hill,1962)
17.Graham,E. and J.Dranoff.I&EC Fundamentals.31(1982)365

FIXED BED PROCESSES:A STRATEGY FOR MODELLING

A.E.Rodrigues and C.A.Costa

Department of Chemical Engineering
University of Porto,Porto,Portugal

INTRODUCTION

Fixed bed operations involving ion exchange resins can be used for separation purposes and as catalytic reactors[1].The dynamics of ion exchange processes depends on:

a- equilibrium law at the fluid-solid interface
b- kinetics (reaction;mass transfer :film diffusion and intraparticle diffusion;heat transfer)
c- hydrodynamics

A realistic model should take into account such contributions. The main point in our strategy is that model parameters are obtained from simple independent experiments as shown in Table 1.

Table 1 - Type of experiment and parameters to be measured.

Type of experiment	Model parameter
Batch adsorber	Equilibrium isotherm
Batch and CSTR adsorbers	Intraparticle diffusivity
Shallow bed	Film mass transfer coefficient
Fixed bed -tracer experiment	Axial dispersion

Model parameters are then fed to the model equations.Computer simulations are compared with experimental breakthrough curves obtained at the laboratory scale to test the validity of the model.However, a good model should be able to predict breakthrough curves obtained at conditions different from those used at the bench scale.So we have to see if model parameters are affected or not by scale-up.This remark applies to hydrodynamic parameters.

Equilibrium models are very common in the ion exchange field |2,3,4|.They give a picture of the limiting (ideal) performance of a fixed bed. If equilibrium model predictions are accurate enough for design purposes is a question to be discussed for a particular problem.

In Figure 1 we compare predicted and experimental breakthrough curves for the removal of tungsten from hydrometallurgical liquors |5|. The ion exchange equilibrium isotherm for tungstate /Duolite A 101 D (strong anionic resin) has an inflection point as shown in Figure 2.The equilibrium model will predict a dispersive wave followed by a compressive wave;this is in reasonable agreement with experiment. However,equilibrium model fails when predicting the behaviour of fixed beds of chelating resins due to kinetic limitations.

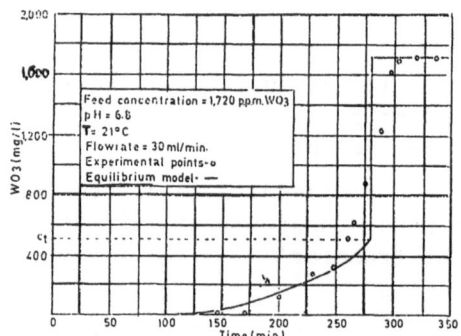

Figure 1- Breakthrough curve for the removal of tungsten from hydrometallurgical liquors with Duolite A 101 D

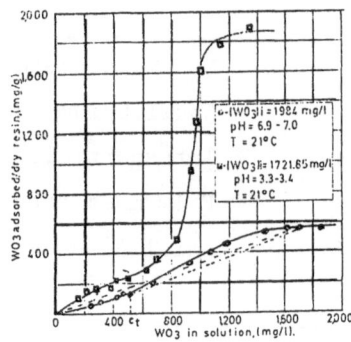

Figure 2- Ion exchange equilibrium isotherms tungstate/Cl⁻

FIXED BED ADSORPTION OF PHENOL IN A POLYMERIC ADSORBENT

Let us illustrate the methodology for modelling through an example: adsorption of phenol in a fixed bed of a polymeric adsorbent (Duolite ES 861).Model parameters were measured by independent experiments as indicated below.

Equilibrium isotherms: Adsorption equilibrium isotherms for phenol in Duolite ES861 were obtained from batch experiments.Experimental results are shown in Figure 3 and were fitted with a Langmuir equation |6|.

Intraparticle diffusivity: Intraparticle mass tranfer is a complex problem which requires a model for the particle structure.Figure 4 shows a picture obtained with a scanning electron microscope.It supports a model in which the bead is viewed as an ensemble of microspheres (gel phase) with pores amongst them (Figure 5).
Four models can be developed for describing intraparticle diffusion in such adsorbents:

a- homogeneous model (gel type particle)
b- pore diffusion model
c- parallel model (pore and gel/surface diffusion in parallel)
d- series model (pore and gel diffusion in series)

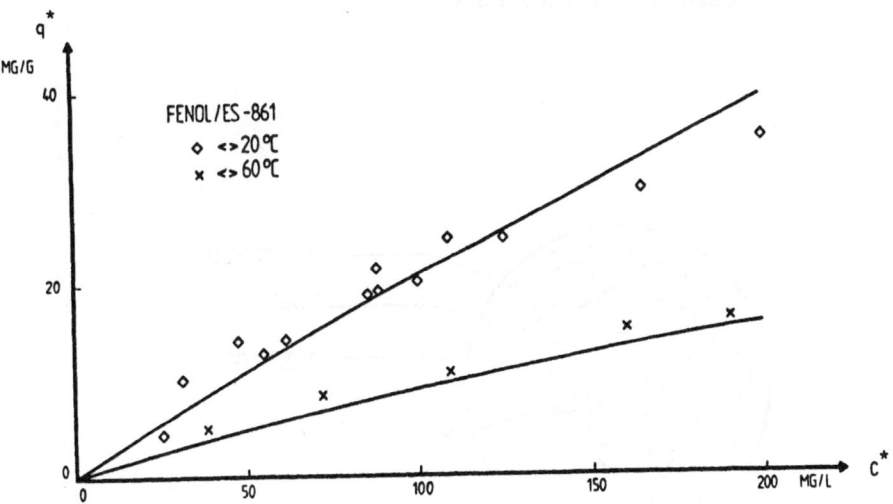

Figure 3 - Adsorption equilibrium isotherms:phenol/Duolite
          ES861

Figure 4- Scanning electron microscope photograph of a
bead of Duolite ES861

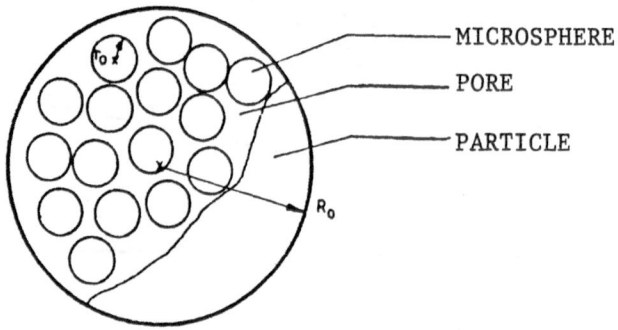

Figure 5- A bidisperse model for Duolite ES861 particles.

Model equations for batch and CSTR adsorbers using the various intraparticle diffusion models were solved by appropriate numerical methods.Batch runs were carried out at three different particle diameters (0.077cm;0.066cm;0.034cm) and initial phenol concentration in solution around 96 mg/l. The pore diffusion model was found to describe experimental results. The intraparticle effective diffusivity was found to be D_{pe} =7.7x10^{-10} m^2/s.

Then experiments in a CSTR adsorber were carried out by using different particle sizes and various flowrates.The predicted responses with the effective diffusivity value calculated from batch runs agree well with the experimental results.This is shown in Figure 6 for CSTR runs at different flowrates.

Film mass transfer coefficient: Correlations are available in the literature to predict film mass transfer coefficients k_f.They are of the form:

$$j_d = f(Re)$$

where j_d =Sh/Re Sc$^{1/3}$ with Sh=$k_f d_p$/D ,Sc=ν/D and Re=$u_o d_p$/ν .Data for Sherwood number as a function of Reynolds number in fixed beds are represented in Figure 7. These data were obtained by using various experimental techniques:dissolution of solid particles (benzoic acid), sublimation of particles (naphtalene),evaporation of pure liquid from the surface of a porous particle,fast catalytic reaction on a nonporous solid or ion exchange accompanied by neutralization.This last method developed by Myauchi |7| is sketched in Figure 8.

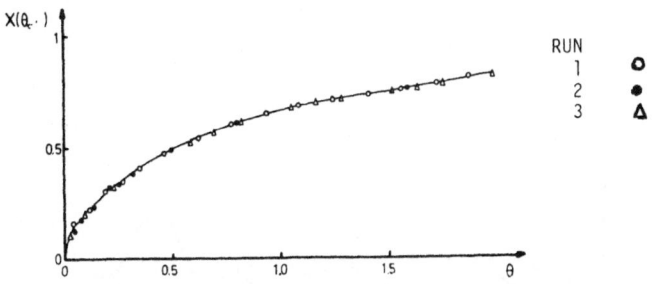

Figure 6-Response of a CSTR to a step input
1- U=108.3ml/min;c_o=92.6mg/1;d_p=0.06cm
2- U= 54ml/min   ;c_o=92.6mg/1;d_p=0.06cm
3- U=77.8ml/min ;c_o=93.8mg/1;d_p=0.06cm

276

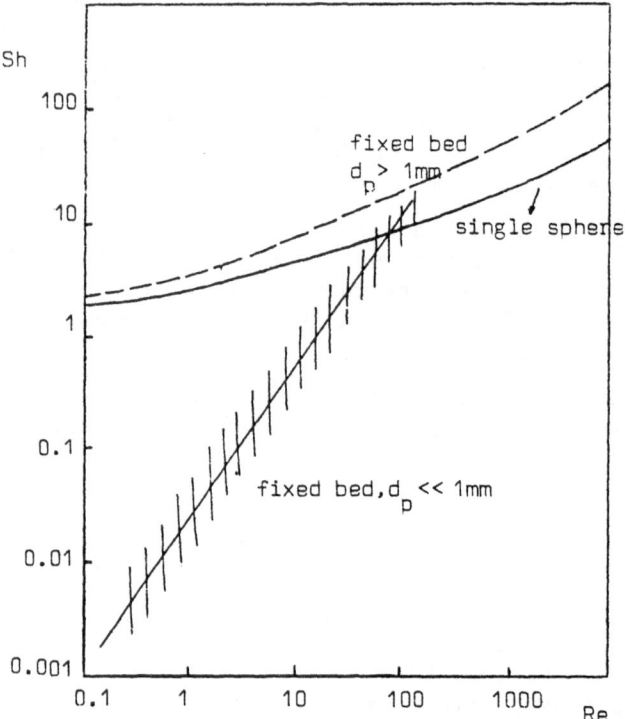

Figure 7- Film mass transfer in fixed beds. Sherwood number versus Reynolds number

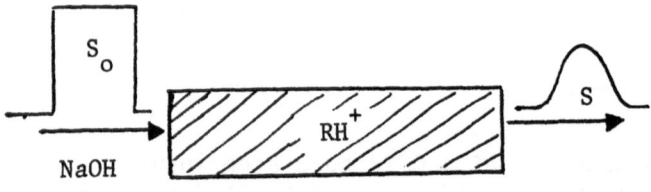

Figure 8- Myauchi method for measuring k_f

One can show that

$$S/S_o = c/c_o = f(Pe,k) \tag{1}$$

where c is the steady state concentration at the bed outlet when the feed is a step change in concentration $c_{in} = c_o H(t)$. The dispersion model leads to:

$$c/c_o = \frac{4m\exp(Pe/2)}{(1+m)^2\exp(mPe/2) - (1-m)^2\exp(-mPe/2)} \tag{2}$$

with

$$m = \sqrt{1+4\ Da/Pe}$$

$$Da = k\tau$$

$$Pe = u_i L/D_{ax}$$

In the above treatment k is equivalent to the film mass transfer coefficient.

In our work we used a shallow bed technique (Figure 9). We also took into consideration deformation of signals due to the sections above and below the shallow bed. If we plot the number of film mass transfer units as a function of time (Figure 10) a straight line is obtained which was extrapollated to zero time to get the true film mass transfer coefficient. This technique due to Tien [8] is supposed to be supported by the fact that at the beginning only mass tranfer resistance in the film is important. However if film mass transfer resistance is important during a finite time we would expect a constant number of film mass transfer units during that time followed by a decrease of N_f as soon as the intraparticle resistance becomes important. Then extrapollation to zero time can lead to significant errors.

We simulate our shallow bed with a model which includes film and intraparticle resistances. We took advantage of the fact shown in Figure 11 [9] that for a given N_f intraparticle diffusion does not affect the response during a certain time . This supports the idea that it is not correct to extrapolate $N_f(t)$ line to zero time. For our system we found $j_d = 7.32\ Re^{-0.569}$.

Axial dispersion: Tracer experiments in fixed beds using either KCl or dye were carried out in order to measure axial dispersion. A dispersion model which took into account intraparticle porosity was used to analyse the experimental results. These are summarized in Figure 12 where the particle Peclet number is shown as a function of the Reynolds number.

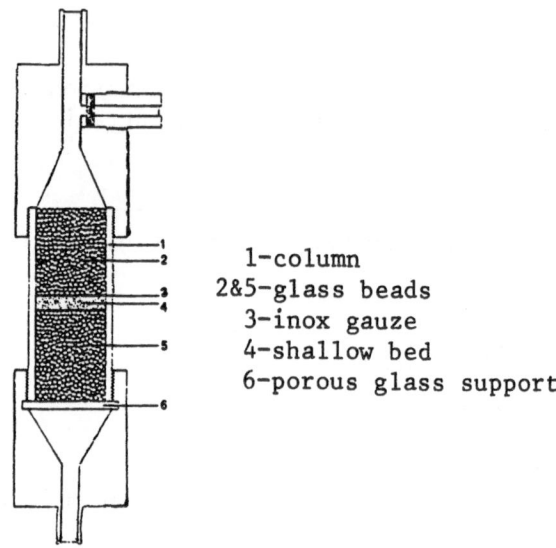

Figure 9- Shallow bed technique for measuring film
mass transfer coefficients

1-column
2&5-glass beads
3-inox gauze
4-shallow bed
6-porous glass support

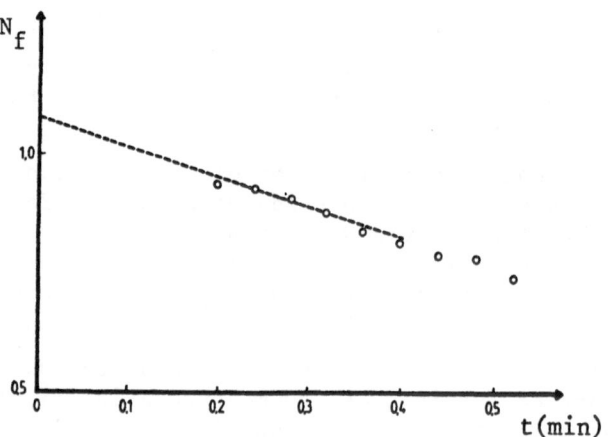

Figure 10- N_f versus time

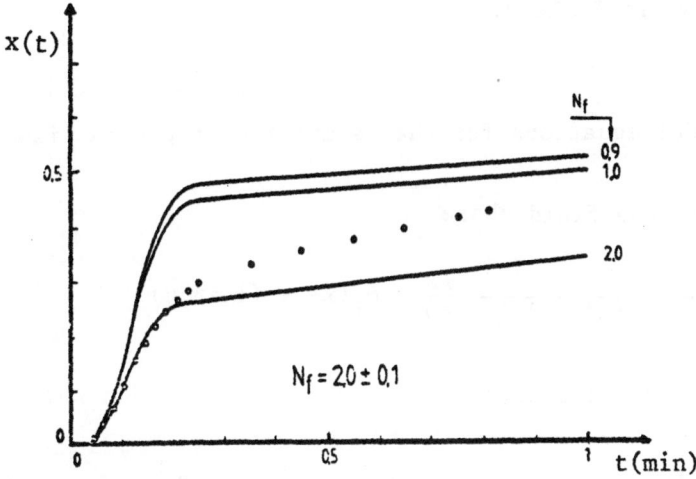

Figure 11- -Response of a shallow bed to a step input in con-
centration:experimental and model results.

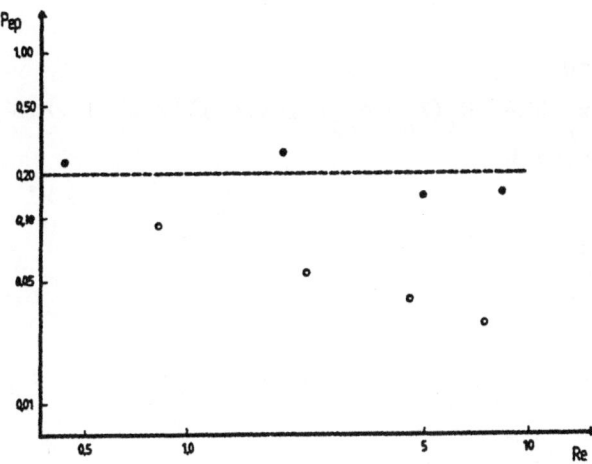

Figure 12- Axial dispersion in a laboratory fixed bed.
Peclet number versus Reynolds number

MODELLING FIXED BED ADSORBERS:SATURATION STEP
Model equations for the saturation (loading) of a fixed bed adsorber are shown in Table 2.

Table 2 - Model equations for the saturation step of a fixed bed

Mass balance for the fluid phase

$$\frac{1}{Pe}\frac{\partial^2 x(z^*,\theta)}{\partial z^{*2}} = \frac{\partial x}{\partial z^*} + \frac{1}{1+\xi}\frac{\partial x}{\partial \theta} + N_f(x - x_p(1,z^*,\theta))$$

Mass balance inside particles

$$\frac{\partial x_p(R^*,z^*,\theta)}{\partial \theta} = \frac{(1+\xi)N_d}{\varepsilon_p + \dfrac{K_L Q \rho_{ap}}{(1+K_L c_o x_p)^2}}\left(\frac{2}{R^*}\frac{\partial x_p}{\partial R^*} + \frac{\partial^2 x_p}{\partial R^{*2}}\right)$$

Initial and boundary conditions

o Particle
$R^*=0$, $\partial x_p/\partial R^*=0$
$R^*=1$, $3\xi N_d \partial x_p/\partial R^* = N_f(K_L Q \rho_{ap}/(1+K_L c_o))(x-x_p(1,z^*,\theta)$
$\theta=0$, $x_p(R^*,z^*,0)=0$

oo Fixed bed
$z^*=0$, $x(0,\theta)=1$
$z^*=1$, $\partial x/\partial z^*=0$
$\theta=0$, $x(z^*,0)=0$

Computer simulation enabled us to analyse the effect of the model parameters [10] .Having set the equilibrium parameters for the system phenol-water/Duolite ES861 the model parameters are:

$\xi=(1-\varepsilon)q_o/\varepsilon c_o$ capacity parameter
$Pe=u_i L/D_{ax}$ Peclet number

$$N_f=(1-\varepsilon)a_p k_f \tau/\varepsilon$$
$$N_d=\tau D_{pe}/R_o^2$$

Our strategy for modelling was then tested.Model parameters already obtained from independent experiments were put in the model equations.Comparison between the experimental breakthrough curves and calculated curves is shown in Figure 13.

MODELLING REGENERATION OF FIXED BED ADSORBERS

Regeneration of adsorbent particles saturated with phenol was carried out by passing sodium hydroxyde through the bed.This is a problem of adsorption coupled with chemical reaction.The reaction taking place is:

$$C_6H_5OH + OH^- \rightleftharpoons C_6H_5O^- + H_2O \tag{3}$$

with K_{eq} (20 C)= 1.3×10^4 1/mole. On the other hand,the adsorption equilibrium follows a Langmuir isotherm;coupling adsorption and reaction leads to an equation relating the adsorbed quantity of phenol as a function of the total concentration (phenol + phenate) in solution. That equation shows the effect of the regenerant concentration or pH on the regeneration efficiency (Figure 14).

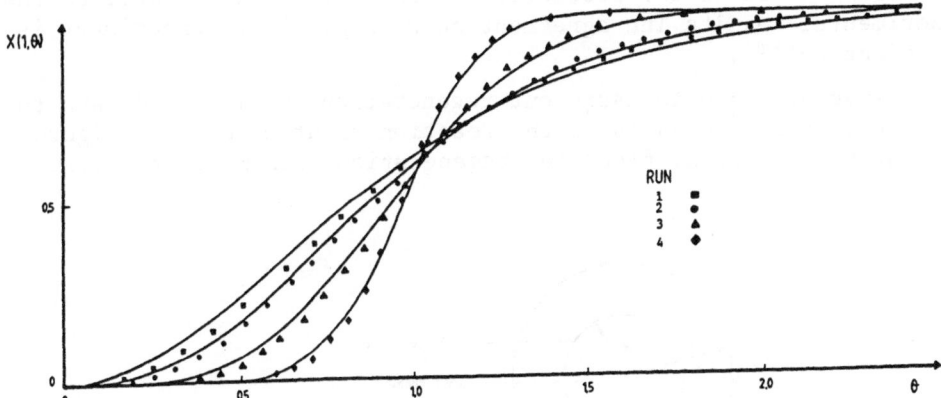

Figure 13- Saturation of a fixed bed adsorber (d_p=0.06 cm; L=40 cm;ID=2.18cm,T=20C)
1- U=158.7ml/min;c_o=82 mg/1
2-   115.2ml/min;   91.6mg/1
3-   54.4 ml/min;   91.6mg/1
4-   16.8 ml/min;   82.4mg/1

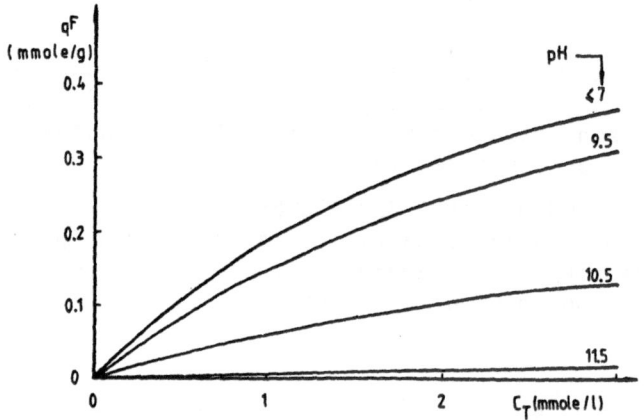

Figure 14- Influence of pH on the adsorption equilibrium isotherm.

In order to understand the regeneration process we run simple dynamic experiments in a CSTR.We tried to explain the experimental results on the basis of an equilibrium model.This gives a qualitative picture of the process. However,a better model was developed - the reaction front model (Figure 15).

Figure 16 shows the predicted response of a CSTR as well as the experimental results.The agreement between model and experiment is satisfactory|11|.

Now we are able to carry out regeneration in fixed beds and to interpret it on the basis of the reaction front model |12|.Figure 17 compares a typical fixed bed regeneration with model results.

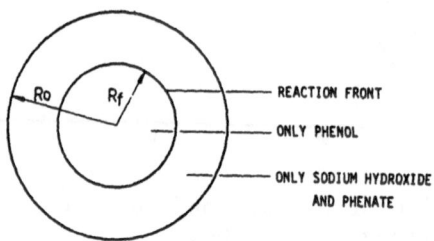

Figure 15- The reaction front model

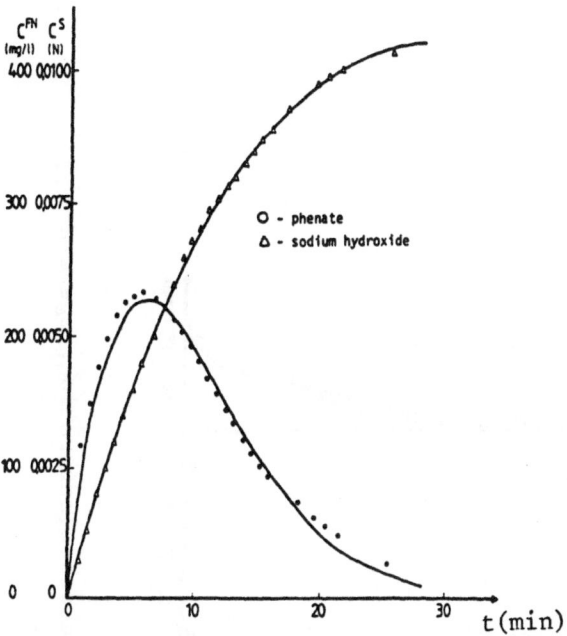

Figure 16- Regeneration in a CSTR (U=77.5ml/min;hydroxyde concentration: 0.0108 N;initial phenol concentration=93.8 mg/1;d_p=0.06 cm; porosity of the CSTR=0.946)

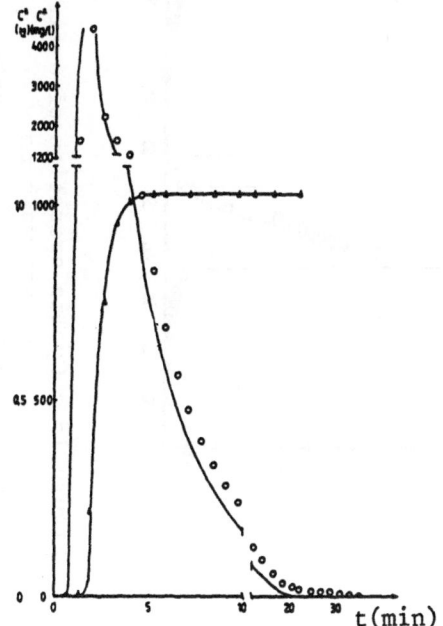

Figure 17-

Regeneration in a fixed bed saturated with phenol at 91.6mg/1.
Regenerant concentration= 1.035 N;flowrate=56.6 ml/min.

284

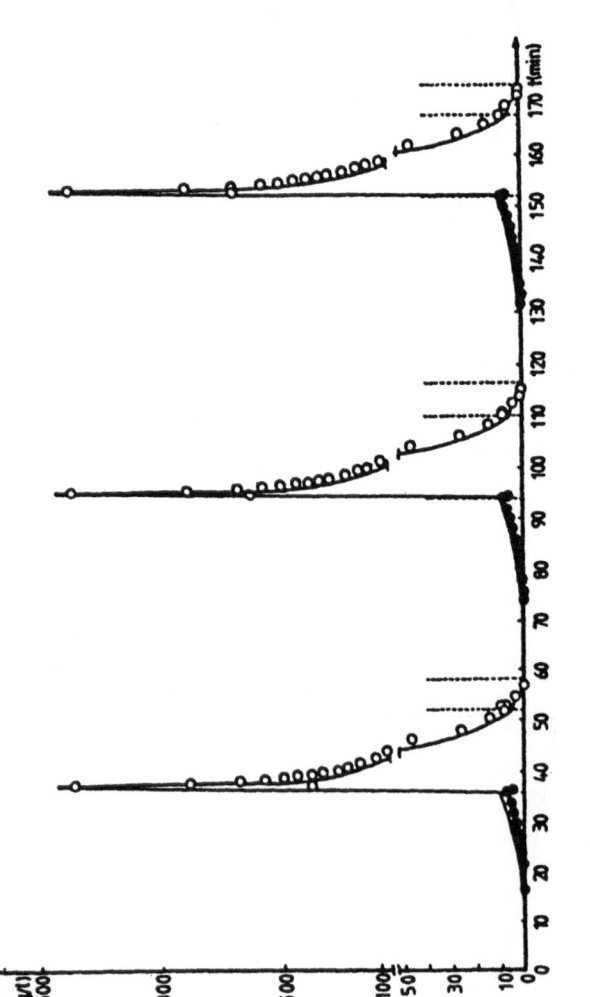

Figure 18 – Cyclic operation of a fixed bed adsorber:saturation,regeneration and washing.

● phenol   o phenate

CYCLIC OPERATION

Adsorption cycles involve three steps: saturation, regeneration and washing. Now we are able to develop a package for cyclic adsorption |12| since we already modelled saturation and regeneration steps.

Experimental and calculated cyclic runs are compared in Figure 18.

CONCLUSION

The design methodology mentioned above is based on the modelling of fixed bed adsorbers. It is used as a learning tool (to understand the effect of the various model parameters) and also as a predictive tool. This means that model predictions should correspond to the real behaviour of a fixed bed either at the bench scale or at the industrial scale.

In order to achieve these objectives model parameters are determined by simple independent experiments. Scale-up will eventually influence hydrodynamic parameters.

It is also important to stress that adsorption processes are cyclic in nature. Few studies on the regeneration step have been done. However, a robust package for cyclic processes should include models for saturation as well as for regeneration.

The same methodology has been extended for multicomponent liquid phase adsorption |13| and to the separation of copper, zinc and lead with chelating resins |14|.

NOTATION

a_p - specific area of the particle
c - outlet concentration
c_o - inlet concentration
d_p - particle diameter
D^{ax} - axial dispersion coefficient
D^e - effective diffusivity
Da - Damkholer number
K_{eq} - equilibrium constant
K^L - parameter in the Langmuir equation
k_f - film mass transfer coefficient
L - column length
N_f - number of film mass transfer units
N_d - number of intraparticle mass transfer units
Pe - Peclet number
$q*$ - adsorbed phase concentration in equilibrium with c*
Q - adsorbent capacity
r_o - radius of microspheres
R - particle radius
$R*$ - reduced radial coordinate for the particle
Re - Reynolds number

Sh $-$Sherwood number
Sc $-$Schmidt number
t $-$time
t_{st} $-$stoichiometric time
u_i $-$intersticial velocity
U^i $-$flowrate
x $-$fluid phase concentration
x_p $-$ fluid concentration inside pores

Greek symbols
ε $-$bed porosity
ε_p $-$intraparticle porosity
τ $-$space time
ξ $-$capacity factor
θ $-$reduced time (t/t_{st})
ρ_{ap} $-$apparent density of the adsorbent

Superscripts
F $-$phenol
FN $-$phenate
S $-$sodium hydroxyde

REFERENCES

1. S.Carrà.Reaction Processes Involving Ion Exchange Resins.NATO
   ASI Ion Exchange:Science and Technology,Troia(1985)
2. Tondeur,D. and M.Bailly.Design Methods for Ion Exchange Processes
   Based on the Equilibrium Theory.NATO ASI Ion Exchange:Science and
   Technology,Troia(1985)
3. Helfferich,F. and G.Klein. Multicomponent Chromatography(Marcel
   Dekker,New York,1970)
4. H.K.Rhee.Equilibrium Theory of Multicomponent Chromatography
   in Percolation Processes:Theory and Applications(Sijhtoff and
   Noordhoff,Alphen aan den Rijn,1981)
5. Martins,J.,Loureiro,J.,Costa,C. and A.Rodrigues.Recovery of
   Tungsten from Hydrometallurgical Liquors by Ion Exchange in
   Ion Exchange Technology edited by D.Naden and M.Streat (Ellis
   Horwood,London,1984)
6. C.Costa. Dynamics of Cyclic Processes:Adsorption and Parametric
   Pumping. Ph.D.Thesis.University of Porto (1983)
7. Myauchi,T. and T.Nomura. Liquid Film Mass Transfer Coefficient
   for Packed Beds in the Low Reynolds Number Region. Int.Chem.Eng.
   12(1972)360
8. Tien,C. and G.Thodos.Ion Exchange Kinetics:The Removal of Oxalic
   Acid from Glycol Solutions. Chem.Eng.Science. 13(1960)120
9. Rodrigues,A. et al. Computational Aspects of the Dynamics of
   Sorption Operations. CACE'79,Montreux(1979)
10. Costa,C. and A.Rodrigues. Design of Cyclic Fixed Bed Adsorption
    Processes. Part I-Phenol Adsorption in Polymeric Adsorbents.
    AIChEJ,September 1985.

11.Costa,C. and A.Rodrigues. Regeneration of Polymeric Adsorbents
   in a CSTR. Chemical Engineering Science.40(1985)707.
12.Costa,C. and A.Rodrigues.Design of Cyclic Fixed Bed Adsorption
   Processes.Part II-Regeneration and Cyclic Operations. AIChEJ,
   September (1985)
13.Costa,C. and A.Rodrigues.Adsorption/Desorption of Phenol and
   m-cresol Mixtures in Fixed Beds of Polymeric Adsorbents:Modelling
   and Experimental Study. Accepted for presentation at the Eng.
   Foundation Conference.Santa Barbara(1986).
14.J.Loureiro. Separation of copper,zinc and lead with chelating
   resins.Ph.D.Thesis .Univ.Porto(1985).

COUNTER-CURRENT ION EXCHANGE.

J.A.Wesselingh and A.P.van der Meer*

Delft University of Technology
2628 BL Delft,Netherlands
*Billiton Research b.v.
PO-box 40,6800 AA Arnhem,Netherlands

1. INTRODUCTION.

1.1. The process and its competitors.

Counter-current ion exchange is a continuous steady state process to separate ions from dilute solutions. The main features are shown in figure 1. A feed stream containing an ion B passes through a loading column. Here it is contacted with a stream of ion exchange particles in a counter-current fashion. The ion exchange particles or resin, contain another ion A which is exchanged. The resin loaded with B is then returned to a regeneration column. There it is contacted with a concentrated solution of A and it releases B at a high concentration. This concentrate can then be further processed by other means. The resin is transported by the process liquids. To avoid contamination between the loading and regeneration sections rinsing sections are required between the two columns.

Continuous ion exchange can be an attractive process if the amounts of material to be removed are substantial, but the concentration very low. If the amounts to be removed are small, conventional fixed bed ion exchange is usually to be preferred. For higher concentrations liquid/liquid extraction (which is a very similar process) usually wins. Continuous ion exchange has the disadvantage of requiring solids handling and rinsing. On the other hand there are no solvent losses due to dissolution in the feed and the process can handle substantial amounts of fine solids suspended in the feed.

290

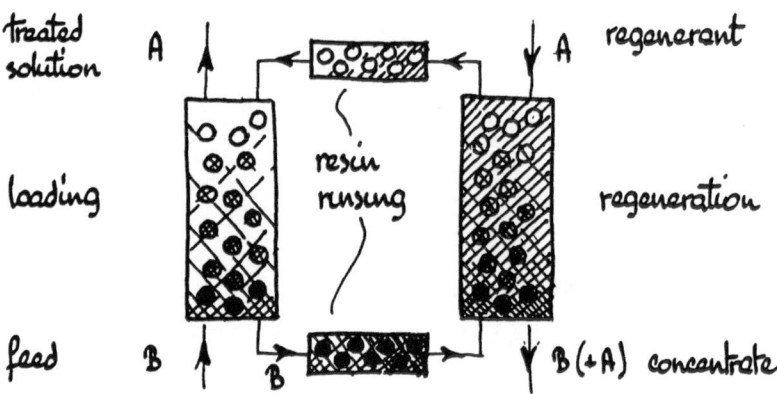

figure 1. Counter-current ion exchange.

Counter-current ion exchange has found a few large applications, of which uranium winning is the most notable (1,2).

1.2. Scales of scrutiny in the equipment.

Designing a separation process involves bringing together a
number of very different subjects. One way of classifying these is
by the scale involved (figure 2).

The largest scale the designer will normally encounter is that
of the complete plant. Dimensions there are of the order of 100
metres. The next lower level is that of the individual units; in
our case separation units. Here dimensions are around ten metres.
In the separation units one can often distinguish stages, with
dimensions of around one metre. These might be stirred vessels,
trays or fluidized beds.

Inside the stages two phases are flowing; the liquid and the
ion exchanger. When viewed from a distance the flow resembles that
of a single phase with jets, turbulence and large scale circu-
lations. These have dimensions smaller than those of the stages,
say between one centimetre and one metre. The local convective flow
around the ion exchange particles is of the same scale as the
particle; around one millimetre. The convective flows die out
towards the particle surface. At distances smaller than about ten
micrometres they become unimportant and transport of ions to and
from the particle is taken over by diffusion. This is governed by
processes on the scale of a few molecules, which is again much
smaller. Inside the particles the main resistance to diffusion can
often be thought as concentrated in a thin layer of some tens of
micrometres. The details of the diffusion processes are again
governed by phenomena on a colloidal or molecular scale (between
100 and one nanometre).

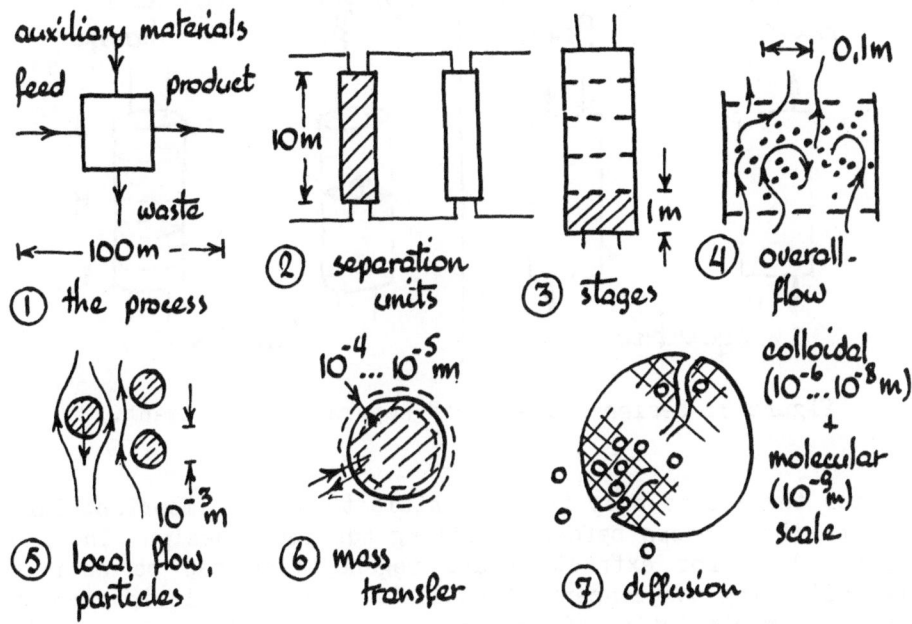

figure 2. Scales of scrutiny in separation equipment.

The largest scale involved has connections with real world
problems such as the materials situation, the economy, the environment and with aspects of plant construction. The smallest
scales are connected to the chemistry and physics of the ion
exchanger. The intermediate scales are the realm of the chemical
engineer. To describe them he draws heavily on the subjects of
phase equilibria, diffusion and fluid flow, supplemented by a
considerable amount of empiricism. It is this area that we consider
here.

1.3. The design sequence.

When designing a separation process the chemical engineer
usually follows a fairly well defined sequence (figure 3).

The first step is to define the feed and the required separation. This is not always as simple as might seem, but here we
will assume this step has already been undertaken. The second step
is the choice of the separation process: here taken to be continuous ion exchange. A very important factor is the choice of the
separating agent (the ion exchanger). Especially the phase equilibria have a large effect on the ultimate process design.

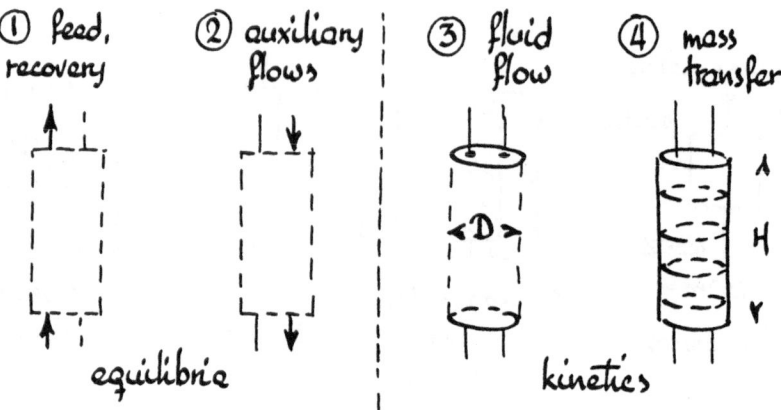

figure 3. Design sequence of separation equipment.

After the second step it is possible to obtain important information on the process before detailing equipment design. The flows
required, both for extraction and regeneration can be estimated
quite well using the concept of equilibrium stages. It is here that
the main characteristics of the process are set. It is important to
realize that this step only requires equilibrium data. These can be
obtained quite easily in the laboratory. The information obtained
now is often sufficient to ascertain the feasibility of a process.
Only if the result looks promising do we go on.

Now comes the overall design of the equipment. This can often
be divided in two parts. In the first we calculate those dimensions
which determine the capacity of the equipment. This is often mainly
a matter of sufficient cross sectional area. In these calculations
the theory of two phase flow plays a role.

The next step is to determine the volume necessary for the required mass transfer between the phases. Together with the cross
sections this fixes the length of the equipment. This step is the
most complicated and uncertain. It requires knowledge of the flow
patterns and diffusional processes in the equipment. It can be
bypassed to a large extent by pilot plant development.

These steps: the calculation of
- the equilibrium flows,
- the hydrodynamics and
- the mass transfer
set all the main dimensions of the process and the equipment. The
remaining part of this article is devoted to separate chapters on
these subjects.

Many other steps will have to follow, such as the design of the
fluid and resin transport systems, the control system, the mecha-

nical details and the layout and structural support of the system. However the previous steps are sufficient for a broad outline of the separation process.

Some aspects of the design cannot be determined well theoretically. They include:
- the chemical and mechanical stability of the resin,
- the accumulation of contaminants in the process which may "poison" the resin and
- fouling of the resin and equipment.
To assess these factors either experience on similar installations or pilot plant work is required. Otherwise continuous ion exchange is a fairly predictable process.

2. EQUILIBRIUM FLOWS.

2.1. Ion exchange equilibria.

The backbone of an ion exchanger is a permeable solid matrix, usually an organic polymer or resin (figure 4). To this matrix are attached fixed charged ionic groups. They are negative for cation exchangers, positive for anion exchangers. These groups make the polymer hydrophilic; it will take up water to about one half of its volume. The counter-ions opposite the fixed charges can diffuse more or less freely inside the particle. The ionic concentration inside the particle is high: two to four eq/l. If such a particle is immersed in pure water a few of the counter-ions will escape. This generates an electric charge which keeps the others inside the particle.

In a dilute solution with the same counter-ions there will also be "co-ions" with the same sign of charge as the fixed groups in the exchanger. These co-ions are to a large extent excluded from

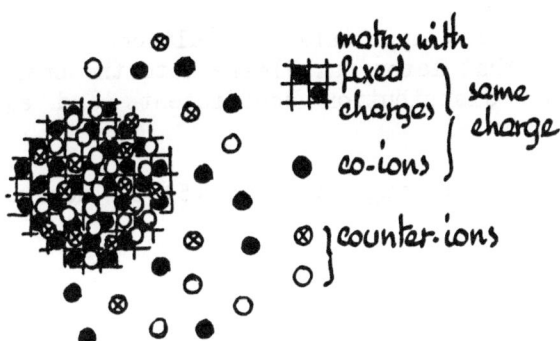

figure 4. Ions in exchanger and surrounding solution.

the inside of the particle by the fixed charges; they play only a minor role in the internal equilibria. Different counter-ions in the liquid however, can exchange with the counter-ions in the particle. It is this property which makes ion exchangers so useful.

At equilibrium the different counter-ions will distribute themselves in different proportions over resin and liquid. The equilibrium can be characterized by a set of distribution coefficients:

- the ratio of volumetric concentrations $\quad m_B = \dfrac{\tilde{c}_B}{c_B}$ (1)

- or the ratio of equivalent fractions $\quad K_B = \dfrac{y_B}{x_B}$ (2)

with

$$y_B = \frac{\tilde{c}_B}{\tilde{c}} \qquad\qquad x_B = \frac{c_B}{c} \tag{3}$$

Here c and $\tilde{c}$ are the total equivalent concentrations of the counterions in the liquid and resin respectively. (Values with a subscript are those of a certain species). The distribution coefficients depend on the liquid composition and (usually weakly) on the temperature. Here we only discuss one of the very simplest models to describe this behaviour.

The exchange of two equally charged species often behaves as a simple chemical equilibrium:

$$B^+ + \overline{A}^+ \rightleftharpoons \overline{B}^+ + A^+ \tag{4}$$

with the equilibrium constant:

$$K_{BA} = \frac{\tilde{c}_B c_A}{c_B \tilde{c}_A} \quad \text{or} \quad \frac{\tilde{c}_B}{c_B} = K_{BA}\frac{\tilde{c}_A}{c_A} \tag{5}$$

This equation is also applicable to multicomponent equilibria; we will make use of that later on. Please note the small difference in notation between distribution coefficients and equilibrium constants.

For a binary rewriting equation (5) in equivalent fractions yields:

$$y_B = \frac{\tilde{c}_B}{\tilde{c}} \qquad\qquad x_B = \frac{c_B}{c}$$

$$y_A = 1 - y_B = \frac{\tilde{c}_A}{\tilde{c}} \qquad x_A = 1 - x_B = \frac{c_A}{c}$$

so

$$\frac{y_B}{1-y_B} = K_{BA} \frac{x_B}{1-x_B} \tag{6}$$

This is a hyperbola (figure 5). The equilibrium constant is a measure of the preference of the resin for component B over component A. The larger its value the more selective the resin is for B. The magnitude of the equilibrium constant can vary considerably; a few rough values are given in figure 5. Some resins have a very high affinity for certain specific ions. This is the case for weak acid and weak base exchangers which have a very strong affinity for H^+ and OH^- ions respectively. With these resins adjusting the pH can cause them to effectively discharge the whole of their content of other counter-ions. Very high equilibrium constants are also encountered for certain metal ions in chelating resins. These high selectivities are not always beneficial; they can make regeneration difficult!

For low concentrations of B equation (6) yields a linear equilibrium:

$$\frac{y_B}{x_B} = K_{BA} = K_B \tag{7}$$

Here the distribution coefficient in equivalent fractions is equal to the equilibrium constant. This is not true in general. In terms of volumetric concentrations:

$$m_B = \frac{\bar{c}_B}{c_B} = K_{BA} \frac{\bar{c}}{c} \tag{8}$$

The value of $\bar{c}$ is a property of the resin, but that of c in the liquid can be varied. This makes changes in the volumetric dis-

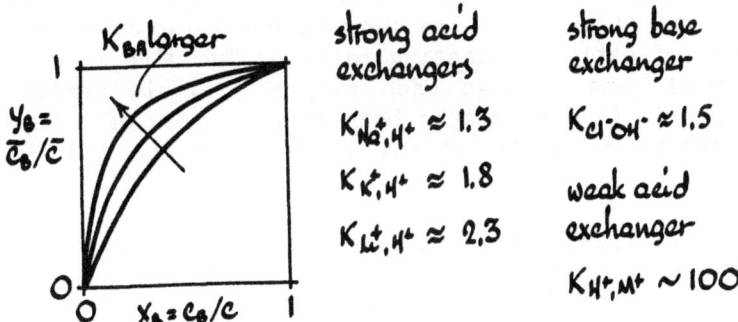

figure 5. Simple ion exchange equilibria.

tribution coefficient possible. We use these for regenerating the resin; it loads B in a dilute solution where m_B has a high value and can discharge it at a higher concentration in a concentrated solution.

Real equilibria of ion exchangers show an enormous variation. (You will find the many reasons for that in the classic work (3) on ion exchange). In general the equilibrium constants are functions of all concentrations in the system and of the temperature. Especially for multicomponent systems data usually will have to be gathered experimentally on the system of interest and in the range of concentrations expected in the plant. These will then be fitted to some equation for interpolation in design calculations. Attempts are being made at deriving general models for predicting equilibrium behaviour (4), but the situation is much less satisfactory than with other types of equilibria such as between gases and liquids. Because ion exchange systems show an enormous variation and are not used so extensively this situation is not likely to change rapidly.

In the next paragraph we will use the very simple description of equilibria given by equation (5) and neglecting composition effects. Use of the general equations for the phase equilibria greatly complicates the mathematics involved in design work. It does not change the principles however.

2.2. An equilibrium loading column.

One of the workhorses of the chemical engineer is the concept of the equilibrium stage; a piece of equipment where the streams leaving are thought to be in equilibrium. Such stages only exist in our mind; in a real stage there is no equilibrium. Equilibrium stages do often approximate the behaviour of real stages however. The utility of the concept lies in the fact that it allows us to make estimates on process and equipment behaviour without considering what happens inside.

As an example of the use of equilibrium stages a continuous loading column (figure 6). The flow is counter-current wise, which is the most effective contacting pattern. Two streams of exchangeable material pass through each stage; a liquid stream L and a solid stream S. In our example both have a value of 1000 eq/s. The streams consist of contributions l and s of the individual components.

The stream leaving the loading column at the right contains a remnant $l_1 = \frac{1}{2} \cdot 1$ of Ca^{2+} ions. The rest has been exchanged by Na^+. The Ca^{2+} equilibrium for low fractions of Ca2+ is described by:

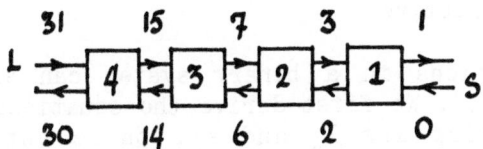

figure 6. Ca^{2+} streams in a loading column.

$$y = 2x \qquad \text{or} \qquad \frac{s_1}{S} = 2\frac{l_1}{L}$$

(we drop the subscript Ca^{2+} for simplicity). So the solid stream leaving the first stage is

$$s_1 = 2\frac{S}{L}l_1 = 2 l_1 = 2$$

But then a mass balance over the stage tells us that

$$l_1 + s_1 = l_2 + s_0$$

The resin entering contains no Ca^{2+}; $s_0 = 0$. So $l_2 = 3$. The equilibrium relation for stage 2 then tells uas that

$$s_2 = 2 l_2 = 6$$

You should be able to construct the other streams yourself. You see that four stages have reduced the feed concentration by a

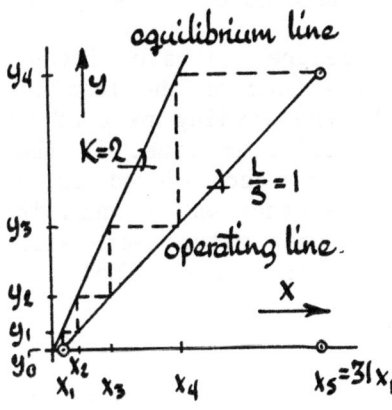

figure 7. Graphical solution to the Ca^{2+} loading column.

factor of 31. It is easily seen that lower values of S make the separation less effective.

Stage compositions in a binary system can also be followed graphicly (figure 7). We first derive the equations for a stage n. The streams leaving are l_n and s_n, those entering l_{n+1} and s_{n-1}. The compositions of the streams between two stages are again related by a mass balance:

$$l_{n+1} + s_0 = l_1 + s_n$$

or

$$\frac{s_n}{S} = \frac{s_0}{S} + \frac{L}{S} \frac{l_{n+1} - l_1}{L}$$

or

$$y_n = y_0 + \frac{L}{S}(x_{n+1} - x_1) \qquad (9)$$

This is a straight line through a point (x_1, y_0) with a slope L/S. So we can construct this line if we know or choose L, S and the compositions on one side of the column.

The compositions of streams leaving an equilibrium stage are related via the equilibrium line:

$$y_n = K x_n \qquad (10)$$

The graphical method follows the previous numerical one closely. Only we compute compositions instead of streams. From the known value of x_1 we read off the y_1 from the equilibrium line. The value of x_2 then follows from y_1 and the mass balance or operating line. This stepping procedure allows one to calculate all compositions between the consecutive stages.

The graphical procedure can be used for non-linear equilibria. Figure 8 gives the behaviour of the loading and regeneration in a continuous system. In the loading part of the cycle the operating line is below the equilibrium line. In the regeneration - where transfer of the component considered is from the resin to the liquid - it is just the other way around. The resin compositons at the ends of the two columns match. Note also the difference of the slope L/S of the operating lines. That of the regeneration will usually have to be higher than in the loading. For the type of equilibrium shown approximate limits to the flow rates are:

- loading: $L/S \lesssim 1$ (slope of diagonal),

- regeneration: $L/S \gtrsim K_{BA}$ (initial slope of equilibrium line)

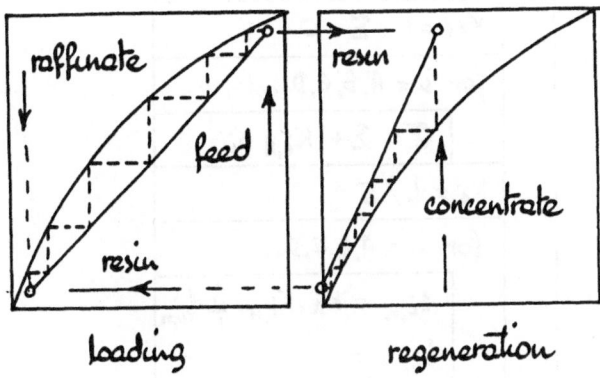

figure 8. Compositions in a binary loading/regeneration cycle.

This means that resins with a high K_{BA} will need a large regenerant flow. Outside these limitations the choice of the flow ratio's is up to the designer. As can easily be seen flows close to the limiting values require many stages. This means a high investment. Larger resin and regenerant flows on the other hand mean more chemicals and higher operating costs. The most economic operation usually turns out to be close to the limiting flows.

Real stages are not equilibrium stages. The compositions usually change less than they would on an equilibrium stage (figure 9). In binary exchange this can be taken into account by a Murphree efficiency. This is a ratio of the change in composition in the real stage over that in an equilibrium stage. The ratio can either be taken over the resin or over the liquid. In general the two values are not equal. We come back to the estimation of these efficiencies in chapter 4.

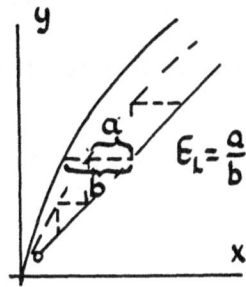

figure 9. Murphree efficiency of a real binary stage.

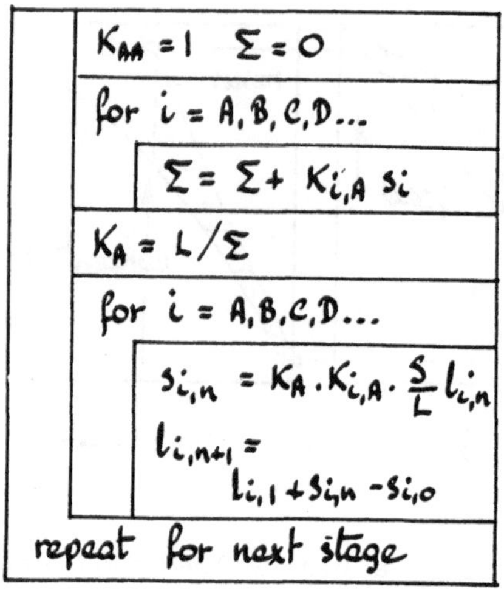

$$K_{AA} = 1 \quad \Sigma = 0$$

for $i = A, B, C, D \ldots$

$$\Sigma = \Sigma + K_{i,A} \, s_i$$

$$K_A = L/\Sigma$$

for $i = A, B, C, D \ldots$

$$s_{i,n} = K_A \cdot K_{i,A} \cdot \frac{S}{L} \, l_{i,n}$$

$$l_{i,n+1} = l_{i,1} + s_{i,n} - s_{i,0}$$

repeat for next stage

figure 10. Stage algorithm for multi-component separation.

2.3. Multi-component systems.

Graphical procedures are not easily extended to multi-component separations, but the numerical methods can be adapted. In most cases the equilibrium calculations are then iterative, because the distribution coefficients are implicit functions of the phase compositions. However for equilibria such as given by equation (5) with constant equilibrium constants (!) a simple non-iterative algorithm exists (figure 10). It is a little tedious to use by hand but even a programmable calculator will suffice.

Using this algorithm we have worked out an example of the behaviour of different components in a hypothetical four component system (figure 11). The components and the equilibrium constants used are:
- the regenerant A (also reference component),
- the main component B to be separated; $K_{BA} = 2$
- a small amount of a component C which has a small affinity to the resin; $K_{CA} = 0.2$, and
- a small amount of a component D which has a large affinity to the resin; $K_{DA} = 20$.

On the whole component B follows the same pattern as in a binary separation, although the separation is a little more difficult. Component C builds up a relatively small equilibrium flow in the resin phase; this stream is completely removed in the regenerator. Component D is a nasty one. Initially it is completely removed in the loading, but not in the regeneration. So in the process it builds up to an equilibrium value where it occupies a substantial part of the resin.

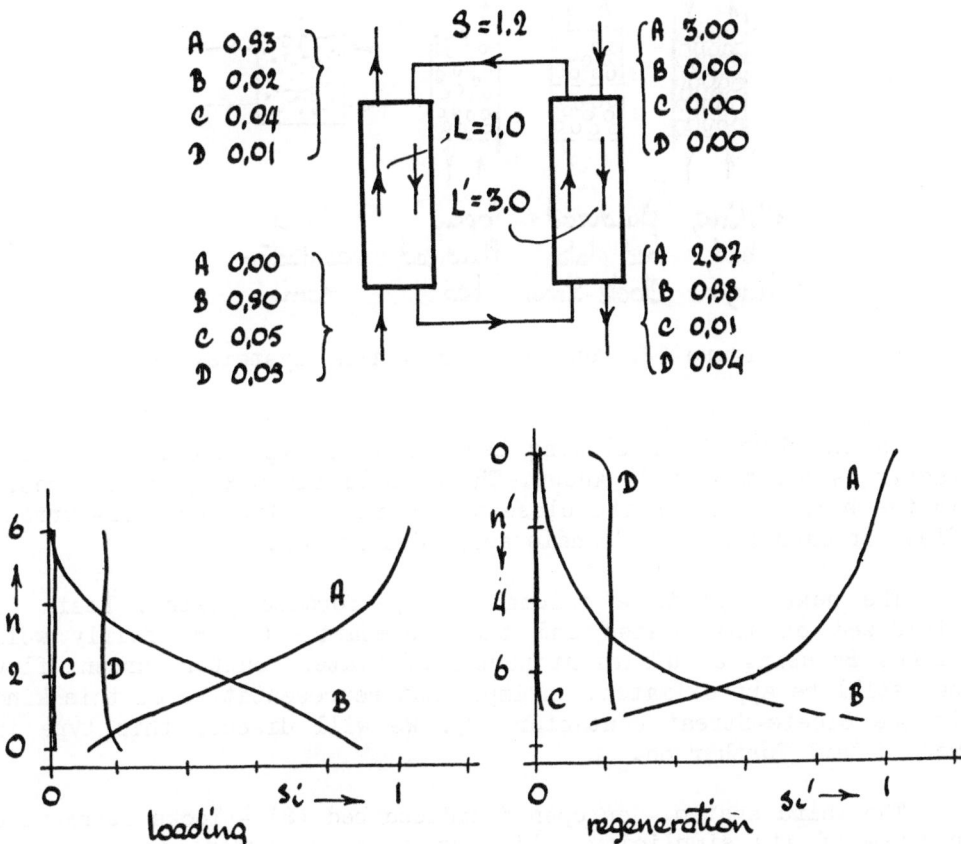

figure 11. Flows in a four component separation.

For the calculations in this paragraph we have used the "stage-to-stage" method. This method requires knowledge of all flows at one extremity of of the system. For a binary exchange these can be chosen independantly by the designer. In multi-component separations this is not so, and estimating a good set of flows is a real problem. Modern computer programs therefore make use of a "component-by-component" method. For further information the reader is referred to (5). The effects of the choice of the streams and the numbers of stages in the two sections is discussed in (6).

3. HYDRODYNAMICS.

3.1. Counter-current systems.

A large number of counter-current systems for continuous ion exchange have been proposed. They can be classified in a few major groups (figure 12).

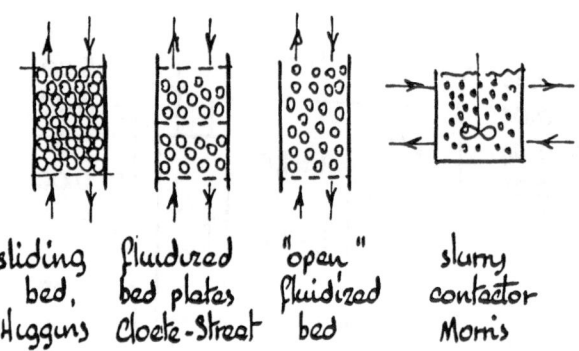

figure 12. Four counter-current systems.

On the left is a sliding bed. Of this type the Higgins contactor is the most well known. There is little mixing of the resin in the bed, so it has the closest approach to true counter-current flow. It cannot however handle suspended solids.

The next type is a column with perforated plates. Resin is fluidized on the plates and these compartments are fairly well mixed. By using a sufficient number of plates counter-current flow can still be approximated. An important representative of this kind is the Cloete-Streat contactor (7). We will discuss this type in more detail further on.

The third system - an open fluidized bed (8) - looks attractive because of its simplicity. This can be an important advantage in fouling systems. However the mixing of the resin phase makes it less suitable for sharp separations. Also the resin mixing is probably rather scale dependant and not yet well predictable.

The last system is a stirred slurry contactor (9). Both phases are continually redistributed over the volume and have a more or less homogeneous composition. To obtain a counter-current flow pattern a number of these systems is placed in series. In between the tanks are devices to separate the liquid and solids and send them in opposite directions. These might be settlers, plate interceptors or hydrocyclones. This system can use fine ion exchange particles, allowing high transfer rates. This is not practical for the fluidized bed types; the low falling velocities of small particles would give them too low throughputs.

3.2. Fluidized plate columns.

In the rest of this chapter we restrict ourselves to the Cloete-Streat type of column. This is not a true steady state system; the solids flow through the column is intermittent. However

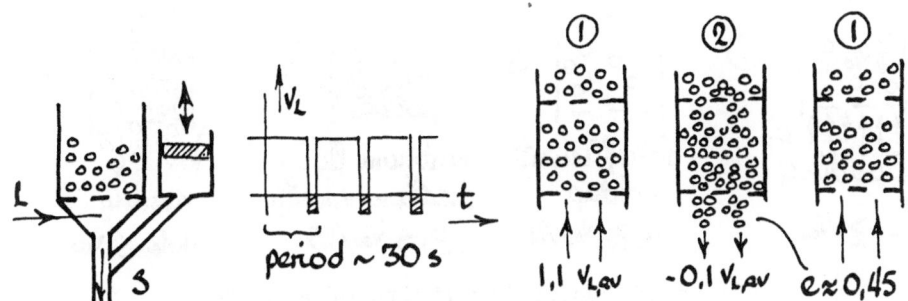

figure 13. Periodic solids removal.

the overall result does not deviate greatly from true steady state counter-current behaviour.

The resin transport is effected by periodically stopping or reversing the liquid flow. This allows part of the resin bed to drain to the next lower plate (figure 13). The resin from the bottom plate is transferred to the regenerator; an equal amount is added to the top plate. In the loading section the volumetric flow of the resin might be only one hundredth to one thousandth that of the liquid. So rapid pulsations are not required; the period is in the range of 10...1000 s.

The transfer rate is determined by the pulse volume and frequency, and the volume fraction of the resin in the transferred material. This last value is usually between 0.4 and 0.5 and almost independant of of conditions on the plate.

After the transfer the resin bed expands again. Any excess of resin on the plate is sucked back into the next higher compartment by the higher liquid velocities near the holes in the plate. This leaves a clear liquid layer of one or two hole diameters above the bed.

To avoid clogging the hole diameter should not be too small; we have often used 12 mm holes. In the laboratory holes as small as 3 mm are used; these do not clog with commercial resins with average particle sizes around 0.7 mm.

3.3 Operational limits.

Four possible limits to the range of operation of a plate are shown in figure 14. The first is "flooding"; entrainment of particles to the next higher tray because the liquid velocity exceeds the falling velocity of a particle. Here this is not a serious problem; the particle holdup on the plate becomes too low before entrainment occurs to any great extent. Only the finest particles –

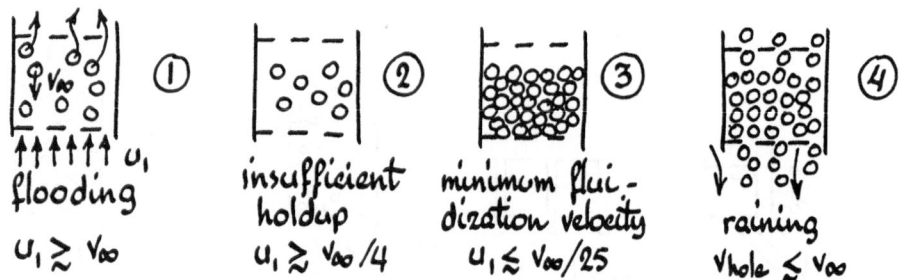

figure 14. Operating limits of a plate.

a small fraction of the resin - will be lost by this mechanism during the starting up of the plant.

The main upper limitation to the flow rate is the requirement of sufficient particle holdup on the plate. This is not a strictly defined criterion; higher velocities mean a smaller cross section but a disproportionally larger column height to accomodate sufficient resin for mass transfer. As a rough guide we suggest not operating at superficial liquid velocities higher than one quarter of the free falling velocity of the average particle.

There are two possible lower limits to the operation: the minimum fluidization velocity and the occurrence of "raining" of particles through the holes in the upflow period. The minimum fluidization velocity is about one twenty fifth of the free falling velocity. As one approaches this point parts of the bed become more or less stagnant.

"Raining" occurs when the the liquid velocity in the holes is smaller than the free falling velocity of the largest particles. Below the raining point the resin content of the column is no longer under control. Raining can be avoided by choosing a small free (hole) area. In our experience the hole area should not be more than about five percent of the plate area. Too small free areas give a poorer fluidization, but we do not know the lower limit.

3.4. Particle and swarm hydrodynamics.

Commercial ion exchange particles are very nearly spherical. Their size distribution is approximately Gaussian with a standard deviation of about one fifth of the arithmetic particle diameter. This means that only a small fraction of the particles falls outside the range $0.5 \bar{d} < d < 1.5 \bar{d}$.

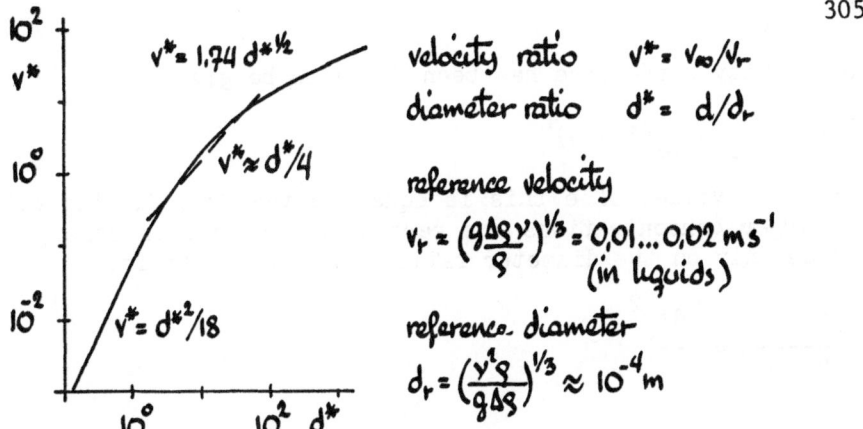

figure 15. Terminal velocity of a single particle.

Free falling velocities of a single particle are given in figure 15. Plotted along the vertical axis is the ratio v^* of this velocity v_∞ to a reference value which only contains physical properties and the gravitational acceleration. For particles in liquids this reference velocity usually has a value between 0.01 and 0.02 m s^{-1}. Along the horizontal axis is plotted a diameter ratio d^*. The reference diameter used there usually has a value of about 10^{-4} m. In the left part of the graph the velocity is proportional to the diameter squared; on the right side it is proportional to the square of the diameter. Ion exchange particles usually fall in the intermediate regime $3 < d^* < 30$ where the particle velocity is proportional to the diameter:

$$v^* = 0.25 \ d^* \tag{11}$$

Free falling velocities are then around 0.02...0.03 m s^{-1}.

The beds on the trays are not homogeneously fluidized; they might be called spouted beds. Even so the holdup relations follow those for homogeneous fluidization quite closely.

In a homogeneous swarm the behaviour is determined by the fluxes or superficial velocities of the two phases:
- u_1 for the fluid and
- u_s for the particles.
(We reckon both positive when they are upwards).

The particles occupy a volume fraction e and the average local velocities of the two phases are:

$$v_1 = \frac{u_1}{1-e} \qquad v_s = \frac{u_s}{e} \tag{12}$$

The velocity difference has been found to be given by

$$v_1 - v_s = v_\infty (1 - e)^{n-1}$$ (13)

(For small values of e this is equal to the free falling velocity; for larger concentrations the particles hinder each other). Here n depends only on the diameter ratio d* of the particle:

$$\frac{5 - n}{n - 2.5} = \left(\frac{d*}{7}\right)^2$$ (14)

For small particles n has a value of 5, for large particles a value of 2.5. (This is basicly a rearrangement of the well known relation of Richardson and Zaki (10) for fluidization; see also (11)).

Equation (13) gives the relation between the holdup and the fluxes. It is a nasty non-linear equation which has to be solved numericly or graphicly. A graphical solution is given in figure 16. As an example of its use: a resin with d* = 10 and $v_\infty = 3 \times 10^{-2}$ m s^{-1}. The particle flux ratio is small; $u_s/v_\infty = 0$. For a liquid flux ratio $u_1/v_\infty = 0.20$ the resin volume fraction is seen to be 0.39. Another example: at the minimum fluidization velocity the resin volume fraction is about 0.6. For the resin above this 0.6 is seen to correspond to a liquid flux ratio of about 0.045.

The origins of this graph are to be found in (13). It works as follows. Inserting equations (12) in (13) yields

$$\frac{u_1}{1-e} - \frac{u_s}{e} = v_\infty (1 - e)^{n-1}$$

or after rearranging:

$$\left(\frac{u_1}{v_\infty}\right) e + \left(\frac{-u_s}{v_\infty}\right)(1 - e) = e (1 - e)^n$$ (15)

The left hand side of the equation is a straight line in the graph. Its intersections with the non-linear curve of the right side are the solutions of the equation.

These data should be sufficient to solve most questions on the hydrodynamics of the plates.

3.5. Stability of the holdup.

The tray contents do not always automaticly adjust to proper values. The reason is shown in figure 17. If a given plate has a

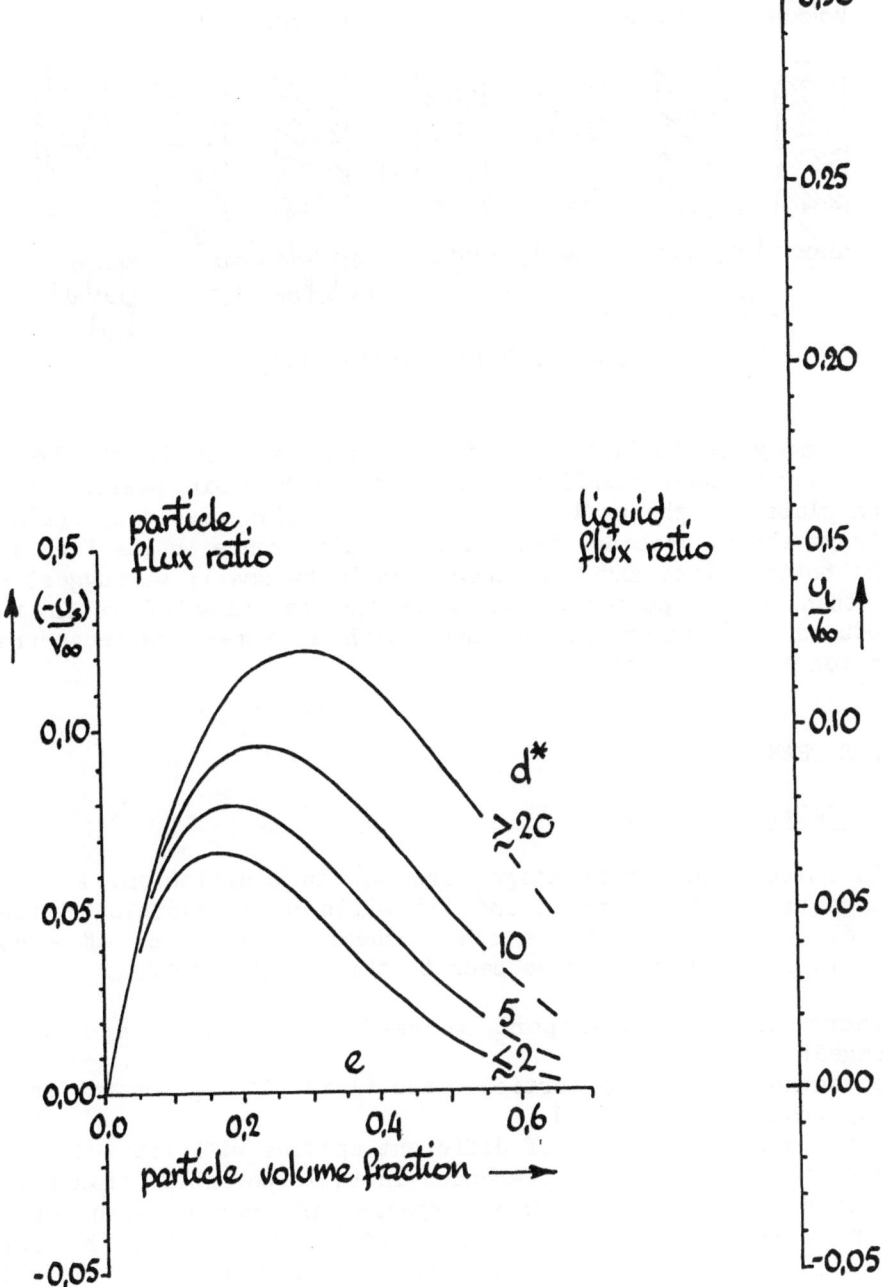

figure 16. Flux/holdup relations in swarms of particles.

larger hole area than the plate above it will discharge more resin than it receives. Such a plate will empty. In a continuous system even minor differences in the plate geometries can lead to un-predictable resin distributions.

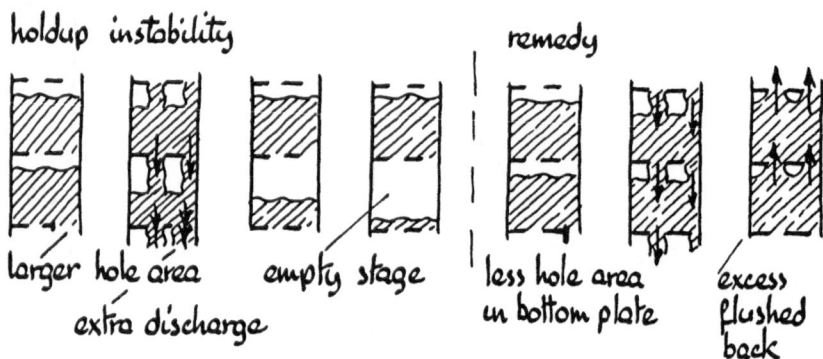

figure 17. Holdup instability.

A remedy is to have slightly less hole area in the bottom plate. During each downflow period too much resin passes to the bottom plate. In the upflow period this is then passed up again to the trays above, keeping them full. The difference in the free area of the bottom plate and the others should be small; we suggest not more than a few percent. Otherwise the backflow of resin this introduces will interfere too much with a proper counter-current operation of the column.

4. MASS TRANSFER.

4.1. Models.

As noted before real stages are not in equilibrium. A finite exchange rate between resin and liquid is only possible if there is a difference between the electrochemical potential of a component in each phase. We come back to this in chapter 4.4.

There are two transport mechanisms for the ions to be exchanged:
- convection, caused by particles and fluid flowing towards or away from each other and
- diffusion; the movement of different species with respect to
each other because of differences in their potential gradients. Convection is not "specific"; all species are carried along at the same rate by the liquid or the particle. Convection is rapid; typical convection rates in ion exchange equipment are of the order of 10^{-2} m s^{-1}. However, unless the ion exchanger is swelling or shrinking there is no liquid convection through the interface between the two phases, and neither inside the particles. So there diffusion must take over. Diffusion velocities are very small; around 10^{-5} m s^{-1} in the liquid and up to several orders of magnitude lower in the resin. Appreciable amounts of material can only be exchanged if a large interphase area is available.

Both convection and diffusion are complicated processes
(13). At present we are not able to describe ion exchange in real
equipment starting from fundamental equations. So the engineer
"models" the phenomena. He looks for a schematic description which
- he hopes - will contain the main mechanisms and phenomena he is
interested in. The model should be so simple that it is tractable
both mentally and mathematicly. Models are not unique; the same
situation can be modelled in many ways. Models are always more or
less incorrect and that also applies to those presented here!

The course of this chapter is to set up models for
- convective transport on a plate,
- the interfacial area between the phases,
- the diffusional processes and
- the diffusional resistances in the two phases.
These models are then combined to allow prediction of the stage
efficiencies. Combining these efficiencies with the equilibrium
calculations of chapter 2 makes an estimate possible of the number
of real plates required. That ends the determination of the overall
dimensions of the separation equipment.

4.2. Convective flow (mixing).

The resin spends a relatively long time on the plate; much
longer than the liquid. If the fluidization is not too bad each
particle will move all over the plate during its stay. So on the
average the resin composition should be the same everywhere on the
plate. This does not apply to each particle separately; a particle
that has just entered will have a different composition from one
that has been on the plate for longer. However these differences
are usually ignored. There are reasons to believe that the errors
involved by neglecting this "segregation" are not of great importance (14), and it simplifies things tremendously. The resin
composition on a plate does change during the upflow/downflow
cycle. This enhances mass transfer compared to a true steady state
(see (15)). However to keep things simple, we shall neglect this
effect.

Small zones on the plate between the holes are stagnant. But
these zones only comprise a few percent of a properly designed
plate. So we will regard the resin phase as well mixed, with the
same composition everywhere.

The behaviour of the liquid is more complicated. On a plate
with a low bed (say 10 cm, but it also depends on the geometry)
spouts of liquid and particles can be seen projecting through the
bed from the holes in the plate. Pulse response experiments
indicate that a considerable part of the liquid passes the tray

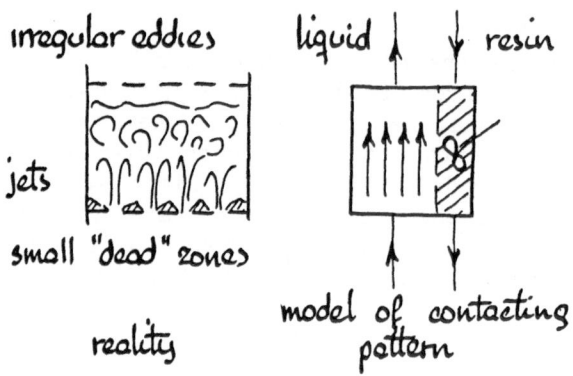

figure 18. Contact pattern on a plate.

rapidly (8). It is not surprising that colouring matter shows that these high velocity areas are the jets above the holes. Bypassing is deleterious for tray performance. Also it is not very predictable. So it is probably best to try to avoid it.

If the bed height is increased - in our measurements to above 0.2 m - the spouts disappear. Pulse response experiments show a more uniform passage time for the liquid. This indicates that the liquid is approaching the ideal pattern for counter-current contacting: that of plug flow (figure 18).

A model in which the liquid on the plate is assumed to be well mixed is not very sound physically. It is however very simple to apply and therefore popular. The predictions using this model are conservative compared to those of the plug flow model.

4.3. The interfacial area.

A spherical particle has a surface to volume ratio

$$\frac{\pi d^2}{\pi/6 \, d^3} = \frac{6}{d} \tag{16}$$

For a dispersion of identical spheres the interface per volume of dispersion is then

$$a' = \frac{6e}{d} \tag{17}$$

where e is the volume fraction occupied by the spheres. If the spheres have a size distribution the diameter to be used in (17) is not the arithmetic diameter but the so-called Sauter diameter. For typical ion exchanger distributions it is between 5 and 10 percent larger than the arithmetic average. Here we are tacitly assuming that all interface is used equally effectively. This is not so, but again the errors involved are thought to be small.

For a typical plate e = 0.3, d = 0.8 x 10^{-3} m and a' $\approx$ 2000 m^2 m^{-3}. A minimum height of 0.5 m between the plates is required if they are to be accessible. The interfacial area per m^2 plate is then 1000 m^2.

4.4. Diffusion.

Convective flows die out near the particle interface, and also do not exist inside the resin. Moving species must then have a velocity difference with their surroundings. The surroundings are mainly water in the liquid phase and the matrix, fixed charges and water in the resin. The driving force for the movements is the gradient of the electrochemical potential:

$$\frac{d}{dz}(\mu_i + Fz_i\phi)$$

which indeed has the dimensions of a force per mole. The contribution from the chemical potential μ_i is related to the concentration gradient. The gradient of the electrical potential is caused by different mobilities of the diffusing ions. Although the electrical potential differences are small they have a profound influence on the transport behaviour of ions.

The work done by the driving force is expended as friction between i and its surroundings. The friction will be proportional to the velocity difference and a friction coefficient f_i:

$$\frac{d}{dz}(\mu_i + Fz_i\phi) = - f_i v_i \tag{18}$$

The velocity is related to the flux by:

$$N_i = v_i c\ x_i \quad \text{or} \quad v_i c\ y_i \tag{19}$$

Note that the friction losses depend on the friction coefficient and the velocity. Friction coefficients are much higher in the resin than in water. On the other hand the velocity for a given flux is highest where the concentration is lowest and this is usually in the water phase. So the potential drop is not necessarily highest in the resin.

These frictional relations have one other unexpected consequence. In binary exchange the two equivalent flows must com-
pensate. The largest frictional loss is then caused by a component
with a high velocity and that is a component with a low concentration.

Equation (18) is seldom used in design calculations. For ideal
solutions it can be transformed quite easily to give:

$$N_i = -D_i c \frac{d}{dz}(x_i + \frac{F z_i x_i \phi}{RT}) \tag{20}$$

These are the Nernst-Planck equations. They contain a diffusivity
D_i which is a property of the component in the solution
considered.

For binary exchange the electrical potential can be eliminated
from the two equations to yield:

$$N_A = - D_{AB} c \frac{dx_A}{dz} \qquad N_B = - D_{AB} c \frac{dx_B}{dz} \tag{21}$$

For $z_A = z_B = 1$ the binary diffusivity is:

$$D_{AB} = (\frac{x_B}{D_A} + \frac{x_A}{D_B})^{-1} \tag{22}$$

and similarly for the resin. As indicated by (22) but also for
other reasons the diffusivity can be a strong function of composition. Fortunately the composition change on one plate is seldom
large, so we can use a constant diffusivity in calculations.

There is no simple equivalent to equation (22) for multicomponent systems. Extensions of our calculations to more than two
exchanging components can probably best be done starting from a
more fundamental viewpoint such as equation (20), but this does not
yet appear to have been done for ion exchange columns.

4.5. Mass transfer coefficients.

Convection in the liquid dies out towards the interface in a
very complicated manner. For design work this is usually modelled
in the following very crude way. In the bulk liquid far from the
interface all transport is assumed to be by convection and the
solution is homogeneous. In a thin layer or "film" next to the
interface, transport is assumed to be solely by diffusion. This
subdivision in two strictly defined zones is artificial but has

been found to be quite effective. Its use depends on the determination of the film thickness δ. For a given value of δ the flux to the interface is predicted as

$$N_B = k_1 c (x_b - x_w) \qquad (23)$$

The subscripts b and w stand for bulk and wall (interface) respectively. The parameter

$$k_1 = \frac{D_{AB}}{\delta} \qquad (24)$$

is known as the mass transfer coefficient in the liquid phase. It has the dimension of a velocity and is indeed closely related to the transport velocities of the species. Empirically it has been found (16) that for single particles of the size we are interested in:

$$k_1 = 0.31 \left(\frac{g \Delta \rho D_{AB}^2}{\rho \nu} \right)^{0.33} \qquad (25)$$

Both the form and constants of this equation agree fairly well with predictions from more fundamental models using what is known as the boundary layer theory. In fluidized beds the constant (0.31) is somewhat lower; its value drops to about 0.20 at minimum fluidization. For our purposes a value of 0.25 is adequate. (For another description see (17)). Values of k_1 for fluidized beds of ion exchanger particles are around 3×10^{-5} m s^{-1}.

If the diffusivities D_A and D_B of the components differ considerably, the assumption of constant total concentration used in deriving (23) no longer holds. Then corrections are necessary (18). These are important for systems with either H$^+$ or OH$^-$ ions (which have very high diffusivities) or with polyelectrolytes (with low diffusivities).

In the resin the diffusion process is time dependant. Also a proper modelling would require regarding the different particles separately. Although possible the results are unwieldly. A simpler, but often adequate description is obtained as follows. Overall the resin particle undergoes a continuous increase in its external concentration as it passes down the column (figure 19). If one assumes a constant concentration gradient it appears there is a semi steady state solution to the concentration profiles in the particle. The average internal concentration lags behind the external one in such a way that

$$N_B = k_s \bar{c} (y_w - y_b) \qquad (26)$$

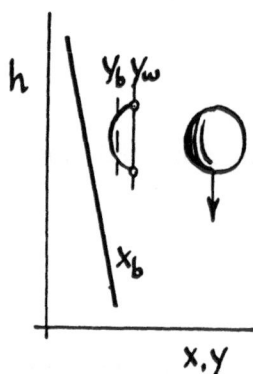

figure 19. Particle falling through a concentration gradient.

where

$$k_s = \frac{10 \, \overline{D}_{AB}}{d} \tag{27}$$

and y_b is the average composition of the particle. The complicated diffusional pattern can be described as if the particle has an internal "film" of one tenth of its diameter with the corresponding mass transfer coefficient. The derivation of the above equation is elementary. More complex situations have been analyzed in (19). As an example; with $D_{AB} = 8 \times 10^{-11} \, m^2 \, s^{-1}$ and $d = 8 \times 10^{-4}$ m we find $k_s = 10^{-6} \, m \, s^{-1}$.

4.6 Combining mass transfer coefficients.

The flux equations for the two phases can be combined to yield a single equation. For small composition ranges the equilibrium line can be approximated by

$$y = K'(x - b) \tag{28}$$

Where K' is the local slope. At the interface we assume equilibrium. Rewriting equations (23) and (26) and adding

$$\frac{N_B}{k_1 c} = x_b - x_w$$

$$\frac{N_B}{K' k_s \overline{c}} = \frac{y_w}{K'} - \frac{y_b}{K'} = x_w - x^*_b$$

yields

$$N_B = k_{ol} c \, (x_b - x^*_b) \tag{29}$$

with an overall mass transfer coefficient

$$k_{ol} = (\frac{1}{k_1} + \frac{1}{m'k_s})^{-1} \tag{30}$$

Here x^*_b is the liquid composition that would be in equilibrium with the average composition of the solid and

$$m' = K'\frac{\bar{c}}{c} \tag{31}$$

is a kind of volumetric distribution coefficient. For strongly curved equilibrium lines a more elaborate treatment is required (20).

4.7. Stage efficiencies.

With the model elements obtained it is now possible to set up estimations of the stage efficiencies. For the simplest flow model – both phases well mixed – this is easy (figure 20). Component B is transferred from the liquid to the resin at a rate:

$$N_B a = k_{ol} a \, c \, (x_n - x^*_n) \tag{32}$$

where a is the interfacial area. But this is also equal to the change in content of the liquid stream:

$$N_B a = L \, (x_{n-1} - x_n) \tag{33}$$

Equating the two right hand sides, adding $k_{ol} a c \, (x_{n-1} - x_n)$ to both and rearranging yields:

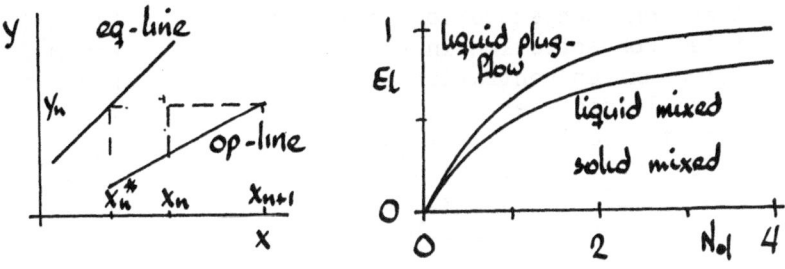

figure 20. Compositions in a mixed stage.
figure 21 (right). Stage efficiencies.

$$\frac{x_{n-1} - x_n}{x_{n-1} - x^*_n} = \frac{k_{ol}ac}{k_{ol}ac + L} = \frac{N_{ol}}{N_{ol} + 1}$$ (34)

where

$$N_{ol} = \frac{k_{ol}a\,c}{L}$$ (35)

is the "number of mass transfer units in the liquid phase".

The difference in equivalent fractions in the nominator is that of an equilibrium stage. That in the denominator is the change obtained in our model of the real stage. So this equation gives an estimate of the liquid phase Murphree efficiency we introduced in chapter 2:

$$E_l = \frac{N_{ol}}{N_{ol} + 1}$$ (36)

Slightly more complicated calculations for the plug flow model yield:

$$E_l = 1 - \exp\left(-N_{ol}\right)$$ (37)

(figure 21). Despite the large uncertainties this way of modelling has been found to give reasonable results in many experiments with laboratory columns (21). The efficiency estimates can now be used to determine the number of real stages required, and with the chosen plate distance the column height. This ends our preliminary design.

SYMBOLS

a	interfacial area	m^2
a'	interfacial area per volume dispersion	$m^2\ m^{-3}$
c	total concentration in liquid	$keq\ m^{-3}$
c_i	component concentration in liquid	$keq\ m^{-3}$
$\bar{c}$	total concentration in resin	$keq\ m^{-3}$
$\bar{c}_i$	component concentration in resin	$keq\ m^{-3}$
d	particle diameter	m
d*	diameter ratio (figure 15)	
D_i	diffusivity of component i	$m^2\ s^{-1}$
D_{AB}	binary diffusivity, eq. (22)	$m^2\ s^{-1}$
e	volume fraction particles	-
E_l	Murphree efficiency	-
f	friction coefficient	$N\ s\ mol^{-1}\ m^{-1}$
F	Faraday constant	$C\ eq^{-1}$
g	gravitational acceleration	$m\ s^{-2}$

K_B	distribution coefficient	–
K_{AB}	equilibrium constant	–
k	mass transfer coefficient	$m\ s^{-1}$
L, L'	total equivalent flow in liquid	$eq\ s^{-1}$
l	same for a component	$eq\ s^{-1}$
m	volumetric distribution coefficient	–
m'	see eq.(31)	–
n	exponent in eq.(15)	–
N_i	flux of component i	$eq\ m^{-2}\ s^{-1}$
N_{ol}	number of transfer units, eq.(35)	–
R	gas constant	$J\ mol^{-1}\ K^{-1}$
S, S'	total equivalent flow in solid phase	$eq\ s^{-1}$
s	same for a component	$eq\ s^{-1}$
T	absolute temperature	K
u	superficial velocity (volumetric flux)	$m\ s^{-1}$
v	velocity	$m\ s^{-1}$
v_∞	terminal velocity of single particle	$m\ s^{-1}$
v^*	velocity ratio (figure 15)	–
x	equivalent fraction in liquid	–
y	equivalent fraction in solid (resin)	–
z	distance	m
δ	film thickness	m
μ	chemical potential	$J\ mol^{-1}$
ν	kinematic viscosity	$m^2\ s^{-1}$
ϱ	liquid density	$kg\ m^{-3}$
$\Delta\varrho$	density difference solid/liquid	$kg\ m^{-3}$
ϕ	electrical potential	V

subscripts

A,B	components
i	any component
l	liquid
s	solid
b	bulk
w	wall (interface)
1,2...n	stage number.

318

REFERENCES

IET stands for the book "Ion Exchange Technology" edited by D.Naden and M.Streat, Ellis Horwood Publishers, Chichester, 1984.

1. Cloete,F.L.D., IET p 661
2. Ford,M.A., IET p 668
3. Helfferich,F.G., "Ion Exchange", McGraw-Hill New York 1962, chapter 5.
4. Soldatov,V.S.; Bichova,V.A., IET p 179
5. King,C.J., "Separation Processes" 2nd ed., McGraw-Hill New York 1980, chapter 10
6. Hartland,S., "Counter-current Extraction", Pergamon Press, Oxford 1970, chapter 4
7. Cloete,F.L.D.; Streat,M., Brit. Patent 1.070.251, 1962
8. Meer, A.P. van der, "On Counter-current fluidized ion exchange columns", PhD thesis, Technological University of Delft, 1984.
9. Naden, D. et al, IET p.690.
10. Richardson,J.F., Zaki,W.N., Trans.Instn.Chem.Engrs. 1954,$\underline{32}$,38
11. Al-Dibouni,M.R.; Garside,J.R., Trans.Instn.Chem.Engrs.1979,$\underline{57}$,94
12. Wallis,G.B., "One-dimensional Two-phase Flow", McGraw-Hill, New York 1970, chapter 4.
13. Helfferich,F.G. in "Mass Transfer and Kinetics in Ion Exchange", eds. Liberti,L.; Helfferich,F.S, p 157, Martinus Nijhoff Publishers, The Hague 1983.
14. Rietema,K. in "Advances in Chemical Engineering, volume 5", eds. Drew,T.B.; Hoopes,J.W.; Vermeulen,Th., Academic Press, New York 1964, p. 237-301
15. Gomez-Vaillard,R.; Kershenbaum,L.S; Streat,M., Chem.Eng.Sci. 1981,$\underline{36}$,306
16. Calderbank,P.H.; Moo Young,M.B., Chem.Eng.Sci. 1961,$\underline{16}$,39
17. Snowdon,C.B.; Turner,J.C.R., "Proceedings of the International Symposium on Fluidization", Eindhoven 1967; Netherlands University Press 1967, p. 599
18. Turner, J.C.R.; Snowdon,C.B., Chem.Eng.Sci. 1968,$\underline{23}$,223 and 1099.
19. Vorstman,M.A.G., Thijssen,H.A.C., "Proceedings of the International Solvent Extraction Confereence", The Hague 1971, Society of Chemical Industry, London 1971, p. 1071
20. Treybal,R.E., "Mass Transfer Operations" 3rd ed., McGraw-Hill, New York 1980, chapter 5.
21. Meer, A.P.van der, et al., IET p 284.

CONTINUOUS ION EXCHANGE TECHNOLOGY

M.Streat

Department of Chemical Engineering & Chemical Tecnology
Imperial College
London SW7 2BY,United Kingdom

INTRODUCTION

A major technological breakthrough in ion exchange has been the establishment of continuous ion exchange (CIX) as a viable economic alternative to fixed bed processes and plant in some selected applications. CIX has advanced ion exchange in a chemical engineering sense and opened up the possibility of even wider applications in the chemical and allied process industries. Modern developments offer advantages hithero claimed for liquid-liquid extraction and afford the possibility of integrated fractional separation processes.

The momentum to develop a truly continuous ion exchange process has been maintained in the uranium industry since the early 1950's. However, there is no doubt that the early incentive came from the nuclear industry where it became obvious that fixed columns were unsuitable for the treatment of highly radioactive solutions. Early work on fluidised bed contactors was directed towards the ion exchange treatment of liquors spontaneously producing radiolytic off-gas. Though this work faltered, it inspired the development of CIX in many process applications. Several ambitious attempts to build CIX plants were carried out in the uranium industry, some designed to treat unclarified liquors and slurries. There were many problems and difficulties and despite the fact that some plants are still in existence today, the major advances have occurred in the late 1960's and early 1970's. Specific advances in the development of CIX will be outlined in this paper.

DEVELOPMENTS IN ION EXCHANGE EQUIPMENT

Ion exchange is widely used in the field of water treatment , effluent treatment and control, hydrometallurgy, sugar processing and biological applications. Invariably, ion exchange materials are used in packed bed reactors, since the equipment is simple to design and fabricate, the ion exchange resin is not subjected to arduous physical conditions and therefore the useful lifetime of the resins is enhanced and thus the operating costs of a fixed bed ion exchange plant are relatively low. The disadvantage of a fixed bed process is that it is cyclic in operation and that at any one instant in time only a relatively small part of the resin hold-up is doing useful work. Operation of a column involves either downflow or upflow; the flow-rate is usually restricted in downflow to about 20 Imp.gal ft^{-2} min^{-1} ($60m^3$ m^{-2} h^{-1}) to avoid excessive pressure drop.

Fixed columns are normally contained in pressure vessels and provided with ancillary valves and pipework to control the flow. Columns are rarely used singly and are normally cascaded in order to give continuous operation as far as the feed solution is concerned. Typical examples of this are in boiler feed water treatment or in the processing of a pregnant uranium leach liquor. In a large uranium mill, there may be several lines of ion exchange columns in parallel and upwards of 20 columns in service is not uncommon in a typical plant (e.g. in South Africa). A large plant will require many automatic valves and timers and therefore the control system is complex. In essence, a continuous process is maintained by keeping the ion exchange resin stationary and moving the pregnant feed point from column to column. Though the capital cost of these installations is large, the operating costs are usually low. Replacement of resin is negligible and the main direct operating cost is the regenerant chemical.

Whereas a fixed bed column is essentially simple, the disadvantage of a fixed bed plant lies in the complexity of the ancillary equipment. Each column in a cascade will require several automatic control valves and associated equipment involving process timers. The hold-up of resin is large and this can be a problem if poisoning is a threat, e.g., in the South African uranium industry. The cyclic variation of product concentration in many applications can be a disadvantage and often leads to extensive tankage and storage of bulked solutions. By analogy with other chemical engineering separation processes, ion exchange in fixed beds is a multiple batch process. There has always been interest in the development of continuous ion exchange processes and early attempts to improve the process technology date back to the nineteenth century. The patent literature abounds with novel and ingenious techniques for the operation of continuous ion exchange equipment. However, advances have occurred only in recent years and these are mainly associated with the expansion of the uranium processing industry. Examples of the development of ion exchange equipment are given in Table 1, and it may be seen that there are three major categories:

> moving fixed bed
> jigged and agitated beds
> fluidised beds

TABLE 1- Examples of ion exchange equipment

Type of Equipment	Example	Application
Fixed Bed Cascade of Fixed Beds		Water Treatment and Uranium Purification
Continuous Ion Exchange Moving Fixed Bed	Porter-Arden Higgins Asahi	Uranium Purification Water Treatment Water Treatment
Jigged Bed Agitated Bed	Weiss et al. Resin-in-Pulp	Uranium Purification Uranium Purification
Fluidised Bed	Swinton/Weiss Cloete/Streat USBM Himsley NIM	Uranium Purification Uranium Purification Uranium Purification Uranium Purification Uranium Purification

The effectiveness of the fixed bed column has been proved in water treatment, e.g. demineralisation, dealkalisation, softening, etc., and it is thus not suprising that early attempts to improve the process technology centred on the concept of a countercurrent moving fixed bed process. The intention was to retain the high efficiency of the ion exchange process, but simplify the number of columns, valving, pipework, controls, etc. The concept of a moving bed contactor is well known and the most notable examples are those invented by Higgins (Figure 1) and developed by Asahi in Japan (Figure 2). The Higgins contactor involves a closed loop in which resin moves intermittently as a sliding bed in countercurrent flow to the process solution, e.g. water(1). Elegant slide-valves are now used to separate the sorption, regeneration and resin backwash stages. The contactor operates in pre-determined cycles and is an ideal process for the treatment of clarified liquors, e.g. removal

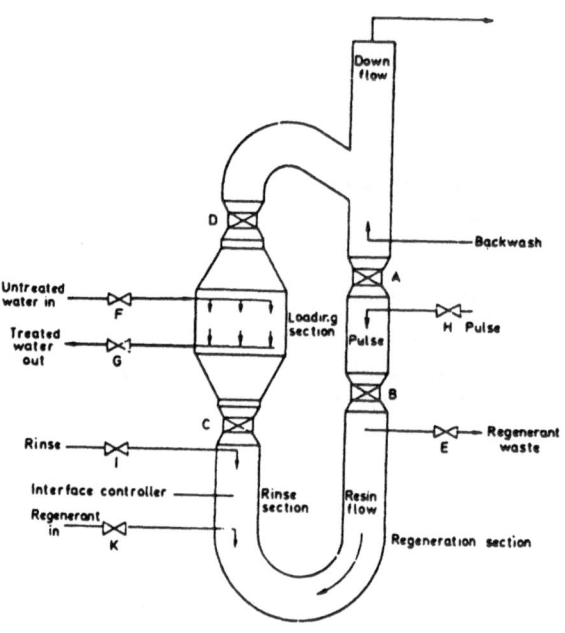

Valve positions during cycles:
(a) Run cycle. Valves A. E. F. G. I. K open; valves B, C, D. H closed
(b) Pulse cycle. Valves B. C. D. H open; valves A. E. F. G. I. K closed

Figure 1
Schematic diagram of the Higgins Moving Bed Contactor.

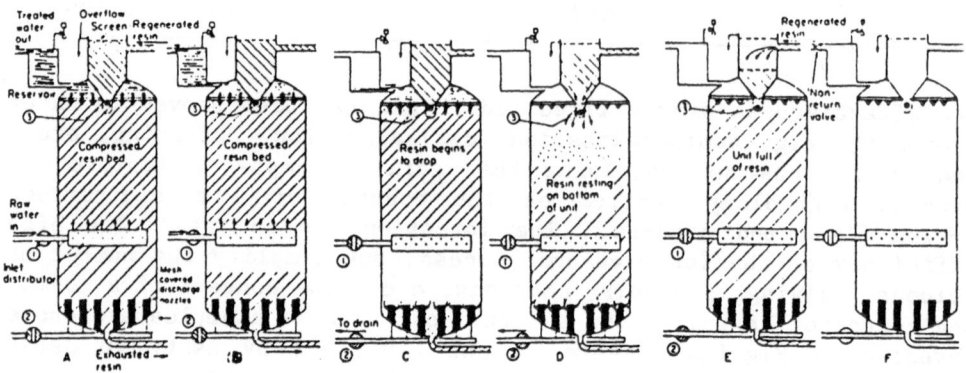

Figure 2
Operating sequence of the Asahi Type Countercurrent Moving Bed Contactor.

of nitrates from water, water softening and dealkalisation. The major uncertainty about the equipment is the lifetime of the resin. Estimates range from 5-30% resin inventory replacement per year due to attrition and breakage in the resin valves. Nevertheless, the Higgins contactor represents a major advance in ion exchange technology. A later development is the Ashai contactor(2). Here, the resin is retained in a conventional pressure vessel containing a resin support grid at the base and a resin retention screen at the top. Process water is fed in upflow through the fixed bed and periodically the liquid contents of the column are allowed to drain rapidly. This allows the resin to flow as a slurry from the bottom of the bed to a comparable regeneration contactor. During this period, fresh resin flows into the column from a resin feed hopper. The latter contains a ball-valve which passes resin in downflow and seals during upflow of process water. The success of the equipment depends greatly on the operation of this valve. The Asahi contactor has been installed in many water treatment projects and is also suitable for effluent treatment applications e.g. copper recovery from rayon waste solution.

The idea of moving fixed beds was also adopted in the uranium industry and an early example of continuous ion exchange was the Porter-Arden process, operated in the Blind River region of Ontario, Canada, in the late 1950's(3). Here, exhausted resin was transported to an empty conventional column by hydraulic conveying. The resin was then backwashed and transported to a bank of elution columns, fresh resin having previously been transported to the vacant extraction column. The movement of resin around the fixed bed installation was essentially countercurrent, though the process was not fully continuous by modern standards. Installations of this type were the forerunners of much of the continuous ion exchange development in the uranium industry. At about the same time, considerable work was in progress on the handling and treatment of uranium leach pulps. The possibility of recovering uranium from unfiltered pulps offered great savings in cost and encouraged some outstanding development work. Clearly, fixed bed equipment was unsuitable for unclarified pulps and thus a radical change in ideas was required. Separate development took place in the USA and Australia, culminating in the concept of jigged or agitated beds of resin. Here, the ion exchange particles were separated momentarily by pulsation, vibration or agitation. Notable developments were the resin-in-pulp contactors. Typical of these are the Anaconda Blue Water Plant in Grants, New Mexico, and the Atlas Mineral Moab Mill in Utah, USA(4). The equipment is crude by modern standards and usually involved a cascade of baskets containing ion exchange resin retained by stainless steel mesh(5). The porosity of the mesh retained the resin, but allowed ingress of a flowing leached uranium ore pulp. The baskets were agitated by reciprocating action to avoid clogging of the bed (see Figure 3).

324

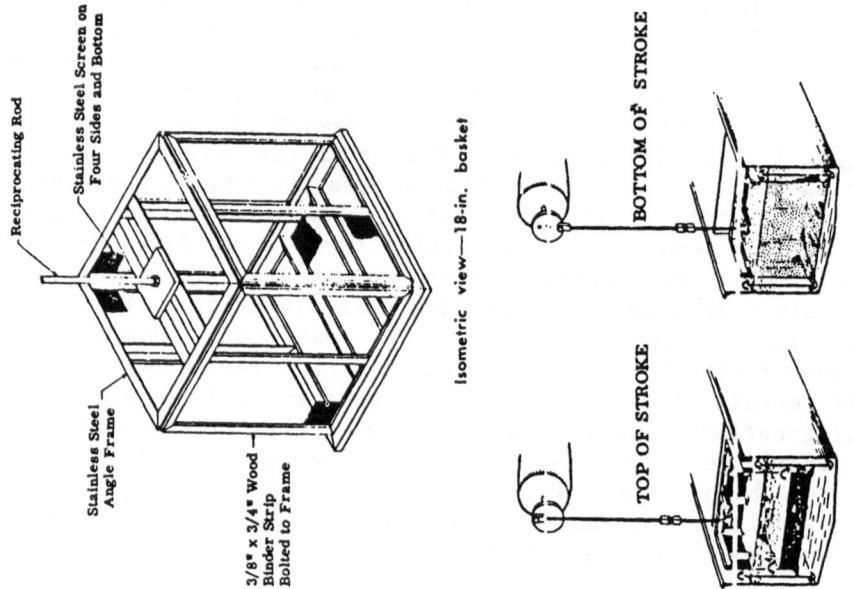

Figure 3.
Concept of the Agitated Basket Resin-in-Pulp
Contactors used for Uranium Recovery from Ore
Pulps.

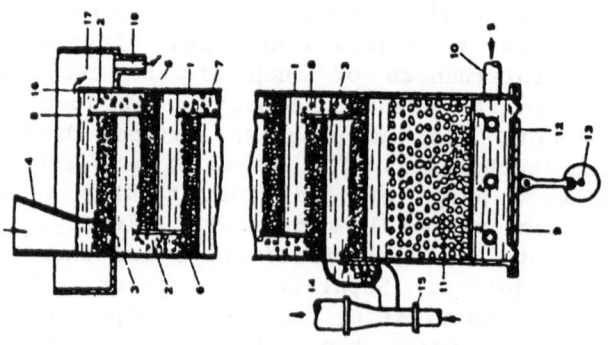

Figure 4
Schematic diagram of the Swinton and Weiss Multistage Fluidised Bed Column.

The process operates by alternating pregnant feed, wash and eluant solutions. Though there is no countercurrent movement of the particles, this process demonstrated that an "expanded" bed contactor could recover and separate uranium from bulk impurities. A similar concept was developed separately in Australia by Swinton, Weiss and co-workers(6). They devised a moving agitated or jigged bed for the treatment of high solids concentration uranium ore pulps. Continuous pulsation of the bed (about 200 cycles/min) caused momentary separation of the particles and this allowed the slurry to pass through the bed in upflow. A screen prevented carry over of the ion exchange particles. In many respects, this contactor was revolutionary, since it attempted to solve the problem of slurry handling and countercurrent flow of resin simultaneously. In the event, the project was dogged with problems associated with the top screen. This screen would blind with wood pulp and other gangue and thus the pressure drop became excessive. The idea did not proceed to full-scale development.

The concept of a multistage plate-type column is well known in chemical engineering and forms the basis of many separation processes in vapour-liquid, gas-liquid and liquid-liquid operations. The development of a vertical multistage fluidised bed column has always been appealing in ion exchange. The earliest attempts used a vertical column containing perforated distributor plates and downcomers. Ion exchange resin was supported on the distributor plate in the no-flow condition and during upward flow, the particles passed over a weir and into the downcomer. Resin moved in downflow from stage-to-stage and could be removed from the base of the column. The concept was feasible in principle, and pioneering work was performed by Swinton and Weiss(7) in 1953, (see Figure 4). However, the operation of this column and others like it was hydraulically unstable. Control of resin holdup and by-passing of the pregnant feed were problems. Start-up was not easy since the pressure drop in the downcomers was invariably less than through the resin bed and thus by-passing was readily initiated. This early work with fluidised beds faded in the 1950's and was ultimately revived in the early 1960's by Cloete and Streat at Imperial College. In essence, the Cloete-Streat ion exchange contactor is a multistage fluidised bed containing perforated distributor plates and no downcomers. The hole size in the plates is greater than the maximum resin particle size. Countercurrent flow of the resin is achieved by controlled cycling of the liquid feed (see Figure 5). In this way, a precisely determined fraction of the resin hold-up per stage can be transferred. In practice, it is desirable to transfer the entire resin hold-up per cycle. This column has been developed for application in the uranuim industry and the greatest advances have occurred in South Africa. Work at Mintek (Council for Mineral Technolgy) in Johannesburg(8,9) proved the effectiveness of the Cloete-Streat column for uranium recovery from acid leach solutions, (see Figure 6). Pilot-plant experiments

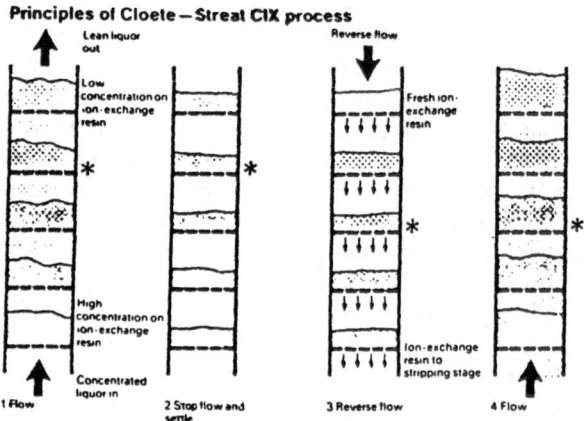

Figure 5
Sequence of operations of the Cloete-Streat CIX Process.

Figure 6
The experimental CIX plant for uranium recovery at Blyvooruitzicht Gold Mine, South Africa. This Plant was designed by NIM and operates on the Cloete-Streat principle.

confirmed the scale-up of the equipment and showed that it was possible to produce uranium concentrate of similar quality to that produced by liquid-liquid extraction. Full-scale plants have now been installed at several gold mines in South Africa. Extraction columns 4.5m diameter containing eight stages of total height about 20m are in operation.

Independent work in the USA led to the development of the USBM contactor(10). This is a multistage countercurrent fluidised bed reactor and operates by controlled cycling (see Figure 7). In many respects, it is similar to the Cloete-Streat contactor and differs only in plate design, mode of solids transfer, and resin withdrawal. This column has all the advantages of a multistage fluidised bed and has found application for uranium recovery at various locations in the USA.

More recently, a slightly modified form of the multistage fluidised bed has been developed by Himsley(11). The Himsley column consists of vertical stages containing resin and a single fluid inlet. Pregnant feed in upflow passes from stage to stage continuously and during operation the feed flow is not stopped (see Figure 8). Resin is transferred sequentially from stage to stage by using an auxiliary pump to transfer particles down the column. An empty stage at the base of the column allows the process to start. Fresh resin is transferred hydraulically to the top stage. The process requires a series of control valves and timers to sequence the operations. Also, a screen in each stage avoids the carry-over of resin particles during transfer. This column is in operation for uranium recovery at Agnew Lake, Ontario, Candada.

Though most fluidised bed development is in the form of vertical columns it is also possible to operate the system in a series of horizontal tanks. Possibly the largest CIX plant in the world operates at Rossing uranium mine in Namibia. This plant, designed by Porter, comprises an array of horizontal interconnected tanks, each acting as an effective fluidised bed. The resin beads are progressively moved from tank to tank by airlifts in countercurrent to the flow of pregnant solution. The total pregnant flow to the plant is about 3500 m^3/hr and each contacting tank is 6m square by 3.5m deep. The initial plant had 4 lines of 5 contacting tanks in each. Careful design of resin traps has been included in the plant; these comprise a sophisticated rotating trommel and this has now permitted a sixth contactor to be installed in each line.
The Rossing process flowsheet incorporates both ion exchange and solvent extraction and thereby employs the considerable advantages of both techniques. CIX is used to upgrade the low concentration feed (0.15g U_3O_8 per litre) and produce an eluate feed of almost constant composition containing 3-4 U_3O_8 per litre. This eluate feeds the solvent extraction plant and produces a final product

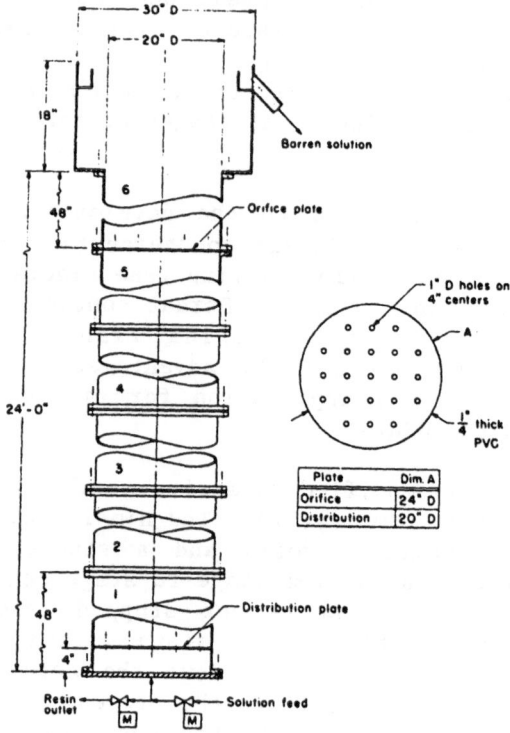

Figure 7
Schematic diagram of the USBM Contactor.

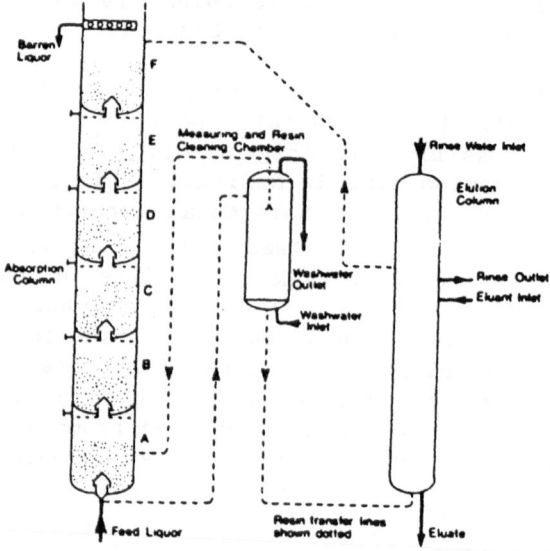

Figure 8   Typical Arrangement of Himsley
continuous ion exchange system

solution containing about 10g U_3O_8 per litre. This solution is precipitated and calcined to give a concentrate containing approximately 97% U_3O_8. The combination of continuous ion exchange and liquid extraction is a powerful process route for the separation and purification of metals, pharmaceutical products and allied chemical products.

FRACTIONAL SEPARATIONS BY CONTINUOUS ION EXCHANGE

Stagewise continuous countercurrent ion exchange processes can be applied to fractional separations. A CIX process does not necessarily give steady-state operation, but it is reasonable to treat the concept as such in trying to develop a coherent separation theory. In this sense, CIX can be considered analogous to liquid-liquid extraction and the same theoretical treatment can be applied. If we assume constant flows of solution S, and resin R, through a cascade of n perfectly mixed ion exchange stages, then the mass transfer of a solute can be predicted by the Kremser-Souders-Brown equation(12). The treatment is explained in Fig.9 and applies where the distribution coefficient of the solute remains constant. In this case, an extraction factor, E, can be defined and thus the loss of the solute in the aqueous raffinate can be found in terms of E and n, the number of ideal stages.

In order to achieve separation of two sorbing components, an intermediate fed contactor is used (by strict analogy with liquid-liquid extraction). This can be explained by considering for example, the separation of ferric and cupric ions in chloride solution. Both ferric and cupric ions form anionic chloride complexes which are sorbed by a typical strong base resin such as Dowex-1 X8. The likely reactions are:

$$\overline{2R_4N^+Cl^-} + CuCl_4^{2-} \rightleftharpoons \overline{(R_4N^+)_2CuCl_4^{2-}} + 2Cl^-$$

$$\overline{R_4N^+Cl^-} + FeCl_4^- \rightleftharpoons \overline{R_4N^+FeCl_4^-} + Cl^-$$

R_4N^+ denotes the quaternary strong base resin and overbars indicate the species in the resin phase. In this case study, the distribution coefficient of ferric ions is greater than for cupric ions. The operating diagram for an intermediate fed process involving both extraction and scrubbing, is shown in Fig.10. Using an aqueous scrub solution of 6M HCl, it is possible to back extract copper and produce a resin phase product containing essentially iron and raffinate solution containing copper. The process given in this example requires 16 ideal stages, 6 for extraction and 10 for scrubbing. The latter is large and not optimum but this is merely an artifact of the operating conditions chosen in the example. Note the pinch condition for iron in stages 4-10 and for copper in stages 11-13.

330

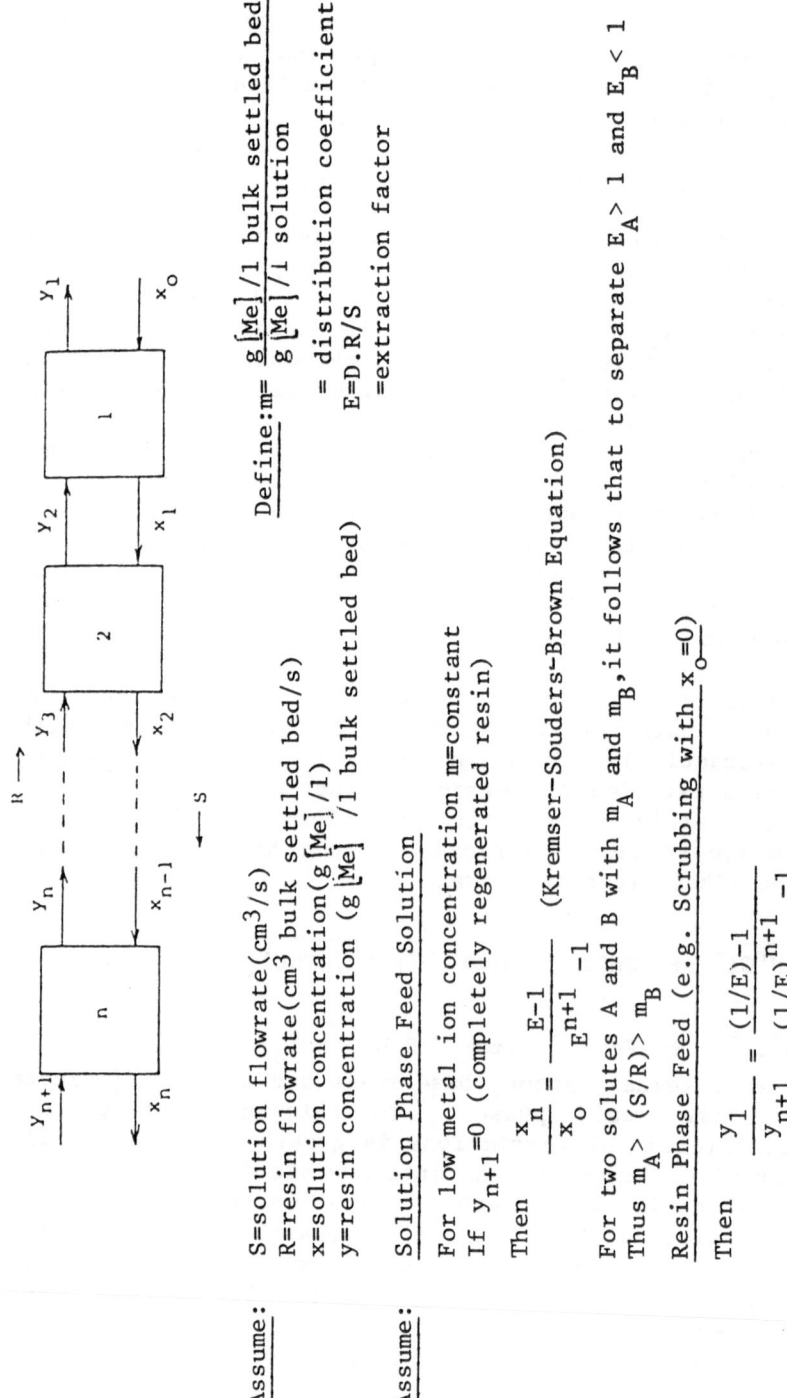

Assume: S=solution flowrate(cm³/s)
R=resin flowrate(cm³ bulk settled bed/s)
x=solution concentration(g[Me]/1)
y=resin concentration (g[Me] /1 bulk settled bed)

$$\underline{\text{Define}}: m = \frac{g[Me]/1 \text{ bulk settled bed}}{g[Me]/1 \text{ solution}}$$
= distribution coefficient
$E = D.R/S$
=extraction factor

Assume: Solution Phase Feed Solution

For low metal ion concentration m=constant
If $y_{n+1}=0$ (completely regenerated resin)

Then $\quad \dfrac{x_n}{x_o} = \dfrac{E-1}{E^{n+1}-1} \qquad$ (Kremser-Souders-Brown Equation)

For two solutes A and B with m_A and m_B, it follows that to separate $E_A > 1$ and $E_B < 1$
Thus $m_A > (S/R) > m_B$

Resin Phase Feed (e.g. Scrubbing with $x_o=0$)

Then $\quad \dfrac{y_1}{y_{n+1}} = \dfrac{(1/E)-1}{(1/E)^{n+1}-1}$

Figure 9– Separation theory in continuous ion exchange

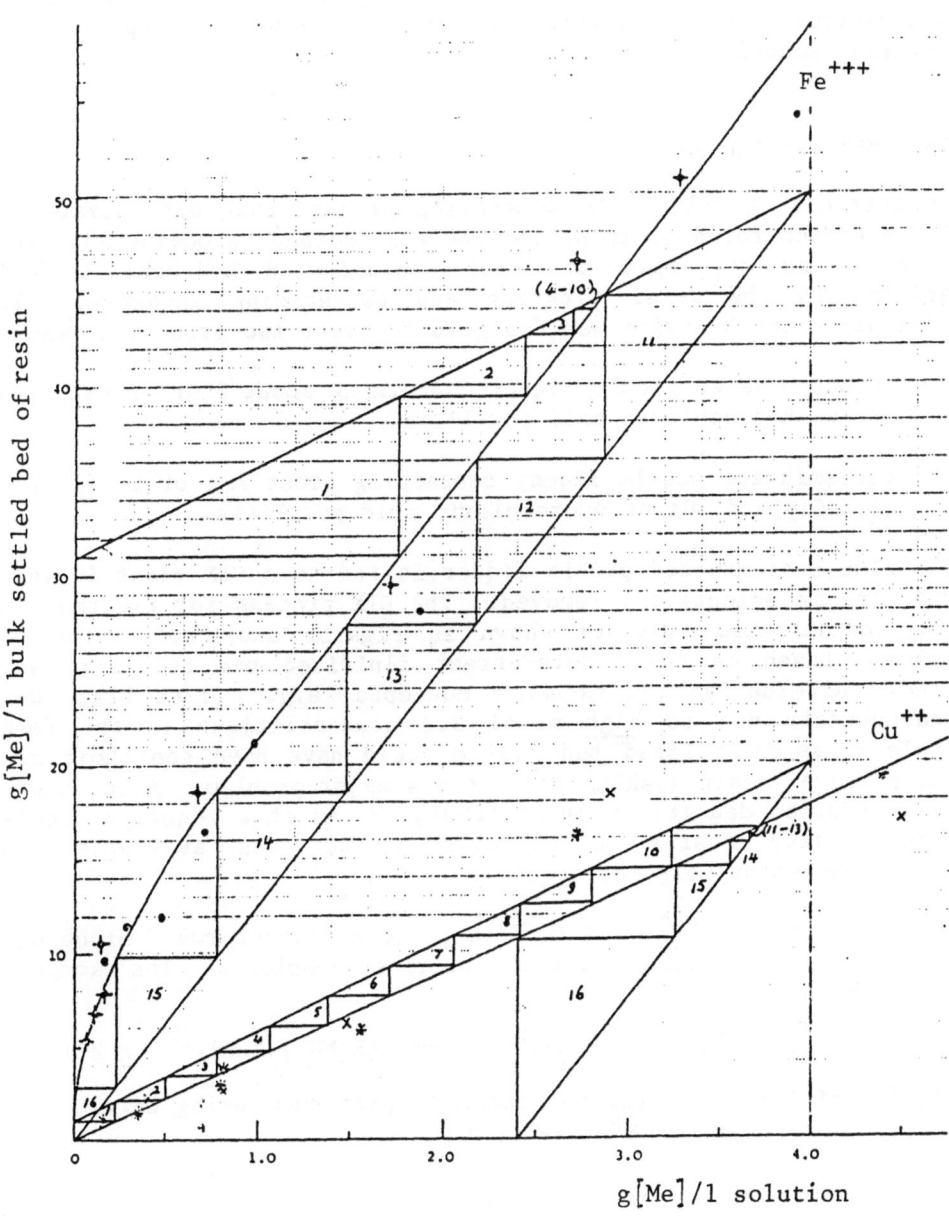

Figure 10- Equilibrium diagram for adsorption of ferric and
cupric ions onto Dowex-1 X8 in 6M hydrochloric
acid.

This example serves to illustrate the theoretical possibility of operating a fractional ion exchange process in a stagewise CIX contactor. In practice, these concepts could be applied to separations of the transition metals or noble metals, especially in chloride solution.

THE "METSEP" PROCESS

An interesting full-scale separation process that was operated in South Africa for a short period was the "Metsep" process(13). This process was developed to treat contaminated hydrochloric acid arising in the metal pickling and galvanizing industry. The process in South Africa was designed to treat two process streams:

a. uncontaminated pickle liquor containing about 220g of $FeCl_2$ and 31g of hydrochloric acid per litre.

b. contaminated pickle liquor containing about 73g of $ZnCl_2$, 230g of $FeCl_2$ and 30g of hydrochloric acid per litre.

The zinc contaminated pickle liquor is fed to a CIX plant in which the zinc is selectively removed. The CIX process for treatment of the contaminated acid is shown in Figs.11 and 12. Three CIX columns based on the Cloete-Streat fluidised bed technique were used. Alternative ion exchange concepts, e.g. fixed beds, were unacceptable due to <u>high</u> metal ion concentrations in the feed. Cycle times in a fixed bed system would have been too short and, furthermore, resin washing presented a major problem. Also, the feed solution density is about $1200kg/m^3$ and this requires special high density resins (a suitable product was avaliable from Resindion-Sybron Corporation at the time).

The separation of Fe^{2+} and Zn^{2+} is quite straightforward since only Zn^{2+} forms a stable anionic chloride complex. The sorption reaction is:

$$\overline{2R_3NH^+Cl^-} + ZnCl_4^{2-} \rightleftharpoons \overline{(R_3NH^+)_2ZnCl_4^{2-}} + 2Cl^-$$

Elution of the zinc from the resin is performed using water, since a weak base resin (R_3NH^+) was used.

$$\overline{(R_3NH^+)_2ZnCl_4}^{2-} \xrightarrow{H_2O} 2R_3NH^+Cl^- + ZnCl_2$$

This process proved highly successful in separating zinc from the acid feed to the spray roaster and furthermore a neat liquid-liquid extraction process step was used to convert $ZnCl_2$ to $ZnSO_4$ using the cation exchange properties of the solvent extractant di-2-ethyl hexyl phosphoric acid.

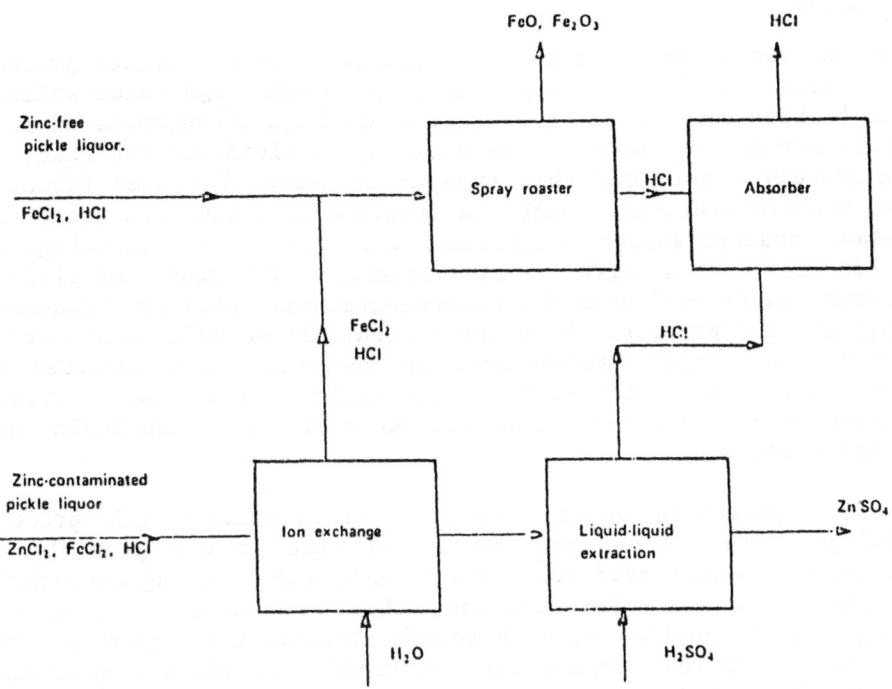

Figure 11-Block flowsheet of the Metsep process.

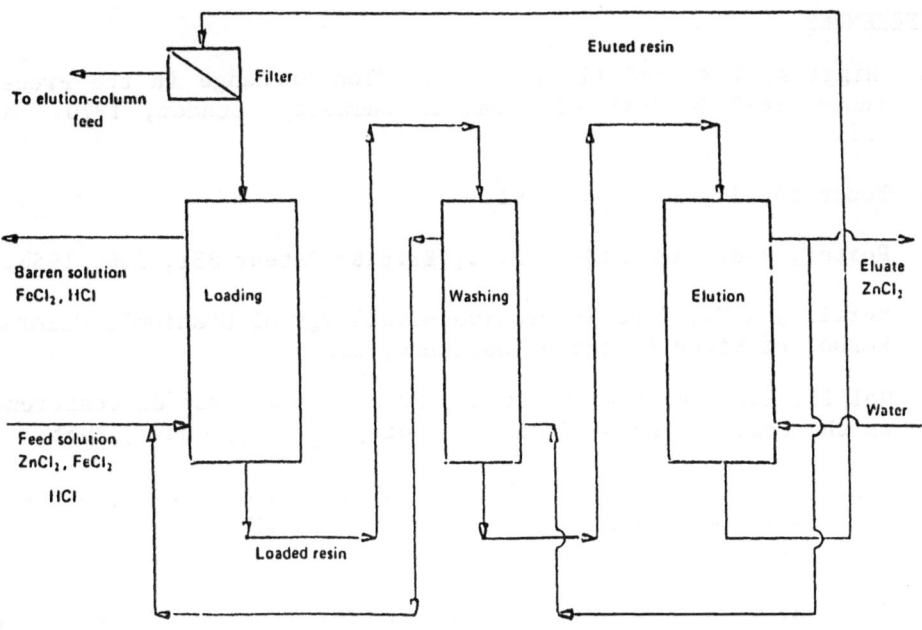

Figure 12- Flowsheet of the ion exchange plant.

CONCLUSIONS

Continuous ion exchange is firmly established in specialist process applications, notably in hydrometallurgy, where high value solutes are to be recovered or alternatively where high throughputs of feed solution are to be treated. The scale up of fluidised bed reactors is relatively simple and this type of contactor has been favoured in the uranium industry. Water is regarded as a low cost commodity in most industrialised countries and thus the technological advances associated with water treatment for power utilities, ultra-pure water treatment for microelectronics, effluent treatment and pollution control are less adventurous. Fixed columns are still acceptable and major improvements in operation have occurred in recent years, e.g. counterflow regeneration. There is no strong incentive to use continuous ion exchange where fixed bed technology is totally adequate.

Separation processes would benefit from multistage CIX process technology. The Metsep process showed that it was possible to treat highly concentrated solutions by CIX, albeit using favourable separation factors. Nevertheless, this example shows that ion exchange can be applied at high solute concentration, provided the resin can be cycled rapidly from extraction to elution reactors. The scope of CIX in separation process development requires more work and it is hoped that this will lead to new areas of potential application.

REFERENCES

1. Higgins, I.R. and Chopra, R.C., "Ion Exchange in the Process Industries" Society of Chemical Industry, London, 1970, page 121.

2. Bouchard, J., ibid, page 91.

3. Porter, R.R., and Arden, T.V., British Patent 831, 206, 1956.

4. Merritt, R.C., "The Extractive Metallurgy of Uranium", Colorado School of Mines Research Institute, 1971.

5. Hollis, R.F. and McArthur C.K., Proceedings First UN Conference on the Peaceful Uses of Atomic Energy, 8, 54, Geneva, 1955.

6. Arden, T.V., Davis, J.B., Herwig, G.L., Stewart, R.M., Swinton, E.A., and Weiss D.E., Proceedings Second UN Conference on the Peaceful Uses of Atomic Energy, 3, 396, Geneva, 1958.

7. Swinton, E.A., and Weiss, D.E., Australian J.Appl Sci. 4, 316, 1953.

8. Haines, A.K., J.South African Inst.Min. and Metal, $\underline{78}$, 303, 1978.

9. Haines, A.K., Craig, W.M., Faure, A., Hendricksz, A.R., Wills, W.J. and Nicol, D.I., Eleventh International Mineral Processing Congress, paper 32, Cagliari, 1975.

10. George, D.R., Ross, J.R., and Prater, J.D., Min.Engng., $\underline{1}$, 73, 1968.

11. Himsley, A., Canadian Patent, 980, 467, 1973.

12. Treybal, R.E., "Liquid Extraction", McGraw Hill 1963, page 350.

13. Haines, A.K., Tunley, T.H., Te Riele, W.A.M., Cloete, F.L.D., J.South African Inst.Min.Metal, $\underline{74}$, (4), 149, 1973.

EFFICIENT FRACTIONATION BY ION EXCHANGE

Phillip C.Wankat

School of Chemical Engineering
Purdue University
West Lafayette,IN 47907 USA

ABSTRACT

Methods for improving the efficiency of large-scale chromatographic ion exchange fractionation will be discussed. For both "migration" chromatography (the band moves at a finite speed) and "on-off" chromatography (the material is eluted only when conditions are changed) operation with smaller diameter particles and faster cycles will increase resin productivity. Other methods for on-off chromatography include using layers of different resins, using beds in series, and using moving bed techniques for preliminary fractionation. The efficiency of migration chromatography can be improved by using recycle or segmented recycle, column switching, moving port methods, simulated moving beds, and two-way chromatography methods.

PARTICLE SIZE EFFECTS

Particle diameter, shape, and the distribution of particle sizes have very important effects on both the hydrodynamics and the mass transfer in packed beds. Since particle diameter is under the designer's direct control, we will look at diameter effects in detail.

Pressure drop in a packed bed can be estimated from a variety of equations and correlations (11,36). For rigid particles the pressure drop can be estimated from,

$$\Delta p = \frac{\mu v L}{K d_p^2} \tag{1}$$

Note that pressure drop is inversely proportional to d_p^2. Thus the usual argument for not using small diameter particles is that the pressure drop will be too high. However, if L/d_p^2 is kept constant the pressure drop remains the same. If, in addition, L/t_{cycle} is kept constant, the column will have the same capacity as the larger column although it is shorter and cycles faster.

Small diameter particles have other effects on the hydrodynamics of packed columns. The packed bed acts as a filter for suspended solids, and with small diameter particles the bed is a much more efficient filter. Average mass transfer rates will be effected by wall effects for column diameters from 30 to 50 times the particle diameter (20). Thus, the use of small diameter particles should minimize wall effects. At the same time, channeling is often worse when smaller diameter particles are used, and is particularly troublesome with larger diameter columns (41). Despite this problem, large diameter columns have been successfully packed with small diameter particles and have been used industrially. Examples include the EcoTec ion exchange system (2,26), the Elf large scale gas chromatography system (13), and the Union Carbide pressure swing parametric pump (37). Successful packing requires accurate sieving so that the size fraction is sharp. This becomes more difficult as very small particles are used. Special packing techniques such as those used in large scale chromatography (13,36,59,67) may be required. To fully utilize the high mass transfer rates of small diameter packings the distribution of fluid must be quite uniform. This may require special distributors (67).

Linear Systems

The advantage of reducing particle diameter is that mass transfer rates usually increase markedly; thus, better separations or shorter columns can be used. For *linear* isotherms where the Van Deemter equation (28,39,62) is valid, the mass transfer effect is easily seen. The simplest form of the Van Deemter equation is,

$$H = A + \frac{B}{v} + Cv \tag{2}$$

The A term is the contribution of eddy diffusion and is directly proportional to d_p,

$$A = \lambda d_p \tag{3}$$

The B term represents the contribution of molecular diffusion and is independent of particle diameter.

$$B = 2\gamma D_M \tag{4}$$

The mass transfer contribution is contained in the C term. Since mass transfer usually controls, a variety of expanded forms for the C term have been developed (28,39,59,67). For a typical liquid system C can be written as

$$C = \frac{c_M d_p^2}{D_M} + \frac{c_{SM} d_p^2}{D_{SM}} \qquad (5)$$

The first term on the right hand side represents film diffusion and the second term diffusion in the stagnant mobile phase contained in the pores. Since pore diffusion usually controls, H is proportional to d_p^2. The number of plates is then

$$N = \frac{L}{H} \propto \frac{L}{d_p^2} \qquad (6)$$

N is proportional to L/d_p^2.

Several solutions for concentration profiles in fixed beds have been developed for systems with linear isotherms (28,43,54,56,63-65,67). As expected they all give essentially the same predictions. All of these models predict that zone spreading or the mass transfer zone is proportional to $\sqrt{L}$, $\sqrt{N}$ or $\sqrt{\text{distance traveled}}$. If L/d_p^2 is kept constant then N is constant even though the column may be much shorter. The fractional length of the mass transfer zone will be constant.

$$\frac{L_{MTZ}}{L} = \frac{N_{MTZ}}{N} = \text{constant} \qquad (7)$$

Thus the fraction of the column which can be used per cycle is constant.

For a linear system the elution step will also scale as L/d_p^2. Thus if we decrease the particle diameter by some factor f, also decrease L by f^2, and decrease the period of the cycle by f^2, we should observe constant separations at the same pressure drop with the same total feed throughput. Since L has been reduced by f^2 the productivity will have been increased by f^2. The conclusion is that for linear systems operating as simple load-elute cycles productivity can be drastically increased by decreasing the particle diameter, using a shorter bed, and cycling faster.

This conclusion is also true for the chromatographic separation of linear systems. In this case a pulse of feed is input and eluted with a less retained elutant. Linear chromatographic theory (39,59) shows that the resolution between two peaks can be calculated as

$$R_s = \frac{2(t_{R2} - t_{R1})}{w_1 + w_2} = \frac{1}{2} \left[\frac{\alpha_{21} - 1}{1 + \alpha_{21}} \right] \frac{\vec{k}'}{1 + \vec{k}'} N^{1/2} \qquad (8)$$

where t_{R1} and t_{R2} are the retention times of the two peaks, w_i are the peak widths, α_{21} is the selectivity, and k' is the relative retention. If decreasing the particle size does not change either the chemistry or the porosity of the system, then

$$R_S \propto N^{1/2} \propto \left[\frac{L}{d_p^2}\right]^{1/2} \tag{9}$$

If L/d_p^2 and t_{cycle}/L are constant, the resolutions of repeated pulse inputs will be constant. This is illustrated in Figure 1. Thus we should obtain higher resin productivities with constant pressure drop and constant resolution.

However, there is one problem buried in the dimensionless abscissa of Figure 1. With smaller particles t_{cycle} decreases and the accuracy of the timing must be much greater when short columns packed with small diameter particles are cycled rapidly. Although still useful for migrational linear chromatography, this scaling procedure will probably be more useful for on-off type cycles where timing is less critical.

Non-Linear Systems

Except at very low concentrations of the solutes, ion exchange will be non-linear. Does the scaling procedure work for non-linear systems? Obviously, the answer must be yes or I wouldn't have spent as much time on this topic. With nonlinear isotherms one will observe shock (constant pattern waves) and diffuse (proportional pattern waves). When these patterns occur, depends on the shape of the isotherms. To be specific we will discuss favorable isotherms where shock waves occur when the column is loaded, and diffuse waves occur when the column is eluted. However, the results are general.

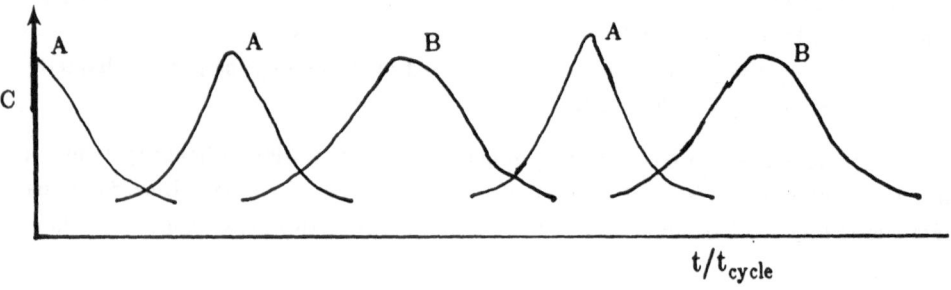

Figure 1. Scaled linear chromatography results.

Shock or constant pattern waves have been extensively studied (42,43,46,54,56,63,64,67). In the asymptotic limit the mass transfer zone has a constant length. Once this limit has been reached, the assumption of a constant pattern greatly simplifies the mathematical analysis. Analytical solutions are easily obtained for a variety of mass transfer expressions. For example, for a lumped parameter mass transfer form,

$$\rho_B(1 - \epsilon)\,\frac{\partial q}{\partial t} = -k_M a_p(c_i^* - c_i) \tag{10}$$

and a generalized Langmuir isotherm,

$$q = \frac{ac}{1 + bc} \tag{11}$$

the mass transfer zone length is

$$L_{MTZ} = \frac{u_{sh}\rho_B(1 - \epsilon)\dfrac{q_F}{c_F}}{k_M a_p}\left\{\frac{-a}{a - \dfrac{q_F}{c_F}}\ln\left[\left(\frac{0.05}{0.95}\right)\frac{a - \dfrac{q_F}{c_F} - 0.95bq_F}{a - \dfrac{q_F}{c_F} - 0.05bq_F}\right]\right.$$

$$\left. + \ln\left[\frac{a - \dfrac{q_F}{c_F} - 0.95bq_F}{a - \dfrac{q_F}{c_F} - 0.05bq_F}\right]\right\} \tag{12}$$

Note that the length of the mass transfer zone, L_{MTZ}, is inversely proportional to $k_M\,a_p$. If pore diffusion controls then $k_M\,a_p$ is inversely proportional to d_p^2, and thus L_{MTZ} is directly proportional to d_p^2.

If we scale so that L/d_p^2 is constant,

$$\frac{L_{MTZ}}{L} \propto \frac{d_p^2}{L} = \text{constant} \tag{13}$$

and the fractional bed use in the loading step of an on-off type system is constant. If we then cycle proportionally faster, we can obtain constant feed throughput (of the loading step) with a much shorter column.

The diffuse wave as a first approximation is controlled by the isotherm shape. Unless mass transfer rates are very low, the width of the diffuse wave is proportional to the distance travelled and is independent of mass transfer rates (d_p has no effect). If we scale as L/d_p^2,

$$\frac{\text{Width diffuse wave}}{L} \propto \frac{L}{L} = \text{constant} \tag{14}$$

Thus the diffuse wave also scales, but for a different reason than why the shock wave scales. The fractional time required for the elution (diffuse wave) step compared to the cycle time is constant, and the total throughput per hour is unchanged.

So far the analysis has inherently been limited to binary ion exchange or single non-linear isotherms for adsorption. The scaling will also work for more complex coupled systems. Coupled systems have been extensively studied using the local equilibrium theory (28,31-33,43,51,52,54). When the system is coupled, one still observes diffuse and shock waves. This is illustrated in Figure 2 where the local equilibrium solution is shown for chromatography of A^+ and B^+. Elution uses H^+ which has less affinity for the resin. Each individual shock wave will scale as L/d_p^2, as will the diffuse waves. The plateau region will also scale appropriately. This can be seen from a simple example. Assume that L, t_F, and t_{cycle} are all reduced by a factor of two. The elutant is now input earlier and the fast wave intersects the shock wave at 1/2 the original height. The dotted line will intersect the plateau at 1/2 the previous height. Thus the plateau will develop at 1/2 the height and 1/2 the elapsed time as in the original system. If z=L intersected the plateau region in the original case, it will in the scaled case and at the same relative time. Although the diagram will look considerably different, chromatography also scales when the elutant has a greater affinity for the resin than both A^+ and B^+ or an affinity between these two ions. Again the same caveat must be made as for the linear systems illustrated in Figure 1: timing becomes critical as cycle times are decreased.

Practical Aspects and Alternatives

Many of the practical aspects of using this scaling technique have already been discussed, but it will be useful to collect these aspects here.

1. Column design. The distributor must be more carefully designed. The amount of dead volume in the design must also be decreased, or the separation will be adversely affected.

2. Packing. The packing material must be carefully sieved so that the relative distribution of sizes does not increase drastically. A commercial source for the smaller diameter resin must be found. Packing procedures which utilize the full power of the smaller diameter packing are required.

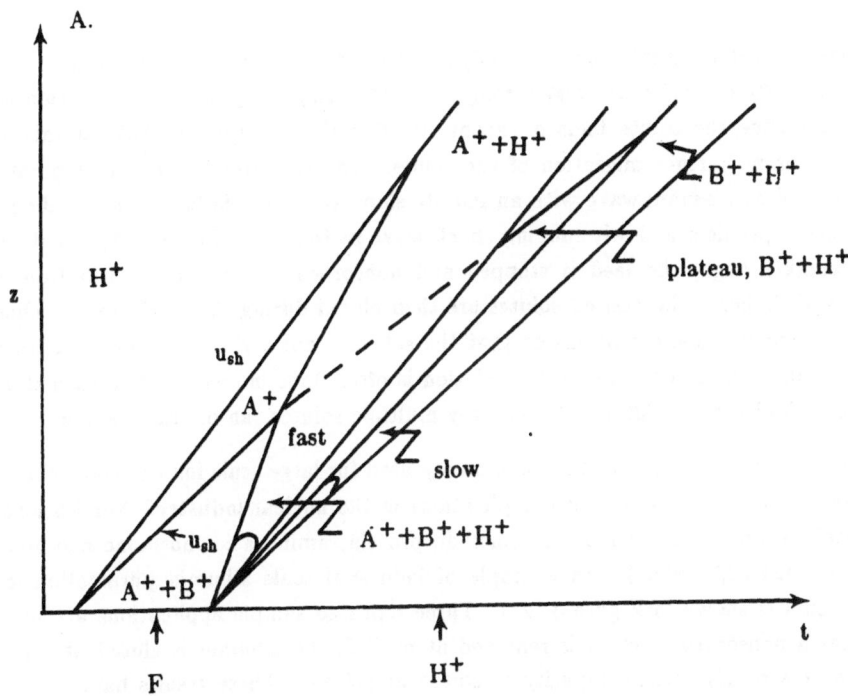

Figure 2. Characteristic solution for local equilibrium theory for coupled ion exchange chromatography.

3. Operation. Fast cycling requires more accurate timing, and if concentration control is used, more rapid concentration measurements are required. These limitations may limit use of extremely small diameter packings. In addition, the entering fluid must have all suspended solids removed.

Some alternate approaches which are a bit more elegant than the brute force method outlined here can be tried. The Graver Company (40) attaches micron sized particles to fibers and then packs these into filters. The large fibers give a reasonable pressure drop while the small particles have very rapid mass transfer rates. Composite resins with large macropores and micron sized regions where the bulk of the exchange capacity is available should also have reasonable pressure drops and high mass transfer rates. Perhaps the ultimate as one keeps decreasing particle diameter is to replace particles entirely with a porous sheet or membrane. Since this will necessarily be quite thin, cycling will have to be rapid.

ON-OFF CHROMATOGRAPHY

In on-off chromatography the thermodynamic conditions (pH, ionic strength, etc.) are set so that desired solutes are very strongly sorbed during the "on" step. Sorption is so strong that once the solute finds a vacant site it sorbs and remains held at this site. Thus there is no further migration of the solute. The isotherm for the on step can be approximated as a square wave with an infinite slope at the origin (67). The local equilibrium theory predicts a slowly moving shock wave as fresh feed fills up additional sites. Before breakthrough, the feed is stopped and nonsorbed material is washed from the column with buffer. The desired solutes are then eluted during the "off" step by changing the thermodynamic conditions so that the solute is not sorbed, is strongly attracted to a group in solution, or is displaced. Elution is often done in steps with a wash step in between each change in elutant. In this way multiple solutes can be fractionated.

On-off chromatography has been commonly used for large scale ion exchange chromatography. Schultz *et al* (55) review applications in the nuclear industry. Applications in biotechnology are quite common and include protein, amino acid, hormone and steroid purification (21,22,36,50,68). An example of industrial scale albumin purification from blood plasma is shown in Figure 3 (21). Three repeated sample applications are shown. Note that a nonsorbed fraction is removed at pH 5.2, the albumin is eluted at pH 4.5, and a more strongly sorbed impurity is eluted at pH 4.0. These results have not been optimized for throughput, probably because purity is of paramount importance.

The on-off chromatography system is very similiar to typical adsorption and ion exchange operations with the addition of multiple elution steps. Thus methods used for adsorption and ion exchange to increase efficiency and increase throughput can often be adapted. For example, counterflow regeneration is often advantageous since the solute clean end of the column is kept clean, complete regeneration of the most strongly sorbed component may not be required, and higher solute concentrations are usually obtained during elution.

It is very desireable to have L_{MTZ}/L be a small fraction so that bed utilization will be high during the loading step. This can be done by decreasing L_{MTZ} (see the previous section) or increasing L. Unfortunately, as L/L_{MTZ} increases past about 2 there is a declining rate of increase in bed utilization (44). This is illustrated in Figure 4 (44) for a simple adsorption system, but the same principles will be valid for more complex separations. Increasing the bed length also increases the pressure drop. For proteins and other large molecules the diffusivities are very low which results in low mass transfer rates and quite long mass transfer zones. Thus high bed utilization may not be possible with a reasonable pressure drop. The use of beds in series or in parallel can be used to increase bed utilization. In the beds in series arrangement shown in Figure 5 the adsorption step is done in two or more beds in series. In this way the mass transfer zone can be retained

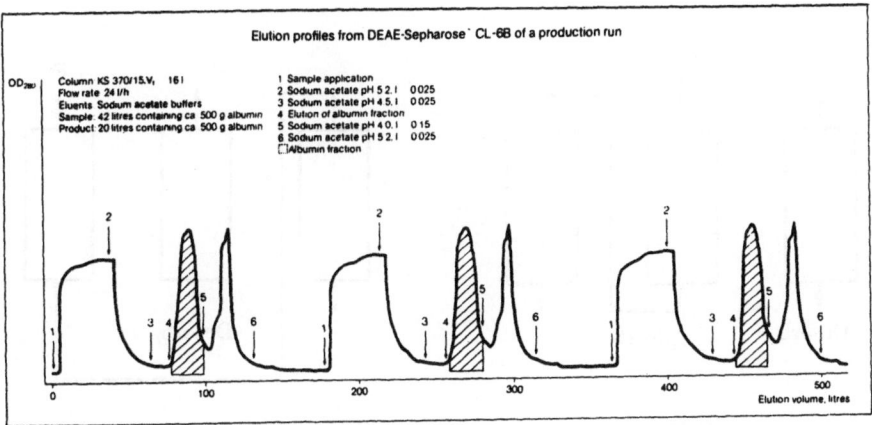

Figure 3. Purification of albumin from blood pressure plasma by on-off ion exchange (21). Reprinted with kind permission of Pharmacia, AB, and J.M. Curling, J.H. Gerglof, S. Erickson and J.M. Cooney.

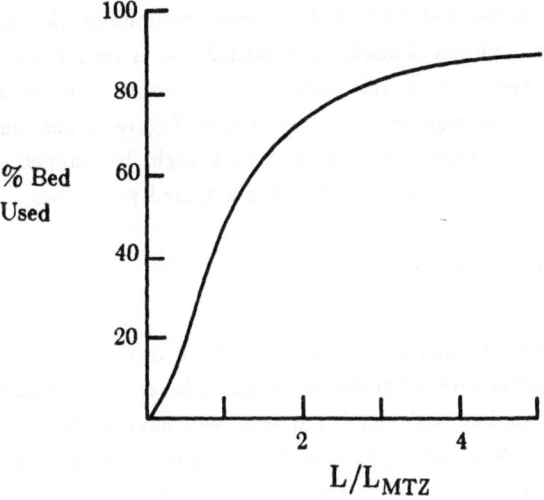

Figure 4. Bed utilization in adsorption (44). Excerpted by special permission from CHEMICAL ENGINEERING, *80*, (13), 111 (June 11, 1973). Copyright 1973 McGraw-Hill, Inc., New York, NY, 10020.

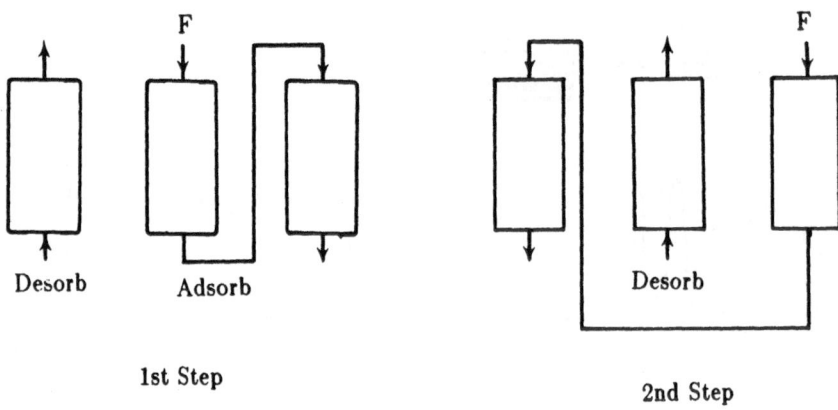

Figure 5. Beds in series system.

in the second bed while it has passed completely through the first bed. Thus the first bed is completely loaded. The loaded bed is then regenerated and a clean bed is added as the last bed in the series. Either co or counter-flow regeneration can be used. Counter-flow regeneration is shown in Figure 5 and has the advantage of helping to remove any suspended material which might be trapped in the bed. The beds in parallel arrangement is less efficient and is not used very often.

Use of Layers of Resins

Layers of different resins can be used advantageously to increase the bed efficiency. For systems with favorable isotherms a layer of large particles followed by a layer of the same resin but with small particles will have some of the advantages of both types of particles. The large particles have a low pressure drop but a large L_{MTZ}. The broad mass transfer zone will be sharpened in the layer of small particles because of the self-sharpening character of shock waves. The result will be a sharp breakthrough curve with a modest total pressure drop.

Use of two different resins can also be useful. Suppose that two different resins are available. One is expensive but has very favorable isotherms which will give very sharp breakthrough curves. If this resin follows a cheaper resin with a less sharp isotherm the mass transfer zone will be sharpened when it reaches the more expensive resin. This procedure can be advantageous even if the resin with the more favorable isotherm is not more expensive. If counter-flow elution is used the diffuse wave formed will spread much less in the resin with a more linear isotherm. The composite bed will then require less regenerant than a single bed of the resin with the very favorable isotherm, and a more concentrated stream will result.

If a resin with an unfavorable isotherm (resin A) and a resin with a favorable isotherm (B) are available, they can also be used in series (38). The characteristic diagram for this situation is shown in Figure 6 for two arbitrary isotherms (67). The diffuse wave formed during the loading step in resin A is sharpened into a shock wave in resin B.

During the regeneration step, a diffuse wave forms in resin B and is sharpened in resin A. The result can be sharp breakthrough patterns during both the loading and the regeneration steps. A similiar result can be obtained with a single resin if its isotherm is favorable under the conditions of loading and unfavorable under the regeneration conditions. Of course, this is what happens in ordinary water softening.

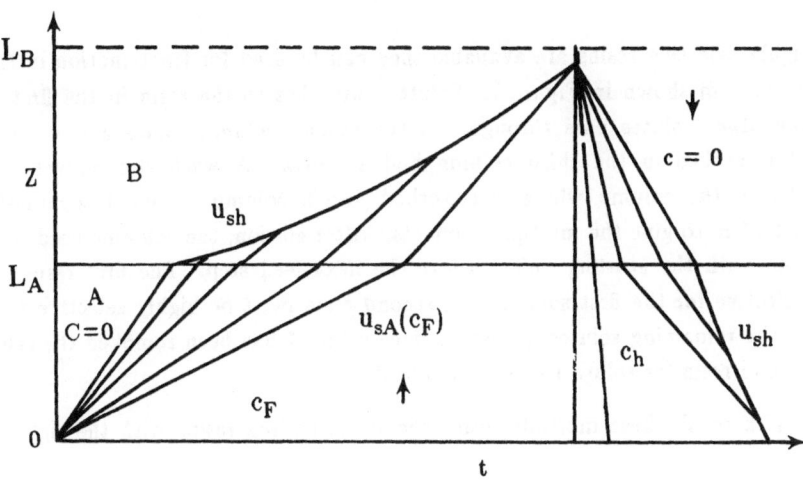

Figure 6. Characteristic diagram for binary ion exchange for resin with unfavorable isotherm (A) followed by resin with a favorable isotherm (B) (67). Reprinted with permission from P.C. Wankat, *Large Scale Adsorption and Chromatography*, CRC Press, Boca Raton, FL, 1985. Copyright CRC Press 1985.

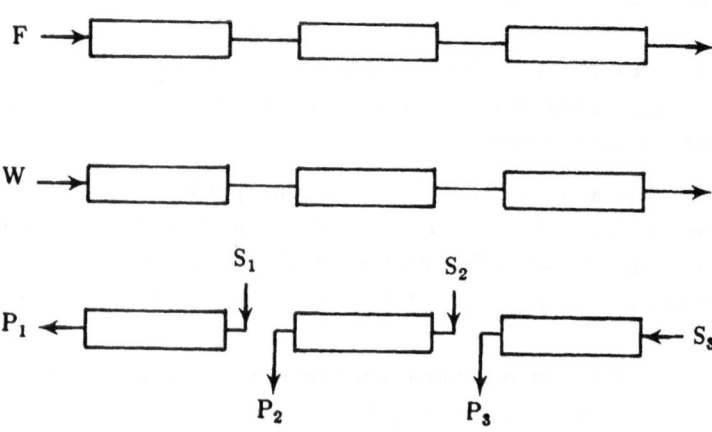

Figure 7. Fractionation scheme with highly selective resins.

If highly selective resins are available they can be used for fractionation in the multicolumn system shown in Figure 7. Solute 1 attaches to the resin in the first column while the other solutes pass through. In the second column solute 2 is sorbed while solute 3 is sorbed in the third column and so forth. A wash step moves unsorbed material into the column where it is sorbed. Each column is eluted separately with different buffers to give the multiple products. After elution, the columns need to be reequilibrated with the starting buffer before the next feed step. The first resin must be highly selective for the first solute. The second resin must be highly selective for solute 2 among the remaining solutes; however, since solute 1 has been removed the selectivity of the second resin for solute 1 is not important.

The trick in all these multiple resin schemes is finding resins with the right properties.

Fluidized and Dense Moving Beds

In on-off chromatography the desired solutes are very strongly sorbed during the on step. Thus only a few equilibrium contacts are needed for adequate separation during the loading step. Fluidized or dense moving beds are a high throughput possibility for a first rough fractionation. These systems have been used commercially for a variety of ion exchange separations and can be scaled to any desired size (3,19,25,34,48,53,57,58,60,67). A complete continuous system is shown in Figure 8. Intermittently pulsed Cloethe-Streat (19,48,57,60,67) or other types of fluidized beds (34,67) could be used for the loading step. The fluidized beds have a high capacity and can be fed an unfiltered feed. This could be a major advantage since the filtration step is bypassed and there will be less loss of desired product on the solids.

The solutes are removed separately in the regeneration columns. Upwards solids flow is shown, but downward flow is also used. Since flow rates in the regeneration columns are usually quite low and since long residence times are desired, moving beds or pulsed

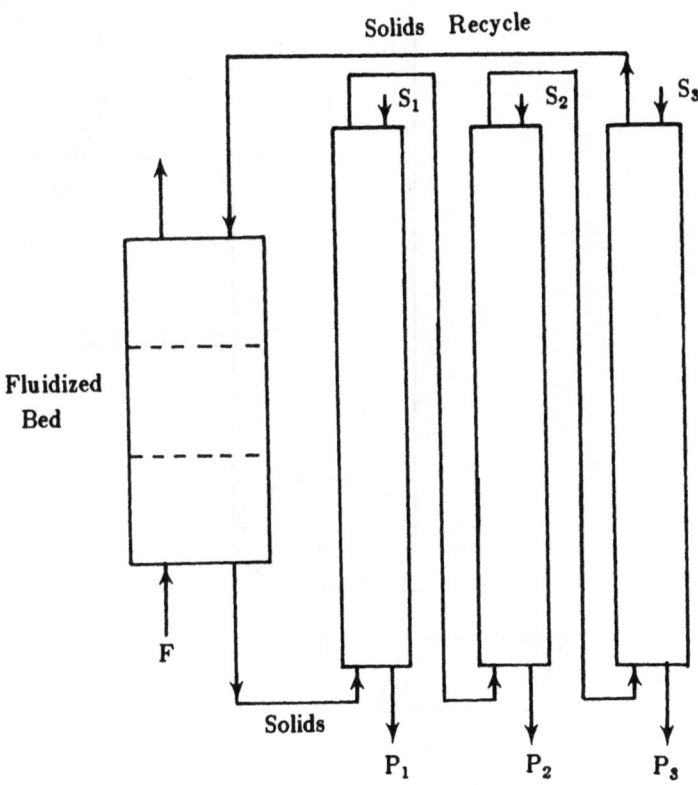

Figure 8. Continuous system for on-off chromatography.

moving beds would be appropriate. Current technology appears to be adequate (34,48,57,60,67), but better results can probably be obtained with magnetically stabilized moving beds (53). A wash step might be desireable between the regeneration columns. A single fluidized bed would probably be satisfactory. Although quite complex this continuous system would have a very high capacity. It could take unfiltered broth directly from a fermenter, remove the desired solutes, and return the broth to the fermenter. Since the cuts are not perfect the products will have to be further purified downstream.

Another fluidized bed system which has been used commercially in pharmaceutical manufacture (8-10) is shown in Figure 9. The raw broth is sent to a high capacity vibrating screen where large solid particles are removed. The feed then fluidizes the

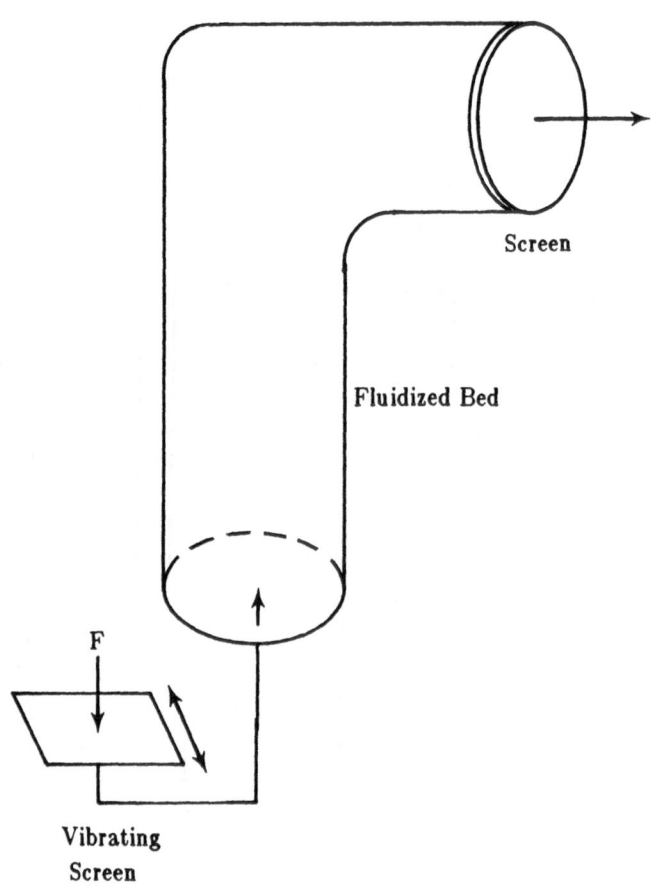

Figure 9. Elbow system for fluidized bed ion exchange.

resin in a fluidized bed held in an elbow. The elbow helps to remove the resin so that it does not clog the screen. The much smaller suspended solids pass through the screen. Usually a stirrer with the blade immediately next to the screen was used to prevent clogging of the screen. If desired the apparatus can be operated continuously with addition and removal of resin, but usual operation was batch. Several columns were placed in series as in Figure 5. When a column is saturated, it is removed from the series. The resin is allowed to settle and the column is regenerated in downflow as a packed bed. Elution can be done in steps to remove different solutes.

Capacity of the fluidized bed systems can be increased by using the newly developed magnetic resins which flocculate when stirring ceases (12,61).

MIGRATIONAL CHROMATOGRAPHY

In migrational chromatography a solute does not stay attached to the solid once it has been sorbed. Instead, the solute will sorb and desorb many times as it moves down the column. The stronger the affinity of the solute for the solid the more time the solute spends attached to the solid and the slower the average velocity of the solute in the column.

The usual large scale operating technique for migrational chromatography is to input a large pulse of feed and then elute different peaks. A second feed pulse is input before the slowest moving component exits the column. Although complete separation can be obtained, this is usually not done since excessive elutant is required. Instead, partially separated portions are recycled or sent to other columns. An example is shown in Figure 10 for the ion exclusion chromatography purification of sucrose from molasses(30). Water is used as the elutant. Figure 10 shows mainly the results of one feed pulse, but the tailing edge of the previous pulse and the leading edge of the next pulse can also be seen. Note that complete separation was not obtained and there are two recycle streams.

Recycle is commonly employed since it allows the next feed pulse to be input sooner which increases throughput, it reduces the amount of elutant required, and it keeps the products from becoming excessively diluted. Recycle operation can be improved by using segmented recycle (5). This is shown schematically in Figure 11 (67). A pulse of the critically important academic chemicals, A and B, is fed to the column. Since the separation is difficult, there is considerable overlap of the peaks. To acheive the desired purity a large fraction can be recycled. However, if this recycle stream is mixed together, the partial separation which has been obtained will be lost. Entropy production can be minimized by segmenting the recycle stream as shown. Stream R_1 is more

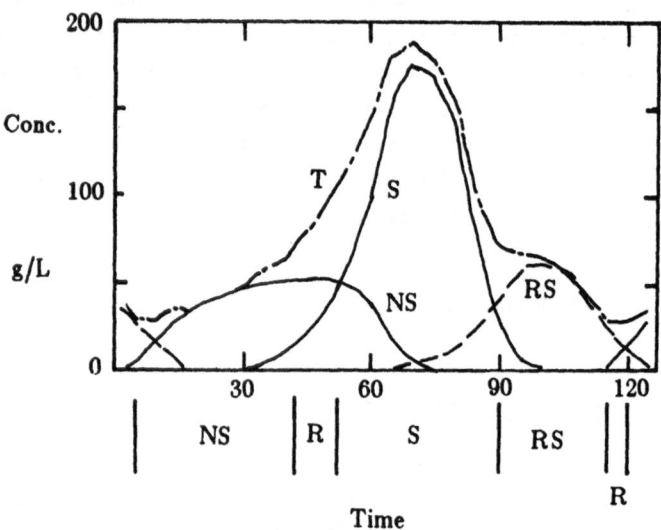

Figure 10. Ion exclusion chromatography of sugars (30). T = total sugars, S = sucrose, RS = reducing sugars, NS = non-sugars, R = recycle streams. Excerpted by special permission from CHEMICAL ENGINEERING, 50, Jan. 24, 1983. Copyright 1983, McGraw-Hill, Inc., New York, NY, 10020.

concentrated in the faster moving A so R_1 should be input at the beginning of the feed. R_2 is approximately at the feed concentration and can be mixed with the feed. Finally, R_3 is concentrated in B and should be input at the tail of the feed pulse. The more subdivisions the better, but each subdivision requires special handling. Essentially infinite subdivisions can be obtained if column switching methods (see next section) are used instead of recycle.

Migrational chromatography results can also be improved by proper choice of elutant. An elutant with a weak affinity for the resin does not rapidly remove slow moving materials and excessive time can be required to remove these compounds. An elutant with a very strong affinity for the resin will hustle the slow moving components out of the column, but will also block sites and reduce the separation of the components. When a single elutant is used the optimum is an elutant with an affinity in-between those of the two key components(18,27). This compromise appears to give the highest productivity, but one may not be able to find an elutant with the correct affinity.

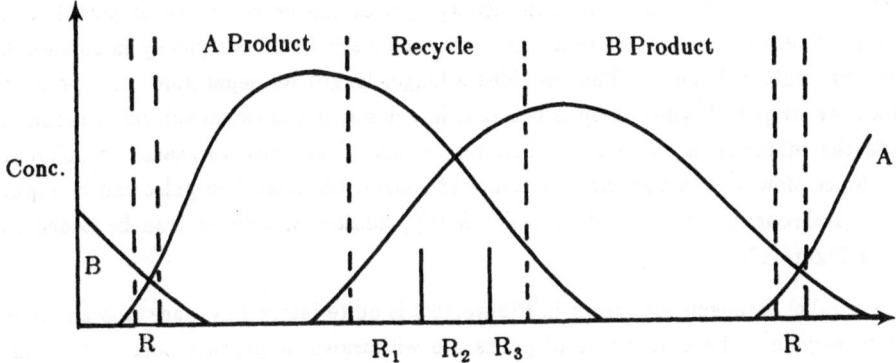

Time or Volume

Figure 11. Illustration of segmented recycle (67). Reprinted with permission from P.C. Wankat, *Large Scale Adsorption and Chromatography*, CRC Press, Boca Raton, FL, 1985. Copyright 1985, CRC Press.

Slow moving compounds can also be removed by programming which is the addition of elutant of changing pH, ionic strength etc. This approach does require that the column be reconditioned before the next feed pulse. Programming is also considerably more complex than isocratic operation, and isocratic operation seems to be much more common for large scale migrational chromatography.

Column Switching and Moving Ports

The productivity of large scale migration chromatographs can be improved by using column switching and moving port methods. One problem with a typical chromatograph is that the leading edge and the trailing edge of the feed pulse rapidly become purified, but they must pass through the entire column before they are withdrawn as products. This causes further zone spreading and dilution, and is an inefficient use of the resin. It is much more efficient to remove purified material as soon as possible. This can be done with one of a variety of column switching methods (13,59,66,67,69). Some of the possible arrangements are shown in Figure 12 (67). The basic column switching system is shown in Figure 12A. The feed is input into column A. A species with a low affinity for the resin will pass through the column quickly and will be completely separated in the first column. This low affinity species can then be collected at product port 1. A species with a high affinity for the resin will move very slowly and will also be completely separated in column A. Thus the high affinity species can be recovered at port 1 at a later time. Species of intermediate affinity will not be separated completely in column A and are sent onto column B. This provides a longer length for separation. If a product is withdrawn at port 1 while there is material in column B additional solvent is required to elute the material in column B. Columns A and B can use the same or different resins. Since slow and fast species are removed sooner, the next feed pulse can be input sooner and throughput will be higher. Obviously, additional columns can be added as shown in Figure 12B.

The parallel arrangement shown in Figure 12C is quite interesting since it is an alternative to recycle. The center cuts of peaks are withdrawn at product port 1. Streams which would normally be recycled are sent to the additional columns. Since there will be very little mixing in a well designed piping system, the streams sent as feeds to columns B,C,and D will retain the partial separation shown in Figure 11. Thus, this parallel arrangement is approximately equivalent to segmented recycle with an infinite number of segments. Figure 12D shows that parallel columns can be combined with recycle. In fact, any of the arrangements can use recycle, and the recycle can be returned to a variety of locations. A good optimization problem!

If some components have a very high affinity for the resin, they will not migrate very far in the column. It is often more efficient to backflush these species than to drive them through the entire length of the column. Backflushing is also useful with nonlinear isotherms since the diffuse wave formed during elution can later be resharpened by the shock wave. Backflushing can be done with the same solvent used in forward flow or with a desorbent which will force the material off the resin. Column switching arrangements can be used in several ways with backflushing. One way to do this and to fractionate the high affinity species is shown in Figure 12E. Once the backflushed material has entered column B, which probably uses either a different resin or different buffer conditions, column A can be reconditioned for the next feed pulse if necessary.

Other variants of column switching methods such as "box-car chromatography" and "moving withdrawal" methods have been developed (67). Column switching methods are also used in "two-way chromatography" (discussed later). The general ideas of column switching methods have a wide variety of applications to improve the efficiency of large scale chromatography.

The column switching methods essentially work on the inefficiencies of chromatography at the product end. Inefficiencies at the feed end can be attacked with the moving feed method (4,2745,67,68). In this approach a system similiar to Figure 12B can be used. However, product ports 1 and 2 are not used, and the ports marked S are used for

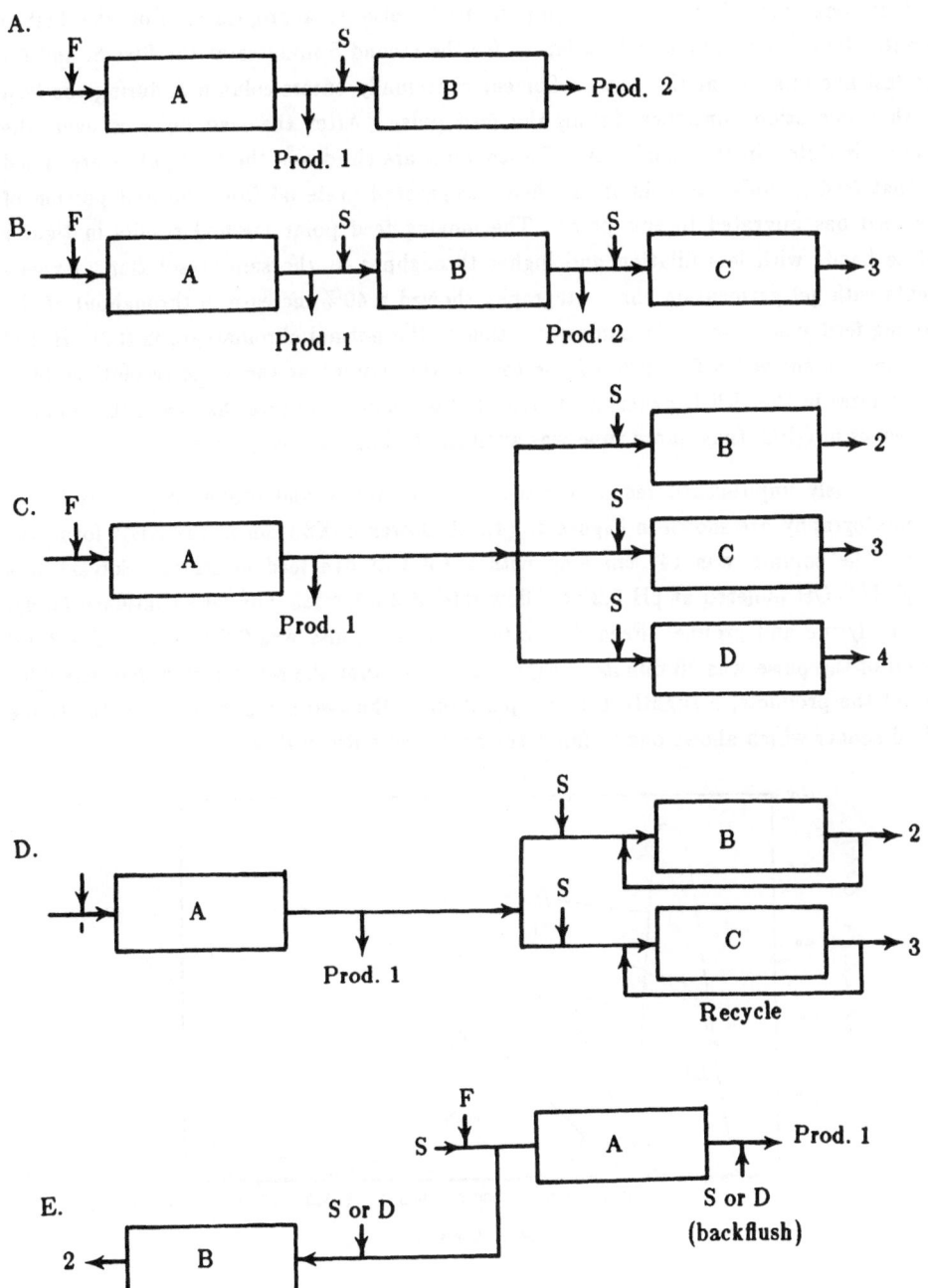

Figure 12. Column switching arrangements (67). Reprinted with permission from P.C. Wankat, *Large Scale Adsorption and Chromatography*, CRC Press, Boca Raton, FL, 1985. Copyright 1985, CRC Press.

356

feed introduction. Suppose a 15 minute feed pulse is appropriate. For the first 5 minutes feed can be input at location F, for the second 5 minutes at the first S, and for the last five minutes at the last S. Solvent continually enters column A during the feed so that migration continues during the feed pulse. After the feed pulse is over, the column is eluted in the usual way. The columns are sized and the feed pulses are timed so that feed is added to column B when unseparated material from the first portion of the feed has migrated to this point. The moving feed point method results in tighter solute bands with less dilution and higher throughput at the same resolution. Experiments with gel permeation chromatography showed a 40% increase in throughput of the moving feed system with the same resolution as the normal chromatograph (68). HPLC experiments showed a 90 to 300% increase in throughput at the same resolutions (45). The solutes in the HPLC experiments travel at a much lower rate than even the smallest molecules in GPC; thus, more time was available to improve the process.

Previously unpublished results for separation of lysine and proline by ion exchange chromatography are shown in Figure 13 (4). A Dowex 50X8 resin in the NH_4^+ form was used. The column was 147 cm long with a total of five feed locations. Solvent was $NH_4OH/AcOH$ buffered at pH 5.2 at a flow rate of 2 ml/min. The feed contained 80 g/l of both lysine and proline. Feed flow rate during the pulse was 0.2 ml/min. The total period of the pulse was 58 minutes. Figure 13 shows that the MFP system had less dilution of the products, a slightly better separation of the two components, and the lysine exited sooner which allows one to input the next feed pulse earlier.

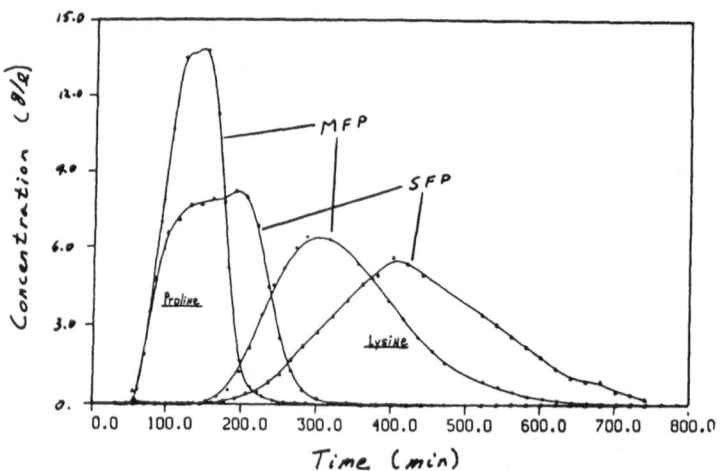

Figure 13. Separation of proline and lysine by ion exchange chromatography (4). MFP = moving feed port, SFP = stationary feed port (normal operation). Used with the kind permission of Dr. Blaine Asay.

Obviously, one can combine moving feed and column switching ideas. One way to do this is to use the apparatus shown in Figure 14 (66). Feed or solvent can be introduced into any of the even numbered ports, and products can be withdrawn at any of the odd numbered ports. Results of a numerical simulation of the separation of naphthalene N, anthracene A, and pyrene P are shown in Figure 15 (47). Feed was introduced at 0,5,10,15, and 20 cm while products were withdrawn at 25, 30,35,40,45, and 50 cm. The intermediate withdrawal ports were opened and closed to give high purity products. This is shown in Figure 15A. Very little anthracene contaminants either of the other products. The sharp drops in the N and P peaks indicate that a valve has closed or opened. Most of the anthracene is recovered at the end of the column as shown in Figure 15B. The numerical simulations predicted a 22% increase in average purity and a 120% increase in feed throughput for the moving port chromatograph compared to normal operation (47). Calculations for concentrated ion exchange chromatography using the local equilibrium theory also show considerable improvement of the moving port methods (27).

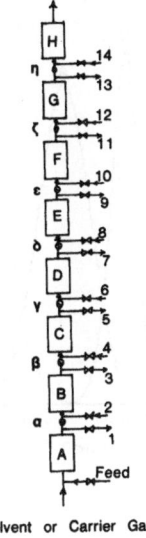

Solvent or Carrier Gas

Figure 14. Moving port chromatography system (66). Reprinted with permission from *Ind. Eng. Chem. Fundam., 23,* 256 (1984). Copyright 1984, American Chemical Society.

358

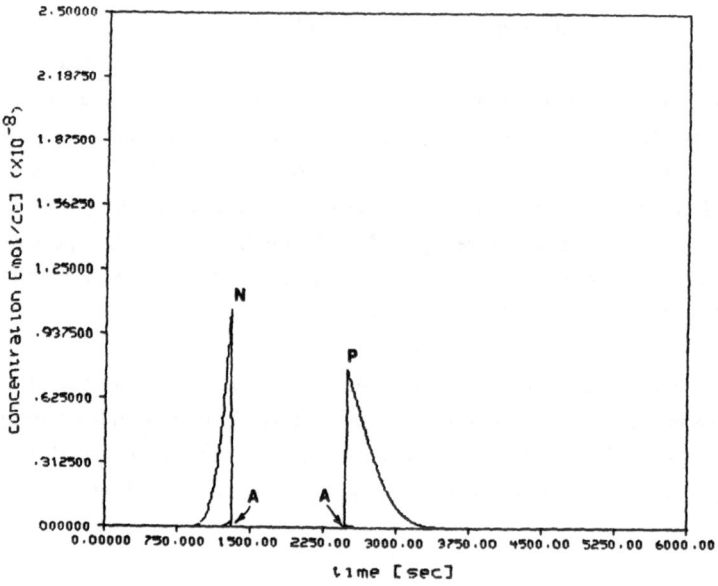

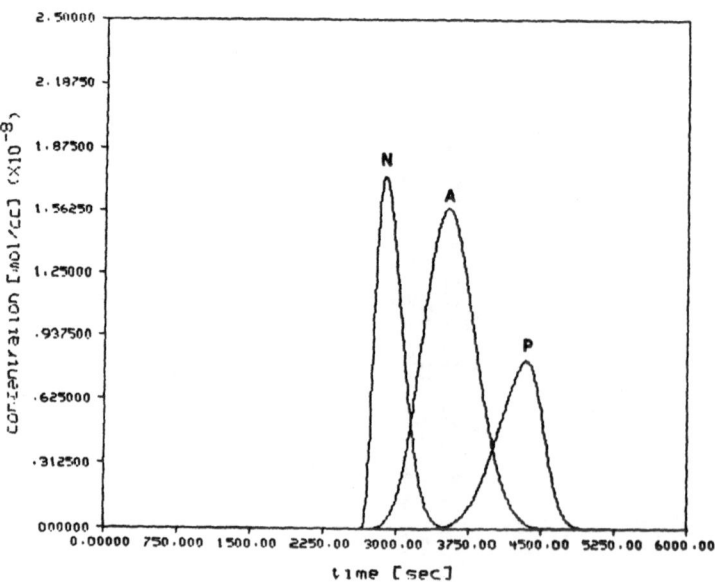

Figure 15. Computer simulation of moving port chromatography (47). A. Intermediate withdrawal at 25 cm. B. Final product withdrawal. Reprinted with permission from *Chem. Eng. Commun., 31*, 21 (1984). Copyright 1984, Gordon and Breach Science Publishers, Inc.

Simulated Moving Beds

The previous systems have all been chromatographs; that is, they are all batch processes which can all do multicomponent separations in a single cascade. For binary separations many commercial systems use simulated moving beds (7,14-17,23,24,35,49,67). The idea is to simulate countercurrent flow of the solid and fluid without actually moving the solid. One way of doing this is shown in Figure 16. The column is packed in a series of sections with plumbing in-between each section. In the arrangement shown in Figure 16, developed by Universal Oil Products, a custom rotary valve is used to control the addition of feed and desorbent, and removal of the two desired products. The column is divided into four zones with an addition or withdrawal between each zone. There may be several small packed sections within each zone. With continuous flow of liquid up the column, every few minutes the rotary valve switches so that all port locations move up one section. An observer located at a product port sees the solid move down when the ports switch. Thus, countercurrent motion of fluid and solid has been simulated.

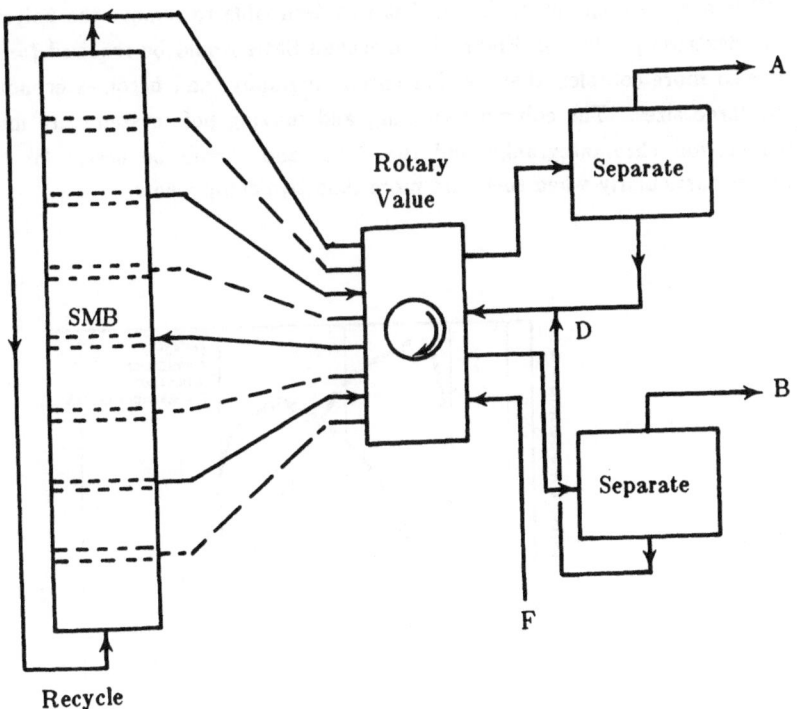

Figure 16. Simulated moving bed system following UOP arrangement.

Other arrangements can be used to simulate countercurrent motion. Simpler system using solenoid valves are used commercially (35,67). In this case separate columns are used for each zone. The simulation of countercurrent motion is not as good as the multisection system shown in Figure 16, but this arrangement is less expensive when sharp separations are not required. For fructose and glucose separations an ion exchange resin in the calcium form is used.

The profiles observed in the column for a pilot scale version of the system shown in Figure 16 are shown in Figure 17 (49). A molecular sieve packing was used. Fructose complexes with the cation held by the zeolite and thus is the more strongly sorbed species. Fructose acts as if it were moving downward in a countercurrent cascade. Dextrose is less sorbed and the polysaccharides are excluded so they both act as if they were moving up the countercurrent cascade. The results shown in Figure 17 show an excellent separation of fructose and glucose, and there are no parts of the column which are not doing a useful separation. This system is being used commercially for sugar separations.

The SMB is more efficient than an elution chromatograph for binary separations. Numerical simulations show that the SMB will require about 1/2 as much desorbent and about 1/4 as much adsorbent as normal elution chromatography (15,16,23,54). However, the SMB is a *binary* separator. Thus, if it was desireable to remove the polysaccharides from the dextrose product in Figure 17, a second SMB would be required for this. The SMB is also more complex than elution chromatography, and becomes economical only at fairly large sizes. The column switching and moving port systems are intermediate between elution chromatography and the SMB, and should be useful for large scale separations particularly when there are more than two components.

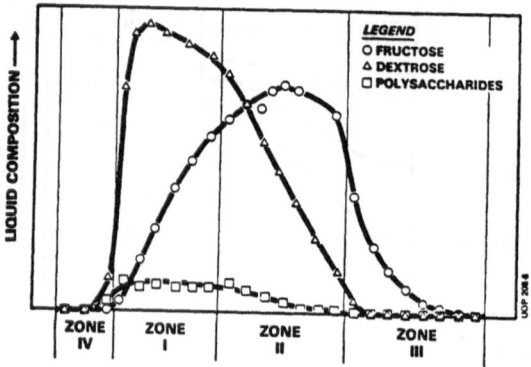

Figure 17. Composition profiles for fructose and dextrose separation in a SMB (49). Reprinted with kind permission of Dr. R.W. Neuzil.

Two-Way Chromatography

Bailly and Tondeur (6) developed a clever series of techniques which utilize flow reversal and intermediate withdrawals to improve chromatography of a feed containing two ions. The processes use flow reversal and the competing nature of the ion exchange system to both recompress the diffuse wave and to concentrate one of the products. Intermediate withdrawal is used to obtain the other product. A variety of different processes were developed. Only the simplest process will be discussed here, and the interested reader is referred to the original paper (6).

The simplest version of the process is single feed, simple reversal, two-way chromatography shown in Figure 18 (6). The particular operation shown is for a case where the elutant H^+ has less affinity for the resin than the two ions in the feed, Na^+ and K^+. K^+ has the highest affinity for the resin. At the start of the cycle the resin is saturated

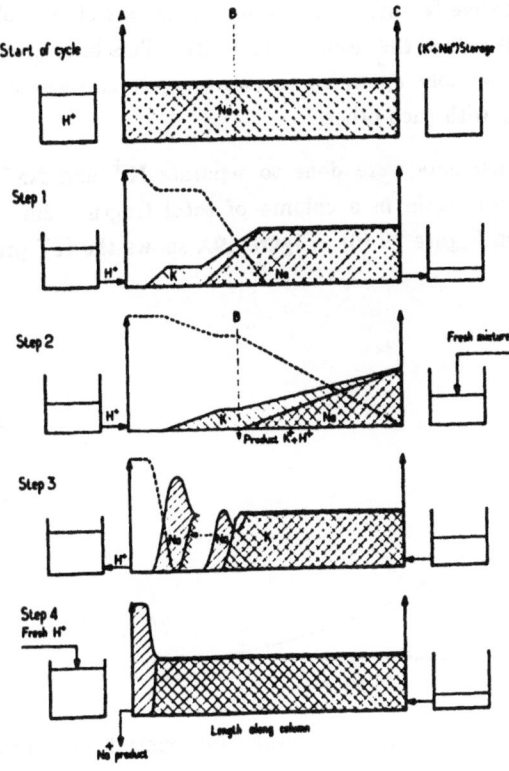

Figure 18. Schematic of concentration profiles for single feed, simple reversal, two-way chromatography (6). Reprinted with permission from *Chem. Eng. Sci., 36,* 455, 1981. Bailly, M. and D. Tondeur, "Two-way chromatography. Flow reversal in non-linear preparative liquid chromatography." Copyright 1981, Pergamon Press.

with Na$^+$ and K$^+$. During the first step elutant is added from the left, and K$^+$ and Na$^+$ are stored in the right reservoir. The K$^+$ lags behind the Na$^+$ and a region of separated K$^+$ diluted in H$^+$ results. In step 2 the region of separated K$^+$ has reached the intermediate withdrawal port. Since this material has been purified it is removed while continuing to elute with H$^+$. The right hand end of the column is stagnant during step 2. Once the desired quantity of K$^+$ has been removed, flow is reversed and the column is flushed with a combination of fresh feed and stored K$^+$ and Na$^+$ from the reservoir. This recompresses the diffuse waves for both Na$^+$ and K$^+$. Pure H$^+$ is eluted from the left end of the column and is stored for future use. Eventually, the Na$^+$ wave will pass through the strongly held K$^+$ wave. When this happens, the Na$^+$ concentration starts to build up since the K$^+$ serves to displace the Na$^+$. The result is a shock wave of concentrated or even pure Na$^+$ (not mixed with H$^+$). In step 4 this Na$^+$ wave is withdrawn as product. When the K$^+$ wave breaksthrough the column has returned to its initial state.

This cycle is interesting since the Na$^+$ is concentrated using the feed solution. The one negative feature of the cycle is that part of the column is stagnant during step 2 and this will reduce the resin productivity. This later problem can be solved by using several column sections operating as two or more two-way chromatographs in parallel, but out of phase with each other.

Experiments were done to separate K$^+$ and Na$^+$ using a Duolite C20, sulfonated polystyrene resin in a column of total length 3.2m. Some of the reported results are shown in Figure 19 (6). Figure 19A shows the K$^+$ product. Note that it contains very

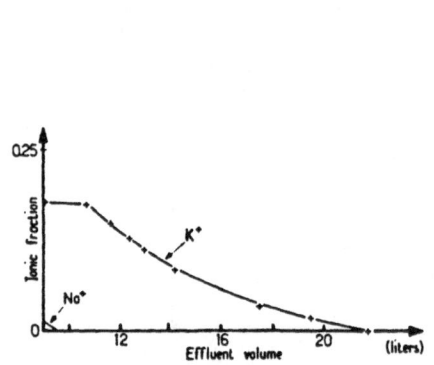
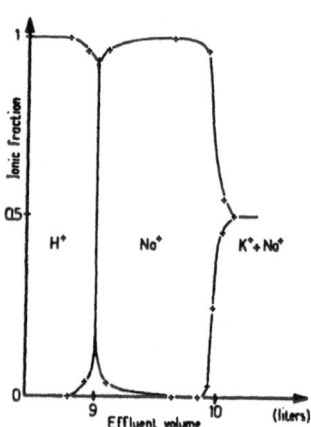

Figure 19. Experimental results for single feed, simple reversal, two-way chromatography (6). A. Effluent concentration history for side stream product (step 2). B. Effluent concentration history of Na$^+$ product (steps 3 and 4). Reprinted with permission from *Chem. Eng. Sci.*, *36*, 455, 1981. Bailly, M. and D. Tondeur, "Two-way chromatography. Flow reversal in non-linear preparative liquid chromatography." Copyright 1981, Pergamon Press.

little Na^+, but is quite diluted by the H^+. Figure 19B shows material exiting the left side of the column during steps 3 and 4. The pattern between H^+ and Na^+ is very sharp which gives a sharp cut of the Na^+ product. The K^+ wave is also sharp and operation would be ceased at breakthrough. The local equilibrium theory was successfully used to predict the general shape of the separations.

ACKNOWLEDGMENT

Much of the review reported here was done while on sabbatical at LSGC-ENSIC in Nancy, France. The hospitality of LSGC-ENSIC was greatly appreciated. Discussions with Dr. John Dodds, Dr. Georges Grevillot and Dr. Daniel Tondeur were very important in sharpening my ideas. Support from NSF/CNRS through the US/France Scientific Exchange program is gratefully acknowledged.

NOMENCLATURE

A,B,C	Parameters in Van Deemter equation (2)
a,b	Equilibrium constants in Langmuir isotherm
c_M, c_{SM}	Constants in Equation 5
c	concentration
c_F	feed concentration
d_p	particle diameter
D_M	molecular diffusivity
D_{SM}	molecular diffusivity in stagnant mobile phase
E_D	Eddy diffusivity
H	height of a theoretical plate
k_i	linear equilibrium constant
k_i'	relative retention, $k_i V_s/V_M$
$\overline{k'}$	average relative retention
K	permeability
L	bed length
L_{MTZ}	length of mass transfer zone
N	number of plates
Δp	pressure drop
q	amount sorbed or on resin
q_F	amount sorbed in equilibrium with feed

R_S	resolution, Eq. 8
t_{R_1}, t_{R_2}	retention time of peaks
v	interstitial velocity
V_M	volume mobile phase in column
V_S	volume stationary phase in column
w_1	width of peak
α_{21}	selectivity $= \dfrac{k_2}{k_1} = \dfrac{k_2'}{k_1'}$
μ	viscosity
λ	constant in Equation (3)
γ	tortuosity factor, Equation (4)

REFERENCES

1. Anderson, R.E., "Ion-Exchange Separations" in Schweitzer, P.A., Ed., *Handbook of Separation Techniques for Chemical Engineers*, McGraw-Hill, MY, (1979), Sect. 1.12.

2. Anon., Ion exchange system, *Chemical Engineering*, 59, Aug. 20, 1984.

3. Arden, T.V., *Water Purification by Ion Exchange*, Butterworths, London, 1968.

4. Asay, Blaine, Private Communication, 1984.

5. Bailly, M. and D. Tondeur, Recycle optimization in non-linear productive chromatography. I. *Chem. Eng. Sci., 37*, (1982), 1199.

6. Bailly, M. and D. Tondeur, Two-Way Chromatography. Flow reversal in non-linear preparative liquid chromatography, *Chem. eng. Sci., 36*, (1981), 455.

7. Barker, P.E., G.A. Irlam, and E.K.E. Abusabah, Continuous chromatographic separation of glucose-fructose mixtures using anion-exchange resins, *Chromatographia, 18*, (1984), 567.

8. Bartels, C.R., G. Kleiman, J.N. Korzun, and D.B. Irish, A novel ion-exchange method for the isolation of streptomycin, *Chem. Eng. Prog., 54* (8), (1958), 49.

9. Belter, P.A., Ion exchange and adsorption in pharmaceutical manufacturing, *AIChE Symp. ser., 80* (233), (1984), 110.

10. Belter, P.A., F.L. Cunningham, and J.W. Chen, Development of a recovery process for Novobiocin, *Biotech. Bioeng., 15*, (1973), 533.

11. Bird, R.B., W.L. Stewart, and E.N. Lightfoot, *Transport Phenomena*, Wiley, N.Y., (1960), p. 196-200..

12. Bolto, B.A., Sirotherm desalination, ion exchange with a twist, *Chemtech., 5*, (1975), 303.

13. Bonmati, R., G. Chapelet-Letourneax, and G. Guiochon, Gas chromatography: A new industrial process of separation. Application to essential oils, *Separ. Sci.*

Technol., 19, (1984), 113.

14. Broughton, D.B., Molex: Case history of a process, *Chem. Eng. Prog., 64,* (8), (1968), 60.

15. Broughton, D.B., Production-scale adsorptive separations of liquid mixtures by simulated moving-bed technology, *Separ. Sci. Technol., 19,* (1984-85), 723.

16. Broughton, D.B., Adsorptive separations (liquids), in *Kirk-Othmer Encyclopedia of Chemical Technology,* 3rd, ed., Vol. 1, Wiley-Interscience, NY, (1978), 563-581.

17. Broughton, D.B., R.W. Neuzil, J.M. Pharis, and C.S. Brearley, The Parex process for recovering paraxylene, *Chem. Eng. Prog., 66,* (9), (1970), 70.

18. Carra, S., M. Morbidelli, G. Storti, and R. Paludetto, Experimental analysis and modeling of adsorption separation of chlorotoluene isomer mixtures, in Myers, A.L. and Belfort, G., Eds., *Fundamentals of Adsorption,* Engineering Foundaton, (1984), 143.

19. Cloete, F.L.D. and M. Streat, A new continuous solid-fluid contacting technique, *Nature, 200* (4912), 1199, Dec. 21, 1963.

20. Cohen, Y. and A.B. Metzner, Wall effects in laminar flow of fluids through packed beds, *AIChE Journal, 27,* (1981), 705.

21. Curling, J.M., J.H. Gerglof, S. Ericksson, and J.M. Cooney, Large Scale Production of Human Albumin by an all-solution chromatographic process, Joint meeting of the 18th Congress of the International Society of Heamatology and the 16th Congress of the International Society of Blood Transfusion, Montreal, Quebec, Canada, Aug. 16-22, 1980.

22. Darbyshire, J., Large scale enzyme extraction and recovery, in Wiseman, A., Ed., *Topics in Enzyme and Fermentation Biotechnology,* Vol. 5, Ellis Horwood, Ltd., Chichester, England, (1981), Chapt. 3.

23. de Rosset, A.J., R.W. Neuzil, and D.B. Broughton, Industrial Applications of preparative chromatography, in A.E. Rodrigues and D. Tondeur, Eds., *Percolation Processes, Theory and Applications,* Sijthoff and Noordhoff, Alphen aan den Rijn, The Netherlands, (1981), 249-281.

24. de Rosset, A.J., R.W. Neuzil, and D.J. Korous, Liquid column chromatography as a predictive tool for continuous countercurrent adsorptive separations, *Ind. Eng. Chem. Proc. Des. Develop., 15,* (1976), 261.

25. Dorfner, K., *Ion Exchangers, Principles and Applications,* Ann Arbor Sci. Pub., Ann arbor, MI, 1972.

26. Eco-Tec Ltd., Eco-Tec Ion Exchange Systems, 1983 and Recoflo - A breakthrough in water deionization systems, 1984, 925 Brock Rd. South, Pickering (Toronto), Ontario, Canada L1W 2X9.

27. Geldart, R.W., Q. Yu, P.C. Wankat and N.-H. L. Wang. Multicomponent moving port chromatography, American Chemical Society Annual Meeting, Chicago, Sept. 1985.

28. Giddings, J.C., *Dynamics of Chromatography, Part I, Principles and Theory,* Marcel Dekker, N.Y., 1965.

29. Glueckauf, E., Theory of chromatography. Part VII. The general theory of two solutes following non-linear isotherms, *Dics. Faraday Soc.*, *7*, (1949), 12.

30. Heikkila, H., Separating sugars and amino acids with chromatography *Chem. Eng.*, 50, Jan. 24, 1983.

31. Helfferich, F., *Ion Exchange*, McGraw-Hill, NY, 1962.

32. Helfferich, F.G., Conceptual view of column behavior in multicomponent adsorption or ion-exchange systems, *AIChE Symp. Ser.*, *80* (233), (1984), 1.

33. Helfferich, F. and G. Klein, *Multicomponent Chromatography*, Marcel Dekker, NY, 1970.

34. Himsley, A. and E.J. Farkas, Operating and design details of a truly continuous ion exchange system, in Streat, M., Ed., *The Theory and Practice of Ion Exchange*, Society of Chemical Industry, London, (1976), 45.7.

35. Illinois Water Treatment Co., IWT Adsep System, *Making Waves in Liquid Processing*, *1* (1), 1, Rockford, Illinois.

36. Janson, J.-C. and P. Hedman, Large-Scale Chromatography of Proteins, in Fiechter, A., Ed., *Advances in Biochemical Engineering*, Vol. 25, *Chromatography*, Springer-Verlag, Berlin, (1982), 43.

37. Jones, R.L. and G.E. Keller, Pressure-swing parametric pumping - a new adsorption process, *J. Separ. Proc. Technol.*, *2*, (3), (1981), 17.

38. Klein, G., Design and development of cyclic operations, in Rodrigues, A.E. and Tondeur, D., Eds., *Percolation Processes, Theory and Applications*, Sijthoff & Noordhoff, Alpheen aan den Rijn, The Netherlands, (1981), 427-441.

39. Knox, J.H., Practical Aspects of LC Theory, *J. Chromatogr. Sci.*, *15*, (Sept. 1977), 352.

40. Kunin, R., A. Tavares, R. Forman, and G. Wilker, New developments in the use of ion-exchangers and adsorbents as precoat filters, in Naden, D. and Streat M., Eds., *Ion Exchange Technology*, Ellis Horwood Ltd., Chichester, England, (1984), 563-578.

41. LeVan, M.D. and T. Vermeulen, Channeling and bed-diameter effects in fixed-bed adsorber performance, *AIChE Symp. ser.*, *80* (233), (1984), 34.

42. Liberti, L. and F.G. Helfferich, Eds., *Mass Transfer and Kinetics of Ion Exchange*, Martinus Nijhoff Pub. Co., the Hague, Netherlands, (1983).

43. Lightfoot, E.N., R.J. Sanchez-Palma, and D.O. Edwards, in Schoen, H.M., Ed., *New Chemical Engineering Separation Techniques*, Interscience, NY, (1962), 125.

44. Lukchis, G.M., Adsorption systems. Part I: Design by mass transfer-zone concept, *Chem. Eng.*, *80* (13), (June 11, 1973), 111.

45. McGary, R.S. and P.C. Wankat, Improved preparative liquid chromatography: The moving feed point method, *Ind. Eng. Chem. Fundam.*, *22*, (1983), 10.

46. Michaels, A., Simplified method of interpreting kinetic data in fixed-bed ion exchange, *Ind. Eng. Chem.*, *44*, (1952), 1922.

47. Miller, G.H. and P.C. Wankat, Moving Port Chromatography: A Method of Improving Preparative Chromatography, *Chem. Eng. Commun.*, *31*, (1984), 21.

48. Naden, D. and M. Streat, Eds., *Ion Exchange Technology*, Ellis Horwood, Chichester, England, (1984).

49. Neuzil, R.W. and R.H. Jensen, Development of the Sarex process for the separation of saccharides, paper 22d, AIChE meeting, Philadelphia, PA, June 6, 1978.

50. Regnier, F.E., High-performance ion exchange chromatography, in Jakoby, W.B., Ed., *Methods in Enzymology*, Vol. 104, *Enzyme Purification and Related Techniques*, Academic Press, Orlando, FL, (1984), 170-189.

51. Rhee, H.-K., Equilibrium Theory of Multicomponent Chromatography, in Rodrigues, A.E. and Tondeur, D., Eds., *Percolation Processes, Theory and Applications*, Sijthoff and Noordhoff, Alphen aan den Rijn, Netherlands, (1981), 285-328.

52. Rhee, H.-K., R. Aris, and N.R. Amundson, On the theory of chromatography, *Phil. Trans. Roy. Soc. London, A267*, (1970), 419.

53. Rosensweig, R.E., J.H. Siegell, W.K. Lee, and T. Mikus, Magnetically stabilized fluidized solids, *AIChE Symp. ser., 77* (205), (1981), 8.

54. Ruthven, D.M., *Principles of Adsorption and Adsorption Processes*, John Wiley & Sons, NY, 1984.

55. Schultz, W.W., E.J. Wheelwright, H. Godbee, C.W. Mallory, G.A. Burney, and R.M. Wallace, Ion exchange and adsorption in nuclear chemical engineering, *AIChE Symp. Ser., 80* (233) (1984), 96.

56. Sherwood, T.K., R.L. Pigford, and C.R. Wilke, *Mass Transfer*, McGraw-Hill, NY, (1975), Chapt. 10.

57. Slater, M.J., Recent industrial-scale applications of continuous resin ion exchange systems, *J. Separ. Proc. Technol., 2* (3), (1981), 2.

58. Slater, M.J., The relative sizes of fixed bed and continuous countercurrent flow ion exchange equipment, *Trans. Inst. Chem. Engr., 60,* (1982), 54.

59. Snyder, L.R. and J.J. Kirkland, *Introduction to Modern Liquid Chromatography*, 2nd ed., Wiley, NY, 1979.

60. Streat, M., Recent developments in continuous ion exchange, *J. Separ. Proc. Technol., 1* (3), (1980), 10.

61. Swinton, E.A., B.A. Bolto, R.J. Eldridge, P.R. Nadebaum, and P.C. Coldrey, "The present status of continuous ion exchange using magnetic micro-resins," in Naden, D. and Streat, M., Eds., *Ion Exchange Technology*, Ellis Horwood Ltd., Chichester, England, (1984), 542-562.

62. Van Deemter, J.J., F.J. Zuiderweg, and A. Klinkenberg, Longitudinal diffusion and resistance to mass transfer as causes of nonideality in chromatography, *Chem. Eng. Sci., 5,* (1956), 271.

63. Vermeulen, T., Separation by Adsorption Methods, in Drew, T.B. and Hoopes, J.W., Jr., Eds., *Advances in Chemical Engineering*, Vol. II, Academic Press, NY, (1958), p. 14.

64. Vermeulen, T., G. Klein and N.K. Hiester, Adsorption and Ion Exchange, in Perry, R.H. and Chilton, C.H., Eds., *Chemical Engineers' Handbook*, 5th ed., Section 16, McGraw-Hill, NY (1973).

65. Vermeulen, T., M.D. LeVan, N.K. Hiester, and G. Klein, Adsorption and Ion Exchange in Perry, R.H., and Green, D., Eds., *Perry's Chemical Engineers' Handbook*, 6th ed., McGraw-Hill, NY, (1984), Section 16.

66. Wankat, P.C., Improved Preparative Chromatography: Moving Port Chromatography, *Ind. Eng. Chem. Fundam., 23*, (1984), 256.

67. Wankat, P.C., *Large Scale Adsorption and Chromatography*, CRC Press, Boca Raton, FL, in press 1985.

68. Wankat, P.C. and P.M. Ortiz, Moving feed point gel permeation chromatography. An improved preparative technique, *Ind. Eng. Chem. Process Design Dev., 21*, (1982), 416.

69. Waters Div., Millipore, New Waters Kiloprep Process Scale Separation Systems, Milford, MA (1983).

PARAMETRIC ION-EXCHANGE PROCESSES (Parametric Pumping and Allied Techniques)

D.Tondeur and G.Grevillot

Laboratoire des Sciences du Génie Chimique du CNRS
1,rue Granville   54000 Nancy,France

1   INTRODUCTION : THE BASIC CONCEPTS

The term "parametric pumping" implies the idea of pumping, that is of transport of material (in our case, ionic species) up along some potential scale (in our case, chemical potential), at the cost of degradation of some flux of energy (to be specified later). The term "parametric" refers to the fact that the energy flux just mentionned is obtained by modulating some intensive thermodynamic parameter such as temperature, pressure, ionic strength, pH, electric potential, ... Typical examples are temperature-swing ion-exchange and pressure-swing adsorption. The concept of parametric pumping is due to Wilhelm and coworkers (1966, 1968). Their original results deal with adsorption from the liquid phase, modulated by temperature. Let us examine how these phenomena are implemented.

1.1   Equilibrium Shift by Temperature Change

Consider the adsorption of phenol from the aqueous phase on a non-functionalized polystyrene-DVB resin (Almeida et al. 1982). Figure 1 shows that for a given concentration of phenol in the solution, the amount adsorbed is larger at a low temperature than at a high one. In other words, when heat is furnished to an equilibrated and cold mixture of adsorbent and solution (represented by point A), phenol will be transferred from the adsorbent to the solution. If the temperature is stabilized at 60°, and equilibration is allowed, the new state of the system is represented by point B. Points A and B are connnected by an operating segment, expressing the overall conservation of phenol, and the slope of which is the ratio of the amounts L and S of the two phases, according to Eqn 1 :

$$S \, \Delta y = - L \, \Delta y \qquad (1)$$

370

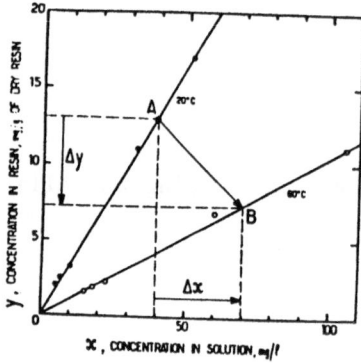

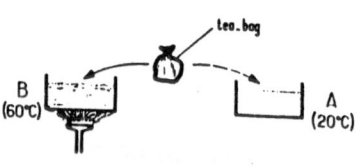

Fig. 2. The simplest form of
parametric pumping : the tea-
bag model.

Fig. 1. Adsorption isotherms
of phenol from water on
Duolite ES 861.

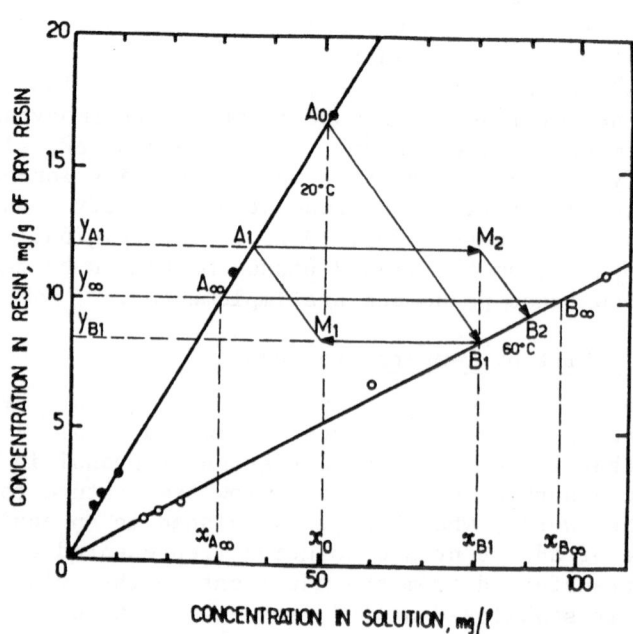

Fig. 3. Successive states of the tea-bag system in
the first cycles and final states A_∞, B_∞.

This example thus illustrates simply that matter can be transported across an interface under the effect of temperature change. Let us now examine how to build a separation process based on that principle ; for that purpose, the two phases must be separated from each other.

1.2. The Simple "Tea-Bag" Model

Figure 2 illustrates what is probably the simplest way to implement the parametric pump concept. Two beakers contain initially the same solution. A bag containing the adsorbent is initially equilibrated at 20° with the solution in beaker A. The bag is then taken out of that beaker, drained and carried into beaker B, where it is re-equilibrated at 60°. Some phenol is then desorbed, so that the concentration of phenol in solution B is now larger that that of solution A. In addition, the adsorbent bag contains less phenol than what corresponds to equilibrium with solution A. When it is carried back into beaker A (after draining and cooling), it will pick up some more phenol, which will be carried back to beaker B, and so forth. The adsorbent bag thus serves as a carrier of phenol from the cold beaker to the warm one, and phenol is pumped from a diluting reservoir to a concentrating reservoir.

However the process just described will not be able in general to transport all the phenol into beaker B : the amount carried at each trip becomes smaller and smaller and finally vanishes. Figure 3 shows a few successive states of the system, and the path followed by the compositions of the two beakers. Let x_0 be the initial concentration of phenol in both beakers. Point A_0 represents the initial state of beaker A. When the bag is carried into beaker B, at 20°, the state of that beaker is also A_0 ; when the temperature is raised to 60°, the state switches to point B_1, on the 60° isotherm. The adsorbent, of concentration yB_1 is then brought back to beaker A where it is contacted with solution x_0 (mixture M_1) ; the re-equilibrium at 60° leads to point B_2, etc.. It is easy to see that these successive points A_i, B_i, M_i tend to converge so as to lie on a same horizontal line $A_\infty B_\infty$. No more phenol is then transferred from A to B, since the adsorbent has a concentration y_∞ which is in equilibrium with both solutions, that is with solution A (xA_∞) at 20°, and with solution B (xB_∞) at 60°.

The limit separation of this "single-stage batch" parametric pump is thus simply related to the distance between the two isotherms. When the latter are linear and characterized by the slopes K_C (cold) and K_H (hot), the ratio of the two limiting concentrations is given by :

$$\frac{x_{B_\infty}}{x_{A_\infty}} = \frac{K_C}{K_H} \quad (\sim 3 \text{ in the present example}) \qquad (2)$$

The tea-bag operating scheme was actually used by Goto et al. (1979)

1.3 The Analogy with Active Transport, with the Heat Pump, and with distillation

In active transport through a membrane (Fig. 4), a complexing carrier (for example H^+ ions) combines with the species to be transported (NH_3) on one side of the membrane. The complex thus formed ($N H_4^+$) diffuses toward the opposite side of the membrane under its own concentration gradient, and is decomposed, releasing the transported species and the carrier which diffuses back. In the present example, there is counterdiffusion of H^+ and NH_4^+ inside a cation exchange membrane. The driving energy is furnished by the protonation of NH_3 : each mole of NH_3 transfered requires addition of a mole of H^+ ions on the right side. Other driving forces may be used : difference in temperature, partial pressure, or in general chemical potential of some species involved in the process across the membrane. It is seen that the carrier ion (H^+) plays at the molecular level a role quite comparable to that of the tea-bag.

Figure 4 summarizes the conceptual analogy of active (carrier mediated) membrane transport, the heat pump, distillation, and the parametric pump. The analogy with distillation may not be easy to trace, because the "carrier" is the mixture to be separated itself. However, it should be understood that in all these systems, an entity (heat or a chemical species) is transported from a depleting "source" to an enriching "well" possibly against its own concentration gradient and thus in the direction of increasing potential, and the driving energy for this transport is furnished by a flux of energy flowing down potential : mechanical energy is furnished to the compressor and degraded in the throttle valve in the heat pump ; acid is added on one side of the membrane and neutralized ; heat flows from the boiler to the condenser in distillation, and from a hot source to a cold well in thermal parametric pumping. The analogy with distillation will be developed further in the following chapter.

2 ANALYZING THERMAL ION-EXCHANGE PARAMETRIC PUMPING

2.1 The Batch Operating Scheme for a Binary Mixture

Figure 5 shows schematically a typical set-up for a parametric pump using an ion-exchange fixed-bed. As in the tea-bag experiment, the operation considered is batch and cyclic, but here the solid "carrier" is fixed, and the solution is the mobile phase. In addition the contacting of the two phases occurs by flow through the resin bed, and this entails the formation of longitudinal continuous concentration profiles which do not exist in the tea-bag. In other words, we have here a "distributed" system instead of a "lumped" system.

Consider the example of separation of the mixture Cu^{++}/Ag^+ on a common sulfonic type cation exchanger. Fig 6 shows the exchange

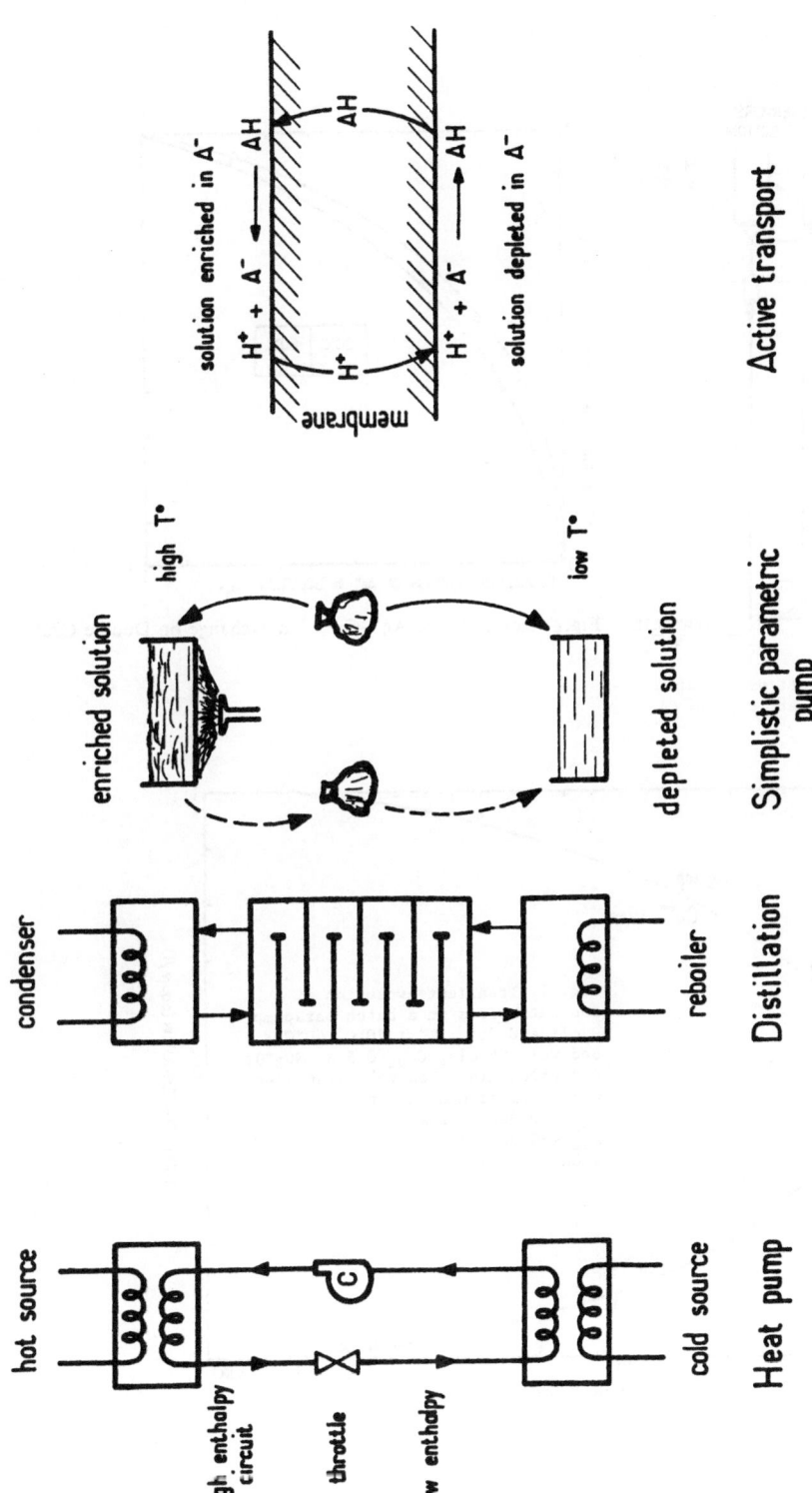

Fig. 4. Analogy of different "active" transport processes.

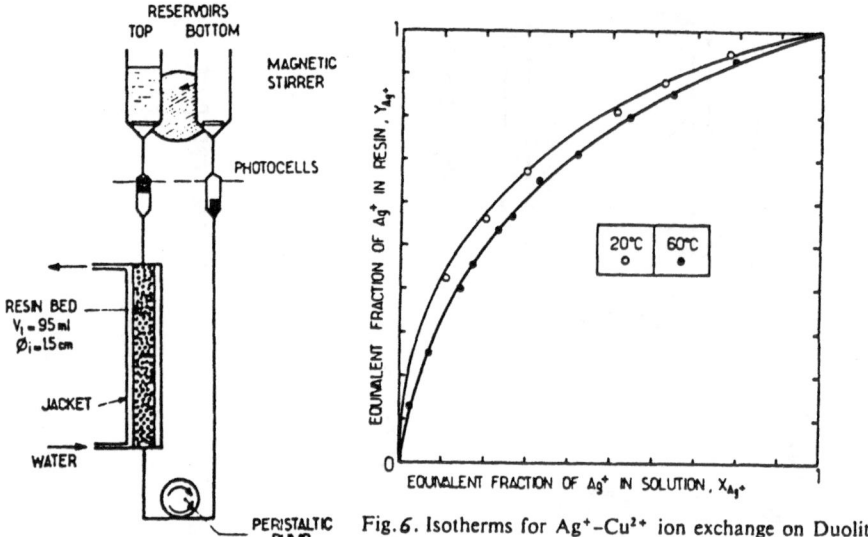

Fig.6. Isotherms for Ag⁺-Cu²⁺ ion exchange on Duolite C265

Fig. 5. Flow-sheet of a simple
packed-bed parametric pump.

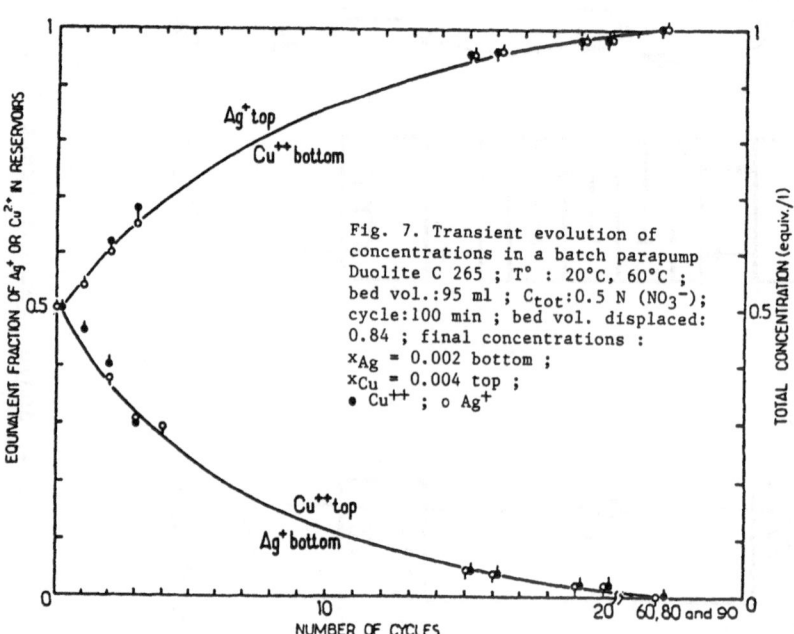

Fig. 7. Transient evolution of
concentrations in a batch parapump
Duolite C 265 ; T° : 20°C, 60°C ;
bed vol.:95 ml ; C_{tot}:0.5 N (NO_3^-);
cycle:100 min ; bed vol. displaced:
0.84 ; final concentrations :
x_{Ag} = 0.002 bottom ;
x_{Cu} = 0.004 top ;
• Cu^{++} ; o Ag^+

isotherms at 20°C and 60°C. It can be seen that Ag^+ has a stronger adsorptivity than Cu^{++}, and that this property is reduced by raising the temperature. Therefore Ag^+ will be relatively retained at 20°C, and will be eluted at 60°C, the opposite being true for Cu^{++}.

A typical cycle will work as follows :

- the column is heated by circulating hot water in the jacket, and when a suitable temperature is reached, the salt solution is pumped upwards from the bottom reservoir through the bed ; Ag^+ being relatively eluted, the effluent collected in the top reservoir is depleted in Cu^{++} and enriched in Ag^+;

- the column is next cooled, and the salt solution is pumped back from the top reservoir to the bottom through the bed ; Cu^{++} being less strongly retained at 20°, a part of it is eluted, and the effluent collected is enriched in Cu^{++} and depleted in Ag^+ with respect to the original solution;

- the process is repeated, and as in the tea-bag experiment, at each cycle, a certain amount of Cu^{++} is transferred from the top to the bottom reservoir, and conversely for Ag^+, until a steady-state is attained.

Figure 7 shows the evolution of the concentration in both reservoirs with the number of cycles, under conditions specified in the legend (Grevillot, 1980). As in the tea-bag experiment, a final cyclic regime is reached in which no more change occurs in the reservoir compositions.

2.2 Modelisation by Staged "Distillation" Approach

This approach has been widely investigated by Wakao, Wankat, Grevillot and Rice. It consists in generalizing the tea-bag model to a large number of tea-bags and of liquid fractions. This can be viewed as a suitable model for actual discontinuous stagewise contacting methods (see Rachez et al for a liquid-liquid example), but possibly also as a model for packed bed pumps, the packed bed being considered as a cascade of theoretical (equilibrium) stages.

Let N be the number of stages, and divide the solution into N+1 equal fractions (Fig 8). The pumping cycle is effected as follows :

- assume that initially the bottom reservoir contains liquid fraction L_{N+1} , and each stage S_k contains the liquid fraction L_k of same index. Solution and solid are assumed to be at equilibrium at the hot temperature T_H ;
1. all liquid fractions are then transferred one stage upwards so that the top reservoir contains fraction L_1 ;
2. the stages are brought to the cold temperature T_C and equili-

376

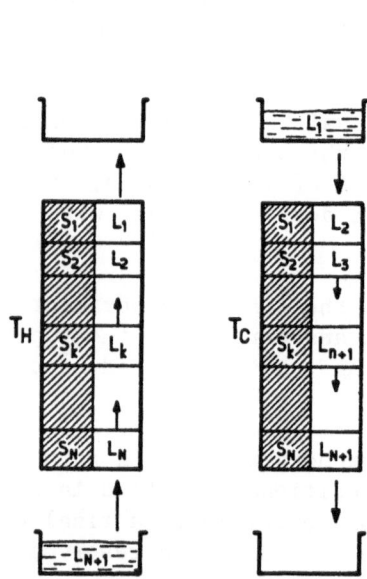

Fig. 8. The staged model of a parametric pump.

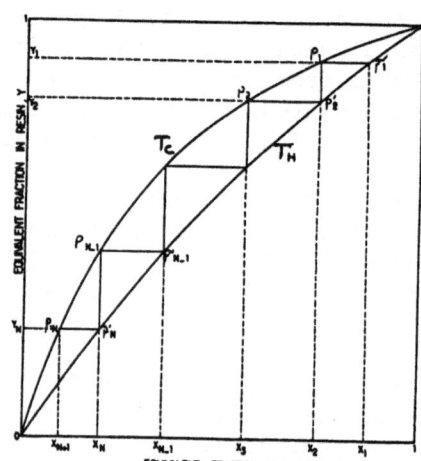

Fig. 9a The limit regime of an N-stage parametric pump.

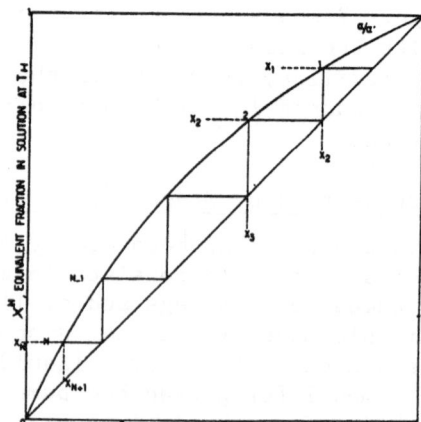

Fig. 9b Representation of Figure a in the $[x^H, x^C]$ plane, illustrating the analogy with the McCabe-Thiele construction for binary total reflux distillation.

brium is allowed to be reached ;

3. next, all liquid fractions are moved back one stage downwards;
4. the temperature is increased to T_H and equilibrium is allowed.

- repeat the four steps of the cycle ...

This mode of operation has the effect of amplifying the effect of the single tea-bag. Actually, it can be shown (Grevillot 1976) that the limiting regime of this system in batch operation (total reflux) is represented by a staircase drawn between the exchange isotherms at the two temperatures (Fig 9a). The meaning of this geometrical property is that no more mass-transfer occurs during the solution movements : the concentration x_k of liquid fraction L_k is in equilibrium with both solid fractions S_{k-1} at T_C (y_{k-1}) and S_k at T_H (y_k) (vertical line P_1P_2' for example on Fig 9a) ; similarly the solid fraction S_k is in equilibrium with both liquid fractions L_{k+1} at T_C and L_k at T_H. If the equilibrium is represented by constant relative adsorptivity α_C and α_H at the two temperatures, then the limiting separation is given by :

$$ SF = \frac{x_T}{x_B} \bigg/ \frac{1 - x_T}{1 - x_B} = \left[\frac{\alpha_C}{\alpha_H}\right]^N \tag{3} $$

$(x_T = x_1 ; x_B = x_{N+1})$
whereas in the linear case, we have simply :

$$ \frac{x_T}{x_B} = \left[\frac{K_C}{K_H}\right]^N \tag{4} $$

These relations generalize Eqn.2 to N stages. Note that Eqn.3 is formally analogous to Fenske's equation of total reflux binary distillation. A representation closer to the McCabe-Thiele diagram is obtained by plotting as ordinate the concentration x^H in the solution fractions that are moving upwards after a hot equilibration, and as abscissa the concentration x^C of the fractions that are moving downwards after a cold equilibration. The relation between the concentrations x^C_{k+1} and x^H_k that are in equilibrium with the same y_k is :

$$ \frac{x^H}{1 - x^H} \bigg/ \frac{x^C}{1 - x^C} = \frac{\alpha_C}{\alpha_H} \tag{5} $$

which indicates the x^H and x^C are related by a "pseudo-isotherm" of coefficient α_C/α_H. The limit regime is then a staircase between this curve and the diagonal (Fig 9b). This is the formal analogue of the McCabe-Thiele diagram for total reflux binary distillation.

We can push the analogy a little further by examining the following correspondence :

Batch parametric pump (limit regime)	Total reflux distillation (steady state)
1. Liquid phase at temperature T_C	- Liquid phase
2. Liquid phase at temperature T_H	- Vapor phase
3. Ratio α_C/α_H of separation factors	- Relative volatility
4. x_H vs. x_C curve on Fig 9 b	- Vapor liquid equilibrium curve (composition of vapor vs. composition of liquid)
5. Solid phase concentration y in a stage	- Temperature T_d on a stage
6. Isotherms α_C and α_H in (x,y) plane	- Boiling curve and dew curve in (composition of liquid or vapor, T_d) plane
7. Reflux : liquid phase withdrawn from top of column at T_H and reinjected at T_C	- Vapor phase withdrawn from top of column and reinjected as liquid phase
8. Reboiling : liquid phase withdrawn from bottom at T_C and reinjected at T_H	- Liquid phase withdrawn from bottom and reinjected as vapor

Let us comment on this analogy. The x_H vs. x_C curve is an equilibrium curve in the following sense. The points of this curve represent the compositions x_H and x_C of two liquid fractions that are in equilibrium with a given y on a given stage, in the same way as the vapor-liquid equilibrium curve represents the compositions of two phases that are in equilibrium at a given temperature on a given stage. The x_H vs. x_C curve may be scaled in terms of y in the same way as the vapor-liquid equilibrium may be scaled in terms of T_d. In the parapump, y plays the same role as T_d in distillation. Each stage of the column is characterized by a constant value of y (respectively T_d in distillation). In this way, x_H appears to play the role of the vapor phase composition and x_C that of the liquid phase. Stated differently, changing the temperature in the parapump is similar to going from vapor to liquid (or reciprocally) in distillation. Note, however, that in the parapump, the liquid fractions of composition x_H and x_C related by the equilibrium curve are never in contact with each other, whereas in distillation, the analogous vapor and liquid phases are actually in contact.

The y $\longleftrightarrow$ T_d analogy entails the analogy between the (x,y) curves in the parapump (isotherms) and the composition vs. T_d curves in distillation (boiling curve and dew curve). In fact, the correspondence between the two diagrams in Fig 9 also exists for vapor-liquid equilibrium (see, for instance, Treybal 1955).

The notion of reflux (or reboiling) may then be analyzed in the following way. In distillation take the fluid exiting at one end of the column, add or substract heat to effect a change of phase (not necessarily a change of temperature), and reinject it (in part or in totality) at the same end of column but in opposite direction. In parametric pumping the description is identical, except that the addition of substraction of heat results in a change of temperature without a change of phase. We may thus speak of a generalized reflux.

The approach presented may be extended to <u>open</u> parametric pumps, with feed and withdrawal of products (Grevillot 1980). Reflux ratios may be defined at the top and the bottom as the ratio of what is "refluxed" from the reservoir to the bed, to what is withdrawn as product. In the x_H vs. x_C diagram, an operating line is constructed for each section, above and below the feed point, and its slope is determined by the corresponding reflux ratio ; the intersection of the operating lines lies on the "q-line" or "feed-line", defined by the "state" of the feed, that is by the ratio of moles fed during the upward half-cycle, to the moles fed during the downward half-cycle. As in distillation, for a specified separation (i.e. a value of SF) one can define a minimum number of theoretical stage (at total reflux), a minimum reflux ratio (for an infinite number of stages), an optimum feed location (for given reflux and number of stages) and an optimum reflux ratio.

In the preceding approach, if the number of stages N and the capacity ratio per stage (eqn 1) are fixed, the amplitude of the displacement of the fluid in the bed is fixed. This constraint can be released by considering more than N+1 solution fractions, and thus several successive transfers of solution fractions in the same direction (Grevillot 1977). However the simplicity of the graphical approach and of the analogy with distillation are then lost. I believe the interest of the staged approach as applied to packed beds is conceptual rather than quantitative : it allows understanding the main trends and the interconnection of the factors, it furnishes a convenient framework for terminology, and for information ; but since no reliable method exists for evaluating the "Height Equivalent to a Theoretical Plate" for such bithermal, non-linear systems, it is hardly realistic to expect quantitative predictions. On the other hand, parametric pumps can actually be operated according to the tea-bag scheme, in a cascade of batches (see Rachez et al 1982 for liquid-liquid extraction P.P.) and then this model becomes quantitatively pertinent.

2.3 Modelisation by Wave-Approach in the Linear Case

This modelisation is based on the classical "chromatographic" approaches presented elsewhere in the present book. As opposed to the staged approach, both the packed bed and the solution flow are considered continuous. We present here a brief account based on the developments by Pigford et al (1969) and Chen and coworkers (1971, 1972, 1973,

1974) which are based on the assumptions of local equilibrium between phases according to linear isotherms.

The velocity at which a value of concentrations, or a front moves through the bed at a given temperature is then given by (see Eqn. 50 in Tondeur, this book) :

$$\frac{dz}{dt} = u_{front} = \frac{u}{1 + K_T} \tag{6}$$

The relation governing the redistribution of solute between the two phases when the temperature is changed from T_C to T_H is :

$$x_C \lfloor 1 + K_C \rfloor = x_H \lfloor 1 + K_H \rfloor \tag{7}$$

or

$$\frac{x_C}{x_H} = \frac{1 + K_H}{1 + K_C} = R < 1 \tag{8}$$

where x_C and x_H are the local concentrations at T_C and T_H respectively For given fluid velocity u, R can be considered as the ratio of the front velocities at the two temperatures. We shall always assume here that $K_C > K_H$. Chen and Hill (1971) define the penetration distances L_H and L_C as the distances travelled by the exchange front (by its stoichiometric point actually) during the hot and the cold half cycle respectively ; these distances are the product of the front velocity (Eqn. 6) and the time of flow, thus :

$$L_H = \frac{u \, t_H}{1 + K_H} \qquad\qquad L_C = \frac{u \, t_C}{1 + K_C} \tag{9}$$

Three regions of operation may be defined depending on the relative magnitude of L_H, L_C and the column height h (Fig. 10).

Batch operation in Region 1 (L_H and h > L_C)

Fig. 11 represents the (z,t) plane for the bottom of the pump (the hot half-cycle corresponds to upward flow). During the hot half-cycle of cycle n+1, the solution in the bottom reservoir, of concentration < x_B >_n, obtained at the end of cycle n, flows upward and the path of the front is represented by characteristic ch1. The concentration upstream of this front (below ch1 on the figure) is thus x_H = < x_B >$_n$, and the local equilibrium is represented by point A in Fig 12. At the end of the half-cycle, the temperature change leads to a new equilibrium represented by point B of abscissa x_C obtained from x_H through Eqn. 8

During the subsequent cold half-cycle, the front motion is represented by characteristic ch2 of slope smaller than ch1 ($L_C < L_H$). Therefore, ch2 does not exit from the bed (there is no breakthrough) and a solution of constant concentration x_C flows into the bottom reservoir. In the absence of mixing with dead volume, the bottom concentration at the end of the cycle is $< x_B >_{n+1} = x_C$, and the change over the cycle obeys :

$$\frac{< x_B >_{n+1}}{< x_B >_n} = \frac{x_C}{x_H} = R < 1 \qquad (10)$$

and therefore, the bottom concentration in cycle n is related to the initial concentration x_0, assumed uniform, by :

$$< x_B >_n = x_0 \, R^n \qquad (11)$$

Fig 12 shows the successive equilibria, and illustrates in which way the bottom concentration decreases from cycle to cycle, and eventually tends asymptotically to zero.

Let us now look at the top reservoir. As seen on Fig 11, at cycle n+2, a new front enters the bottom of the bed and follows characteristic ch4, identical to ch1. Between that front and that of the previous cycle, that is between ch3 and ch4, a band of solute, of concentration $< x_B >_n$ has a net upward movement of distance $\Delta z = L_H - L_C$ at each cycle, and therefore solute is transfered from the bottom to the top reservoir. The analytical expression giving the top reservoir concentration is heavy and can be found in Chen and Hill (1971).

Clearly, the operation in region 1 is interesting in that it allows complete removal of a solute from the bottom reservoir ($x_B \to 0$); this is impossible in the other regions, which imply breakthrough of the downward moving fronts. Further refinements of the linear theory, for equilibrium and non-equilibrium pumps, may be found in the publications of Rice (1975, 1976) and Foo (1973, 1975).

Operation of open (partial reflux) pump

Figure 13 shows an "open" pump (as opposed to the closed, or batch, or total reflux case considered above). While different modes of operation are possible, determined by the feed location and the continuous or discontinuous character of the feeding and product withdrawal operations, we consider here a relatively simple situation of practical interest : the top product is withdrawn during the upflow period, feed is added at the top, and bottom product is withdrawn during the downflow period. The volume flux into the column is then Q on upflow and $(1 + \phi_B)Q$ on downflow (see Fig 13 for notations), while the feed volume is $Q(\phi_T + \phi_B)$. If the same linear velocity u is used throughout the cycle, the half-cycle times t_H and t_C will be such that $t_C = (1 + \phi_B)t_H$ and the penetration distances L_H and L_C (Eqn.9) are modified accordingly. The pump will operate in Region 1 if

382

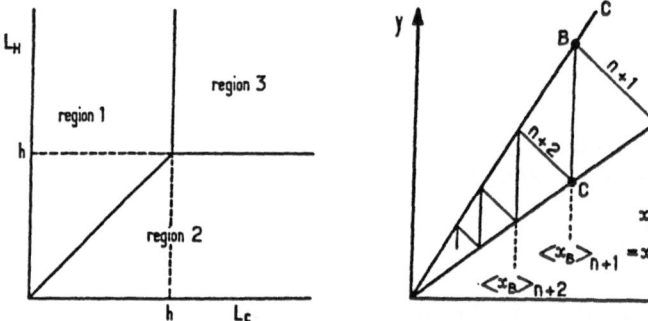

Fig. 10. The three regions of operation
of a linear parapump, defined by the
penetrations L_H and L_C.

Fig. 12. The equilibrium diagram
corresponding to Figure 11.

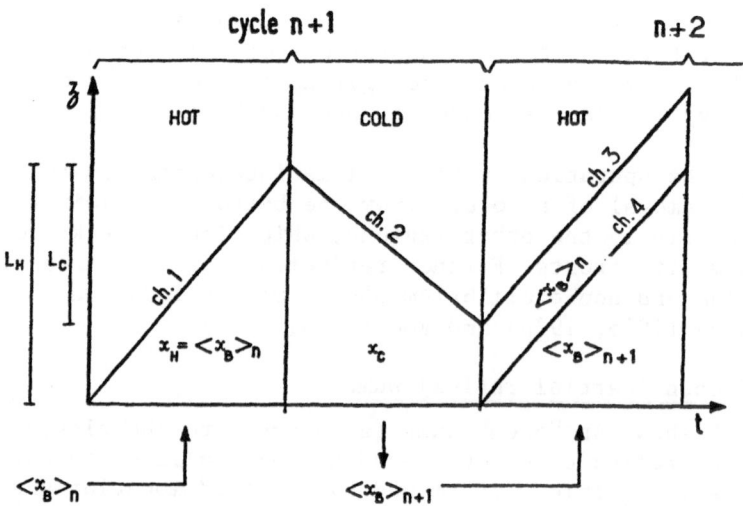

Fig. 11. Characteristics in (z,t) representation for the bottom
of the linear pump.

383

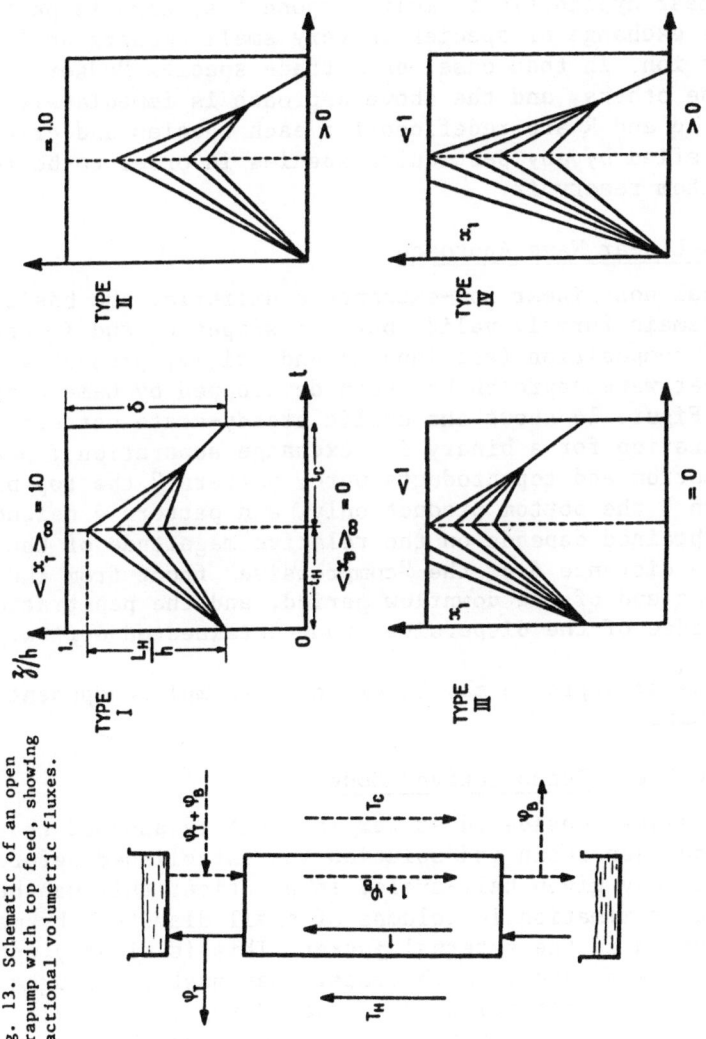

Fig. 14. Cyclic steady-state patterns of characteristics in batch parapump.

Fig. 13. Schematic of an open parapump with top feed, showing fractional volumetric fluxes.

$$\phi_B \leqslant \frac{1 - R}{R} \qquad \text{and} \qquad L_C < h \qquad (12)$$

and complete removal of a solute from the bottom reservoir is then possible ; the limit regime concentrations are :

$$x_B^\infty = 0 \quad ; \quad x_T^\infty = 1 + \frac{\phi_B}{\phi_T} \qquad (13)$$

Muticomponent separations

The linear hypothesis in multicomponent systems is pertinent only for the exchange of species in very small amounts against a common major ion. In that case, each trace species "adsorbs" independently of the others, and the above approach is immediately transposable : L_C, L_H and R are redefined for each species and criteria (12) must be satisfied by any particular species in order to be removed from the bottom reservoir.

2.4 The Non-Linear Wave Approach

For usual non-linear ion-exchange equilibria, the basic equations (6) to (9) remain formaly valid, but the slopes K_T and K_H become functions of composition (see Tondeur and Bailly, present volume). The non-linear wave approach has been developped by Camero and Sweed (1976), and Figure 14 shows the cyclic steady-state patterns obtained in batch operation for a binary ion-exchange separation : pattern 1 gives both bottom and top products pure, pattern 2 the top product only, pattern 3 the bottom product only, and pattern 4 neither. Which pattern is obtained depends on the relative magnitude of the column length h, the distance δ of the "compressive" front from the top of the bed at the end of the downflow period, and the penetration L_H of the leading edge of the dispersive front obtained on upflow.

This type of approach may be extended to multicomponent systems or to open pumps.

2.5 The So-Called "Recuperative" Mode

The processes considered so far implicitly assumed that thermal equilibrium between resin and solution was established before solution was pumped, in any given half-cycle. In practice, this can be achieved with a good approximation in columns of small diameter, by means of fluid circulation in the external jacket. This is clearly unpractical for large diameters, and in such cases, heat must be brought or removed by preheating or precooling the treated solution itself, by means of heat exchangers located between the column and the reservoirs. This mode of operation lends itself to an optimal utilization of calories, hence the term "recuperative" (Rolke and Wilhelm 1969), but it also implies some constraints.

From a theoretical point of view, a thermal wave is formed at the bed entrance at each temperature change, and moves at a velocity

$$u_T = \frac{u}{1 + \mu} \tag{14}$$

where μ is the heat capactiy ratio of the bed

$$\mu = \frac{1 - \epsilon}{\epsilon} \frac{\rho_s C_s}{\rho_f C_{p_f}} \tag{15}$$

An equilibrium shift accompanies the passage of the thermal wave, and the concentrations on each side of the wave are related, in the linear case, by :

$$\frac{x_C}{x_H} = \frac{K_H - \mu}{K_C - \mu} = \frac{u_H^{-1} - u_T^{-1}}{u_C^{-1} - u_T^{-1}} = R_T \tag{16}$$

where u_H and u_C are the concentration front velocities at the two temperatures.

The recuperative mode resembles the direct heating mode when the thermal wave moves faster than the concentration fronts, and when thermal breakthrough occurs :

$$L_T = \frac{u\,t}{1 + \mu} > h \tag{17}$$

The higher u_T , the closer the two modes are of each other.

Sweed and Rigaudeau (1975) have modeled recuperative parapumps operating in region 1 and with thermal breakthrough. The bottom reservoir concentration is given by

$$< x_B >_n = x_0 \left[\frac{h}{L_T} + (1 - \frac{h}{L_T})\, R_T \right]^n \tag{18}$$

which is to be compared to Eqn. 11. Wankat (1978) investigated the various situations that may result from different relative magnitudes of the thermal wave velocity and the concentration front velocities. A "focussing" effect of solute may be obtained when u_T is between u_H and u_C.

3 EXPERIMENTAL EXAMPLES OF THERMAL ION-EXCHANGE PARAPUMPS

3.1 Effect of Temperature on Ion-Exchange Equilibria

Figure 15 shows a series of measured binary isotherms (Marques 1978) and Fig 16 the evolution with temperature of the mass action coefficient for the exchange of various ions against H^+ (Kraus and Raridon, 1959) ; both figures concern conventional sulfonic cation exchangers. Some general properties may be stated in view of these

386

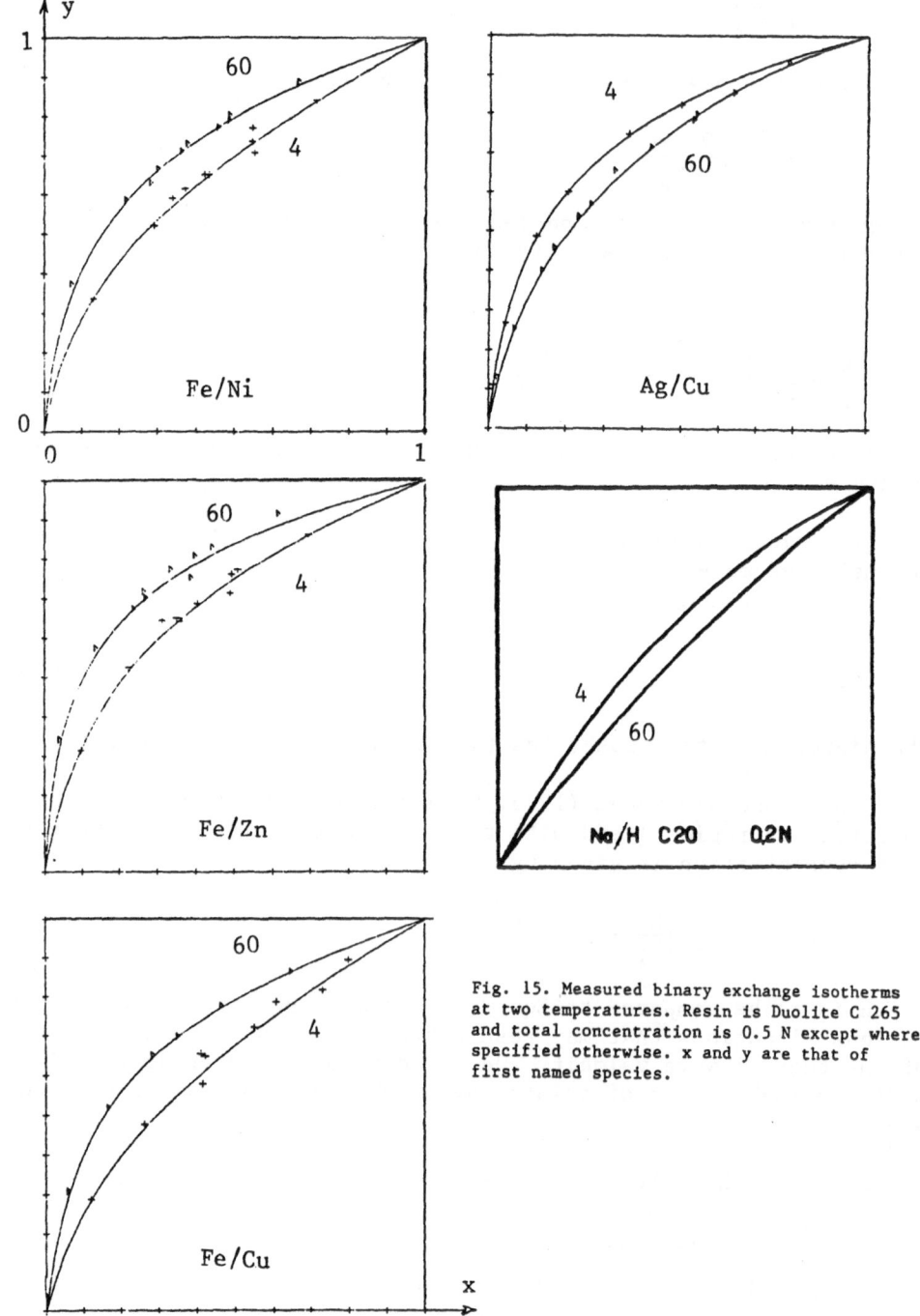

Fig. 15. Measured binary exchange isotherms at two temperatures. Resin is Duolite C 265 and total concentration is 0.5 N except where specified otherwise. x and y are that of first named species.

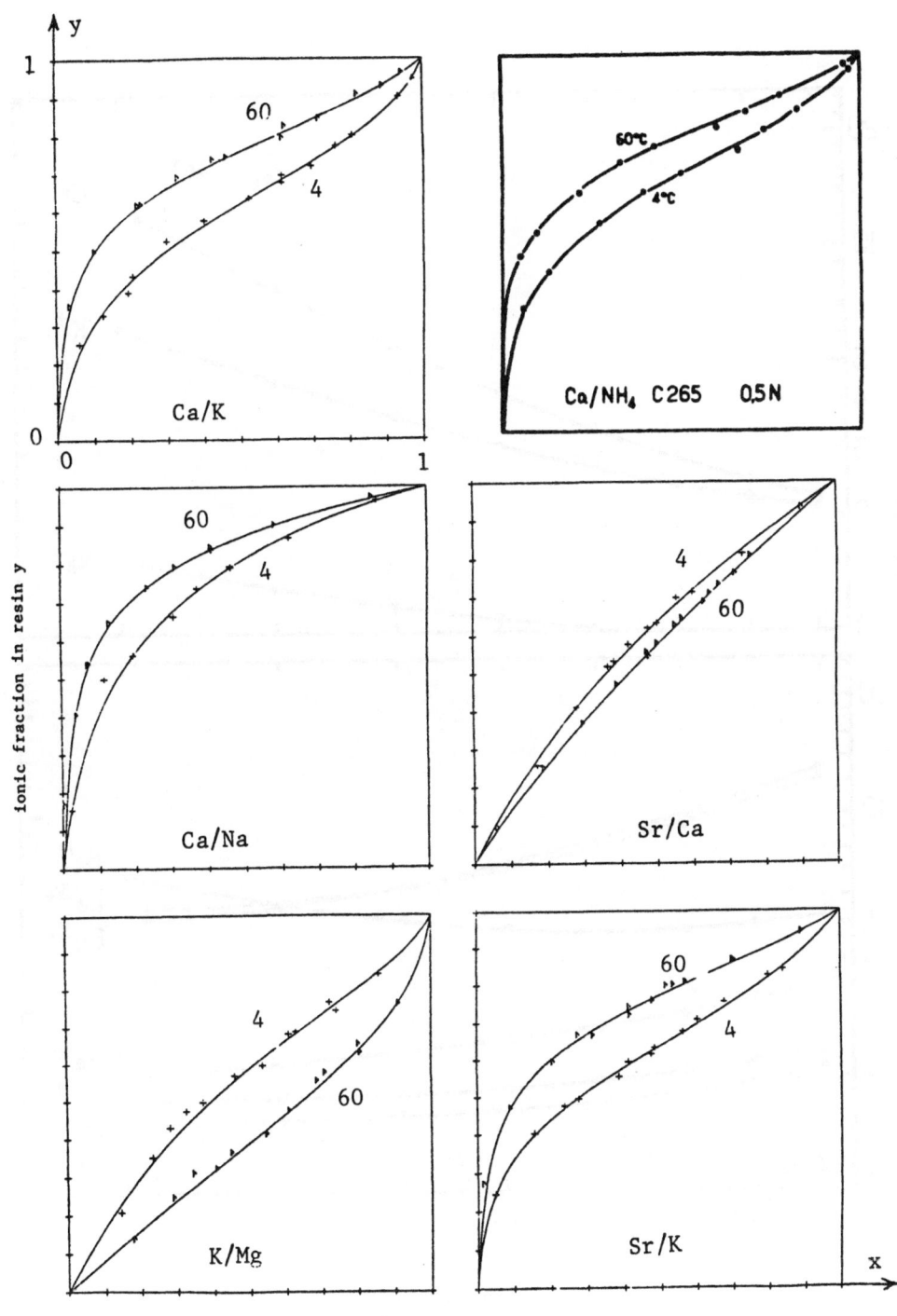

ionic fraction in solution x

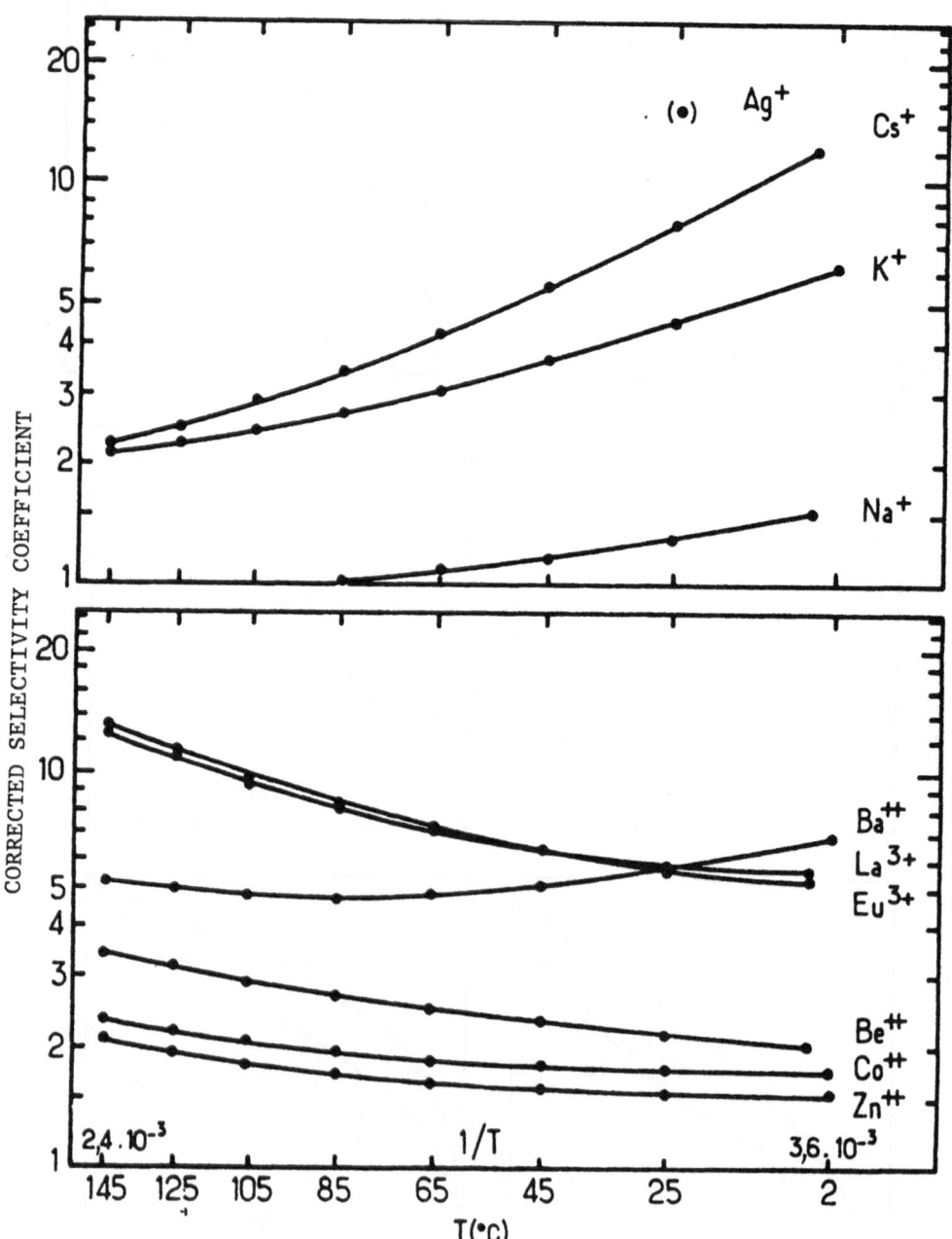

Fig. 16. Effect of temperature on Metal/H$^+$ equilibrium coefficients at low metal concentration on Dowex 50-X 12 (after Kraus and Raridon, 1959).

figures and of other data :

1. The effect of temperature is usually larger for couples of ions of different valences than for ions of equal valences.

2. In some cases, the relative adsorptivity can undergo a reversal with temperature (see K/Mg).

3. Assuming the usual Arrhenius dependence of the mass action constant on temperature :

$$K = K_0 \, \exp(-\frac{\Delta H}{RT}) \tag{19}$$

the slope of the curves of Fig 16 (Ln K vs. 1/T) is a measure of the heat of exchange - ΔH. A high heat of exchange is thus a criterion for searching efficient parametric pumping systems.

4. It can be seen that increasing the temperature decreases the adsorptivity of the monovalent ions (both with respect to H^+ and to divalent ions).

5. By contrast, the polyvalent ions are more prefered when the temperature increases (with respect to H^+ or to ions of lower valence). An exception is Ba^{++}.

Physical explanations for these trends remain to be found.

3.2 Desalting Experiments

The first parametric pumping experiments on ion-exchangers were published by Sweed and Gregory (1971, 1972) who concentrated NaCl from aqueous solution using an ion-retardation resin (resin carrying both anionic and cationic sites). NaCl is more retained at low than at high temperatures. Fig. 17 shows a typical result which illustrates features common to most desalting experiments:a large number of cycles is needed to reach steady-state ; the number of bed volumes displaced needs to remain small if a large concentration factor is sought.

Variations of total concentration during normal ion-exchange parametric pumping were observed by Butts et al (1973), who (erroneously) attributed it to water absorption by the resin related to swelling. It was later shown (Grevillot et al. 1984) that Donnan electrolyte uptake was responsible of this phenomenon, being in some cases strongly affected by temperature changes. Some electrolytes may thus be concentrated on monofunctional ion-exchange resins ; for example a 0.5 N silver nitrate solution can be split in about 20 cycles into a 0.86 N fraction and a 0.13 N fraction, on a conventional sulfonic cation-exchanger in Ag^+ form. A 0.5 N copper nitrate solution was split into fractions at respectively 0.025 N and 1.19 N, but in

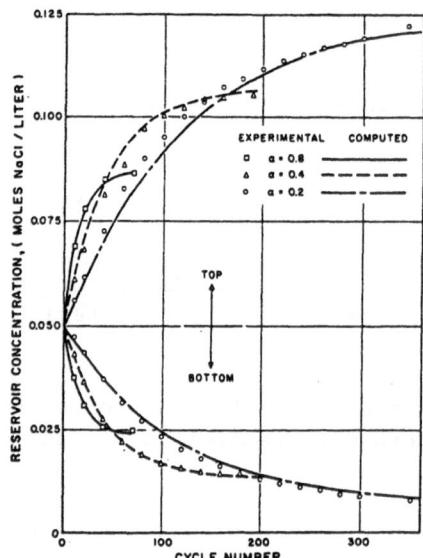

Fig. 17. Desalting NaCl solutions on ion retardation resin in batch parapump (after Gregory and Sweed, 1971). α : bed volumes displaced.

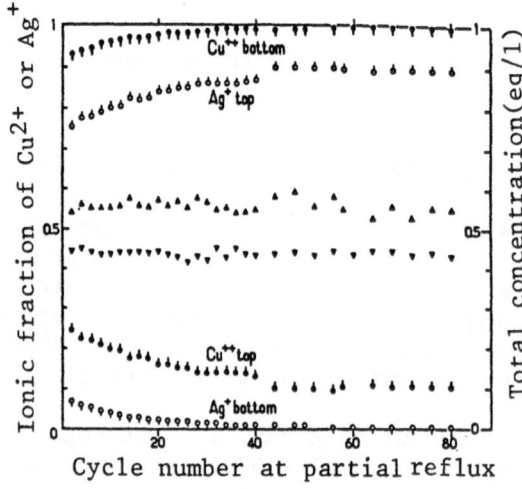

Fig. 18. Separation of silver and copper in two-bed open parapump. Duolite C 265 ; T° : 20°C, 60°C ; C_{tot} : 0.5 N (NO_3^-) ; bed capacity : 0.34 equiv (76 % in enriching top section) ; reflux ratio (top and bottom) : 7.2 ; production : 11 ml/cycle (50 % top, 50 % bottom product) x_{Ag_∞} (bottom) : 0.004 ; x_{Ag} (top) : 0.89 total concentration (bottom ▽, top △).

170 cycles ! It was observed that the Donnan uptake equilibrium is affected by temperature in the same direction as the adsorptivity (Fig 16) ; for example, the uptake of salts of polyvalent cations is favored by a temperature increase, and the opposite is true for mono-valent cations.

Other results on desalting with ion-retardation resins were re-ported by Rice and Foo (1977, 1981), on dual-bed batch and open sys-tems. Some spectacular results were obtained in batch systems with a large number of long cycles (concentration factors above 3000), which were not conserved in open system. Goto (1979) experimented with staged tea-bag type systems ; Ginde and Chu (1972), Shih and Pigford (1977) used the "cycling zone" method (see next chapter).

Probably the only industrial scale application of thermal ion-exchange parametric pumping is the "Sirotherm" desalination process developped by CSIRO in Australia (Bolto et al, 1970, 1975) which uses specially developped bi-functional resins, and applies to brackish waters, with a temperature shift between ambient and about 90°C. The development of the process was accompanied by extensive studies on the properties of weak resins. The operation of the process depends critically on the initial state of the resin after washing with hot water, on the pH, the alkalinity and the hardness of the water trea-ted. It will typically produce 10 to 25 BV of water containing 100 to 500 ppm salt, from a feed at 1000 ppm. Regeneration will require 3 to 5 BV of the feed water heated to 80° or 90°C, to produce an effluent of 3000 to 5000 ppm salt. Various flow sheets have been de-vised including moving bed systems. The reader is referred to the aboundent litterature published by the CSIRO workers, (Weiss, Bolto et al, 1965 to nowadays).

Bifunctional resins appropriate for desalination have been made commercially available by other manufacturers (Ackermann, 1976).

Let us mention finally the Westinghouse process of boric acid concentration control in nuclear reactor coolant systems, based on thermally cycling an anion exchange resin (Van der Schoot, 1977).

3.3 Experimental Separation of Solutes

Butts et al (1973) experimented total reflux separation of the ternary mixture $K^+/Na^+/H^+$ on a cation-exchanger. It was found that K^+ accumulated in the top reservoir, H^+ in the bottom reservoir while Na^+ disappeared from both reservoirs and accumulated in the bed. This result illustrates a general property later interpreted by Camero and Sweed (1976): in a batch operation, the solutes will tend to arrange themselves in successive bands from top to bottom, in a way similar to displacement development. The order is determined by the order of adsorptivity ratios α^C/α^H.

Other binary and ternary ion-exchange separations were studied
by Grevillot (1984), Bailly (1980) in batch and open pumps, in fixed
and moving beds (see below). The problem of removing Cu^{++} from silver
refining electrolytic baths (containing sometimes also Pd^{++}) with a
minimum of silver loss, was investigated. Figure 18 shows the startup
period of a partial reflux pump comprising an enriching and a stripping section, and intermediate feed. Other problems investigated were
separating Fe^{++} and Cu^{++}, in acid solutions, and obtaining pure sodium
chloride from concentrated calcium brines . The table below shows a
sample result , all obtained at high reflux ratios and after long
transient regimes.

The separation of glucose from fructose in water on a cation-
exchange resin in calcium form has been investigated by Chen and
D'Emidio. The sorption of glucose is practically insensitive to temperature, so that the system behaves as a one-solute system. When the
pump is operated in Region 1 for fructose (see Fig 10), the bottom
product contains only water and glucose.

4   OTHER FORMS OF PARAMETRIC PUMPING

4.1   Cycling-Zone Adsorption

Cycling-zone sorption, as defined by Pigford et al (1969) is a
"one-way" version of parametric pumping : the solution is percolated
continuously through a series of beds (zones) the temperature of which
is alternately cold, hot, cold, hot ... Periodically, the temperature
of all beds is changed (Fig 19). When the timing of the temperature
changes is properly adjusted, the separation factor increases with
the zone number in the same way as in "two-way" batch parametric pumping, the separation factor increases with the number of cycles. As
in parametric pumping, two modes exist to effect the temperature
changes : direct heating mode, and "recuperative" or travelling thermal wave mode.

Cycling-zone adsorption has been modelled by both the staged
approach (Wankat 1973) and the local equilibrium approach (Baker and
Pigford, 1981 ; Wankat, 1977). Ion exchange desalting experiments have
been published by Ginde and Chu (1972) and Shih and Pigford (1977).
Other experimental results concern adsorption from the liquid or from
the gas phase, and liquid-liquid extraction. The reader is referred
to the review articles given in the bibliography.

4.2   Moving-Bed Parametric Pumping

Parametric pumping may be effected in continuous countercurrent
type of equipment. For example, Bailly (1980) separated calcium and
potassium chlorides in a two-bed system in which one column is always
maintained hot, while the other is always maintained cold. The resin
and the solution flow countercurrently ; feed is added at some point
and products withdrawn at one end of each bed. The technology used

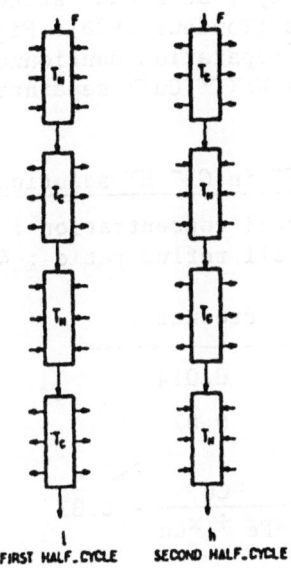

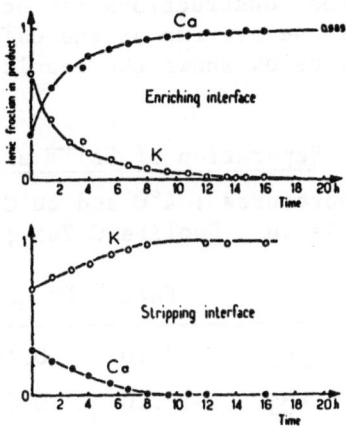

Fig. 20. Startup of Ca^{++}/K^{+} separation in dual temperature moving-bed system (after Bailly and Tondeur, 1980).

FIRST HALF.CYCLE   **SECOND HALF.CYCLE**

Fig. 19. Operating scheme of Cycling-Zone-Adsorption.

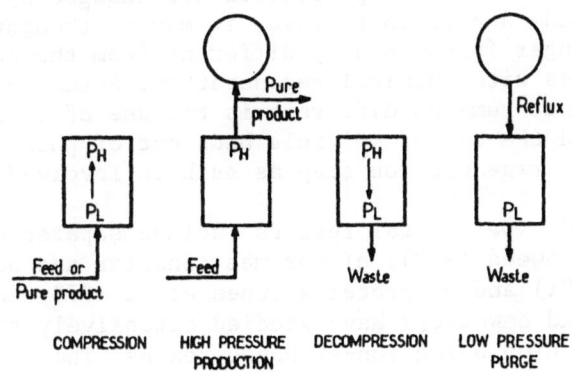

COMPRESSION   HIGH PRESSURE   DECOMPRESSION   LOW PRESSURE
PRODUCTION PURGE

Fig. 21. Operating scheme of a two-column Pressure-Swing-Adsorption.

is inspired by the moving-bed processes in use for water treatment. A pilot plant based on this principle has been in operation in Australia for desalting with Sirotherm resins. The theoretical analysis of such processes has been developped by Bailly ; at steady-state, McCabe-Thiele type constructions may be used (Tondeur 1978). Fig 20 shows the result of the startup of the Ca^{++}/K^{+} separation mentionned above, and the table below shows the result of a Fe^{+++}/Cu^{++} separation in acid solution

Separation of Fe^{+++} and Cu^{++} in Cl^{-} H^{+} solution

Temperatures : 4°C and 60°C ; Total concentration : 0.5 N
Resin : Duolite C 265 ; overall reflux ratio : 40

	Feed	Product 1	Product 2
x_{Fe}	0.10	0.14	0.014
x_{Cu}	0.10	0.0034	0.07

$$\frac{x_{Fe}}{x_{Fe} + x_{Cu}} = 0.975 \qquad \frac{x_{Cu}}{x_{Fe} + x_{Cu}} = 0.83$$

4.3 Using Parameters Other Than Temperature

pH, ionic strength and electric potential have been used to produce parapump effects in adsorption and ion-exchange from the liquid-phase. pH and ionic strength clearly affect the adsorption or ion-exchange equilibria. These parameters are changed by passing chemicals through the bed, and in that sense, it may be thought that the operation is no longer fundamentally different from the common ion-exchange cycles with chemical regeneration. Actually, the characteristic feature that remains different is the use of reflux, in two-way parapumps, and the use of multiple beds out of phase in cycling zone adsorption. No regeneration step as such is involved.

Published experimental results include separation of ions (Sabadell and Sweed 1970), of enzymes (Shaffer and Hamrin 1975 ; Chen et al 1981) and of proteins (Chen et al 1979, Hollein et al 1982). Chen and coworkers have studied extensively the heamoglobin-albumin separation using ion-exchange resins. The two pH levels are chosen so that they frame the isoelectric pH of the proteins to be separated. Thus for example at a low pH, one protein will be anionic, thus retained by an anion-exchange resin while the other is cationic and unretained. At a higher pH, the retained protein becomes neutral or cationic and is thus released. The combined use of pH and ionic strength (Chen et al, 1981) and of pH and electric potential (Hollein et al, 1982) have been experimented. In the latter case, the electric field was applied axially along the bed, and modulated in synchronization with changes in pH and flow direction ; an electromigration process of the proteins is thus superimposed on the chromatographic effect.

4.4. Pressure-Swing Adsorption (P.S.A.)

Although this process applies to gas separation on adsorbents, and is thus outside the scope of this course, it is worth mentioning because it is the only parametric pumping process in abundant use on a large scale in industry. Actually P.S.A. can be traced back to 1954 and is thus anterior to Wilhelm's concept of parametric pumping.

Fig 21 shows a typical and relatively simple P.S.A. scheme, using two columns. The cycle comprises four steps ; compression with mixture, isobaric percolation with production (carried out until breakthrough), decompression, and low pressure purge by reflux of part of the high pressure product. The classical ingredients of parametric pumping are present : the driving force stems from the modulation of a thermodynamic parameter (pressure), no foreign chemicals are used for regeneration, reflux and flow-reversal are used. The only difference with thermal P.P. is that pressure is modified by adding or substracting material from the beds, and the parameter variation is thus coupled with the mass balance. In that sense, it resembles somewhat pH or ionic strength P.P.

P.S.A. processes have reviewed recently by Kenney and Kirkby (1984) and Tondeur and Wankat (1985).

4.5. Ion-Exchange Combined with Other Separation Processes

A form of parametric pumping can be recognized in some coupled processes where ion-exchange is in combination with for example evaporation or a membrane process. Let us consider an example which has received some industrial or semi-industrial developments. Sea-Water or brine is being softened by ion-exchange prior to evaporation to avoid scale formation in the evaporator. The concentrated brine from the evaporator, which contains essentially Na^+, is used to regenerate the ion-exchange resin. We can thus consider that there is a "reflux" of a part of the product of the ion-exchange ; the increase in ionic strength shifts the equilibrium to make it more favorable to elution of Ca^{++}. The temperature may also be brought in as a parameter ; Ca^{++} being more strongly retained at a high temperature, the sea-water can be preheated prior to ion-exchange, in particular using heat-exchange with the concentrated brine, which should be cooled to enhance Ca^{++} elution.

In a similar fashion, ion-exchange can be interestingly coupled with reverse osmosis or electrodialysis.

REFERENCES

Review Articles

- Chen H.T., 1979, "Parametric pumping", Section 1-15, p. 467-486 in "Handbook of separation techniques for chemical engineers", P.A. Schweitzer (Ed.), McGraw-Hill, New York
- Grevillot G., 1985 ; "Principles of parametric pumping" chapter 36, p. 1283-1321, in "Handbook for heat and mass transfer", N.P. Cheremisinoff (Ed.), Gulf Publ., West Orange, N.J., U.S.A. (to be published)
- Rice R.G., 1976, "Progress in parametric pumping", Separ. Purif. Methods, 5 (n° 1), p. 139-176
- Wankat P.C., 1974, "Cyclic separation processes", Separ. Science, 9 (n° 2), p. 85-116
- Wankat P.C., 1978, "Cyclic separation techniques", p. 443-515 in "Percolation processes, theory and applications", A.E. Rodrigues and D. Tondeur (Eds), Sijthoff and Noordhoff, Alphen aan den Rijn (Netherlands).

References Quoted in Text

- Ackermann G.R., Barrett J.H., Bossler J.F., Dabby S.S., 1976, "Industrial deionization with Amberlite XD-2 ; a thermally regenerable ion-exchange resin", p. 107-111 in "Water 1976 - Physical, chemical wastewater treatment", A.I.Ch.E. Symp. Series 73 (n° 166).
- Almeida F., Grevillot G., Costa C., Rodrigues A., 1982, "Removal of phenol from wastewater by recuperative mode parametric pumping", p. 169-178 in "Physicochemical Methods for Water and Wastewater Treatment", L. Pawlowski, Editor, Elsevier, Amsterdam.
- Bailly M., Tondeur D., 1980, "Thermal fractionation by moving-bed ion-exchange : principles and experiments", J. Chromatog., 201, p. 343-357.
- Baker B., Pigford R.L., 1971, "Cycling-zone adsorption : quantitative theory and experimental results", Ind. Eng. Chem. Fundam., 10 (n° 2), p. 283-292.
- Bolto B.A. et al, 1970, "Thermal regeneration of weak-electrolyte resins", p. 270-279 in "Ion-exchange in the process industries", Society of Chem. Industr. Editor, London 1970.
- Bolto B.A., 1975, "Sirotherm Desalination", Chem. Tech., May 1975, p. 303-307.
- Bolto B.A., Swinton E.A. et al, 1984, "The present status of continuous ion-exchange using magnetic micro-resins", p. 542-562 in "Ion-exchange technology", D. Naden and M. Streat (Eds), Ellis Horwood Ltd, Chichester, 1984.
- Butts T.J., Sweed N.H., Camero A.A., 1973, "Batch fractionation of ionic mixtures by parametric pumping", Ind. Eng. Chem. Fundam. 12 (n° 4) p. 467-472.
- Camero A.A., Sweed N.H., 1976, "Separation of non-linearly sorbing solutes by parametric pumping", A.I.Ch.E. Journal 22 (n° 2), p. 369-376.

- Chen H.T., Hill F.B., 1971, "Characteristics of batch and continuous equilibrium parametric pumps", Separ. Sci., 6 (n° 3), p. 411-434.
- Chen H.T., Rak J.L., Stokes J.P., Hill F.B., 1972, "Separations via continuous parametric pumping", A.I.Ch.E. Journal, 18 (n° 2), p. 356-361.
- Chen H.T., Reiss E.R., Stokes J.D., Hill F.B., 1973, "Separations via semi-continuous parametric pumping", A.I.Ch.E. Journal, 19 (n° 3), p. 589-595.
- Chen H.T., Manganaro J.A., 1974, "Optimal performance of equilibrium parametric pumps", A.I.Ch.E. Journal, 20 (n° 5), p. 1020-1022.
- Chen H.T., Lin W.W., Stokes J.D., Fabrisiak W.R., 1974, "Separation of multicomponent mixtures via thermal parametric pumping", A.I.Ch.E. Journal, 20 (n° 2), p. 306-310.
- Chen H.T., D'Emidio V.J., 1975, "Separation of isomers via thermal parametric pumping", A.I.Ch.E. Journal, 21 (n° 4), p. 813-815.
- Chen H.T., Wong Y.W., Wu S., 1975, "Continuous fractionation of protein mixtures by pH parametric pumping", A.I.Ch.E. Journal, 25 (n° 2), p. 320-327.
- Chen H.T., Ahmed Z.M., Rollen V., 1981, "Parametric pumping with pH and ionic strength : enzyme purification", Ind. Eng. Chem. Fundam., 20, p. 171-174.
- Foo S.C., Rice R.G., 1975, "On the prediction of ultimate separations in parametric pumps", A.I.Ch.E. Journal, 21 (n° 6), p. 1149-1158.
- Foo S.C., Rice R.G., 1977, "Steady-state predictions for non-equilibrium parametric pumps", A.I.Ch.E. Journal, 23 (n°1) p. 120-123.
- Ginde V.R., Chu C., 1972, Desalination, 10, p. 309.
- Goto S., Sato N., Teshima H., 1979, "Periodic operation for desalting water with thermally regenerable ion-exchange resin", Separ. Sci. Technol., 14 (n° 3), p. 209-217.
  See also Matsuda et al, 1971, Ibid., 16 (n° 1), p. 31-41.
- Gregory R.A., Sweed, N.H., 1981, "Parametric pumping : behavior of open systems. II : Experiment and computation", The Chem. Eng. J., 4, p. 139-148.
- Grevillot G., Tondeur D., 1976, "Equilibrium-staged parametric pumping. I : Single transfer step per half-cycle and total reflux. The analogy with distillation", A.I.Ch.E.Journal, 22 (n° 6), p. 1055-1063.
- Grevillot G., Tondeur D., 1977, "Equilibrium-staged parametric pumping. II : Multiple transfer steps per half-cycle and reservoir staging", A.I.Ch.E. Journal, 23 (n° 6), p. 840-851.
- Grevillot G., 1980, "Equilibrium-staged parametric pumping. III : Open systems at steady-state-McCabe-Thiele diagrams", A.I.Ch.E. Journal, 26 (n° 1), p. 120-131.
- Grevillot G., Dodds J., Marques S., 1980, "Separation of silver-copper mixtures by ion-exchange parametric pumping. Total reflux separation", J. Chromatogr. 201, p. 329-342.

398

- Grevillot G., Marques S., Tondeur D., 1984, "Donnan partition parametric pumping", Reactive Polymers, 2, p. 71-77.
- Grevillot G., Tondeur D., 1984, "Silver-copper separation by continuous ion-exchange parametric pumping", p. 653-660 in "Ion-exchange technology", D. Naden and M. Streat (Eds), Ellis Horwood Ltd., Chichester, 1984.
- Hollein H.C. et al, 1982, "Parametric pumping with pH and electric field : protein separations", Ind. Eng. Chem. Fundam., 21, p. 205-214.
- Kenney C.N., Kirkby N.F., 1984, "Pressure-Swing Adsorption" in "Zeolites : science and technology", Ribeiro et al (Eds), Martinus Nijhoff, The Hague.
- Kraus K.A., Raridon R.J., 1959, "Temperature dependence of some cation exchange equilibria in the range 0 to 200°C", J. Phys. Chem., 63, p. 1901.
- Marques S., 1978, internal report, Lab. Sciences Génie Chimique-CNRS, ENSIC-Nancy (France).
- Pigford R.L., Baker B., Blum D.E., 1969, "An equilibrium theory of the parametric pump", Ind. Eng. Chem. Fundam., 8, p. 603-604.
- Pigford R.L., Baker B., Blum D.E., 1969, "Cycling-zone adsorption ; a new separation process", Ind. Eng. Chem. Fundam., 8 (n° 4), p. 848-851.
- Rachez D., Delaveau G., Grevillot G., Tondeur D., 1982, "Stagewise liquid-liquid extraction parametric pumping - Equilibrium analysis and experiments", Separ. Sci. Technol., 17 (n° 4), p. 589-619.
- Rice R.G., 1975, "The effect of purely sinusoïdal potentials on the performance of equilibrium parapumps", Ind. Eng. Chem. Fundam., 14 (n° 4), p. 362-365.
- Rice R.G., 1975, Ind. Eng. Chem. Fundam., 14, p. 362.
- Rice R.G., 1976, "Progress in Parametric Pumping", Separ. Purif. Methods, 5 (n° 1), p. 139-188.
- Rice R.G., Foo S.C., Gough G.G., 1979, "Limiting separations in parametric pumps", Ind. Eng. Chem. Fundam., 18 (n° 2), p. 117-123.
- Rice R.G., 1981, "Adsorptive distillation", Chem. Eng. Communic., 10, p. 111-126.
- Rice R.G., Foo S.C., 1981, "Continuous desalination using cyclic mass-transfer on bifunctional resins", Ind. Eng. Chem. Fundam., 20 (n° 2), p. 150-155.
- Rolke R.W., Wilhelm R.H., 1969, "Recuperative parametric pumping", Ind. Eng. Chem. Fundam., 8 (n° 2), p. 235-245.
- Sabadell J.E., Sweed N.H., 1970, "Parametric pumping with pH", Separ. Sci., 5 (n° 3), p. 171-181.
- Shaffer A.G., Hamrin C.E., 1975, "Enzyme separation by parametric pumping", A.I.Ch.E. Journal, 21 (n° 4), p. 782-786.
- Shi T.T., Pigford R.L., 1977, "Removal of salt from water by thermal cycling of ion-exchange resins", p. 129-150 in "Recent Developments in Separation Science", Li N.N. (Ed), CRC Press, Cleveland (USA)
- Sweed N.H., Gregory R.A., 1970, "Parametric pumping : modeling direct thermal separation of sodium chloride-water in open and closed systems", A.I.Ch.E. Journal, 17 (n° 1), p. 171-176.

- Sweed N.H., Rigaudeau J., 1975, "Equilibrium theory and scale-up of parametric pumps", A.I.Ch.E. Symp. Series, 71 (n° 152), p. 1-5.
- Tondeur D., 1978, "Dual-step countercurrent processes", p. 517-538 in "Percolation processes", A.E. Rodrigues and D. Tondeur (Eds), Sijthoff and Noordhoff, Alphen aan den Rijn (Netherlands).
- Tondeur D., Wankat P.C., 1985, "Gas purification by pressure-swing adsorption", Separ. Purif. Methods (to be published).
- Treybal R.E., 1955, "Mass Transfer Operations", McGraw Hill, New York.
- Van der Schoot M.R., 1977, "Boron thermal regeneration system", U.S. Pat. 4,017,358 (April 12, 1977).
- Vermeulen T., 1977, "Process arrangements for ion-exchange and adsorption", Chem. Eng. Progr., oct 1977, p. 57.
- Wakao N.H., Matsumoto K., Suzuki M., Kawahara A., 1968, "Adsorption separation of liquids by mean of parametric pumping", Kagaku Kogaku, 32, 169 (in Japanese).
- Wankat P.C., 1973a, "Liquid-liquid extraction parametric pumping", Ind. Eng. Chem. Fundam., 12 (n° 3) p. 372-381.
- Wankat P.C., 1973b, "Cycling-zone extraction", Separ. Sci., 8 (n°4), p. 473-500.
- Wankat P.C., 1974, "Cyclic separation processes", Separ. Science, 9 (n° 2), p. 85-116.
- Wankat P.C., 1977, "Fractionation by cycling-zone adsorption", Chem. Eng. Sci., 32, p. 1283-1287.
- Wankat P.C., 1978, "Continuous recuperative mode parametric pumping", Chem. Eng. Sci., 33, p. 723-733.
- Weiss D.E., Bolto B.A. et al, 1965-1982, "An ion-exchange process with thermal regeneration".
  Part I : J. Austr. Instn. Engrs. 1965, 37, p. 193.
  Parts II to VI : Aust. J. Chem., 1966-1968.
  Parts VII to XVI : Desalination, 1970-1982.
- Wilhelm R.H., Rice A.W., Bendelius A.R., 1966, "Parametric pumping : a dynamic principle for separating fluid mixtures", Ind. Eng. Chem. Fundam., 5 (n° 1), p. 141-144.
- Wilhelm R.H., Rolke D.W., Sweed N.H., 1968, "Parametric pumping : a dynamic principle for separating fluid mixtures", Ind. Eng. Chem. Fundam., 7, p. 337-349.

PART IV
INDUSTRIAL APPLICATIONS

ION EXCHANGE IN INDUSTRY

P.Grammont,W.Rothschild,C.Sauer and J.Katsahian

Rohm and Haas Company
Duolite International SA
BP 48, 02300 Chauny,France

A range of applications are given in summary form to show the great versatility of the use of ion exchangers in industry.

1.ENERGY PRODUCTION

1.1.Condensate polishing with 3 components mixed bed

Specifications of feed water quality are given by the architects and are functions of the steam pressure in the boiler:high pressure (135 to 140 bars),very high pressure (140 to 220 bars),supercritical (above 224 bars).Turbines convert the energy in the steam to shaft energy.Then steam is condensed and reused to feed the boiler.Due to various sources of contamination there are other ions in solution: corrosion products (Fe^{2+},Cu^{2+},Co^{2+}),condenser leakage (Na^+,Ca^{2+},Mg^{2+}, Cl^-,SO_4^-,HCO_3^-,SiO_2),make-up water (Na^+,Cl^-,SiO_2),air in leakage(CO_2). The condensate has to be polished before reuse.The standard method of obtaining very high quality water (with a conductivity of less than 0.1 microSiemens) is to use a mixed bed of strong cation and strong anion resins.The main limiting factor for high quality water production was the difficulty of separating exhausted resins before regeneration and so limiting the regeneration level of resins.To overcome the difficulty,a 3 component mixed bed system has been developed.Its distinguishing feature is that one of its three components is an inert substance.The particle size distribution and bulk density of anion resin,cation resin and inert substance are selected such a way that after used mixed,all components can be perfectly separated by an upflow expansion.

The layers in the container separate out in the following order: at the top strong anion resin,in the middle particles of inert substance at the level of intermediate regenerant strainer and at the bottom strong cation resin.The layer of inert substance obviously reduces the risk of having a layer of mixed ion exchange resins as

it reduces the effect of reagent diffusion on the resin in the vicinity of the regenerant strainer.The type of equipment used is shown
in Figure 1.

A - Raw water inlet

B - Distribution grid

C - Air outlet

D - Inspection window

E - Manhole

F - Strainer floor with nozzles

G - Treated water outlet and backwash
    water and regenerant inlet

H - Anionic resin

I - Inert resin

J - Cationic resin

K - Regenerant outlet

L - Regenerant inlet grid

M - Drainage and sampling outlet

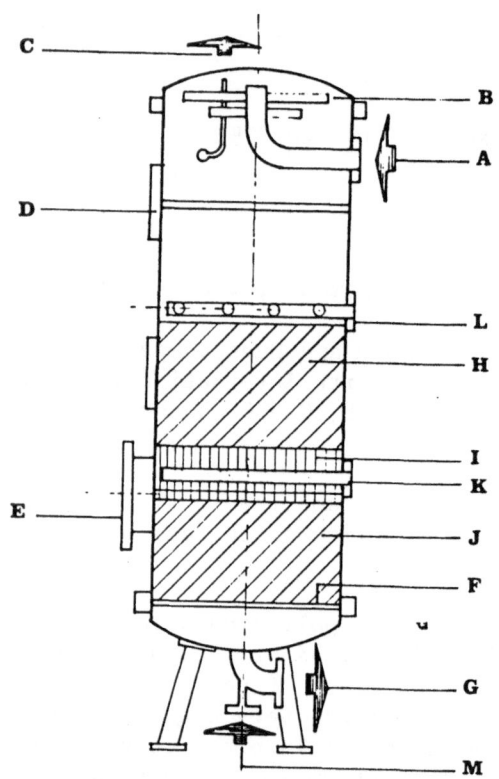

Figure 1 - A three component mixed bed

This technology was implemented in 1976 at DOEL (a nuclear power station in Belgium) by DIA-PROSIM actually DUOLITE International.It was
called TRIOBED.Today the Rohm and Haas AMBERSEP technology is similar.
The 3 component system works either in H/OH or NH_4/OH cycle.The H/OH
cycle allows the following levels to be achieved:
    - Na level below 0.1 ppb usually 0.01 to 0.02 ppb
    - Cl level always < 0.1 ppb
    - SO_4 level below 0.1 or 0.2 ppb
These are results from Fawley Power Station,CEGB,UK.Triobed was installed March 1979 (first external regeneration system).They have to
be compared with former results with conventional mixed bed:
    - Na : 1 to 1.5 ppb
    - Cl : 1 to 2 ppb
    - SO_4: 1 to 2 ppb
In conclusion,this new process has proved itself capable of reducing
permanent leakage rates generally encountered with traditional mixed
beds.It has to be recommended for modern techniques requiring ultra
pure water.

1.2.Condensate polishing with precoat filters

Problem to solve: corrosion and erosion products have to be removed from recycled condensate as well as possible leakage of saline water in the condensate at condenser level.

Filtration with inert materials is well known through precoat filters (ex.SOLKAFLOC) but inert materials never remove ions.In order to solve this problem,mixed ground ion exchange resins (anions+ cations) have been used to replace inert materials.The standard technology was using very finely ground anion resin and cation resin which were premixed first before coating with addition a polyelectrolyte to decrease the size of the floc and to obtain uniform coating. This technology had some disadvantages including premixing time,bad smell of anion resin,possible bad mixing ratio and polyelectrolyte added.

Since 1979 a new technology was proposed which involved the use of a powder with finely ground anion and cation resins in different ratios depending on the use required.

MB series microresins (Table 1)
Chemical purity of microresins:
Regeneration level –   cation H^+ minimum   99%
                        cation NH_4^+ minimum   98%
                        anion OH^- minimum   90%

The description of the power station for the example of condensate polishing on microresins is given in Table 2.

Comparison of traditional microresins (separated) with premixed (MB 400) for condensate polishing is shown in Table 3.

Table 4 presents some data on the performance obtained in the filtration of condensate.

Taking into account the specificity of 3 components mixed bed polishing and precoat filter polishing with premixed resins,the change in water quality required for condensate polishing,one can assess a big future for combination of the two technologies.

2.ULTRA PURE WATER IN ELECTRONIC INDUSTRY

2.1.Theoretical aspect and problem to solve

Pure water has a theoretical resistivity at 25C of18.24 $M\Omega/cm$. This raises the question as to whether is water with a resistivity of 18.24 $M\Omega/cm$ pure?Certainly not,because resistivity takes first into account charged moving "particles".The resistivity of a given water at a given temperature is a function of concentration,valence and mobility of all "ionic" species.Non-ionic species exist also which do not affect resistivity but affect purity.The production of ultra pure water will also include filtration of particles,the size of which varies from 0.001μ to 100μ (0.1 mm).

Figure 2 gives specifications of the water quality required by different companies involved in electronic industry.

TABLE 1 - MB SERIES MICRORESINS

NAME	COMPOSITION	CHIEF USE	TOTAL ION EXCHANGE CAPACITY eq/Kg of DRY MIXTURE (anionic capacity)
MB 200	50% cation H^+ form 50% anion OH^- form	Important demineralization (neutral or slightly alkaline condensate)	1.9
MB 250 NH_4	60% cation NH_4 form 40% anion OH^- form	Important demineralization (condensate with a pH of 9 to 9.6)	1.1
MB 300	66% cation H^+ form 34% anion OH^- form	Demineralization and filtration (neutral or slightly alkaline condensate)	1.3
MB 400	75% cation H^+ form	Colloïd filtration or suspension matter (neutral or slightly alkaline condensate)	0.9
MB 400 NH_4	75% cation NH_4 form	Colloïd filtration or suspension matter (condensate pH 9 - 9.6)	

TABLE 2- <u>EXAMPLE CONDENSATE POLISHING ON MICRORESINS</u>

<u>(DESCRIPTION OF THE POWER STATION)</u>

Nuclear power-station   :   Scandinavia

Type   :   BWR

Power   :   2 unities of 580 MW

<u>Filters</u>

Number	: 12
Filtering surface on 1 filter	: 70 m2
Type of cartridges	: wound candles
Porosity of the cartridges	: 2 μ
Nature	: nylon or polypropylene

<u>Working characteristics</u>

Coating velocity	: 4 to 5 m/h (m3/m2/h)
Service flowrate	: 160 to 170 Kg/s = 8 to 8.7 m/h
Criterion for the end of service	: 1.8 bar

TABLE 3 - EXAMPLE OF CONDENSATE POLISHING:COMPARISON OF TRADITIO_
NAL MICRORESINS (SEPARATED) WITH PREMIXED (MB 400)

CONDENSATES TO BE TREATED

Fe : 10 to 25 ppb   t° C   : 60° C
Cu : 0.5 to 2 ppb   Flowrate   : 160 to 170 Kg/s
Zn : 0.1 to 0.5 ppb   Linear speed : 8 to 8.7 m/h

TRADITIONAL MICRORESINS

860 g/m2

MB 400

860 g/m2

LENGTH OF SERVICE

(Average on 7 months) 12 days
stop at 1.8 bar

Nature of treated condensates

Fe   : 0.12 ppb
Cu   : 0.033 ppb
Zn   : 0.030 ppb
Conductivity : 0.05 µS

LENGTH OF SERVICE

(Average on 7 months) 20 days
stop at 1.8 bar

Nature of treated condensates

Fe   : 0.10 ppb
Cu   : 0.027 ppb
Zn   : 0.030 ppb
Conductivity : 0.05 µS

TABLE 4- <u>PERFORMANCE IN FILTRATION OF CONDENSATE</u>

Suspended matter size	Deep bed polishing	Precoat filters
	- Conventional mixed bed - 3 components mixed bed	Microresins on wound candle
> 0.45 μ	Efficiency = 40 to 80% retained	Efficiency = 95%
< 0.45 μ	Efficiency = less than 10%	Efficiency = 50 to 80%

Figure 3 gives an overview of the size of particles which may be encountered in natural and industrial waters.

2.2.Practical aspects

2.2.1.Chemical purification of water:The ion exchange technology is well known and by using mixed-bed polishing after conventional demineralization one can obtain water which does not contain more than 1 ppb of Na^+,1 ppb of Cl^- and 1 ppb of $SO_4^=$ .By using a 3 components mixed bed system (Triobed or Ambersep) one can drop each ion level below 0.1 ppb and so produce water with a resistivity above 16 MΩ/cm.

2.2.2.Filtration aspect: A lot of technologies are available for separations.Figure 4 gives the possibilities of different technologies to remove particles as a function of their size.One can see that ultrafiltration has the widest range of applications.This technology will be choosen due to its ease of implementation and its lower running costs compared to reverse osmosis.The combination of ultrafiltration followed by reverse osmosis will give improved filtration characteristics.Comparative information about the two systems is given below:

	Reverse Osmosis	Ultrafiltration
Maximum operating temperature	30 C	45 C
Maximum operating pressure	45 kg/cm^2	5 kg/cm^2
Maximum transmembrane pressure	28-42 kg/cm^2	1.75 kg/cm^2
pH	4-11	1.5- 13
Operating cost(¢ /1000 gal)	60	33
Capital equipment cost ($)	800,000	380,000

	IBM	TI	GI	RCA	DR	Mean
Resistivity MΩ/cm	>18	>15	>16	>15	>10	> 16
Electrolyte total concentration expressed as NaCl (ppb)	<25	<30	<30	<30	<45	< 35
Particles content N°./cm3	<150	<150	<100	<150	<100	< 130
Maximum particle size in μ	<0.5	<0.5	<0.5	<0.5	<0.5	< 0.5
Dissolved gas (ppm)	-	<200	-	<200	<200	<200
Living aquatic organisms N°/cm^3	-	<8	<10	-	<10	< 9
TOC ppm	<1	<1	<1	<1	<1	< 1

Figure 2- Specifications of the water quality required by different companies.

Figure 3- Particle size scale

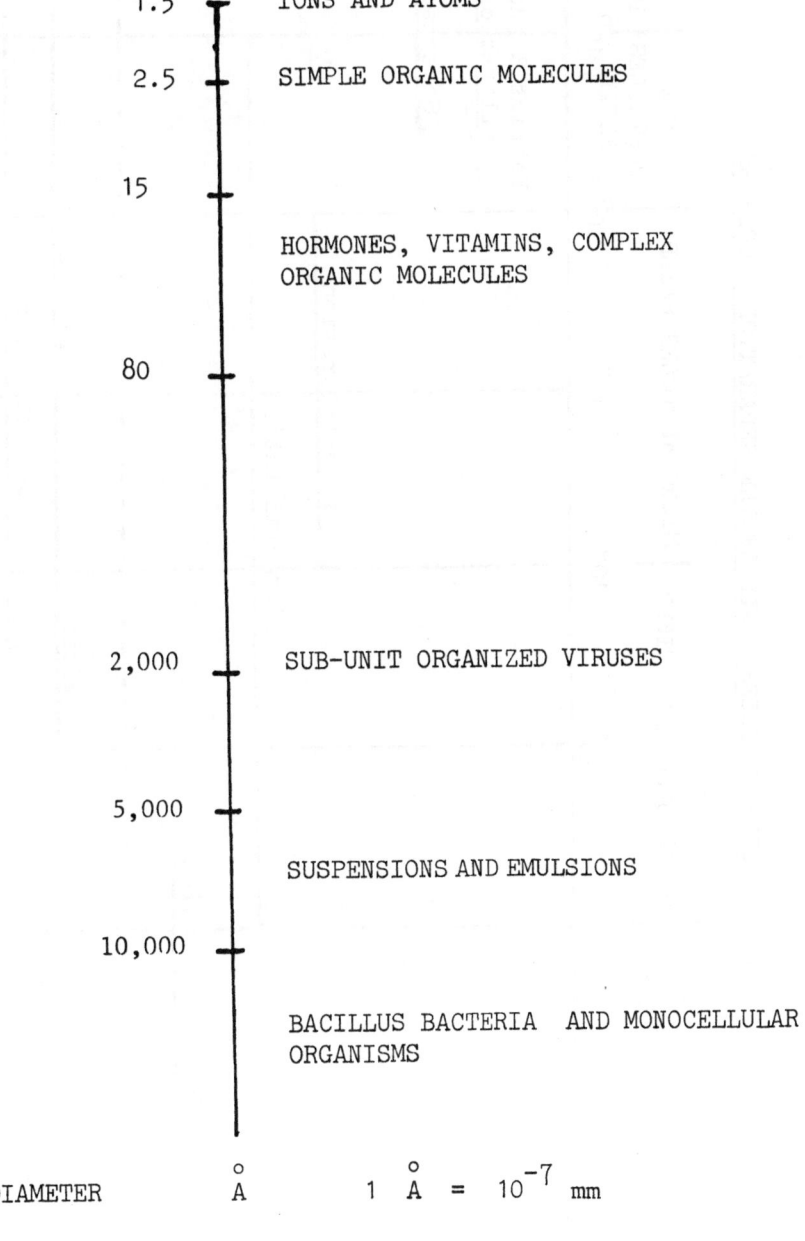

412

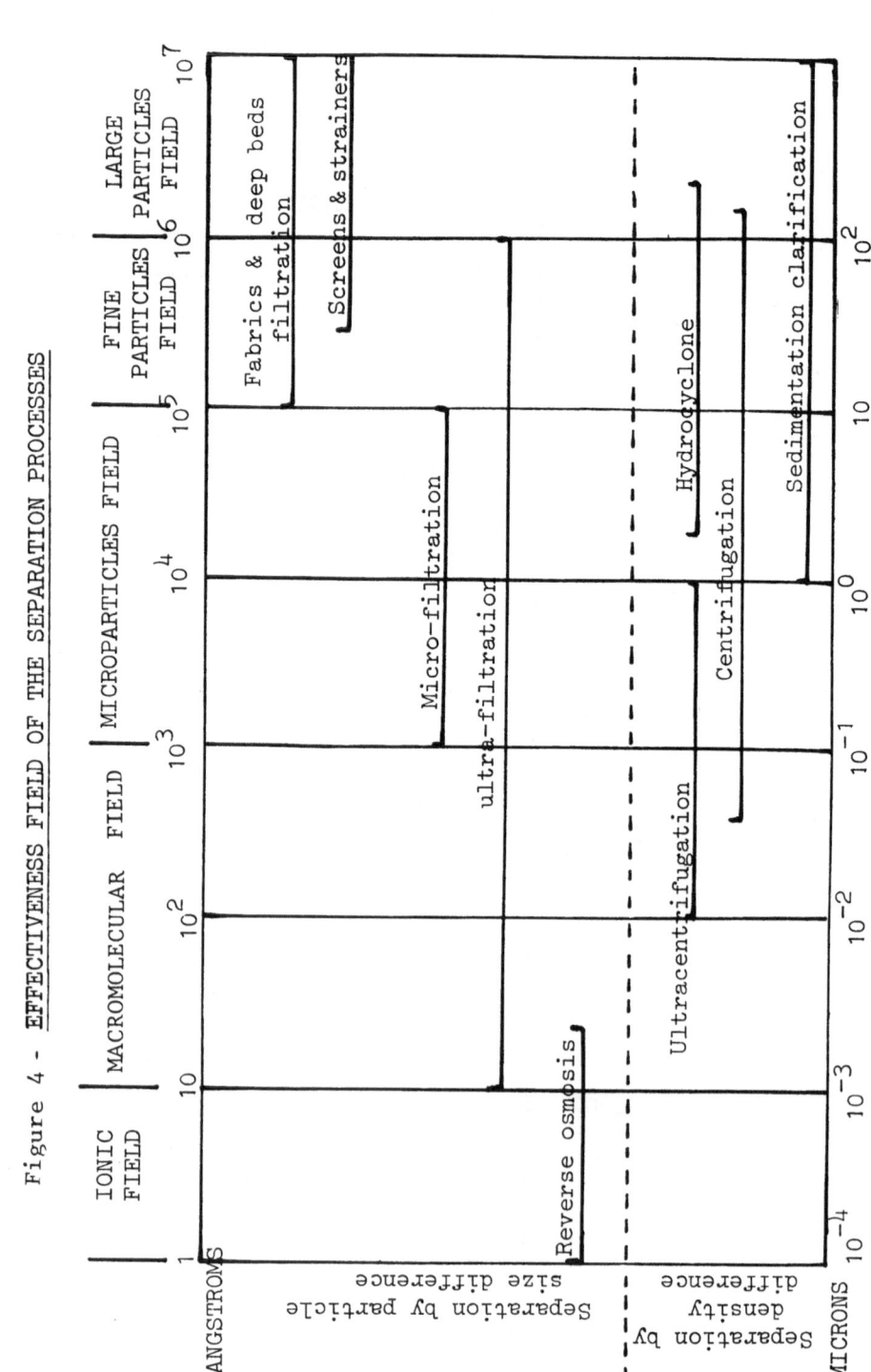

Figure 4 - EFFECTIVENESS FIELD OF THE SEPARATION PROCESSES

2.2.3.Bacteriological purification.Technologies are available from
the pharmaceutical and potable water production industries.There are
two aspects:bacteria filtration and sterilisation.Before demineralization,the water may be chlorinated and then filtered over activated
carbon to remove free chlorine which could damage the resin.Bacteria
have sizes in the order of 1000 Å and viruses of 100 Å.So,the filtration problem may be solved the same way as already discussed.Sterili-
sation will be done either by ozonization or by UV.If ozonization is
used,the water will have to be stored before treatment through mixed
bed resins.As resins are also a filtration media,periodical desinfec-
tion with 1% formol solution is needed.

2.2.4.Storage aspect of ultra pure water. Contamination can occur
from pipes,valves,storage tanks,air contact,etc.So,in practice con-
tinual recycling of the water may be used.The service water will be
delivered directly from a polishing cartridge containing a mixed-bed
of fine particles of nuclear grade ion exchange resins followed by
a filtration cartridge (0.2 μ).A diagram of the different steps is
given in Figure 5.One important aspect is the quality,which is moni-
tored on a continuous basis.

3.FOOD INDUSTRY
     There is a need to remove scaling ions before evaporation of a
liquor and to recover a product which crystallises.Softening is the
name for such a process.
3.1.Lactoserum production.Due to the high selectivity of strong cationic resins over non scaling ions,the regeneration phase of the
softening process requires an excess of NaCl as regenerant.That
creates a further waste water problem.To minimize this constraint
end-users have tried to reuse,during the regeneration step,the mother
liquor which is a concentrated solution of sodium salts.One question
to be answered is :"How can one analyse the feasibility and estimate
the economics of a softening unit based on this approach?

     Removal of Ca from lactoserum.

     In this study Duolite C26 (Na$^+$ form) was used together with a
lactoserum (6% dry matter).The operating conditions were defined as:
     1- end of cycle at Ca leakage 10% of influent content
     2- flow rate 15 bed vol/hour
     Analysis of influent lactoserum: Ca^{2+}=780 mg/l (39 meq/l),Mg^{2+}=
=96 mg/l (8 meq/l),Na^+=450 mg/l (19.5 meq/l),K^+=1240 mg/l (32 meq/l).
     The average useful capacity measured over 3 cycles was 0.9 eq/l.
     The exhausted resin when 10% Ca leakage occured,had the follow-
ing composition:0.9 eq/l Ca^{2+}, 0.23 eq/l Mg^{2+} , 0.61 eq/l K^+ and
therefore the total capacity Na^+ form was approximately 1.8 eq/l.
     Analysis of effluent lactoserum: Ca^{2+}=2.8 mg/l (0.14 meq/l),
Mg^{2+}=1.6 mg/l (0.79 meq/l),Na^+=2019 mg/l (87.8 meq/l),K^+=400 mg/l
(10 meq/l) and TS=98.98. This means that the majority of the calcium
and magnesium was removed.
     This decalcified solution was then concentrated 10 times prior

414

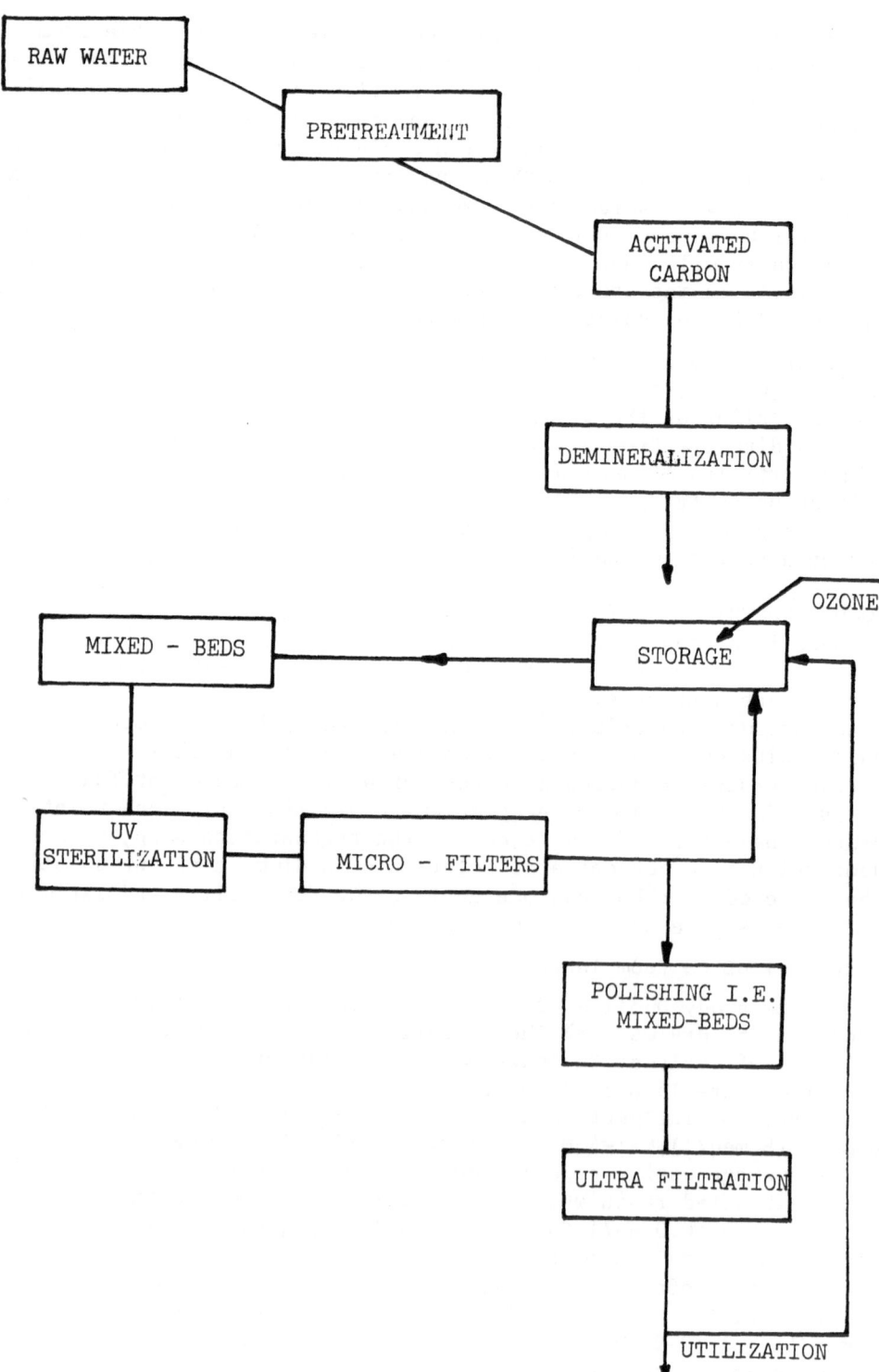

Figure 5- Typical scheme for preparation of ultra-pure water.

to crystallisation.The mother liquor can be recovered and reused for regeneration.The mother liquor contained about 1000 meq/l Na^++K^+which is equivalent to 5.8% NaCl solution.The design procedure is based on Figures 6 to 9.

The regeneration curve of Duolite C26 initially saturated with calcium up to a leakage of Ca of 10 mg/l is shown in Figure 7.Two terms are used in the regeneration curve:

1.The regenerated equivalent fraction of Ca^{2+}: $F_{R\,Ca}$ which is the ratio of Ca^{2+} eq/l eluted to the quantity formerly present into the resin before regeneration:

2.The regeneration level$\tau_R^!$ which is the ratio of equivalents of regenerant passed through the resin to the equivalents of Ca^{2+} present at the beginning step.

The regenerant conditions are: 5 bed vol/vol/hour of 10% NaCl (countercurrent);temperature :20-22 C.

The maximum capacity obtainable is 0.9 eq/l.This means that 23 volumes of lactoserum could be softened per litre of resin.

If we expect to treat Y volumes of lactoserum just by autoregeneration of the resin,the useful capacity will be:

$$Y\,[Ca^{2+}]$$

with $[Ca^{2+}]$expressed in eq/l of lactoserum to be treated.

We can now calculate:

$$\tau_R^! = \frac{Y \cdot TS}{900} = \frac{Y \times 98.5}{900}$$

$$F_{R\,Ca} = \frac{Y\times Ca}{900} = \frac{Y \times 39}{900}$$

$$\tau_R^! = 23\times98.5/900=2.5$$

From Figure 6 we get at $\tau_R^!=2.5$ a value for $F_{R\,Ca}=0.67$ and a useful capacity C_u of 15.5 volumes.

Next cycle we will have a $\tau_R^!=1.69$ (Table 5) and a useful capacity of 0.55 eq/l or 12.7 volumes of lactoserum softened by liter of resin. We can proceed further but the limiting factor is the volume of mother liquor available in regard of the starting volume of resin.So,we are obliged to use new regenerant each time.If we want to treat 20 bed volumes per cycle,the useful capacity will be 0.7 eq/l so we need a $\tau_R^!$ =5.5. We have to add 150 g of NaCl per liter of resin.Implementing such an approach gives an estimated useful capacity of 0.72 eq/l which corresponds to 18 volumes of serum treated.Without recovery of regenerant we would have needed 230 g NaCl/ l resin.The saving is 80 g of salt per litre or about 35% . Figure 9 gives an idea of the range of achievable economies in relation to the treated bed-volumes of lactoserum.

3.2.The NRS process -an IMACTI process for the regeneration of resins used to soften sugar juice.

Problem to solve: we have seen that regeneration of resins used to soften a liquor needs excess of regenerant and creates a waste water problem.To regenerate strong cationic resins used in softening

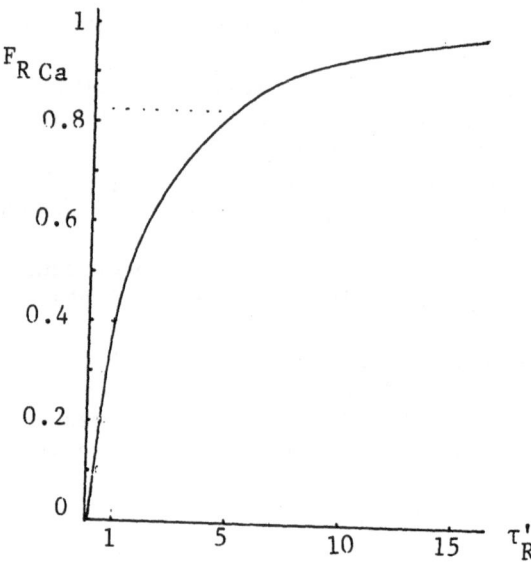

Figure 6- Total regeneration curve (NaCl 10% ;5 v/v/h ;
I.E.R.:Y_s Mg=0.13, Y_s K=0.34 , Y_s Ca=0.53)

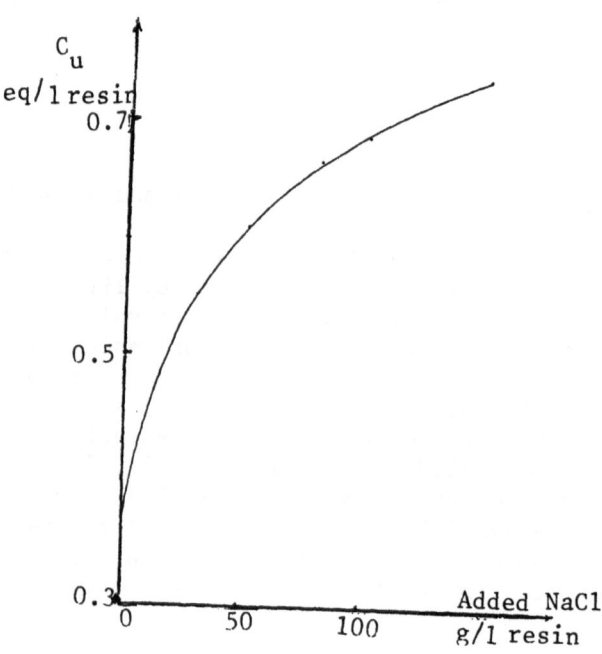

Figure 7- Useful capacity C_u as a function of added NaCl

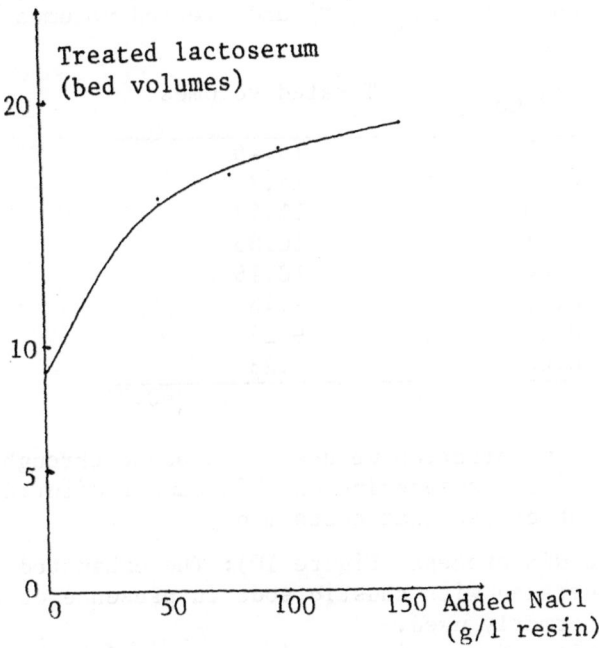

Figure 8- Treated lactoserum (bed volumes) as a function of added NaCl

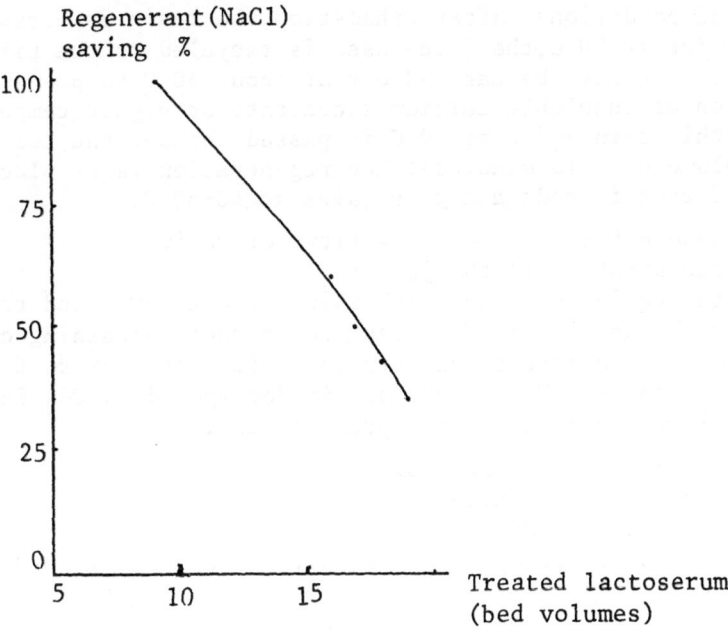

Figure 9- Regenerant saving in terms of treated lactoserum bed volumes

Table 5 - Relation between $F_{R\,Ca}$, τ'_R and treated volumes

τ'_R	$F_{R\,Ca}$	Treated volumes
2.5	0.67	15.46
1.69	0.55	12.7
1.39	0.50	11.53
1.26	0.47	10.85
1.187	0.44	10.15
1.11	0.41	9.46
1.03	0.40	9.23
1.01	0.40	9.23

$$C_u = 0.36$$

thin-juice prior to concentration we need to proceed through two steps of sweetening off and sweetening on.This causes dilution and losses of sugar and of course that costs money.

Principle of the NRS process (Figure 10): The exhausted resin (Ca^{2+} form) is regenerated with caustic soda in presence of sugar instead of NaCl as formerly used.

An aqueous solution of caustic soda in contact with the exhausted resin would,however,cause immediate precipitation of brine.In the presence of sugar a calcium saccharate complex is formed which remains in solution.

Operating conditions: After exhaustion,the resin is backwashed with thin juice at 90 C,the juice used is recycled to the filtration stage.Regeneration must be carried out at about 50 C to prevent possible formation of insoluble calcium saccharate at higher temperatures. To achieve this,thin juice at 40 C is passed through the resin bed (1/4 bed volume over 15 minutes).Then regeneration takes place with a mixture of caustic soda and thin juice at 40-50 C.

NaOH consumption: 40 g/litre of resin
NaOH concentration in the juice: 4%
After that,the resin is rinsed with thin juice at 90 C and the column is ready for the next cycle.The spent regenerant containing calcium saccharate and the excess of caustic (35%) is sent back to the second carbonatation step.The saccharate is decomposed by CO_2 into sucrose and calcium carbonate which precipitates.

Advantages of the NRS process:
  - short regeneration time
  - extremely simple operation
  - no fluctuation in operating conditions of the evaporator due to sweetening off and on
    - no risk of corrosion by chloride
    - no waste water
    - no sugar losses
    - no dilution of thin juice and so energy saving during operation.

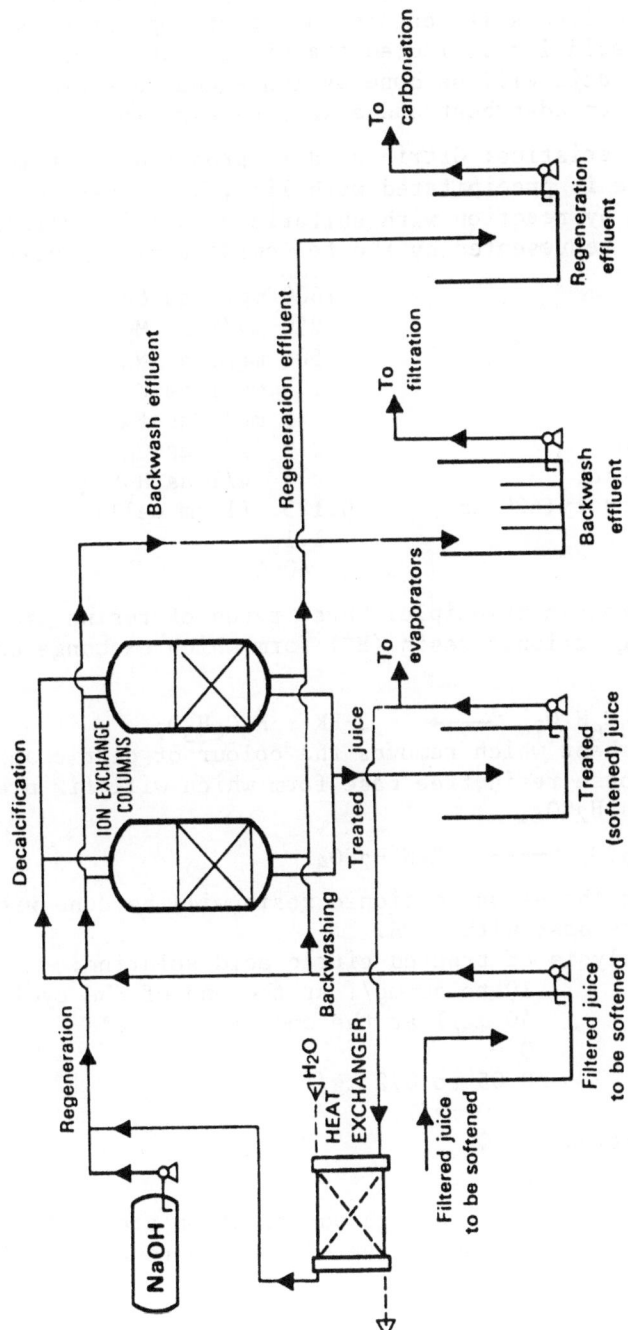

Figure 10– **SUGAR JUICE SOFTENING NRS PROCESS**

3.3. Demineralization and decolourization of citric acid.

Citric acid used in the food and pharmaceutical industries has to comply with quality requirements. When produced by the fermentation of sugar molasses it contains a lot of impurities:mineral salts, free sulfuric acid,iron,coloured organic products and so on.Purification of this acid will be done by ion exchange resins: a strong cationic resin,an adsorbent and a weak base resin.

Process characteristics: Citric acid is produced by fermentation. Calcium citrate is precipitated with lime,then separated and citric acid liberated by reaction with sulfuric acid.This sulfuric acid mixture can be represented by the following typical analysis:

Calcium	1600 mg/l as Ca
Magnesium	250 mg/l as Mg
Sodium	500 mg/l as Na
Potassium	3100 mg/l as K
Iron	5 mg/l as Fe
Citric acid	210 g/l as $C_6H_8O_7$
Sulfuric acid	5 g/l as H_2SO_4
Optical density(420 nm)	0.125 (1 cm cell)
Density	1.1
Temperature	40 C

Purification process principle: Three types of resins are used.

1. A strong cationic resin (H^+) form which exchange cations impurities,

$$3RH + K_3C_6H_5O_7 \longrightarrow 3 RK + H_3C_6H_5O_7$$

2. An adsorbent which removes the colour of citric acid solution.

3. A weak base resin,free base form which will fix preferentially strong acids as H_2SO_4,

$$R_3N + H_2SO_4 \longrightarrow R_3N.H_2SO_4$$

Regeneration of the strong cationic resin will be done with HCl,adsorbent and weak base with NaOH.

Typical analysis of treated citric acid solution:

Na	10 to 50 mg/l at the end of the cycle
K	40 mg/l at the end of the cycle
Ca	0
Mg	0.05 to 0.2 mg/l
Fe	$\approx$1.0 mg/l
Decolourization	> 80%
$SO_4^=$	0

The losses of citric acid through purification are 2-4%: 2% on the weak base and 2% during sweetening off and on regeneration of resins.

Operating conditions :

Duolite	C 26	S 761	A 561
Cycle flow rate Bv/v/h	1 - 5	1.5 - 5	2 - 5
Useful capacity eq/l/resin	1	-	1.1
Regenerant	HCl	NaOH	NaOH
Type of regenerant	counter current	- co-current -	
Regenerant quantity g/l (100%)	60 to 100	50 to 80	60 to 100
Regenerant concentration % (weight basis)	5 - 10	2	2
Slow rinse	2 bed/vol Raw water or decationized water	2 bed/vol Soft water or decationized water	2 bed/vol Soft water or decationized water
Neutralization		HCl at 0.5%	
Rinse	Raw water	Soft or decationized water	Soft or decationized water

Scheme of the process

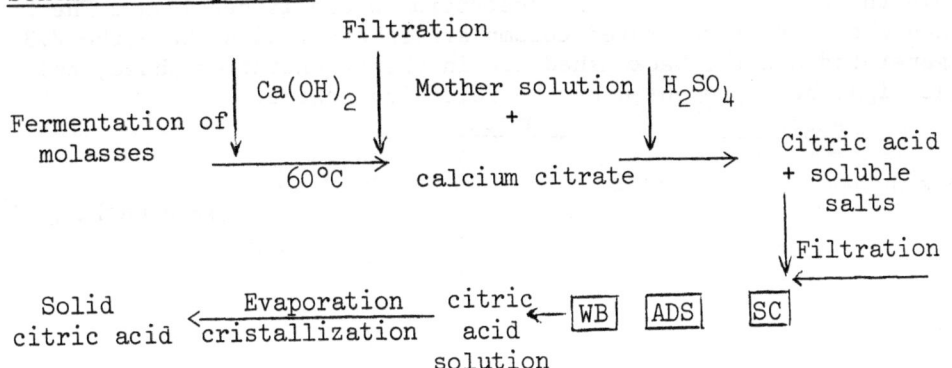

SC= strong cationic resin ,ADS= adsorbent, WB= weak base.

3.4.Production of aminoacids

Aminoacids are classified according to their isoelectric pH.The isoelectric point defines the pH above which the ionic form of the amino acid is anionic and below which is cationic.

Problem to solve:Extraction or separation

Lowering the pH of the solution to convert all species into cationic species and fix them on a cationic resin;raising the pH to convert all the species into anionic species and fix them on an anionic resin.In both cases elution will be done selectively with buffered solutions.

Lysine production: Lysine is produced industrially by fermentation from various carbohydrate raw materials:molasses,corn hydrolysates,acetic acid,etc.Its isoelectric point corresponds to a pH=9.7 . An acidic solution of lysine will contain the cationic species which can be fixed on a cationic resin.Furthermore,the selected strong cationic resin has a better affinity for lysine than for Na^+,NH_4^+,K^+ or Mg^{2+} as well as the others amino acids present in the broth.The strong cationic resin is selected not only for its affinity for lysine over the other cationic species present but also for its resistance to osmotic shock and attrition. Duolite C20 N 2014 is the selected product for this application.

Process description: Three phases:adsorption,backwashing,elution.

Adsorption: The filtration broth is acidified and passed through three columns of strong cationic resin NH_4^+ form.

Backwashing: This involves passing a counter current of deionized water through the first of the 3 columns used previously for lysine adsorption.

Regeneration and lysine elution: This involves passing NH_4OH, cocurrent,through the three exhausted columns.

One unit line contains 7 columns.During the cycle,3 are in the adsorption phase,3 in elution phase and one in backwashing phase.Following cycle the first column of the adsorption phase is backwashed.The 2, 3 and the first regenerated column are in adsorption phase;the 2,3 regenerated and the backwashed are in the regeneration phase,etc. That might be represented by the following scheme:

```
            Adsorption          elution
               ↓                   ↓
Cycle 1     1 → 2 → 3          4 → 5 → 6     7
                    ↓                  ↓      ↑Backwashing
                                    lysine
Cycle 2        2    3    4    5    6    7    1

Cycle 3        3    4    5    6    7    1    2

Cycle 7        1    2    3    4    5    6    7
               ‿‿‿‿‿‿‿‿    ‿‿‿‿‿‿‿‿
               Adsorption    Elution    Backwashing
```

Exhaustion of the resin is followed up by pH measurement.

3.5.Production of sweeteners

In this area,in addition to classical applications such as deco-lorization and demineralization,ion exchange resins are also used as enzyme supports and for chromatographic separation packing.

Enzyme supports: The support has,of course,to have affinity for the enzyme.Its "internal ph" environment has to fit with pH stabili-ty of the enzyme.It also has to be "permeable" to the substrate.Once loaded with the enzyme,activity has to be retained enough time for reasons of economy.

Immobilization of glucoisomerase on Duolite ES 562 : CPC Inter-national in a D.K. Pat.Appl. GB 2.040.949 (1980) gives information about procedures for the preparation and use of ion exchangers in the isomerization of glucose to fructose.Enzyme binding ranged from 90 to 100%.In tests for the isomerization of 97% DE hydrolysate at 50 % d.s. at 60 C with an initial flowrate of 6 BV/hour,half-lives ranged from 30 to 40 days.

Immobilization of β galactosidase on Duolite S. 761 : Lactose hydrolysis to produce glucose and galactose can be done by:
1- Mineral acids
2- H^+ form strong cationic resins
3- Enzymes   a) free:lactase
             b) immobilized : β galactosidase
Mineral acids create column problems. Free enzyme costs too much. So only two useful processes remain : immobilized enzymes and H^+ form strong cationic resin.
Comparative performances : from a lactose syrup at 65% d.s.

	Enzyme immobilized	Ion exchange resin
Lactose %	8	12
Glucose %	28	25
Galactose %	28	26
Fructose %	0.5	0.5
Other sugars	0	1.32 %

Immobilized enzyme are used as ion exchange resins by operating via percolation.

Lyfe time :For economical reason it is better to do the hydroly-sis of a preconcentrated syrup.Figure 11 shows the evolution of con-version percentage as a function of time from a lactose syrup con-taining 100 g/l of lactose.In the example Duolite S 761 was loaded with 300 g of TAKAMINE brand fungal lactase.

Cycle condition: Lactose solution at 100 g/l ,pH=4.5, flowrate= =20 BV/vol/h, temperature=40 C.

3.6. Glucose and fructose separation

Problem to solve: How to select the best resin and the optimum operational parameters for fructose enrichment?

Guidance is given below about the ion exchange resin.Neverthe-less,the equipment side is also an important parameter,but today,

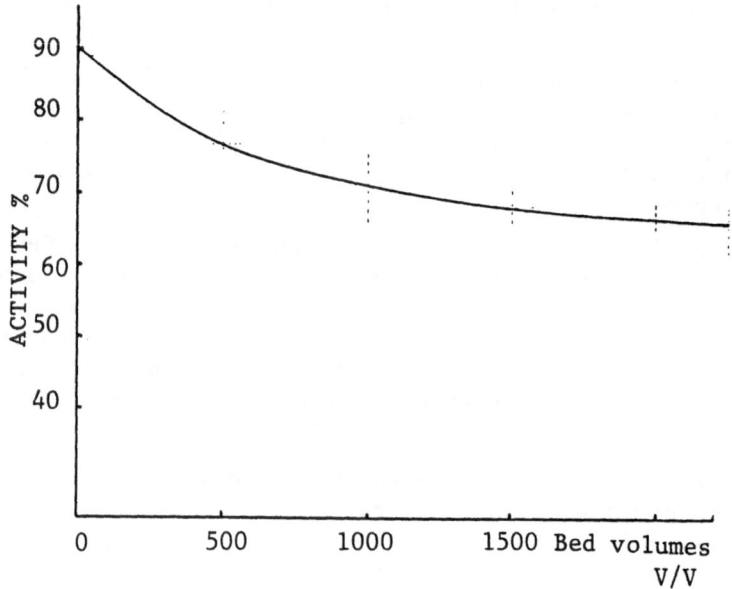

Figure 11- Activity lasting as a function of time
(Takamine 300 g/l in Duolite S 761;20
bed volume/hour;lactose 100 g/l ,pH=4.5
T=40 C)

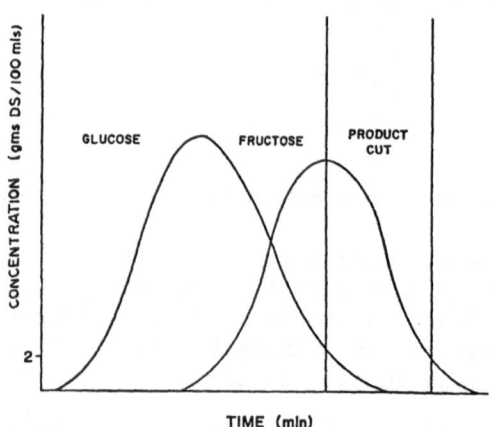

Figure 12- Product cut in glucose/fructose separation

most of industrial equipments are based on pseudo moving bed system. The columns are always completely filled with resin in order to eliminate unnecessary syrup dilution and to prevent resin from moving. Usually the bead is unclassified by resin bead size.

What do we need to know to select optimized resin? First,we need to know which purity of fructose cut is requested and to define the cut.The cut is defined as that in the Figure 12 and the purity as 90%.The operational conditions are fixed as follows: syrup bead level - 0.1 bed volumes, development flowrate- 0.4 bed vol./vol/hour, syrup concentration to separate- 40% D.S. , column temperature- 60 C. These operational parameters are very important as they influence the separation efficiency.For example,for a given resin (Ca form, 4% DVB,50 to 100 mesh size) Illinois Water Treatment Co has published a laboratory study from batch chromatography shown in Figure 13. The quantity of sugar placed in the resin is most critical for performance.

How to define the needed resin? The controlling mechanism in sugar chromatography is the intra-particle diffusion of the sugar into and out of the resin,that is particle diffusion control.Current chromatographic theory accurately describes analytical separations but they operate at very low loading and obtain perfectly separated peaks.Plant conditions do not use analytical conditions so we will just consider for the following experiments purity and yield of fructose in the selected cut.For the resins we have used two variables: particle size and percent moisture (calcium form).

RESIN ANALYSIS

Resins	A	B	C	D	E	F
Vol.cap. Ca eq/l	1.68	2.19	1.80	1.97	2.19	1.66
Moisture Ca %	52.4	40.9	52.1	47.6	42.4	53.2
Particle size, Ca						
+ 25	1	0	0	0	0	0
+ 30	1	1	0	0	0	0
+ 35	26	6	1	1	1	0
+ 40	44	17	12	6	2	0
+ 45	25	36	30	29	15	4
+ 50	3	24	28	31	25	17
+ 60	0	11	16	21	29	27
+ 80	0	4	10	11	25	47
+ 100	0	1	3	1	3	5
Mean particle diameter μ	460	368	338	328	288	247
Standard deviation	72	84	83	70	71	54

426

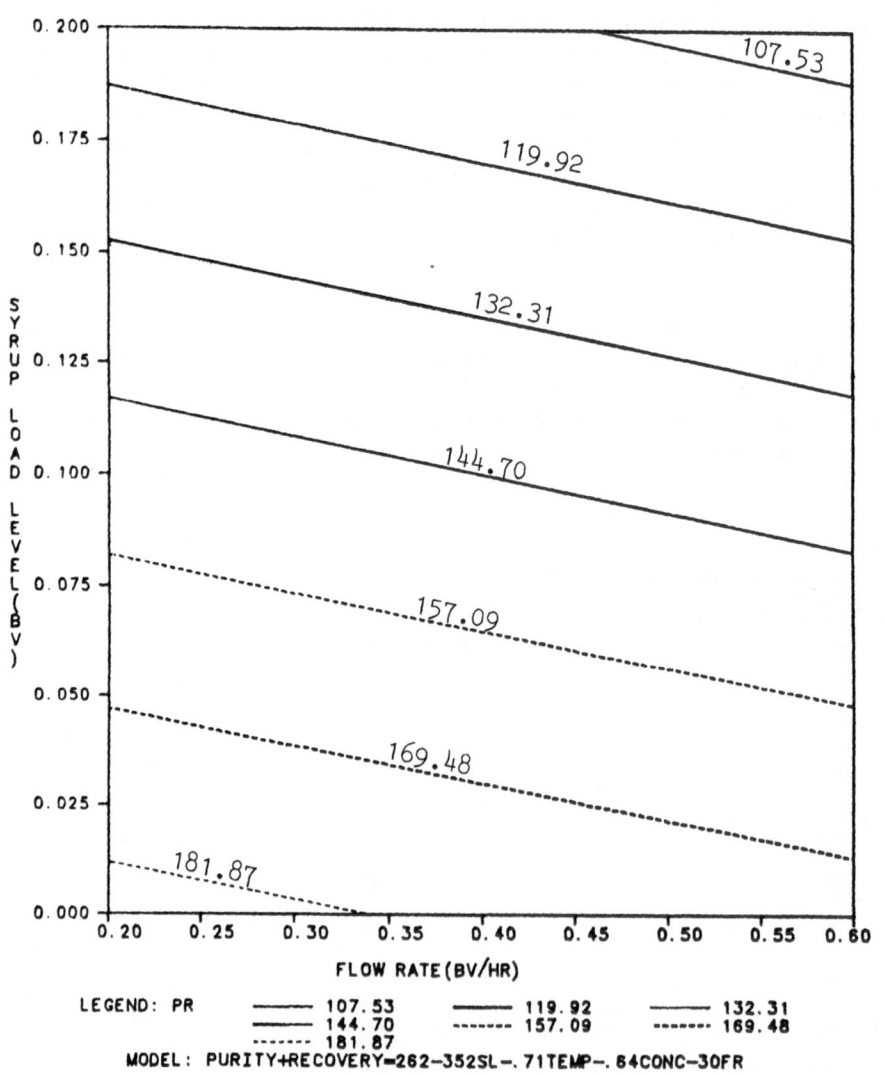

Figure 13- Syrup load level versus flowrate

ELUTION RESULTS

Elution data	A	B	C	D	E	F
Flow rate ml/mn	4.07	4.16	4.18	4.03	4.01	4.33
Elution time glucose, mn	119	96	109	109	98	110
Elution time fructose, mn	137	116	144	134	124	137
tf-tg mn	18	20	35	25	26	27

PRODUCT CUT ANALYSIS

	A	B	C	D	E	F
Glucose g/100 ml	0.8	1.4	1.3	0.8	1.9	2.5
Fructose g/100 ml	10.1	8.8	9.0	10.7	9.9	11.1
Total dry substance g/100 ml	10.9	10.3	10.3	11.5	11.8	13.6
% fructose purity	92.7	85.4	87.4	93.0	83.4	81.6
Yield fructose %	48.0	46.9	44.6	47.2	54.2	60.2

Definition of variables

The product cut is defined as the period beginning at maximum concentration of fructose and ending when total dry substance eluted reaches 2 g/100 ml.

Yield of fructose is expressed as weight percentage and is the amount of fructose recovered in the product cut divided by the total fructose in the feed.

Fructose purity is also expressed as a weight percentage, and is the amount of fructose recovered in the cut divided by the total dry substance in the cut.

Modelling of results

Data was fed to a response surface regression analysis program to determine the effect of each factor on separation.

Model equation used was :

$$\text{Purity, yield} = B_0 + B_1 X_1 + B_2 X_2 + B_{11} X_1^2 + B_{22} X_2^2$$

Where X_1 = mean particle diameter

X_2 = moisture

B_n = coefficient values

REGRESSION ANALYSIS RESULTS

Coefficient	Purity	Yield
B_0	− 417.84527	212.63670
B_1	0.07323	− 0.71407
B_2	20.50009	− 1.00618
B_{11}	− 0.00004	0.00093
B_{22}	− 0.21572	0.00730
Correlation coefficient	+ 0.96976	+ 0.99873
Residual variance	3.4	0.217

Discussion: The yield of fructose is inversely correlated with particle size (see Figure 14) but is only marginally affected by moisture at smaller particle size (240-300 μ).In the region of 320 to 460 μ resin moisture is inversely correlated with yield (diffusion pathway = capacity effect?) as shown in Figure 15. Moisture has a noticeable influence on purity.The most desirable moisture content is one which is high enough so that migration of the sugars in and out of the resin is not impeded,but low enough so that sufficient calcium is present to react with sugars.The separation efficiency is illustrated in Figure 16.

With respect to resin selection a purity of 90% for fructose is required.As yield is inversely correlated with particle size a mean bead size between 275 and 300 microns is chosen in order to maximize yield.As for moisture content it is possible to choose between 46% and 50%.The lower moisture is selected as it corresponds to the resin with the highest capacity.So the selected resin will have around 5.5% DVB,mean particle size 270 to 300 μm.

Pressure drop is also a function of particle size but resin uniformity coefficient to a smaller extent.Figures 17 and 18 illustrate the fine pressure drop results obtained in a one inch by four feet column and at flowrates of 0.3 and 0.6 BV/hour.

4.PETROCHEMICAL INDUSTRY

MTBE synthesis with Duolite C 276

MTBE is synthesised according to the following exothermic reaction.

$$CH_3\diagdown \atop CH_3 \diagup C= CH_2 \ + \ CH_3OH \ \xrightarrow{\ H^+\ } \ CH_3 - \overset{\overset{\displaystyle CH_3}{|}}{\underset{\underset{\displaystyle CH_3}{|}}{C}} - O - CH_3 \tag{1}$$

This is an example of an acid catalysed alcohol addition reaction. It is reversible and consequently the equilibrium and the final composition is a function of temperature.

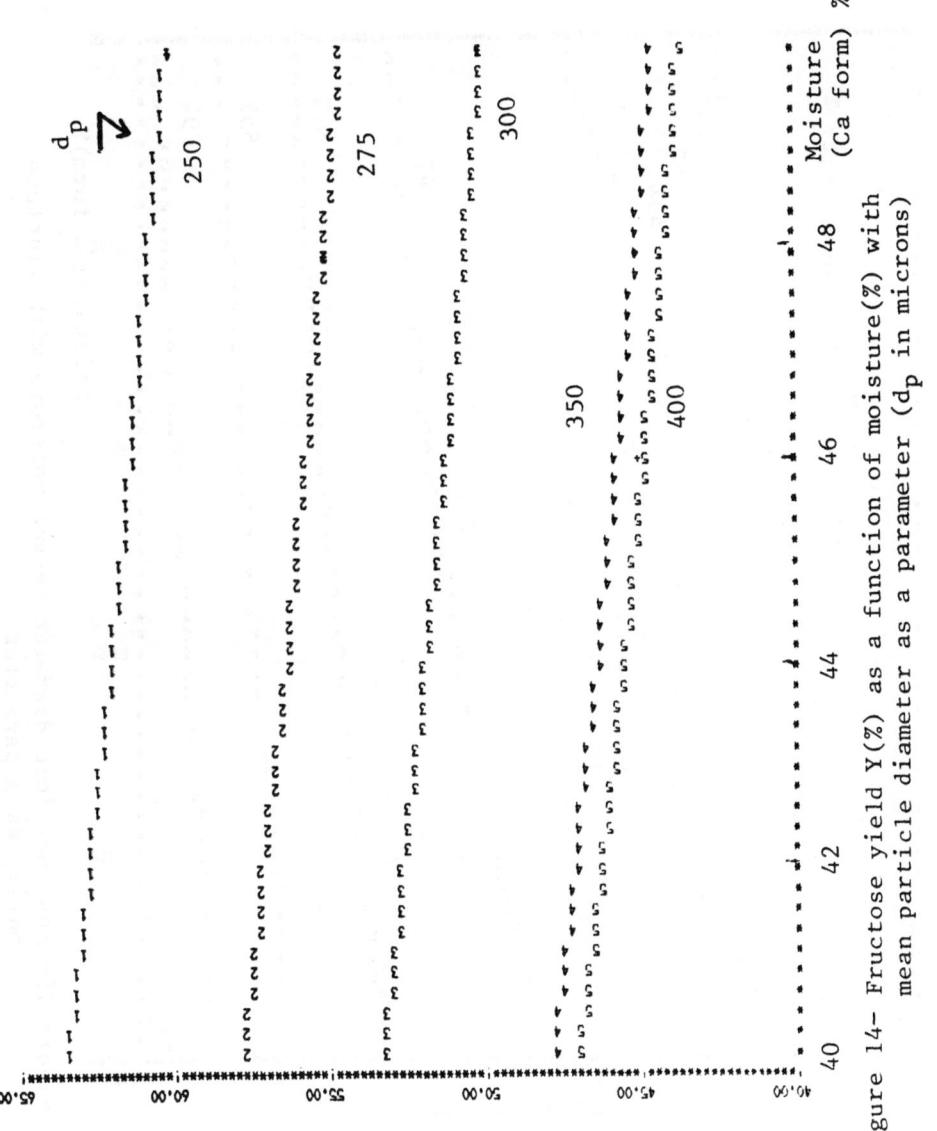

Figure 14— Fructose yield Y(%) as a function of moisture(%) with mean particle diameter as a parameter (d_p in microns)

430

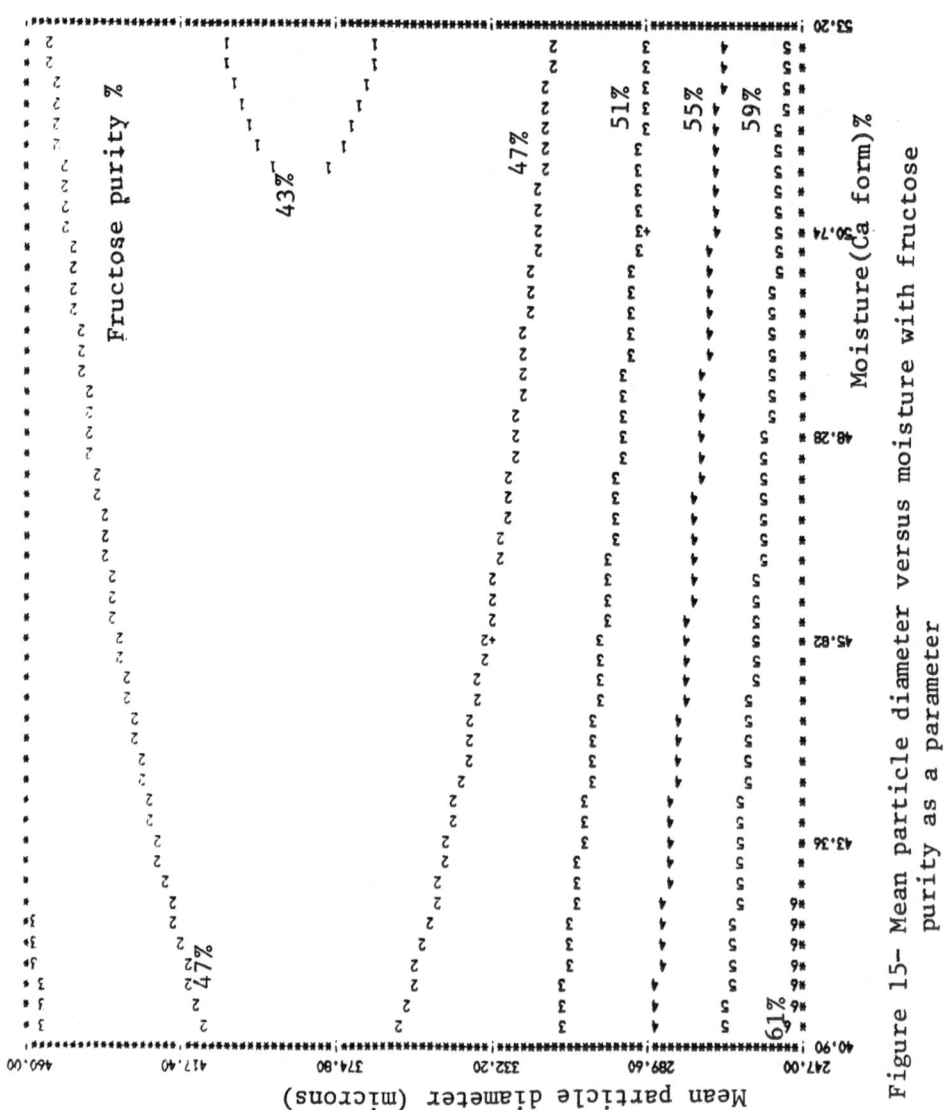

Figure 15— Mean particle diameter versus moisture with fructose purity as a parameter

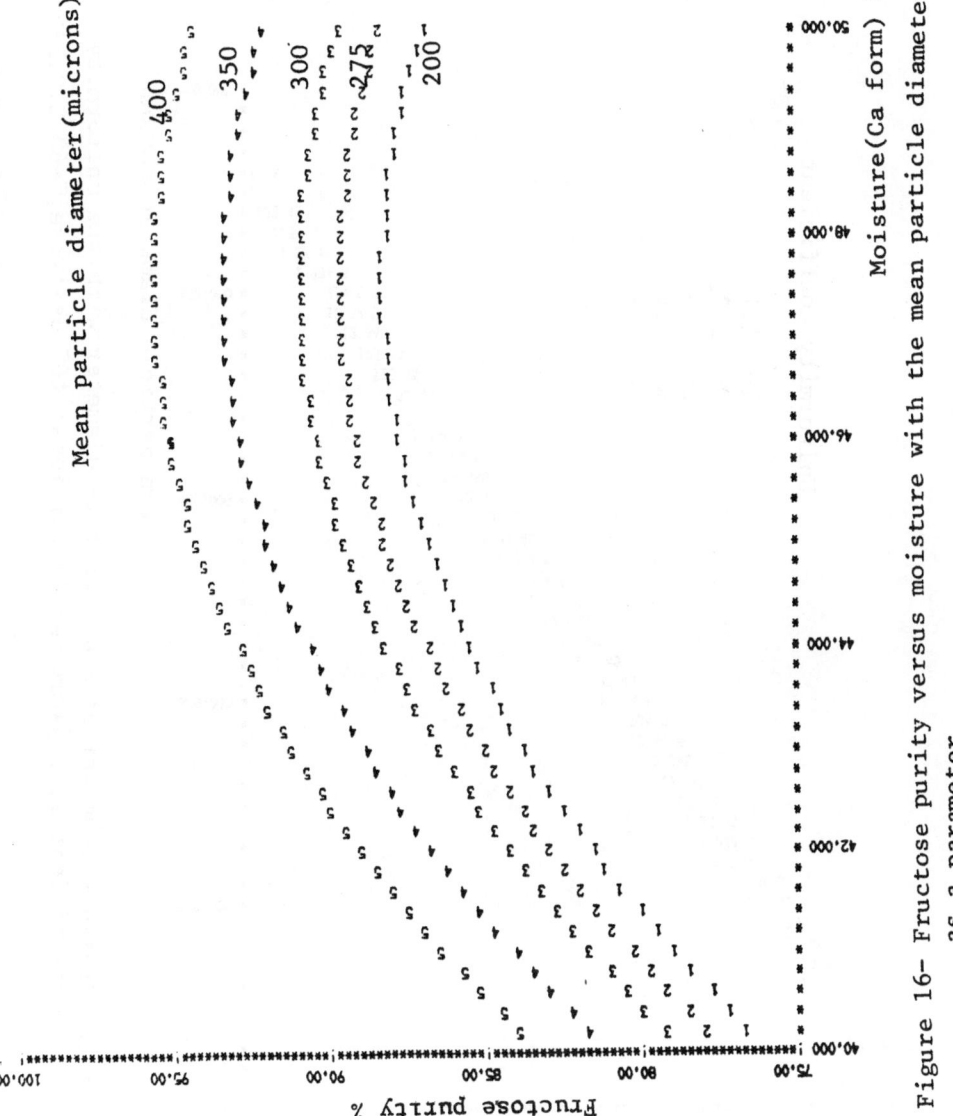

Figure 16- Fructose purity versus moisture with the mean particle diameter
as a parameter.

431

432

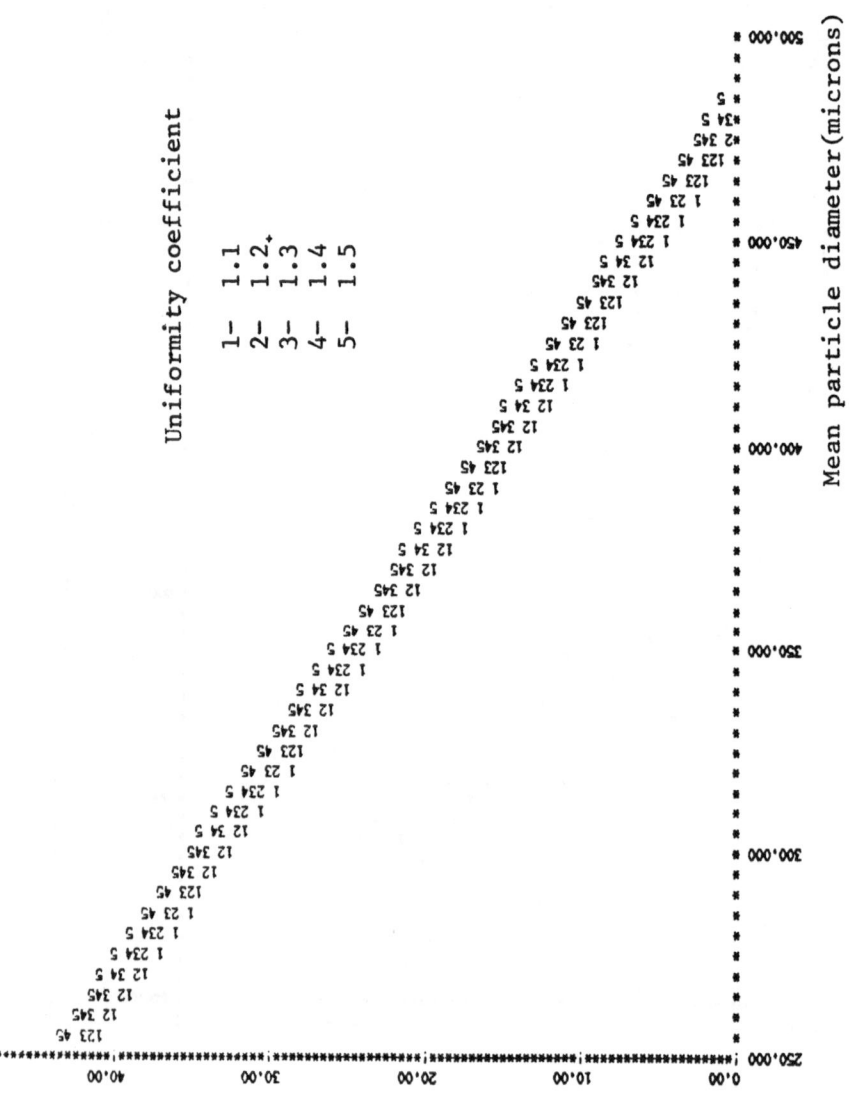

Figure 17- Pressure drop versus mean particle diameter with the uniformity coefficient as a parameter (conditions: 40% dextrose,T=60 C,0.3BV/h)

433

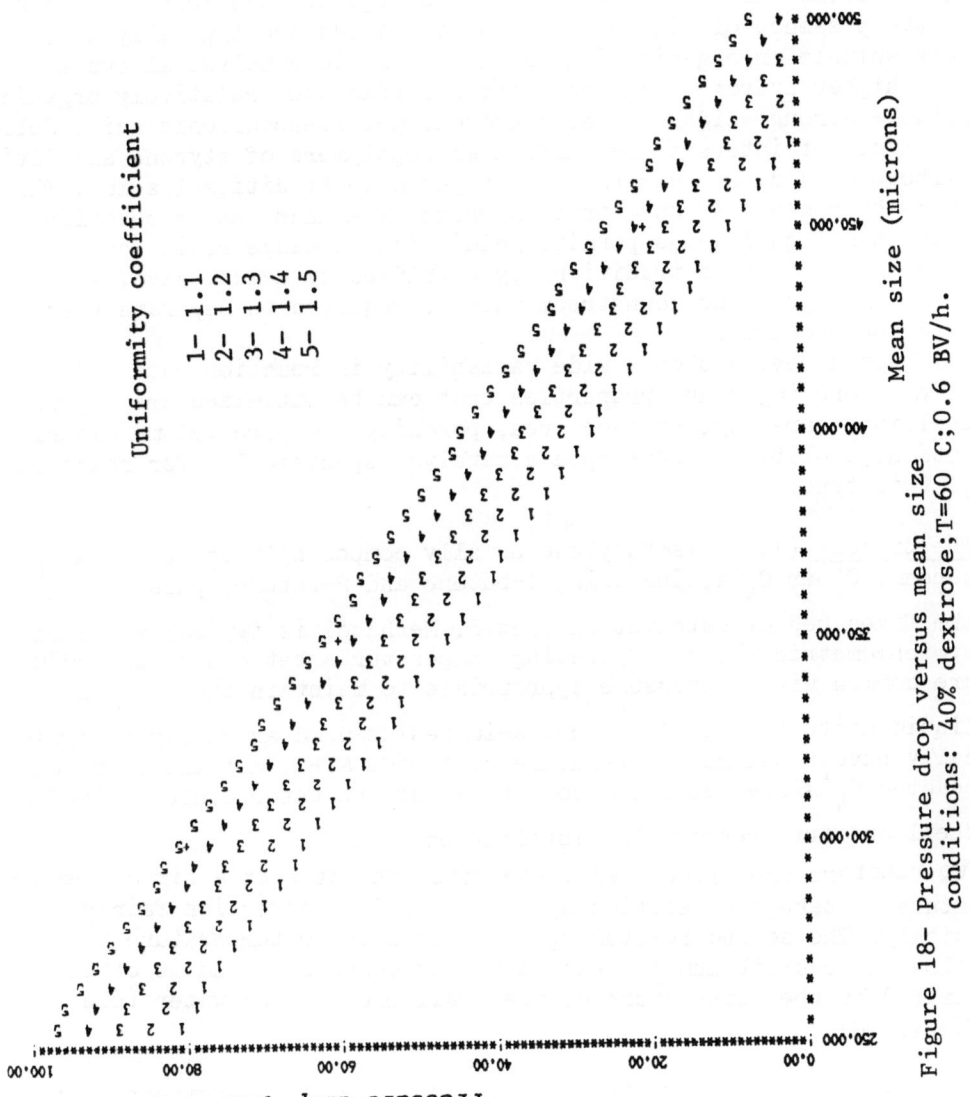

Figure 18- Pressure drop versus mean size
conditions: 40% dextrose;T=60 C;0.6 BV/h.

There is a classical equilibrium-rate conflict .At low temperatures, MTBE is favoured in the quilibrium, but the reaction rate is slow. At high temperatures, the reaction rate is fast, but not as high as conversion of isobutylene is attainable. Figure 19 shows how the equilibrium conversion changes with temperature when the initial stoichiometric ratio of methanol to isobutylene is 1.2 and the C_4 stream contains 45% isobutylene.

This reaction and other acid catalyzed organic reactions will work in the presence of liquid acids. Strong acids are typically not very soluble in organic liquids and the solid catalyst allows a much higher hydrogen ion concentration than even relatively organic soluble strong acids such as anhydrous p-toluenesulfonic acid. Solid catalysts of interest are sulfonated copolymers of styrene and divinylbenzene and have several advantages over traditional acids. This large hydrogen ion concentration promotes a much faster reaction rate than liquid acids permit. Acidic ion exchange resins can catalize most reactions traditionally catalyzed by acids. Also, with solid catalysts, no downstream unit is required to separate the acid and products.

Sulfonated resins show a wide variability in reaction rates. Some of the more important properties that can be optimized are : percent swelling, capacity, surface area, porosity and pore volume distribution. Duolite has developed a catalyst specifically for reactions of this type.

The process (1) : Isobutylene usually composes 10 to 50% of a C_4 stream. Other C_4s, including 1-butene and 2-butene, pass through the fixed bed of catalyst unreacted. Methanol is typically fed in stoichiometric excess. Operating temperatures between 40 and 90°C are common with a pressure appropriate to maintain the C_4's in a liquid state (150 psi). Strong acid resins used as catalysts typically have a maximum temperature of 140°C. MTBE, methanol, and unreacted C_4's are separated downstream of the water cooled, fixed bed shell and tube reactor by distillation.

Two reactors are often used : the first operates at a higher temperature to take the reaction up to around 90% conversion fairly quickly. The second reactor operates at a lower temperature to allow the equilibrium to reach a higher conversion and is often limited by the temperature of the available cooling water (see Figure 20).

Side reactions : There are several possible side reactions, but none occur appreciably under normal circumstances. Reaction 2 shows the formation of an isobutylene dimer and is the only one of concern. Polymers can also be formed that poison the catalyst's activity by presumably accumulating in the pores and lowering the available active sites and surface area. Dimethyl ether (reaction 3) and tert-

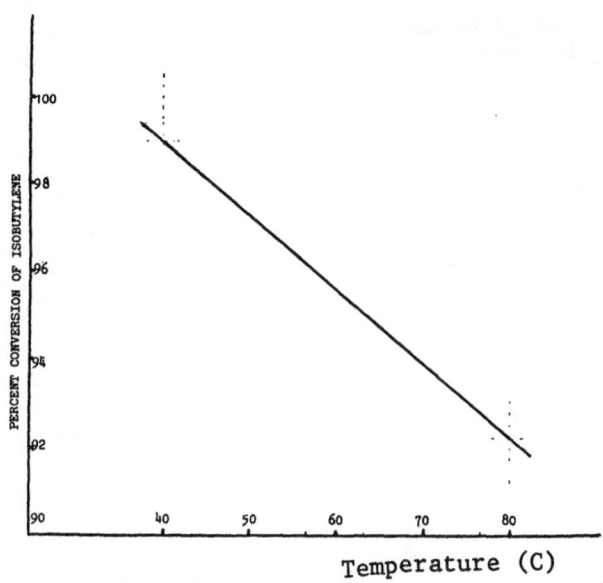

Figure 19- Isobutylene conversion versus temperature
at equilibrium.
(starting reagents: 1179 g C_4 containing
45% isobutylene and 460 ml methanol).

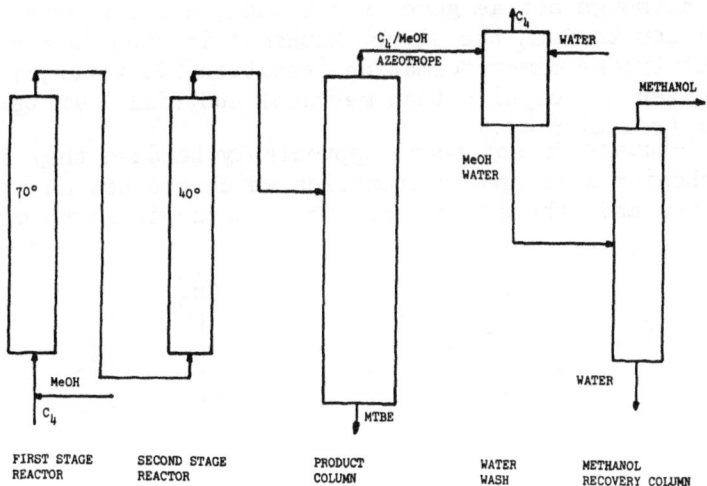

Figure 20- Simplified flowsheet of the MTBE process.

butyl alcohol (reaction 4) are also formed in trace amounts only and do not usually cause problems. The water by-product of reaction 3 can cause reaction 4. Tert-butyl alcohol has good antiknock properties itself.

$$2 \quad \underset{CH_3}{\overset{CH_3}{\diagdown}} C = CH_2 \rightarrow CH_3 - \underset{CH_3}{\overset{CH_3}{\underset{|}{C}}} - CH = C \underset{CH_3}{\overset{CH_3}{\diagup}} \qquad (2)$$

$$2 \quad CH_3OH \rightarrow CH_3 - O - CH_3 + H_2O \qquad (3)$$

$$\underset{CH_3}{\overset{CH_3}{\diagdown}} C = CH_2 + H_2O \rightarrow CH_3 - \underset{CH_3}{\overset{CH_3}{\underset{|}{C}}} - OH \qquad (4)$$

<u>Proposed mechanism</u> : In the MTBE reaction, the primary function of the catalyst seems to be to donate a hydrogen ion to isobutylene (protonate) to form a tertiary carbonium ion. Methanol acts as a nucleophile and attacks the carbonium ion to form the ether and another hydrogen ion (reaction 5). Isobutylene is also an acceptable nucleophile, although not as good as methanol, and can react with the carbonium ion to form the dimer. Methanol is used in excess to avoid the isobutylene dimer formation (reaction 2). Water, if present, is a better nucleophile than methanol and will form tert-butyl alcohol (reaction 4).
1-butene and 2-butene do not react appreciably because they form secondary carbonium ions upon protonation which are not as stable as tertiary ones and, therefore, are not as susceptible to nucleophilic attack.

$$CH_2 = \underset{CH_3}{\overset{CH_3}{\underset{|}{C}}} + H^{\oplus} \rightarrow CH_3 - \underset{CH_3}{\overset{CH_3}{\underset{|}{\overset{\oplus}{C}}}} + CH_3OH \rightarrow CH_3 - \underset{CH_3}{\overset{CH_3}{\underset{|}{C}}} - O\,CH_3 + H^{\oplus} \quad (5)$$

Batch rate studies indicate that the reaction is first order with respect to isobutylene and independent of catalyst and methanol concentrations. This coincides with the proposed mechanism and implies the rate determining step is the protonation of isobutylene to form the carbonium ion. The attack of the nucleophile, methanol, is much quicker. If isobutylene is in stoichiometric excess, the rate is dependent on methanol, but the MTBE reaction is not normally run under these conditions.

It has been shown that the reaction is not mass transfer limited in the liquid phase (2). This is convenient because it allows the use of the rate law as determined from a batch reactor to be easily used to design a continuous process. The fact that the liquid to solid catalyst ratio is much larger in a batch reactor is not important because the reaction is not limited by the liquid phase.

Experimental procedure : Methanol was placed in a 4-liter, isothermal, stirred batch reactor and C_4's containing isobutylene were weighed in for each catalyst tested. Methanol was used in stoichiometric excess of 1.2 moles per one mole of isobutylene.The solution was brought to temperature, and at time zero, a known quantity of dried catalyst was injected into the reactor with a nitrogen blanket to start the reaction. The catalyst was initially regenerated in a similar manner. Two molar sulfuric acid was passed through a column of catalyst for several hours, and then rinse water was introduced to remove the acid. The catalyst was dried to a constant weight in a vacuum oven at 100°C.

Liquid samples were taken periodically through a dip tube and expanded into a sealed vial. A gas sample was then evaluated using gas chromatography. Plots of percent conversion of isobutylene versus time were made for each experiment and the data was fit to the rate law to determine the rate constants (see Figure 21).

Role of a catalyst : A catalyst speeds up a reaction, but does not effect the equilibrium. The best catalyst offers the fastest reaction rates. A better catalyst allows for a smaller reactor to yield the same conversion and capacity.

In this reaction, a better catalyst may allow the second reactor to operate at a lower temperature and, therefore, reach a higher equilibrium conversion.

The reaction rates of two catalysts can be compared by their rate constants at a given temperature.

Catalyst	Rate constant (hr^{-1}) at 70°C
Duolite C 276	2.95
Duolite C 26 H	2.15

Catalysts of interest expand considerably from their dry state, depending on the solvent. Many MTBE producers buy catalysts in the wet form because they are less expensive,when the water is displaced by methanol, the catalyst contracts. It is the expanded volume in water that determines how much can be placed in a fixed bed. Filling the bed with catalyst in the expanded water state has

438

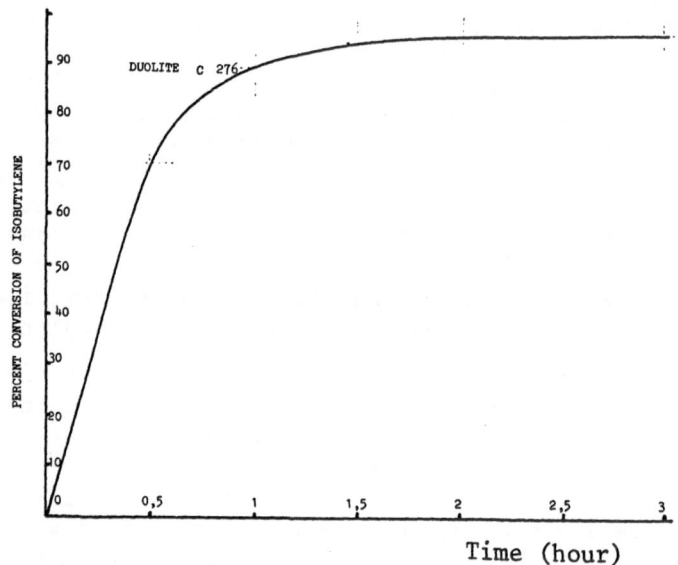

Figure 21- Isobutylene conversion versus time.
(conditions: 460 ml methanol;1179 g C_4
containing 45% isobutylene; 70 C).

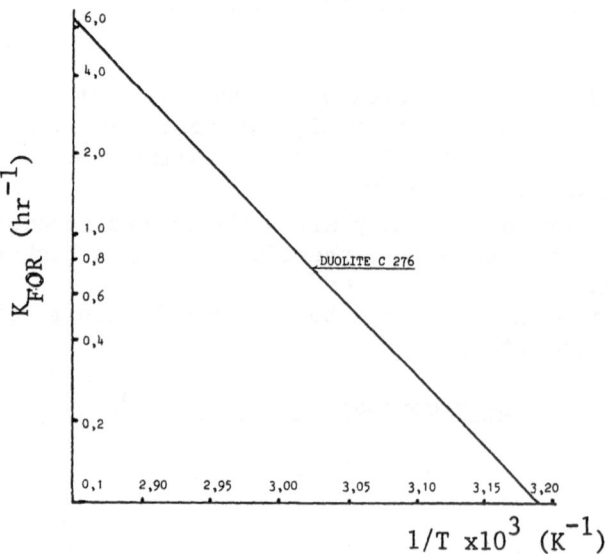

Figure 22- Arrhenius plot : K_{FOR} versus 1/T

the additional advantage of allowing the catalyst to be regenerated inside the reactor.

Over a period of time, the acidic active sites of the catalyst can get bound with cation impurities (trace metals, salts, or ammonia) and part of the regeneration procedure involves a water wash. If the reactor was filled with catalyst expanded in methanol, a smaller reactor would be required, but if water was introduced, the catalyst would expand and damage the reactor.

The expansion and contraction of the resin can cause loading problems and excessive pressure drops across the fixed catalyst bed. The less a catalyst expands from its dry state to its wet state, the less it expands and contracts during its producing and regeneration cycle. Listed below is the percent swelling of Duolite's catalyst from their dry to wet state.

Catalyst	Percent water	Swelling in methanol
Duolite C 276	33	23
Duolite C -26H	120	79

Rate expression : The reaction rate is the moles of isobutylene consumed per time per unit of catalyst and is equal to the rate of formation of MTBE. The rate is independant of methanol concentration under typical operating conditions and independent of catalyst concentration. Since the reaction is also reversible, a rate law of this form is suggested.

$$-r_{IB} = k_{FOR}C_{IB} - k_{REV}C_{MTBE}$$

Putting the concentration of isobutylene and MTBE in terms of initial isobutylene concentration, Co, and fraction of isobutylene reacted, X, yields :

$$-r_{IB} = k_{FOR}Co\,(1-X-KX)$$

K is equal to k_{REV}/k_{FOR}, and at equilibrium, the reaction rate is zero giving :

$$K = \frac{1}{X_{eq}} - 1$$

X_{eq} is the equilibrium conversion of isobutylene and is mostly a function of temperature (see Figure 19). The equilibrium conversion is also a function of composition which makes it dependent on the stoichiometric ratio and isobutylene concentration in the C_4 stream.

k_{FOR} is a function of temperature and fits the Arrhenius equation:

$$k_{FOR} = Ae^{-E/RT}$$

The activation energy, E, is approximately 24,250 cal/g mole and the frequency factor, A, is about 8.24×10^{15} hr^{-1} for Duolite C276; 6.00×10^{15} for Duolite C-26H. (see Figure 22).

5. PHARMACEUTICAL INDUSTRY

The pharmaceutical industry is concerned with fermentation which permits production of antibiotics and vitamins.The majority of antibiotics are extracted from broths either with ion exchange resins or adsorbents.Some examples are cephalosporine C, streptomycine,framycetine,gentamicine,kanamycine,etc.

5.1.Streptomycine production :Streptomycine is produced by fermentation of molasses in presence of micro-organisms.The final liquor containing streptomycine to be extracted contains fine particles of mycelium and mineral salts:calcium,magnesium sulfates and chlorides and of course the streptomycine in its sulfate form at a concentration of 10 to 15 g/l.

Problem to solve:
- To find the right extractant
- To eliminate colour
- To eliminate salts

Extractant selection: Streptomycine contains two strongly basic guanidine groups and one weak base glucosamine group.So a cationic exchanger has to be selected.Due to the neutrality of the streptomycine sulfate in the broth,one can use a weak acid resin which presents high selectivity for streptomycine and high total volume capacity. One can expect a high degree of loading on to the resin.In practice using Duolite C 464 useful capacities of 0.5 to 0.7 eq/l are encountered (MW of streptomycine= 581.58).The resin is operated in its Na$^+$ or NH$_4^+$ form.After loading and regeneration,streptomycine elution is done either with HCl or H_2SO_4. Duolite ES 401 (pellet form of C 464) is also used for operating advantages due to its form which gives higher void volume in a column and less clogging when operating by percolation.But its total capacity,its kinetic and its ability to separate isomers are less than with C 464 which operates only in fluidized bed system.
Colour removal: Experience has demonstrated that a strong base type I resin performs well in this job.The Duolite selected resin is A 143 which permits operating on streptomycine solutions of 100 to 150 g/l to get 90 to 95% decolouration.
Calcium and magnesium salts removal: This is a demineralization process but the strong cationic resin has to be selected in order to have a better selectivity for Ca and Mg over streptomycine.The anion exchanger is normally a weak base one.The resins can be separated or

mixed.Duolite C 264 and A 368 are recommended.
Process scheme: Extraction with a weak acid resin Na$^+$ or NH$_4^+$ form.
Elution wit HCl 2N then neutralization before next loading.This step
can be done on pulsed bed,fluidized bed or fixed bed columns.The
fluidized bed and fixed bed systems contain generally three columns
in series.In the second step decolouration is carried out with strong
base resin in fixed column.The resin can work in OH$^-$ or Cl$^-$ forms;
during the cycle the flow rate is 1 BV/vol/h. Working life is 100 to
150 cycles depending on the nature of the liquor outlet of step 1.
The expected life is 1 year. In the third step most often the two
resins are used mixed due to less streptomycine fixation and pH can
so be kept in the range of 5-7.Strong cationic resin H$^+$ form and
weak base in the free base form should be chosen.

5.2.Cephalosporine C production: Cephalosporine C (amphoteric anti-
biotic) is produced by fermentation from a mixture containing proti-
des,glucides and lipids.The broth can contain from 5 to 15 g/l cephalosporine C.Its pH is about 6.8 - 7.

Problem to solve:
   -Extraction
   -Decolouration
   -Purification and concentration
Extraction: The broth is filtered,acidified to pH 2.5-3 and then
passed through an adsorbent resin selective for cephalosporine C.
Decolouration: The eluate from the extraction step is treated with
an acrylic weak base resin,acetate form.Decolouration can also be
performed with an adsorbent resin before extraction.
Purification and concentration: Cephalosporine acetate coming from
decolouration step is precipitated with zinc powder and cephalosporine zinc salt is precipitated and washed.
Process scheme for extraction: We will review a 2 columns (in series)
process.The first column is loaded with an adsorbent which has a
better specificity for the coloured matter of the broth than the cephalosporine C.The second column is loaded with an adsorbent which
is hyghly specific for cephalosporine C.The first column is loaded
with Duolite ES 873 equilibrated with H$_2$SO$_4$ pH=2 (2 vol/vol. of adsorbent).The second column is loaded with XAD 16 also equilibrated
with H$_2$SO$_4$ pH=2.In order to prevent cephalosporine C leakage the volume of adsorbent into column 2 is 5 times the one of column 1.When
column 1 has been treated 7 bed volumes of broth,the two columns
are treated in series with water (2.5 volumes of 1).This column 1
is regenerated to eliminate fixed colour. Other conditions are:
regenerant NaOH 4%(2 bed volumes);rinse -water 1 bed volume;pretrea-
tment- H$_2$SO$_4$ pH=2 (2 bed volumes).During that time,column 2 is washed
with water (1 bed volume) and cephalosporine is eluted with 1.7 bed
volume of an organic solvent.After solvent displacement with water,
pretreatment (H$_2$SO$_4$ pH=2),the system is ready for an other cycle.
Every 10 cycles,both columns are treated with NaOH 4% for depollution,
Useful capacity for XAD 16 is 23-24 g of cephalosporine/litre of
adsorbent.

6. CHLOR ALKALI PRODUCTION

Old technology:mercury cells.The production of chlorine from mercury cells operated plants consumes mercury which ends up in waste stream Problem to solve: To fit with stringent regulation on mercury disposal in waste streams (ppb level) and reuse the removed mercury. Solution proposed: AKZO process with resin IMAC TMR.

AKZO process : In waste water mercury can occur as metallic and as ionic mercury.The AKZO process consists of the following steps: oxidation/pH adjustement;filtration;dechlorination;ion exchange.

IMAC TMR resin: It is known from literature that organic compounds having -SH groups form very strong bond with Hg and that the solubility products of such compounds are very low.The TMR resin is a thiol resin having -SH group.The following dissociation equilibria of mercuric chloride complexes exist in brine:

$$Hg^{++} + HCl \rightleftharpoons HgCl^+ + 3 Cl^- \rightleftharpoons HgCl_2 + 2 Cl^- \rightleftharpoons$$
$$HgCl_3^- + Cl^- \rightleftharpoons HgCl_4^=$$

The vast majority of the mercury is then present as $HgCl_4^=$ complex ions.The affinity of IMAC TMR resin is so strong for $HgCl^+$ and Hg^{2+} that dissociation equilibria is shifted towards left and the resin is able to reduce the mercury level in the treated stream to below 5 ppb.With a feed containing 20-50 mg/l the resin can treat 8000 bed volumes producing an effluent below detection level of Hg. Figure 23 illustrates equilibrium curve of IMAC TMR and Figure 24 the AKZO IMAC TMR process.

New technology:membrane cells. To remove mercury pollution a new technology has been developed which uses membrane cells.
Problem to solve: Manufacturers of membrane cells require brine of specified quality to guarantee life time and the running costs of the system.Calcium content of the brine has to be less than 20 ppb and magnesium less than 5 ppb to prevent membrane fouling.In fact, most of them require that total content of $Ca^{2+} + Mg^{2+} + Si^{2+} + Ba^{2+}$ < 50 ppb.
Solution proposed:Duolite ES 467.
Duolite ES 467 is a chelating resin with aminophosphonic functional groups.Its isoelectric point is at pH=3.
In brine treatment some of the metals are present as anionic complex species.So,it is not so easy to predict the behaviour of the resin for a given ion.That will vary with pH and brine concentration and of course the nature of the ions to be removed.In the case of membrane cells,the brine concentration is in the range of 280 to 300 g/l and pH in the range of 10-11.

The affinity order of the resin was determined experimentally at a temperature of 60-80 C,flowrate of 10 to 30 BV/vol/hour,inlet concentration of impurities of 1 to 10 mg/l.The order was established to be : $Mg^{2+} > Ca^{2+} > Si^{2+} > Ba^{2+} > Na^+ > Al^{3+}, Ni^{2+}$.
Process description: The product ES 467 is sensitive to free chlorine,so in front of the unit,a dechlorination unit is needed but most

Mercury concentration in
resin (g Hg/litre)

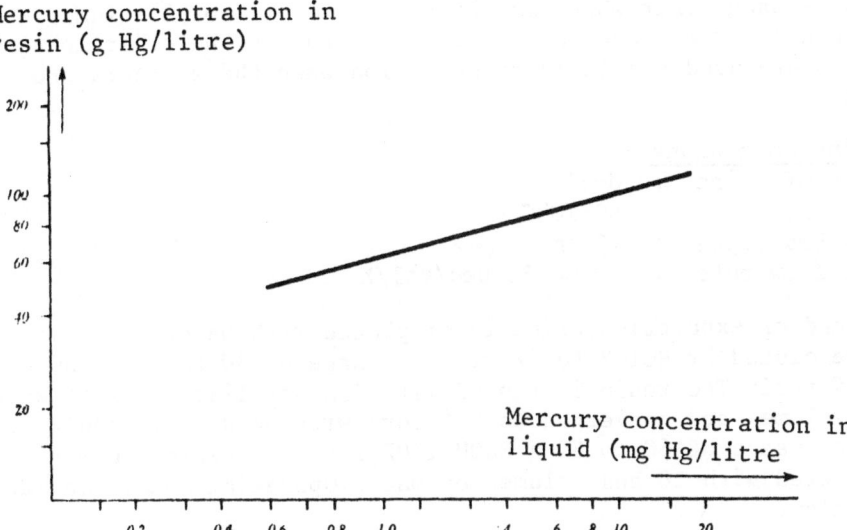

Figure 23- Equilibrium curve of Imac TMR

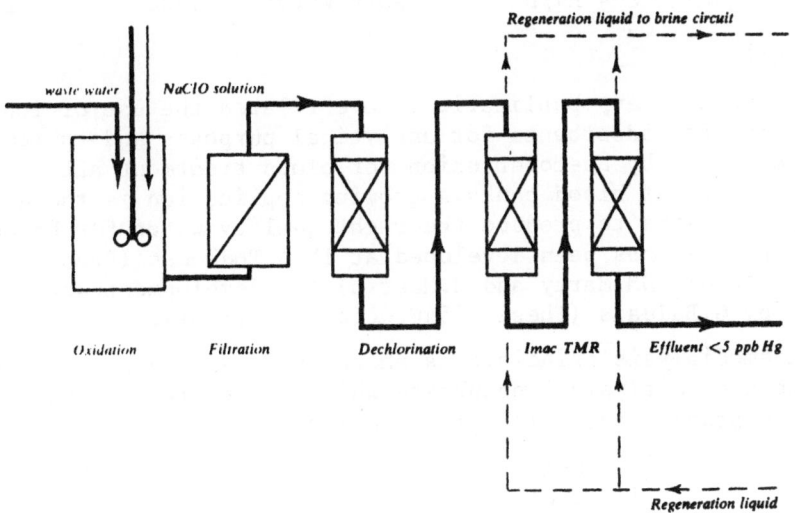

Figure 24- The Akzo Imac TMR process

often an activated carbon is used.Brine decalcification is done in the ion exchange unit,which may be consisted of 2 or 3 columns.When only two columns are used one is in regeneration,the other in service.When 3 are used one is in regeneration when the 2 others are in service.

<u>Operating conditions</u> :
pH range of brine : 8-11
Temperature   : 60-70°C
Minimum bed depth : 75 cm
Service flow rate : 10 to 30 bed/vol/h

At the end of exhaustion,brine is displaced with water,Ca^{2+},Si^{2+},Mg^{2+} Ba^{2+} are eluted by HCl 3 to 7% with a dosage of 80 to 100 g HCl 100%/litre of resin.The resin is rinsed with demineralized or soft water (2 bed volume) and converted to Na^+ form with NaOH (solution 4 to 8%) at a dosage of 60 to 80 g NaOH 100% litre of resin.A new rinse is conducted with 10 bed volumes of water.Operating conditions are shown below:

Operations	Flow rate	Solutions	Time (mn)	Quantity
Service	20 Bv/h	brine	–	–
Brine displacement	4 Bv/h	soft water	60-90	4-6 Bv
Backwash	8-12 m/h	" "	30	–
Regeneration	2-6 Bv/h	HCl, N	30	100-200g/l
Rinse	2-4 Bv/h	soft water	30-60	2 Bv
Na^+ conversion	2-4 Bv/h	NaOH, N	30	60-80 g/l
Rinse	2-4 Bv/h	soft water	30-60	2 Bv

7.BIOMEDICAL

In this field many publications have related the use of ion exchange resins and adsorbents for analytical purposes and to remove toxins from blood by haemoperfusion for blood treatment after collection and storage at blood banks.A growing application is the use of ion exchange resins to produce the right quality water for haemodialysis.The process has been developed at INSA Toulouse (French patent 77.32009 A.Abadie,DM.Marty and M.Mustin).The development work has been done by G.Baluais (Thesis 2540,October 20,1981).

The haemodialysis principle is shown in Figure 25.Typical analysis of the haemodialysis concentrate which is required to be diluted 35 times to produce haemodialysis solutions is as follows:

Na^+ 4830 meq/l
Ca^{2+} 122.5 meq/l
Mg^{2+} 52.5 meq/l
Cl^- 3780 meq/l
CH_3COO^- 1225 meq/l

Water quality required: French legislation since 1977 requires the following water quality for haemodialysis:

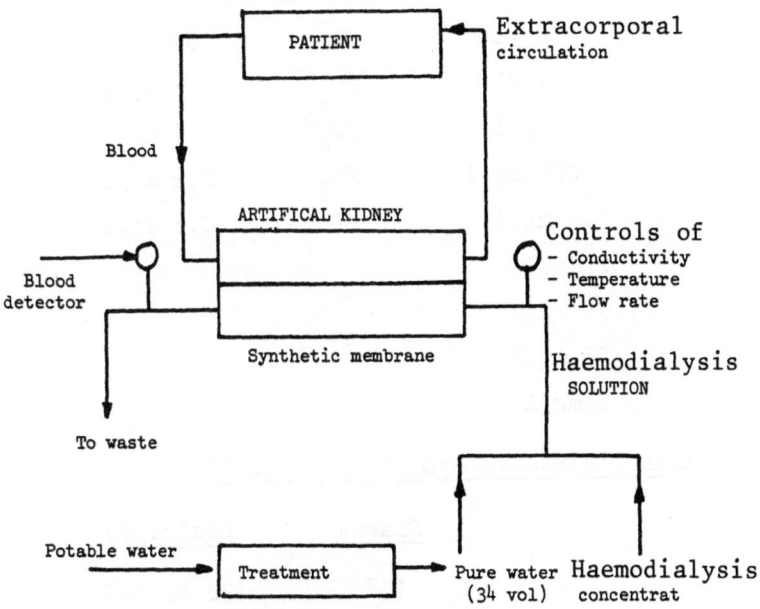

Figure 25-Haemodialysis principle

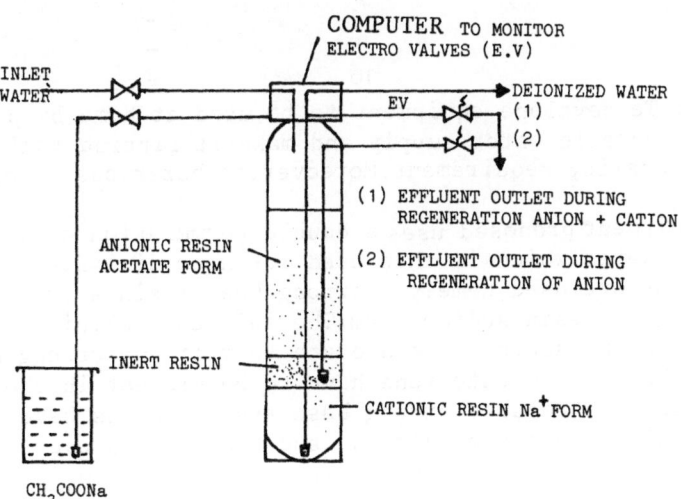

Figure 26- A scheme for haemodialysis for use in homes

Cations		Anions	
NH_4^+	0.2 mg/l	Cl^-	50 mg/l
Heavy metals	0.1 mg/l	SO_4^{2-}	5 mg/l
Zn^{2+}	5 mg/l	PO_4^{3-}	5 mg/l
Hg^{2+}	0.004 mg/l	NO_3^-	0.2 mg/l
Sn^{2+} and $^{4+}$	0.1 mg/l	F^-	0.5 mg/l
Ca^{2+}	2 mg/l		
Mg^{2+}	1.2 mg/l		
Na^+	50 mg/l		
K^+	2 mg/l		

Plasma/Dialysatecompositions (meq/l)

	Plasma	Dialysate
Sodium	138	130 to 145
Potassium	4	0 to 3
Calcium	5	3 to 3.5
Magnesium	3	0 to 1.5
Cloride	99	113
Bicarbonate	27	–
Acetate	–	30 to 38
Phosphate	2	–
Sulfate	1	–
Organic acids	5	–
Proteines	16	–

Problem tc solve: To develope equipment to be used at home by pati‐
ents,to use local potable water supply and make it fitting with hae‐
modialysis water quality requirement.Moreover,no hazardous chemicals
can be used.
Solution: The equipment proposed uses a double permutation of the
available potable water which is then used for haemodialysis.Two
ion exchange resins are used namely a strong base resin acetate form
and a strong cationic resin sodium form.The influent salinity is
transformed into sodium acetate.The acetate form of the strong base
resin was chosen because acetate ions have to be present in the hae‐
modialysis solution and because strong base type I resins have better
selectivity for sulfate,nitrate,chloride and bicarbonate over aceta‐
te.Only fluoride has lower affinity.The strong cationic resin is ope‐
rating as standard softening medium.Both resins can be used in the
same column and regenerated with the same sodium acetate solution
(but not in series to prevent precipitation in the cationic resin).
In practice for inlet water of total salinity in the range of 1 to
10 meq/l the useful capacities are: 0.5 eq/l for strong base resin
and 1.2 to 1.5 eq/l for strong acid resin. Regeneration is done with

4 to 6 times the stoichiometric quantity of sodium acetate.Products used are: nuclear grade strong base resin Duolite ARA 9366 and "Food grade" strong cationic resin Duolite C 20.A scheme for haemodialysis for use in homes is shown in Figure 26.

CONCLUSIONS

The subjects mentioned above illustrate the usefulness of ion exchange as a unit operation in industry.New processes will be developed for new applications.

NOTATION

A - frequency factor (hr^{-1})
C_{IB} -concentration of isobutylene (g mole/1 wet resin)
C_{MTBE} -concentration of MTBE
C^o -initial concentration of isobutylene
E^o -activation energy (cal/g mole)
F_o -initial molar flow rate of isobutylene (g mole/h)
K_{FOR} -forward rate constant (hr^{-1})
K_{REV} -reverse rate constant (hr^{-1})
K - ratio K_{REV}/K_{FOR}
$-r_{IB}$ -rate of isobutylene consumption (g mole/hr x 1 wet resin)
R -ideal gas constant
T -temperature (K)

REFERENCES

1. Scheeline,H. et al. SRI International Process Economics Program Methyl Tertiary Butyl Ether and Tertiary Butyl Alcohol.Report 131,August 1979.
2. Ancillotti,F. et al.Ion Exchange Catalysed Addition of Alcohols to Olefins.J.Mol.Catal. 46,1(1977),45-57.

APPLICATIONS OF ION EXCHANGE IN HYDROMETALLURGY

M.Streat

Department of Chemical Engineering and Chemical
Technology,Imperial College
London SW7 2BY,England

INTRODUCTION

A list of metals that have been recovered and purified commercially
by ion exchange is given in Table 1. In some cases, the scale of
operation is relatively small, e.g. the rare earth elements, the
transuranic elements, the platinum group metals, though the value
of metal recovered is usually extremely high. Ion exchange is
particularly suited to high cost, low throughput purification
processes. Alternatively, the recovery of trace amounts of metals
from waste streams is carried on a large scale, e.g. cadmium and
mercury from industrial effluents, chromium from spent metal
plating solutions and copper and zinc from the waste arising in the
synthetic fibre industry.

The largest single application in hydrometallurgy is the recovery
of uranium from naturally occurring ore-bodies and as a by-product
in the production of gold. Also, more recently, it has been shown
that uranium can be recovered as a by-product during the treatment
of wet process phosphoric acid.

Modern advances in the application of ion exchange in
hydrometallurgy depend greatly on the synthesis and development of
new synthetic organic ion exchange resins possessing selectivity in
complex aqueous solutions. This paper will review recent trends in
the design of polymer based chelating ion exchange materials and
discuss their use in several selected process applications.

ION EXCHANGE RESIN DEVELOPMENTS

It is interesting to note that separation of the first row

TABLE 1   Metals Recovered and Purified Commercially
            by Ion Exchange

> Uranium
> Thorium
> Rare Earths
> Transuranic Elements
> Gold
> Silver
> Platinum Group Metals
> Chromium
> Copper
> Zinc
> Rhenium
> Nickel
> Cobalt

transition metals has been achieved with selective liquid
extractants containing nitrogen and oxygen containing ligands, e.g.
hydroxyoximes. However, these elements will also form weak anionic
complexes, especially in chloride media, and this is also true of
the second and third row transition metals, making possible
separations amongst the precious metals, i.e. gold and silver, and
the platinum group metals. There is a considerable economic
incentive to develop solid phase ion exchange processes for the
recovery of precious or alternatively toxic metals, such as
mercury, from low grade dilute solutions. Therefore, the
literature concerning the development of selective ion exchange
polymers is expansive and cannot be fully discussed here.
Warshawsky (1) has presented a comprehensive review article
recently which discusses the trends in the synthesis of selective
chelating polymers containing oxygen, nitrogen, phosphorus and
sulphur ligands.

Chelating Resins With Nitrogen Containing Pendant Groups

Tertiary and quaternary polymeric amines are produced in large
quantities and are effective for some specialist separation
processes. The most notable is the recovery of uranyl sulphate or
uranyl carbonate from acid or alkali leach solutions. Some
selectivity is also found amongst the first row transition metal
chloride complexes. The incorporation of a chelating ethyleneimine
group - $(NH - CH_2 - CH_2 - NH)_n$ increases the stability of the
polymeric ligand-metal complex markedly. For example, crosslinked
chlormethylated styrene divinylbenze copolymers containing di, tri
or tetraethylene imine groups show high affinity for Au(III),
Hg(II) and Cu(II). Jones and Grinstead (2) have prepared more
complex ethylenediamine derivatives and suggested their use for
Fe(II) and Cu(II) separation.

Chelating Resins With Nitrogen and Oxygen containing Pendant Groups

Amino-diacetic acid ion exchange resins have been available for commercial use since the early 1960's. Resins of this type are useful for the separation of the first row transition metals, though the selectivity is very dependent on pH value. More recently, attempts have been made to synthesise more elegant polymer supports and ligands.

Many workers have proposed chelating ligands based on poly(vinylimidazole) or poly(vinylimidazole) decarboxylate groups (3). It has been suggested that the loading rates for base metals, Cu(II), Ni(II), Co(II) are enhanced with this type of structural modification.

Hydroxyamine polymers have also been extensively explored. Vernon and Eccles (4) have prepared several hydroxyoxime type polymers and recommended their use for copper separation from iron at low pH value. Amidoxime type functional groups are found to chelate uranium at low concentrations in sea water and to offer a potential solution to the problem of separating heavy metals from acidic solutions.

Chelating Resins With Sulphur Containing Pendant Groups

Sulphur ligands are known to complex or precipitate most of the heavy transition metals. Early developments in this field arose out of pollution control and analytical applications. Typical polymer derivatives are based on macroreticular polymethacrylate beads containing pendant mercapto groups. Such resins are reactive for Ag(I), Hg(II) and Au(III). Thioglycolate resins have been developed and used for the laboratory separation of Ag(I), Bi(III), Sn(IV), Sb(III), Hg(II), Cd(II), Pb(II) and U(VI). Similar resins are dithiocarbamates and their derivatives. Recently, the platinum group metals (PGM's) have been separated in chloride media using a combined ion exchange and liquid-liquid extraction process(5). The adsorption step involves a weak base isothiouronium group capable of extracting the chlorocomplexes of the PGM's. Industrial exploitation of this technique has already occurred.

Chelating Resins With Phosphorus Containing Pendant Groups

Phosphoric acid, esters and phosphine oxides are very effective extractants for uranium, gold and the first row transition metals. One commercially available chelating resin contains aminophosphonic groups attached to a crosslinked polystyrene matrix. Though developed for the decalcification of brines, this resin shows high selectivity for the separation of trace amounts of uranium in wet process phosphoric acid (6). A general review of phosphorus containing polymers is given by Efendiev (7).

Chelating Resins With Oxygen Containing Pendant Groups and
Macrocyclic Structures

Phenolic ion exchangers derived from a phenol-formaldehyde
condensation reaction appeared in the first generation of ion
exchange polymers. More recently, styrene-divinylbenzene copolymers
incorporating azo subsituted cresol and salicylic acid, catechol,
hydroquinone and benzoquinone have been described. The quinone
type polymers selectively sorb Hg(III) and the catechol resins sorb
Cr(VI).

The complexation of metal salts by neutral macrocyclic ligands is
well known (8). Polymeric crown ethers are an expanding group of
functional ion exchangers capable of selective sorption of alkali
metals such as K, Cs, Na and Li. The crown ether may be derived
from a conventional chlormethylated hydrocarbon backbone which is
converted to a polybenzylated catechol. Crown ethers are highly
reversible and possess rapid reaction kinetics thus allowing an
interesting thermal elution procedure, whereby a species is sorbed
at 20^0C and eluted at 60^0C.

Solvent Impregnated Ion Exchange Resins

The idea of developing solvent impregnated ion exchange resins was
to combine the selectivity and specificity of conventional liquid
extractants with the advantages of a discrete polymer support
material, thus tailor making adsorbents for a specific separation
process, usually in the field of hydrometallurgy. Though it is now
possible to functionalise polymers as outlined above, it is still
difficult to overcome some steric problems and thus it is
interesting to consider the potential use of liquid extractants
immobilised within a polymer matrix. This can be achieved by
physical impregnation of the reagent onto a polymeric or other
porous support without chemical binding of any sort.
Alternatively, copolymerisation of a monomer (e.g. styrene)
crosslinking agent (divinylbenzene) in the presence of a reagent
(e.g. tri-n-butyl phosphate) will produce a polymer "encapsulated"
product. Typical of these products are the Levextrel resins
developed by Bayer (9) An exhaustive review of extraction with
solvent impregnated resins has been published by Warshawsky (10).
The principal difficulty in the use of these materials is the slow
diffusion of the reagent out of the polymer matrix. Though this
can be overcome by re-impregnation, the possible environmental
implications and cost would make large scale commercial use
unlikely.

NEW TYPES OF ION EXCHANGE MATERIALS

Ion exchange materials are normally synthesised in granular form
and in most cases spherical particles of precise size range

distribution are supplied. In some applications, e.g. the treatment of unclarified liquors, the sorption of slow diffusing species, the use of fluidised ion exchange particles, it is desirable to modify the properties of the ion exchange materials. Increasing particle size or increasing the relative density will improve the hydraulics in a fluidised bed, whereas a reduction in particle size might improve kinetics for a very slow diffusing species. Alternatively, the use of ion exchange fibres and woven fabrics have been prepared for these and similar applications. Sorption of a solute onto fibres is inherently rapid due to the large surface area of reactive sites that can be exposed at any one instant in time. However, supporting the fibres or fabric and contacting the ion exchanger with the liquid phase calls for novel engineering design. The usual idea involves the use of an endless belt and this has been tried for the recovery of copper from a dilute aqueous solution using phosphorylated cotton towelling (11). A similar idea involved the polymerisation of a quaternary ammonium resin onto cotton cloth and this was used to remove chromate ions from a dilute aqueous solution (12).
Separately, Vernon and Shah (13) have synthesised a poly(amidoxime)- poly(hydroxamic acid) fibre and shown that this will sorb significant amounts of uranium from sea water. It is suggested that this fibre could be produced as an endless belt and thus sustain continuous retrieval of uranium by continuous countercurrent operation.

Specialist ion exchange resins have been developed in recent years in an attempt to overcome typical problems encountered in ion exchange process technology. Bolto, Weiss et al (14) have described the synthesis of novel magnetic micro resins suitable for application in desalination, water treatment and hydrometallurgy. These resins are usually manufactured in the form of beads, typically in the size range 100-500 m and containing about 10-15% by volume of a magnetic material such as gamma iron oxide as an inert core. Reactive sites are produced by shell graft polymerisation of organic monomers onto the inert core.

Since the discrete particles are small, they react faster than conventional ion exchange resins, but they can be used successfully in fluidised bed systems at economic flow rates because the micro beads agglomerate magnetically into large flocs when agitation ceases. These flocs have hydraulic properties similar to conventional ion exchangers. The recent status of continuous ion exchange using magnetic micro resins has been presented by Bolto et al (15).

A slightly different approach to the preparation of composite materials has been tried in an attempt to synthesise high density ion exchange materials possessing rapid kinetics and thus suitable for fluidised bed application in hydrometallurgy, especially for uranium recovery from unclarified solutions. Usually an inert porous core material such as alumina or silica gel is impregnated with a reactive monomer, e.g. a substituted vinyl pyridine and crosslinking agent such as divinylbenzene and the resultant impregnate is reacted to polymerise the organic reagent within the pores of the inorganic matrix. Very high density sorbents have been synthesised using a stannic oxide core, though the ion exchange capacity is relatively low and the material only moderately stable over the entire pH range (16). An interesting idea is the impregnation of liquid extractants such as Alamine 336 (a long chain tertiary amine) into an inert porous support such a crushed fireclay or firebrick. It was claimed that materials of this kind were selective for uranium sorption but suffered from instability due to the leaching of the reagents and poor hydraulic behaviour (17). None of these or similar ideas have yet proved to be commercially viable.

ION EXCHANGE PROCESSING OF URANIUM

Uranium is recovered from the host mineral by leaching and subsequently the pregnant solution is clarified and purified using either solid ion exchange, liquid–liquid extraction or both processes in series. Pregnant solutions usually contain between 100-1000ppm uranium as U_3O_8 and therefore ion exchange treatment at the head end of the process is preferred. By-product uranium can be obtained in gold recovery and in the treatment of wet process phosphoric acid. Here, the pregnant solution concentration is about 100ppm uranium as U_3O_8 and is therefore ideally suited to continuous ion exchange processes. Uranium has been effectively recovered from wet process phosphoric acid, using liquid–liquid extraction and considerable research has shown the feasibility of liquid membrane technology, though neither process is economically viable in view of the depressed state of the uranium market. However, ion exchange affords an alternative separation process which appears to be extremely attractive from an economic stand point and this will be discussed below.

Acid leaching of uranium bearing ore-bodies tends to dissolve a wide range of metal impurities, in particular, iron and vanadium. Other impurities such as silica are liberated which tend to poison anion exchange resins. Likewise, the cyanide complexes of cobalt will poison anion exchange materials in by-product recovery of uranium from gold cyanide liquors. This, however, is largely overcome by adopting the reverse-leach technique whereby uranium is recovered prior to cyanide treatment of the ore to leach gold. If the host mineral contains acid consuming material, e.g. dolomite, then leaching with an alkaline reagent is preferred. In fact, most in-situ and solution mining is performed with sodium carbonate/bicarbonate solutions.

Uranium will dissolve in sulphonic acid or in sodium carbonate in the hexavalent state and it is customary to provide an appropriate oxidant. The following reactions are typical in acid and alkaline leaching:

$$UO_2^{2+} + SO_4^{2-} \rightleftharpoons UO_2\,SO_4 \tag{1}$$

$$UO_2\,SO_4 + SO_4^{2-} \rightleftharpoons UO_2\,(SO_4)_2^{2-} \tag{2}$$

$$UO_2\,(SO_4)_2^{2-} + SO_4^{2-} \rightleftharpoons UO_2\,(SO_4)_3^{4-} \tag{3}$$

$$UO_2 + \tfrac{1}{2}\,O_2 \longrightarrow UO_3 \tag{4}$$

$$UO_3 + 3Na_2CO_3 + H_2O \rightleftharpoons Na_4\,UO_2(CO_3)_3 + 2NaOH \tag{5}$$

Reactions (1-3) relate to the acidic complexes and show that both di- and quadrivalent anionic sulphate complexes are formed. The quadrivalent complex predominates at pH values of 0.5-1.5 and about 0.2M sulphate concentration.

The alkaline reactions (4-5) show the formation of a quadrivalent anionic complex provide the pH is adjusted by the presence of sodium bicarbonate to avoid the precipitation of uranium. Sorption of these complexes onto conventional weak and strong base anion exchange resins is highly selective, since most impurities (except iron) do not form anionic species.

$$2\left[R^+\right]_2 SO_4^{2-} + UO_2(SO_4)_3^{4-} \rightleftharpoons \left[R^+\right]_4 UO_2\,(SO_4)_3^{4-} + 2\,SO_4^{2-} \tag{6}$$

$$2\left[R^+\right]_2 CO_3^{2-} + UO_2(CO_3)_3^{4-} \rightleftharpoons \left[R^+\right]_4 UO_2\,(CO_3)_3^{4-} + 2\,CO_3^{2-} \tag{7}$$

$\left[R^+\right]$ denotes the resin matrix of a typical anion exchange resin of macroreticular of polyelectrolyte gel type.

Reactions 6 and 7 are readily reversed using either hot or cold mineral acids. Sulphuric acid is the preferred eluant for the sulphate loaded resin and sodium nitrate is used to elute the carbonate loaded resin because acid would cause spontaneous evolution of CO_2 gas.

Digestion of phosphate ores with sulphuric acid results in the production of phosphoric acid containing traces of cationic uranium species. A macroporous polystyrene-divinylbenzene copolymer containing amino-phosphonic functional groups (Duolite ES467) selectively sorbs uranium from other trace metal impurities provided uranium is in the correct valency state(18). The chelating reaction is shown below:

$$R-CH_2-NH-CH_2-\overset{\overset{\text{O}}{\|}}{\underset{\underset{\text{ONa}}{|}}{P}}-ONa + UO_2^{2+}$$

$$\longrightarrow \quad R-CH_2-NH \overset{CH_2-P=O}{\diagup} \overset{\diagdown}{\underset{O}{\diagdown}} \overset{O}{\underset{\diagup}{}} + 2Na^+ \qquad (8)$$

The sorption of uranium is readily reversed using either concentrated phosphoric acid or alternatively an ammonium carbonate solution. The product solutions can be further refined and concentrated prior to precipitation. A pilot-plant has been operated in Israel and it is claimed that this technology can produce uranium for as little as \$15/lb U_3O_8(19). Thus, the potential for recovering uranium from this source is vast, since it is estimated that about 6 million tonnes of uranium exist in phosphates throughout the world(20).

ION EXCHANGE PROCESSING OF GOLD

Gold exists as an anionic auro cyanide complex $(Au(CN_2)^-)$ in cyanide leach liquors and can be recovered from solution by conventional anion exchange resins. The sorption of gold by a protonated weak base resin is given by the following equation:

$$(R_3N:H^+)_2 \ SO_4^{2-} + 2 \ Au(CN)_2^- \rightleftharpoons 2(R_3N: H^+ Au(CN)_2^-) + SO_4^2 \quad (9)$$

Resin selectivity for gold is adequate, though the cyanide complexes of Ag, Co, Cu, Fe, Ni and Zn are also sorbed and will therefore affect the purity of the eluted product. Elution of gold is usually performed with sodium hydroxide solution according to the following equation:

$$R_3N: H^+ Au(CN)_2^- + OH^- \longrightarrow R_3N: + Au(CN)_2^- + H_2O \qquad (10)$$

The free base form of the eluted resin is treated with dilute sulhuric acid to protonate the resin functional groups prior to the extraction cycle:

$$2(R_3N:) + H_2SO_4 \longrightarrow (R_3N: H^+)_2 \ SO_4^{2-} \qquad (11)$$

A strong base resin contains quaternary ammonium groups and can absorb gold over the entire pH range.

$$(R_4N^+)_2 \ SO_4^{2-} + 2 \ Au(CN)_2^- \rightleftharpoons 2(R_4N^+ Au(CN)_2^-) + SO_4^{2-} \quad (12)$$

The elution of gold is slightly more complex since it is necessary to break the strong ligand-complex interaction by using acidified thiourea.

$$R_4N^+ Au(CN)_2^- + 2CS(NH_2)_2 + 2\ HCl \rightarrow R_4N^+ Cl^- + Au\ CS(NH_2)_2^+ Cl^-$$
$$+ 2\ HCN \quad (13)$$

It is also possible to elute the auro cyanide complex by treatment of the resin with a strongly preferred counter ion such as $Zn(CN)_4^{2-}$, though this will require further elution steps in order to recycle fresh resin to the extraction cycle (equation 12). The cost of the reagents consumed during various operations suggests that weak base resins are more suitable because the elution procedure is significantly cheaper. Also, the elution of weak base resin is easier requiring less labour and lower temperatures[21]. Commercial weak base resins possess some process disadvantages and there is considerable independent research in progress to synthesise custom designed weak base resins based on an imidazole structure[22]. No doubt, novel selective ion exchangers for gold will be available in the near future. Separation of gold from acid leach liquors is possible using ion exchange resins containing weak ester groups, eg. Amberlite XAD-7[23]. The mechanism of extraction is either solvation

$$R-CO_2 + AuCl_4^- \rightarrow R-CO_2\ AuCl_3 + Cl^- \quad (14)$$

or by ion exchange

$$R-CO_2 + H_2O \rightarrow R-COOH^+ + OH^- \quad (15)$$

$$R-COOH^+ + AuCl_4^- \rightarrow R-COOH^+ AuCl_4^- \quad (16)$$

Elution of the gold is performed using a mixture of hydrochloric acid and acetone and it is necessary to provide a distillation step if the eluant is to be recycled.

ION EXCHANGE PROCESSING OF BASE METALS

The separation of base metals from chloride solution is particularly attractive. A diagrammatic representation based on the periodic table of the elements has been prepared by Kraus and Nelson[24] which clearly establishes the behaviour of metal chloride complexes at trace ionic concentration in the presence of a strong base anion exchange resin. This indicates that the alkali, alkali earth and rare earth elements do not interact, whereas the transition metals and noble metals can form anionic chloride complexes with varying affinity for an anion exchange resin. The concept of separating metals from chloride solution by continuous ion exchange has been described by Streat and Gupta[25]. The tendency of the transition metals to form anionic chloride complexes in hydrochloric acid solution is given in Table 2.

TABLE 2   Complex Formation of Transition Metal Chlorides in
Hydrochloric Acid

Acid Concentration

< 2M	4M	6M	> 6M
Zn^{2+}	Zn^{2+}	Zn^{2+}	Zn^{2+}
Fe^{3+}	Fe^{3+}	Fe^{3+}	Fe^{3+}
	Cu^{2+}	Cu^{2+}	Cu^{2+}
		Co^{2+}	Co^{2+}
			Fe^{2+}
			Mu^{2+}
			Ni^{2+}

Separation and recovery of the transition metals is possible by
careful control of the ambient hydrochloric acid concentration in
solution. For example, the 'METSEP' process is highly effective at
low acid concentration, less than 4M, since Zn^{2+} forms a strong
anionic chloride complex, whereas Fe^{2+} does not. Hence, separation
is possible in the extraction cycle of an ion exchange process.
Alternatively, separations can be achieved by fractional elution of
the anionic chloride complexes. Zinc, copper and iron(-ic) are
strongly sorbed in strong hydrochloric acid (6M) and can be
separated by fractional elution using 4M HCl to strip Zn^{2+}, 2M HCl
to strip Cu^{2+} and 0.05M to strip Fe^{3+}. The full potential for
separating base metals from chloride solution using standard
commercial anion exchange resins has not yet been realised.

The recovery of copper from sulphuric acid solutions has been
widely researched and it has been found that commercially available
chelating resins containing imino-diacetic acid groups show good
selectivity for copper over iron (-ous) at pH2 (26).

$$R-CH_2-N \begin{matrix} CH_2\ COOH \\ CH_2\ COOH \end{matrix} \quad + \quad Cu^{2+} \quad R-CH_2-N \begin{matrix} CH_2-C-O \\ CH_2-C-O \end{matrix} Cu^{2+}$$

Elution of the resin is possible using sulphuric acid solutions
(0.5-2N).

An alternative technique for copper recovery has been pioneered by
Warshawsky(27). He advocates the use of solvent impregnated ion
exchangers, whereby liquid extractants such as the commercially
available hydroxyoxime reagents such as LIX63, LIX65N, etc., are
supported within the pores of a macroreticular hydrocarbon bead.
Vernon and Eccles (28) have also shown that macroreticular beads

containing hydroxy-oxime, hydroxamic acid and hydroxyquinoline can sorb copper selectively from aqueous sulphuric acid. Solvent supported systems do have a role to play in hydrometallrgy, but the physical properties of the material will require much improvement to overcome the gradual leaching of the reagent.

CONCLUSIONS

The role of ion exchange in hydrometallurgy has been reviewed in this paper. Apart from the platinum group metals, which is reviewed elsewhere, it will be seen that the major areas of interest are uranium, precious metals, particularly gold and the base metals. This paper has identified the process options available using commercially available ion exchange materials. However, it is quite clear that major advances will occur with the design and development of speciality ion exchange resins capable of selective sorption of these and other important metal values. Sufficient work has not yet been done to create selective ligands attached to a polymer backbone capable of sorbing a metal species and then readily releasing the species during elution. It is found, not surprisingly, that the more effective a ligand is for sorbing a metal ion, the more difficult is the elution step. This may be overcome by better design of the stereochemistry of the ligand group or the polymer backbone structure. Work on the design and synthesis of selective ion exchange materials should be encouraged and it is to be hoped that industry will not desert this important field of progress. Present indications are encouraging based on the number of patents and publications to be found in the open literature.

REFERENCES

1. Warshawsky., A, "Selective Ion Exchange Polmers," Angew.Makvomol. Chem., 109-110, 171, 1982.

2. Jones, K.C., and Grinstead, R.R., Chemistry and Industry 637, 637, 1977.

3. Green, B.R., and Jaskulla, E., "Poly(vinylimidazole) - a versatile matrix for the preparation of chelating resins" in Naden, D., and Streat., M, Editors, "Ion Exchange Technology" Ellis Horwood, 1984, page 490.

4. Vernon, F., and Eccles, H., Anal.Chim.Acta 77, 145, 1975.

5. Warshawsky, A., "Hydrometallurgical processes for the separation of platinum group metals (PGM) in chloride media", in Naden, D., and Streat, M., Editors, "Ion Exchange Technology", Ellis Horwood, 1984, page 604.

6. Gonzalez-Luqe, S, and Streat. M., Hydrometallurgy, 11, 207 and 227, 1983.

7. Efendiev, A.A., Issled Obl. Kinet Model Optim Khim Protsessov 247, 1974.

8. Warshawsky, A., et al. J.Am.Chem.Soc. 101, 4249, 1979.

9. Kauczor, H.W., and Meyer A., Hydrometallurgy, 3, 65, 1978.

10. Warshawsky, A, "Extraction with solvent-impregnated resins" in Marinsky, J.A., and Marcus., Y, Editors "Ion Exchange and Solvent Extraction" volume 8, Marcel Dekker, 1981.

11. Muendel, C.H., and Selke, W.A., Ind.Eng.Chem. 47, 374, 1955.

12. Brown, D.A., et al. Applied Polymer Symp. No 29, 189, 1976.

13. Vernon, F., and Shah, T., Reactive Polymers, 1, 301, 1983.

14. Blesing, N.V., Bolto, B.A., Ford, D.L., McNeill R., Macpherson, A.S., Melbourne, J.D., Mort, F., Siudak, R., Swinton, E.A., Weiss, D.E., and Willis. D., in "Ion Exchange in the Process Industries", Society of Chemical Industries", Society of Chemical Industry, 1969.

15. Swinton, E.A., Bolto, B.A., Eldridge, R.J., Nadebaum, P.R., and Coldrey, P.C., in Naden D., and Streat, M. Editors, "Ion Exchange Technology", Ellis Horwood 1984, page 542.

16. Streat, M., British Patent, 1,456, 974, 11 Dec. 1973.

17. Lloyd, P.J.D. South African Patent, 70/4209.

18. Gonzalez-Luque, S., and Streat, M., in Naden D., and Streat Editors, "Ion Exchange Technology", Ellis Horwood, UK, 1984, page 679.

19. Ketzniel, Z., and Volkman, Y., "Recovery of Uranium from Wet Process Phosphoric Acid by Ion Exchange", IAEA Symposium, Vienna, 1983.

20. Cathard, I.B., "Uranium in Phosphate Rock" USGS Open File Report, 1975.

21. Mehmet, A., and Te Riele, W.A.M., in Naden D., and Streat. M., Editors, "Ion Exchange Technology", Ellis Horwood, UK, 1984, page 637.

22. Green, B.R., and Potgieter, A.H., ibid, page 626.

23. Edwards, R.I., Haines, A.K., Te Riele, W.A.M., "The Separation of Gold from Acidic Leach Liquors with Amberlite XAD-7" in "The Theory and Practice of Ion Exchange", Society of Chemical Industry, London, 1976.

24. Kraus, K., and Nelson, F., Proceedings Int.Conf. on Peaceful Uses of Atomic Energy, 7 113, 1956.

25. Gupta, A.K., and Streat., M., I.Chem.E. Symp.Series, No.42, 21.1, 1975.

26. Naden, D., and Willey, G., "Reduction in Copper Recovery Costs using Solid Ion Exchange" in "The Theory and Practice of Ion Exchange", Society of Chemical Industry, London, 1976.

27. Warshawsky, A., Inst.Min.Metal, 83, C101, 1974.

28. Vernon, F., and Eccles, H.A., "Some Hydrometallurgical Applications of Hydroxy-Oxime, Hydroxyquinoline and Hydroxamic Acid Solvent Impregnated Resins" in "The Theory and Practice of Ion Exchange", Society of Chemical Industry", London, 1976.

SOME OF THE USES OF ION-EXCHANGERS IN HYDROMETALLURGY

Michael J. Hudson

Department of Chemistry, University of Reading,
Whiteknights, P.O. Box 224,
Reading, Berks, RG6 2AD, UK

INTRODUCTION

In the other section [1] on hydrometallurgy the general outline of the uses of ion-exchangers in hydrometallurgy were discussed. In this section some of the points raised will be discussed in greater chemical detail and the extraction of some toxic,
precious and platinum group metals will be discussed. Attention
is drawn to the section on "Introduction to the Coordination
Chemistry of Ion-Exchangers" [2].

Extraction of Metals from Chloride Media

The extraction of metals from chloride media has been
studied in some detail [3]. Some metals are able to form
anionic complexes in HCl solutions and this ability changes
according to the oxidation state of the metal as well as the
molarity of the acid (See Figure 1). Some of the metals are able
to form anionic species over a wide range of molarity of HCl
and consequently they can be extracted by an anion exchange resin.
Noteable examples are the precious metals; the zinc group (Zn,
Cd, Hg); the trivalent state of the gallium group (Ga, In, Tl);
iron(III). There are, however, some noteable exceptions such as
Mn(II) (with little LFSE) and Ni(II) which has a high LFSE for
the octahedral aquo-ion and is unable to form $NiCl_4^{2-}$ in aqueous
solution. Groups 1A, 2A, 3A do not form anionic chlorocomplexes. Anion exchangers can be used to separate metals which
form anionic complexes from those which are unable to do so. Thus,
for example, it is possible to separate Fe and Ni; precious metals
from nickel and cobalt from nickel. However, the extraction of
nickel from cobalt is more difficult [3].

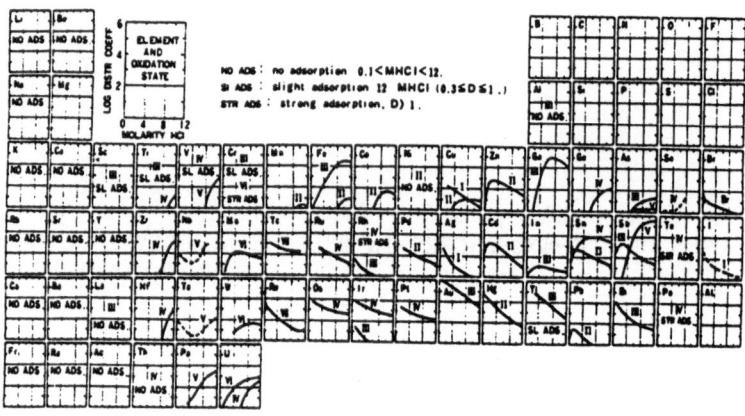

Figure 1. Extraction of Some Elements from
Hydrochloric Acid Solutions Using Anion
Exchange Resins (after Kraus and Nelson).

Extraction of Platinum Group Metals

The platinum group metals are extracted from chloride media
and a typical flow sheet of the INCO process is illustrated in
Figure 2. [4,5]. The copper-anode slime concentrate is given a
total leach (Cl_2/HCl) to give the chloro-anions. Although this
process emphasizes solvent extraction, the use of anion-exchangers
for iridium and rhodium separation [6-9] is discussed below. One
feature of the extraction chemistry of the platinum group metals
(PGM's) is that they can form stable ion-pairs. For this reason
anion-exchangers [9,10] can separate PGM's from base metals such
as Fe(III) even though $(FeCl_4)^-$ also forms an ion-pair with an
anion-exchanger. This anion can be eluted by those of the PGM's.
The stability of ion-pairs of the PGM's has not really been
satisfactorily explained. However, for a given anion exchange resin
"stability" of the ion-pair seems to be related to the overall
charge on the anion. Thus anions with a charge of 1 or 2 appear
to be extracted more than those with a -3 charge. This idea
may be used to separate rhodium and platinum. Rhodium(III) may
be considered as   forming the anion $(RhCl_6)^{3-}$ although some
aquo species are formed [11,12]. The anion exchangers have a
higher affinity for the anion $(PtCl_6^{2-})$, which is formed by
Pt(IV) and the exchangers largely reject rhodium.

Similarly Rh(III) and Ir(IV) can be separated on the basis
that Rh(III) forms $(RhCl_6^{3-})$ and Ir(IV) forms $(IrCl_6^{2-})$.
Hitherto anion-exchangers have reduced Ir(IV) to Ir(III)

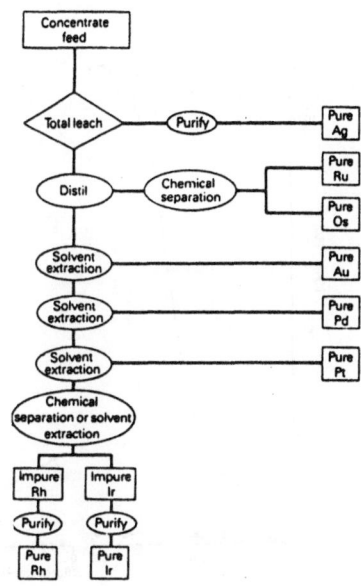

Figure 2. The INCO Process for the
Extraction and Separation of Platinum
Group Metals (PGM) from Copper Anode
Slimes (after Barnes and Edwards)

which makes the separation difficult. However, the increasing
industrial demand for rhodium, principally for use in catalysis,
ensures that its very high market value is retained and that the
precious metals industry continues to seek improved methods for
refining the metal from process concentrates and solutions [5,10].

It has been shown that copolymers can be used to concentrate
and to separate precious and platinum group metals from base metals.
A new separation, which has potential commercial application for
the separation of iridium from rhodium has been reported [6].
Whereas previous attempts to use amines [13,14] have usually
resulted in the reduction of Ir(IV) to Ir(III), this does not
appear to be a significant factor with the copolymer poly(4-
vinylpyridine)-6%divinylbenzene [6]. There are no signs of
decomposition of the copolymer, which has the added advantage
that there is a high concentration of amine groups of which
approximately 60% appear to be protonated. The remaining unpro-
tonated groups are potential ligands and it was considered that
a separation might be achieved by exploiting the different
coordinating and ion-pair formation properties of iridium(IV) and
rhodium(III). The copolymer (Figure 3) which was in a granular form
(250-500μm.) with approximately six per cent of divinylbenzene as
the cross-linking agent, was preequilibrated with hydrochloric acid

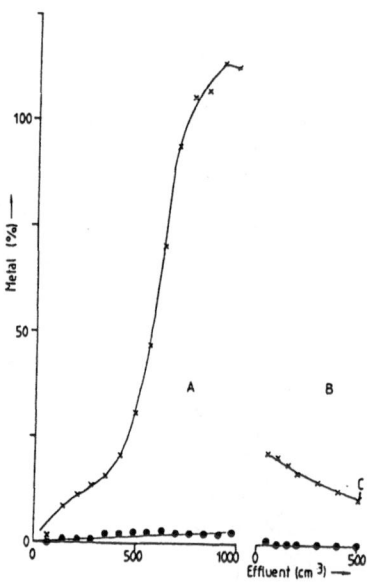

Figure 3. Metal concentration in the effluent, expressed as a percentage of output/input. Second portion B refers to elution of rhodium(III) and iridium(IV) with 4 m hydrochloric acid. (Rh x, Ir ●,) (After Barnes, Ellis and Hudson).

(4 molar) for twelve hours. The approximate bed volume was 22 cm^3 but the column shrank by ten per cent or so during the extraction process. The total metal concentration in the synthetic feed liquor was 0.02 molar and the Rh/Ir ratio was 3/1. The flow rate through the column was approximately 3.5 cm^3 min^{-1}.

In figure 3 part A the rhodium and iridium concentrations in the effluent (expressed as percentage of initial concentration) are plotted against effluent volume and show the preferred uptake of iridium(IV) over rhodium(III).

The effect of eluting the column (part B) with further quantities of hydrochloric acid (4 M) after 40-50 bed volumes of feed solution suggests that rhodium is further eluted whilst iridium remains bound to the copolymer. The final loadings of the column at point C are Rh 0.2 mmol g^{-1} and Ir 3.5 mmol g^{-1} with respect to dry polymer. The copolymer had a water regain value in 4 M HCl of 3.5 g .g^{-1}.

Figure 4 illustrates the separations achieved using an authentic, industrial process solution containing rhodium(III), copper(II), platinum(IV) and iridium(IV) as well as traces of arsenic, selenium and tellurium etc. which were not analysed. The solution was saturated with chlorine gas to ensure that iridium and platinum were fully oxidised and remained in the IV oxidation state. It can be seen that iridium and platinum were effectively retained by the column whilst rhodium and copper were hardly extracted at all. The integrated percentages of extraction are copper 3.3, rhodium 5.2, platinum 98.9 and iridium 93.3. The selectivity coefficients based on the concentrations of the metals per gram of dry copolymer were Ir/Rh(253), Ir/Cu(410) and Pt/Ir(6.5).

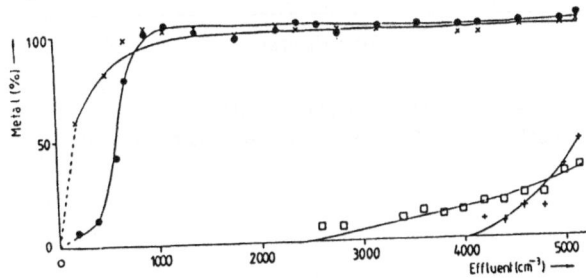

Fig.4- Metal (% as in Figure 3) (Cu X, Rh ●, Ir ▢, Pt +).

There is evidence to support the idea that the iridium remains predominantly as Ir(IV) on the polymer. The reflectance spectrum showed bands at 582, 490 and 437 nm., compared with 575, 489, 431 and 414 nm for $[(IrCl_6]^{2-}$ in solution [15]. There may, of course, be some aquo-containing species [11,12] on the copolymers. However, it appears that the ion exchange mechanism rather than coordination predominates as there is no evidence for any amine species by comparison with the spectrum of (Irpy$_2$.Cl$_4$) [16]. This is to be expected in the strong HCl solutions but approaching pH3 the coordinating mechanism may predominate.

Separation of Copper and Rhodium Using the Copolymer PIED

The active sites of these two metals on this copolymer were discussed in the other section in this volume on coordination chemistry (the structure is in Figure 2.6). The esr at points 1,2, and 3 (Figure 5) have been discussed. The competitive extraction [7] of the two metals is shown in Figure 5. In the

468

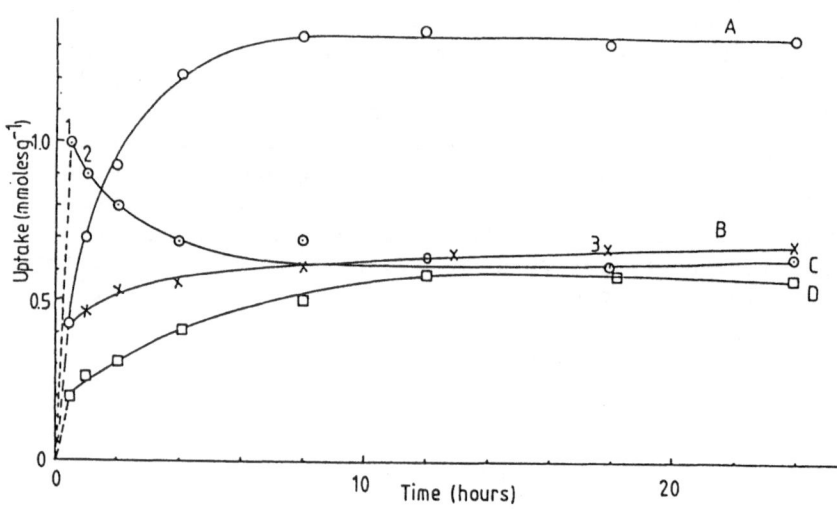

Figure 5   Competitive Extraction of Copper and Rhodium.
A. Cu(ua) B. Rh C. Cu D. Rh (ua)   (ua=no added acid)

unacidified solution A(Cu) D(Rh), copper is preferentially extracted because the aquo groups are more easily replaced with copper
than with rhodium.   (The original solutions were Cu (0.1 m)
Rh (0.1 m) - shaken with PIED (0.2 g) which contained 24% S).   The
maximum separation ratio was only 80 as the metals are bound to
different sites.   Copper is bound to groups with donor S and
N with different geometries whilst rhodium is bound octahedrally
to six sulphur atoms (see text in other section).   There is little
selectivity in the acidified solutions although copper (C) does
appear to be more rapidly extracted than rhodium (D).   The
copolymer PIED appears to be a scavenger for precious and base
metals.   The selectivity of this copolymer for cadmium over zinc
has been discussed previously and is reconsidered later in this
section.

Treatment of Effluent from a Nuclear Power Station.

        Selective ion exchangers are being studied [17] to see
whether they are capable of removing radionuclides such as ^{106}Ru
from simulated nuclear wastes. One such study concerns ^{106}Ru($t_{\frac{1}{2}}$
368 days) which used to be released to the environment by BNFL

(or BNF plc). It has been shown [7,8] that solid selective ion-exchangers with covalently bound $-NCS_2H$ (dithiocarbamate groups) can concentrate platinum group metals (Ru, Os, Rh, Ir, Pd, Pt). These resins were classified as bidentate (S,S) ligands which strongly bind the metals but it has been shown [7] that copper is N_2S_2 coordinated. Ruthenium like the other PGM's, forms strong bonds to donor sulphur groups which may be present in the selective ion-exchanger as bidentate or monodentate groups (e.g. a thiol) [18].

Nuclear fuel reprocessing uses a nitric acid medium and consequently the octahedral nitrosyl-containing species $[Ru-N-O.X_5]$ (X and n vary) are always present. Even so the site *trans* to the NO group is active to nucleophilic attack, particularly by charged sulphur-containing groups. Three simulated waste solutions were studied and these compositions are given in Table 1. The extractions of the nitrosyl complex are listed in Table 2. In particular, the copolymer PIED (number 8) has a very high K_D value for extraction of ruthenium from solution 3 which approximates to pond water. The thiadiazole (number 9), which is possibly an S, N donating reagent [18] is particularly effective for the acidic solution A in which the zeolites are useless.

Table 1. Composition of simulated waste solutions.

	A	Solution B	C
NaOH	—	—	0.005 M
Na_2CO_3	—	—	0.0028 M
HNO_3	0.15 M	0.08 M	—
$NaNO_3$	5 M	2.5 M	0.01 M
Ru-NO[a]	3.63×10^{-6} M	3.63×10^{-6} M	3.63×10^{-6} M

[a] Ru-NO signifies species which contain this group

Table 2. Extraction of ^{106}Ru with solid reagents. Apparent K_D values $(cm^3 g^{-1})$.

Sample (physical form)	A	Solution B	C
(1) 1,3,4-Thiadiazole-2,5-dithiol (powder)	800	80	10
(2) An aminophosphonic acid copolymer (beads)	100	100	36
(3) A thiol copolymer (beads)	800	530	70
(4) An iminodiacetate copolymer (beads)	480	300	1.1
(5) An amideoxime copolymer (beads)	70	95	103
(6) A pyrrolidone copolymer (powder)	65	1.5	—
(7) Amidinothiourea (powder)	200	512	2 460
(8) A dithiocarbamate copolymer powder	250	250	13 800
(9) A thiadiazole copolymer (beads)	1550	260	16
(10) Zeolites – Na,X (beads)	<1	20	994
(11) – Na, clinoptilolite	<1	<1	334
(12) – Linde, A-51	—	—	1 762

Table 2 (after Dyer, Keir, Leung and Hudson)

Selective Extraction of Silver

Silver from photographic effluents with thiosulphate may be
extracted onto an anion exchanger because the Ag(I) - thiosulphate
complex is anionic [18,21] Research is in progress to improve
the rate of extraction and capacities especially for the colour-
bleach effluent [21,22]. When silver is present as the Ag(I)
aqueous ion it may be extracted by a thiol [18] which has the
possibility of acting as an S-N coordinating reagent. The synthesis of the copolymer is outlined in Figure 6 and the extraction
curves for Ag(I) and the base metals are given in Figure 7. The

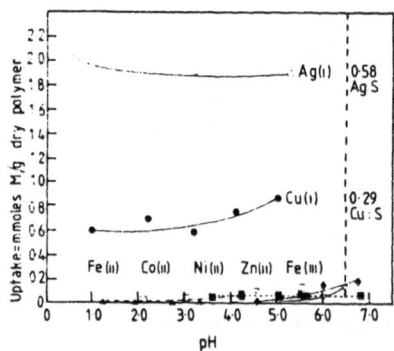

Fig.6-Preparation of poly(2-S-vinyl-1,3,4-thiadiazole-5-thiol).

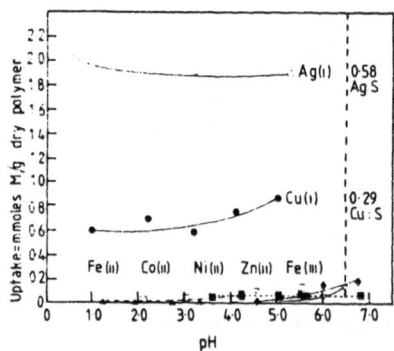

Fig. 7 Uptake of metal ions by the polymer, which was equilibrated with separate solutions containing the metal ions (i.e. these were not competitive studies).

(After Hudson and Shepherd)

copolymer rejects base metals as the main bonding thiol group has
little affinity for these metals. This, and related aspects, were
discussed in the section on "The Coordination Chemistry of Selective Ion Exchangers". An alternative list of properties which
need to be considered [18] is:

A. Properties of the metal ion

1. Ability of the metal ion to bond to the coordinating groups (e.g. Class A or Class B character).

2. Ability of the metal ion to form a species which is capable of bonding to thiol groups.

3. Oxidation state of the metal.

4. Stereochemical requirements of the metal ion.

5. Concentration of the metal ion.

B. Properties of the polymer

1. Stability of the polymer e.g. resistance to oxidation.

2. Ability of the polymer to complex with metal ions.

3. Nature of the co-ordinating group

   (a) charged or uncharged

   (b) Monodentating or chelating

   (c) Number of types of donor atoms.

4. Hydrophillic or hydrophobic nature of the polymer.

5. Amount of cross-linking in polymer.

6. Ability to be reused after elution of the metal.

C. Properties of the metal-polymer complex

1. Rate of formation of metal polymer complexes.

2. Solubility.

3. Stability.

4. Strength of metal-ligand bonds.

5. Stereochemistry of the metal.

6. Oxidation number of the metal.

The Effect of Particle Size

The rates of extraction of metal ions by selective ion-exchangers is greatly influenced by the size of the particles. The metals which are covalently bound to the copolymer limit the rate of diffusion of unbound ions. This limit arises partly because of the size of the bound metal species but is particularly serious when the bonds formed between the metal and the copolymer introduce cross-linking. This increase in cross-linking may further cause contraction in volume of the resin by as much as twenty five per cent [23].

The rate of extraction of rhodium(III) (0.1 m, 4 m HCl, 25 C), is illustrated in Figure 8. For optimum rates of extraction and

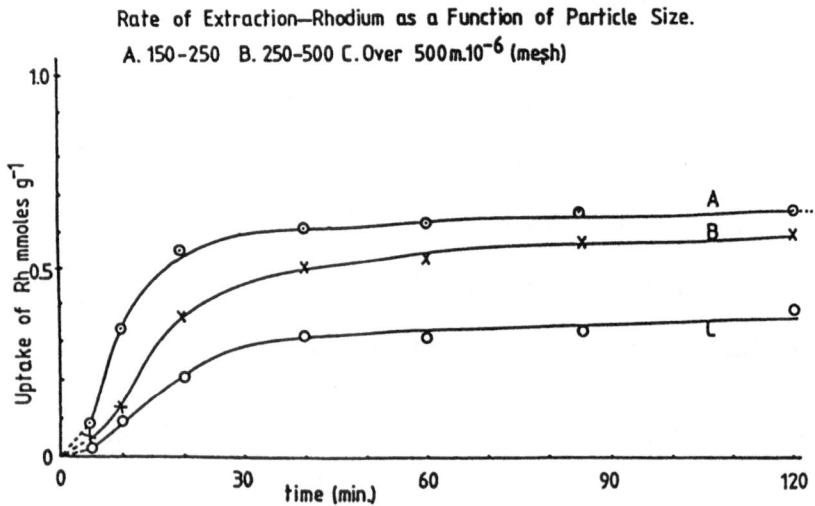

Rate of Extraction of Rh(III) from chloride medium (4 m HCl) as a function of the particle size of the copolymer PIED.

Figure 8.

maximum capacity the particle size should be kept as small as possible. The loaded powders are often readily filterable because they are heavier and denser than the unloaded powders. This is a distinct advantage that copolymers have over low molecular weight precipitating agents [24]

SHORT BED ION-EXCHANGE TECHNOLOGY

The principal disadvantages with the selective ion-exchange resins are that they are expensive and have slow kinetics. The expense may be difficult to justify. These disadvantages can be partially off-set by using small amounts of powdered material, recycling the exchanger and by using specially designed equipment. The Recoflo ion-exchange system [25] uses resins columns 5-60 cm long which are only slightly longer than the exchange zone. This is much shorter than normal ion-exchange columns. The system is designed for fine mesh resins, low ion-exchanger capacities, counter current regeneration, little free space above the resin, liquid stratification for rinsing and short cycle times. It is claimed that such equipment allows for reduced equipment size, lower installation costs, lower operating costs and longer resin life. These features could enable the advantages of the selective ion-exchangers to be exploited.

COPPER RECOVERY USING SHORT BED ION-EXCHANGE TECHNOLOGY

The picolyamine resin, Dowex XFS 43084 has a selectivity for copper over iron [26,27]. The reasons for this selectivity are being investigated [24] but appear to be connected with a combination of the right type of ligand for copper and the formation of a tetrahedral/square planar complex. Copper requires four donor atoms but iron(III) appears to require six which may be sterically hindered by the polymer chain.

The flow-sheet for separation of Fe(III) and Cu(II) is shown in Figure 9. Using the numbered stages in this Figure 9.

    1. Pregnant leach is pumped to the top of the resin. Copper is extracted onto the column with some of the iron.

    2. Iron is scrubbed with dilute acid with some copper which replaces the iron on the active sites of the resin.

    3. Water reuse of residual acid and iron.

    4. Spent electrolyte elutes copper.

    5. Barren (copper-free) sulphuric acid leaches residual copper.

    6. Elution in preparation for next cycle.

474

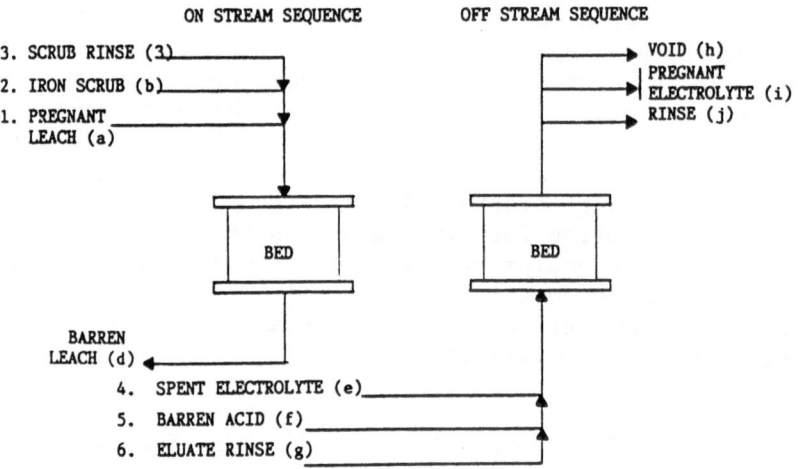

Figure 9.

The operating cycle of the 'Recoflo' System which uses short bed Ion Exchange Technology. (After Brown).

A pregnant feed (Cu:Fe/1:2) can be processed to give a final loaded electrolyte (Cu:Fe/20:1). The resin inventory was twenty-three times lower than a continuous counter current ion-exchange process [28].

ACTIVE SITE IDENTIFICATION IN COPPER EXTRACTION WITH N-DONORS

It has been shown [29] that copper can be extracted from EDTA solutions using polymers with covalently polyamine groups. Magnesium may be added to the solution to bind the EDTA. Copper is at the maximum part of the Irving-Williams series (see related section) and coordinating polyamines are able to give acceptable rates and capacities for copper(II). Consequently, an understanding of the way the metal is bonded is important.

The stereochemistry of the copper which is bound to the copolymer may be determined using spectroscopic techniques. The electron spin resonance (esr) spectrum of IRA45 with copper(II) sulphate is shown in Figure 10. The well defined peaks lead to determination of $A_{//}$ and $g_{//}$ values which can in turn be related to the geometry of the nitrogen donor atoms around the copper. For this sample, the $A_{//}$ value 165±5, $g_{//}$ 2.204 (A) are probably

Hydrometallurgy

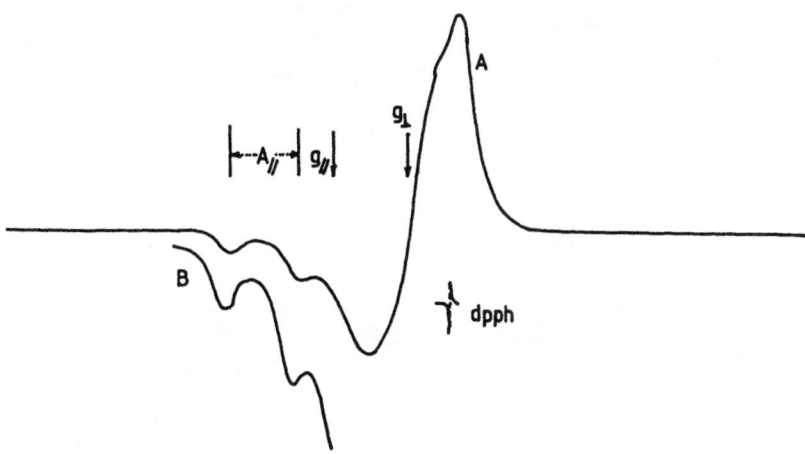

Figure 10.

Esr spectrum of Cu(II) on IRA-45. The field set is 3125G, scan range ± 1000G. A has receiver gain 500 and B 2000.

indicative [24] of a square planar coordination.

Interestingly the Cu-IRA 45 copolymer which was loaded from copper in hydrochloric acid (4 m HCl) has a higher $g_{//}$ value of 2.27 [24] which implies that copper adopts different arrangements of nitrogen ligands according to the groups which are being displaced. In the aquo-ion copper is surrounded by six water molecules (octahedron with a tetragonal distortion) in 4 m HCl the $[CuCl_4]^{2-}$ anion, which is tetrahedral, is the dominant species.

It is important to appreciate that the esr parameters change when the ligand changes. For $[Cu(py)_4]^{2+}$ (py = pyridine) the square planar marker values are $A_{//}$ 150 ± 10G, $g_{//}$ 2.3 and $g_\perp$ 2.1. These values can be used for other nitrogen donor atoms which are also in aromatic ring systems. The esr parameters for a polyvinyl-imidazole PI 1(586) [24,30] loaded with Cu^{2+} from chloride solution (dried 60 °C in vaccuo are rather similar $A_{//}$ 150, $g_{//}$ 2.31, $g_\perp$ 2.07 indicating this also probably has a square planar arrangement of nitrogen atoms. The nitrogen atoms are coordinating and have displaced the chlorine atoms. It is possible to correlate $A_{//}$ and $g_{//}$ (Figure 11) for similar ligands [31]. Broadly speaking

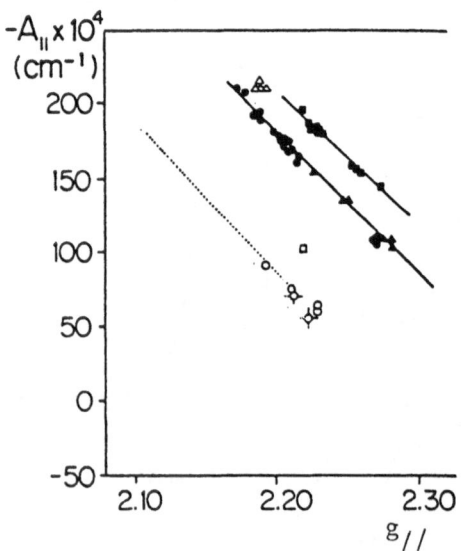

Relationships between $A_{//}$ and $g_{//}$ for copper(II) complexes with (●) pyrrole-2-carboxaldimines, (▲) dipyrromethenes (Δ), tetra-phenylporphyrins and (▬) salicylaldimines. The points O are for selected copper proteins.

The point ⬛ is for CuN_2S_2 in Cu-PIED (high loaded) (17).

Figure 11 (After Addison and Yokoi).

high A$_{//}$ values and low g$_{//}$ imply square planar or tetragonal geometry and (for similar ligands) low A$_{//}$ and higher g$_{//}$ implies a more tetrahedral geometry for copper(II). However, care must be taken to ensure that when such comparisons are being made that the coordination number is the same and the ligands are the same or comparable. It is not possible to make direct comparisons, for example, between aliphatic and aromatic N-donor atoms as they have, as indicated above, different esr parameters. The g$_{//}$ value is particularly diagnostic of geometry. Comparison with the g$_{//}$ value for the aliphatic nitrogens in IR45 above shows that a value of 2.2 is probably diagnostic of a square planar configuration whereas 2.3 is the indicative of a similar stereochemistry for aromatic nitrogens. It is easy to draw false conclusions from esr data because it is necessary to compare similar systems. There is no evidence for metal-N bonds in the iminodiacetate system even though this has been claimed to be the case from esr and spectral data [39]. When comparisons are valid [31] it is possible to estimate the angle (ω) (Figure 12)

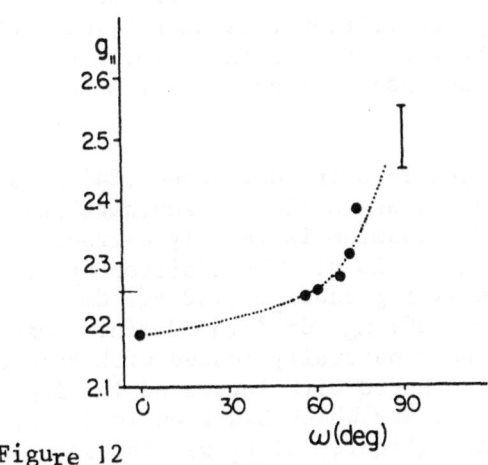

Figure 12

Estimated relationship between g$_{||}$ and ω,

(After Addison and Yokoi)

which approximates to a distortion from the square planar configuration towards a tetrahedral one. The selective ion-exchangers, which are known to extract copper well, seem to generally give a square planar configuration. Thus copper extracted by IRA45, PI 1(485), XES 43084 [24] poly(4-vinylpyridine) normally has this square planar arrangement. The reason for this must be connected with the fundamental coordination chemistry of copper.

The Extraction of Toxic Metals e.g. Cadmium.

Toxic metals include cadmium, mercury and lead. Mercury can be removed to low concentrations using thiol reagents [32] or cation exchangers. In industrial societies, water re-use and the quality of drinking water depends on efficient treatment plant which inevitably leads to large quantities of sewage sludge. In EEC countries together, annual production of sludge exceeds 5×10^6 tonnes dry sludge solids and is increasing [33]. Cadmium is a principal factor limiting the use of sludge on agricultural ground. The metal is a cumulative poison and displaces zinc in enzymes. The metal occupies an interesting position between zinc and mercury in the periodic table. The general chemistry seems to be closer to zinc than to mercury. Nevertheless, cadmium does appear to have some affinity for sulphur ligands and it is this aspect which will be considered. In the form of Cd^{2+}(aq.) the metal can be removed by cation exchangers but frequently the metal is bound to humic acids. Liming out processes are unsatisfactory but a new precipitation method shows some promise using bismuthiol [24]. Two equilibrium constants K_1(ca. 10^6) and K_2 ca. 10^4 were related to SN coordination in a single complex and SN coordination to form an insoluble polymer with the metal as the cross-linking agent.

Starch xanthate gel is being developed [34] as a reagent for the removal of cadmium as are cellulose xanthate and the copolymer PIED (vide infra). The cadmium is rapidly extracted by PIED [35] at pH 5,5 with $t_{\frac{1}{2}}$ of about 20 s. The addition of (ca. 1 g) to a solution (100 ml) containing cadmium (102 mg. dm^{-3}) reduces the concentration down to 0.001 mg. dm^{-3} in 10 min. Some samples, which had been previously partially loaded with cadmium, were studied to see how $t_{\frac{1}{2}}$ changed with increased loading of the copolymer. For the copolymers which had been loaded by 10% (or 41%) of the maximum capacity (Figure 13) $t_{\frac{1}{2}}$ was the same as above, but with 75% loading the kinetics are slower ($t_{\frac{1}{2}}$ = 60 s). Even so this rate is acceptable for an industrial process.

Zinc and Cadmium Separation.

The zinc salt of PIED was used with different loadings of cadmium. The two loading were:-

a) Low load (0, 10 mmol Cd. g^{-1} Zn-1)

b) High load (0, 50 mmol Cd. g^{-1} Zn-1)

The lower the loading of cadmium the faster the kinetics, but the presence of zinc does not influence the kinetics of cadmium uptake on the copolymer, which is selective for cadmium in the presence of zin

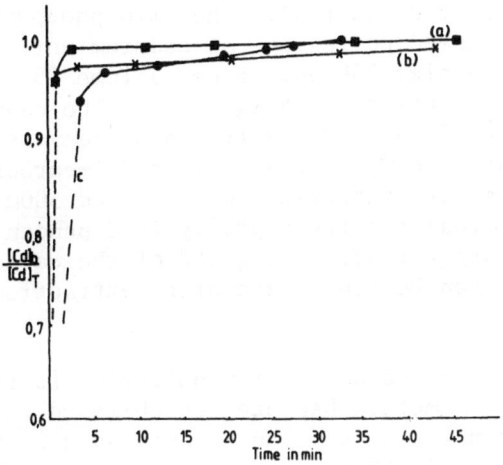

Figure 13. (35)
Uptake of cadmium by PIED pH 5,5 as a function of time.

(a) 0.09
(b) 0,72 mmol Cd added/g of PIED

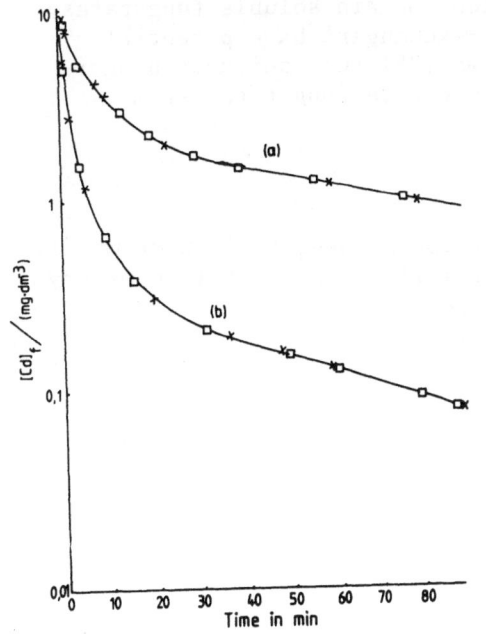

Figure 14. (35)
Residual cadmium concentration in solution after treatment with the zinc salt of PIED (X): initial $[Cd]$ = 11,2 mg. dm^{-3}; ($\square$); initial $[Cd]$ = 11,2 $mg. dm^{-3}$ + $[Zn]$ = 13,0 mg. dm^{-3}. Cd load, (a): 0,10 and (b): 0.50 mmol Cd/g zinc salt of PIED pH = 2. (H_2SO_4)

Cadmium can be removed from marine (harbour) sludges by dilute HCl. The resulting solution of $CdCl_2$ can be treated with a cation exchanger. Selective ion-exchangers such as the aminophosphonic acid may be used to recover cadmium [24]. The aminophosphonic acid in the calcium form showed no leakage of cadmium until 310 bed volumes. (Particle size 250-500 μm bed volume 25 cm^3, flow rate 5 $cm^3 . min^{-1}$, concentration 75 mg dm^{-3}). The cadmium can be eluted with EDTA or HCl and the resin can be recycled. Cadmium can also be extrated using a poly(acrylamide)-carboxylic acid resin [24]. The rates of extraction for 100μ and 500μ mesh are acceptable ($t_{\frac{1}{2}}$ ca. 2 mins) but the capacity is dependent upon mesh size with the larger beads having 85% of the capacity of the former. The resin can be reutilized after extraction with 3 m HCl.

One of the principal sources of cadmium pollution is related to cadmium in phosphates. A method has been established whereby cadmium can be removed from strong phosphoric acid solutions [36] which are used in the industrial manufacture of superphosphates.

The Application of Ion Exchange to the Extraction of Tungsten.

Tungsten has an extremely high melting point [37] and, there-fore, does not really lend itself to pyrometallurgical methods. The mineralogy of tungsten is essentially simple [37] and is related to the occurence of tungstates. Consequently, anion-exchangers should be able to extract tungstate from aqueous solution. There are many lakes which contain soluble tungstates (e.g. Searle's Lake) [37] and anion-exchangers have potential for these sources. It has been shown [38] that anion-exchangers such as Duolite A101 D are able to separate tungstates from phosphates.

Concluding Comment.

The continued development and use of ion-exchangers in hydrometall-urgy should place more emphasis on relating fundamental chemistry and engineering to industrial practice.

References.

1. Streat, M. (this volume). "Applications of Ion-Exchange in Hydrometallurgy.

2. Hudson, M.J. (this volume)."Introduction to the Coordination Chemistry of Ion-Exchangers".

3. Kraus, K. and F. Nelson. Proceedings Int. Conf. on Peaceful Uses of Atomic Energy 7 (1956) 113.

4. Barnes, J.E. and J.D. Edwards. Chemistry and Industry (London) (1982) 151.

5. Rimmer, B.F. Chemistry and Industry (London) (1974) 63.

6. Barnes, J.E., A.F. Ellis and M.J. Hudson. Chemical Communications (London) (1985) (in press).

7. Ellis, A.F., M.J. Hudson and A.A.G. Tomlinson. Chemical Transactions (Dalton), (1985) (in press).

8. Hudson, M.J. and J.F. Thorns. Hydrometallurgy 11 (1983) 289.

9. Cleare, M.J., P. Charlesworth and D.J. Bryson. J. Chem. Tech. Biotechnol., 29 (1979) 210.

10. Warshawsky, A. "Selective Ion-Exchange Polymers," Angew. Makromol. Chem., 171 (1982) 109.

11. Chang, J.C. and C.S. Garner. Inorg. Chem. 4 (1965) 209.

12. Evers, A.P., R.I. Edwards and M.M. Frieberg. British Patent (1978) 22 Nov. 1533373.

13. Kanert, G.A. and A. Chow. Anal. Chim. Acta 69 (1974) 355.

14. Kanert, G.A. and A. Chow. Anal. Chim. Acta 78 (1975) 375.

15. Jorgensen, C.K. "Absorption Spectra and Chemical Bonding in Complexes", Pergamon, London (1962).

16. Inamura, Y. and Y. Kondo. J. Chem. Soc., Japan,, Pure Chem. Section 72 (1951) 787.

17. Dyer, A., D. Keir, M.J. Hudson, B.K.O. Leung, J. Chem. Soc., Chem. Comm. (London) (1984) 1457.

18. Hudson, M.J. and M.J. Shepherd, Hydromet. 9 (1983) 223.

482

19. Simon, G.P. "Stability of Ion-Exchangers in Ionising Radiation", in Ion-Exchange for Pollution Control. C. Calmon, H. Gold and R. Prober (editions). C.R.C. Press Florida (1979) 55.

20. Hudson, M.J. and A. Dyer. Proceedings International Conference, Solvent Extraction and Ion-Exchange in the Nuclear Fuel Cycle, Harwell (1985) September.

21. Hatano, T., H. Iwano, S. Matsushita, and Shirasu K. J. Appl. Photograph. Eng. 2 (1976) 65.

22. Marsh, D.G. J. Appl. Photograph. Eng. 4 (1978) 17.

23. Ellis A.F. Ph.D. Thesis, Reading, England (1985).

24. Hassan Mohamed Bashir., Ph.D. Thesis, Reading, in preparation for 1986.

25. Brown C.J., Conference Society of Mining Engineers of AIME, Denver, Oct. 1984 – available through Eco-Tec Ltd., 925, Brock Road South, Pickering, Ontario, Canada L1W 2X9.

26. Grinstead, R.R. J. of Metals, 31 (1979) 13.

27. Grinstead, R.R. International Conference "Ion-Exchange Technology", Cambridge, IEX 84, Ellis Horwood (1984) 509.

28. Jones, K.C. and R.A. Pyper. J. of Metals 31 (1979) 19.

29. Matejka,Z., personal communication (1984).

30. Green, B.R. and Jaskulla E. (IEX 84 – see ref. 27)(1984) 490.

31. Addison, A.W. and Yokoi H., Inorg. Chem. 16 (1977) 1341.

32. Calmon,C., H. Gold and R. Prober (editors), Ion-Exchange for Pollution Control, CRC Press Florida (1979) 55.

33. Davis, R.D., Cadmium on Complex Environmental Problem, Experientia 40 (1984) 117.

34. Tiravanti, G., D. Marani, M. Mezzana, R. Passino. Int. Conference on Heavy Metals in the Environment, Amsterdam, 1981.

35. Ellis, A., M.J. Hudson and G. Tiravanti, Die Macromol. Chem. (1985) in press.

36. As reference 24 but restrictions due to patent applications.

37. Hudson, M.J., Chemistry in Britain (1982) 438.

38. Martins, J., J. Costa, J., Loureiro, Rodrigues A IEX 84
    (International Conference on Ion Exchange) Cambridge
    (1984), 714, Ellis Horwood Publishers for the Society
    of Chemical Industry (London).

39. Sahni, K. and J. Reedjik, Coord. Chem. Revs. 59 (1984) 1-139.

REACTION PROCESSES INVOLVING ION-EXCHANGE RESINS

Sergio Carrà

Dipartimento di Chimica Fisica Applicata
Politecnico di Milano,Milano,Italy

INTRODUCTION

Many chemical reactions can be catalyzed by ion exchangers. Industrial interest in the application of ion exchangers as catalysts has been directed towards inorganic zeolites, sometimes substituted with metals such as manganese, iron or vanadium. The more common reactions in the liquid phase catalysed by ion exchangers include esterification, ester hydrolysis, sucrose inversion, dehydration of alcohols, condensations and so on.

A summary of some significant examples of the application of synthetic ion exchange resins as catalysts, is given in table 1[1].

As it is well known the characteristic behaviour of the ion exchangers is due to the peculiar features of their structure. An ion exchange material is constituted by a poly-ionic framework, held together by lattice energy in inorganic crystals, or chemical bonds in macromolecular systems, which carries a surplus electrical charge, positive or negative, as shown in fig. 1. This charge is neutralized by counter-ions of opposite sign, free to move in the mentioned framework [2] .

The catalytic activity of ion exchangers on liquid and solutes is explained in terms of the counter-ion. In other words the cation exchangers, in the H^+-form, and the anion exchangers, in the OH^--form, catalyze processes that are accelerated by acids and alkalis . In this respect the process is analogous to the homogeneous-phase catalysis by dissolved electrolytes.

Exchangers loaded with metallic ions exhibit the catalytic

Table 1

Examples of the use of synthetic ion exchange resins as catalysts

Reaction	Exchanger; loading	Temp.
Hydrolysis of different acetates	sulfonated phenol-formaldehyde resin	
Saponification of ethylacetate	Dowex 2; OH$^-$	20°
Hydrolysis of proteins	Dowex 50; H$^+$	100°
Hydrolysis of nitril	Amberlite IRA-400, OH$^-$	
Esterification of glycerol with ethanol	Zeo-Karb; H$^+$	115°
Esterification of n-butyl alcohol and acetic acid	Amberlite IR-100, H$^+$	114-116°
Conversion of glucose and fructose	Amberlite IRA-400; OH$^-$	
Acyloin condensation from benzoin from benzaldehyde	Amberlite IRA-400; CN$^-$	80°
Cyanohydrin synthesis	different exchangers	
Hofmann decomposition	Amberlite IRA-400; OH$^-$	18°
Hydration of olefins	sulfonated coal	
Polymerization of unsaturated hydrocarbons	sulfonic acid exchanger	
Hydration of propylene	sulfonic acid exchanger	150°
Decomposition of diazoacetate	sulfonic acid exchanger, H$^+$	130°
Production of polyesters	sulfonic acid exchanger, H$^+$	140°
Condensation of Methyl-styrene with formaldeyde	Kationit KU-2, H$^+$	
Esterification of methacrylic acid	Kationit KU-2, H$^+$	125°
Deydration of hydroxamic acid	Wofatit KPS	
Aromatic nitrogen bases	Amberlyst 15	100°

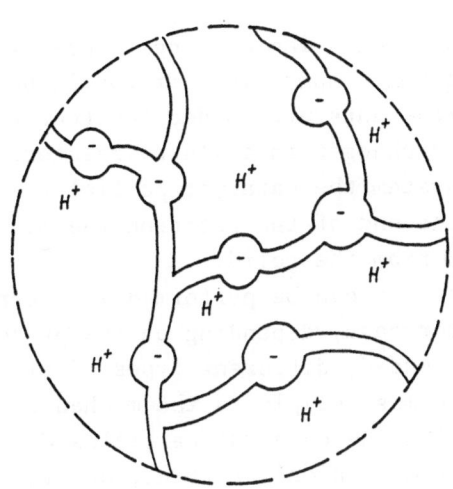

Fig. 1 - A cation exchanger containing counter ions H^+

behaviour of the ions; then the exchanger network acts only as a catalyst support.

Ion exchange resins can be easily transformed into the acid and basic form. Besides their chemical stability, particularly in acid environment, is high .Their porosity and swelling characteristics allow even large organic molecules to penetrate into the resins. The most serious drawback in the use of organic resins as catalysts is their thermal instability and their tendency to oxidative decrosslinking in reactions involving strong oxidation agents.

The employment of ion exchangers as catalysts offer a series of advantages over homogeneous catalysis by dissolved electrolytes, these are:

1 - The catalyst can be easily separated from the reaction products, for instance through decantation or filtration.

2 - It is possible to perform continuous operations in slurry reactors ,in fixed bed columns or continuous processes.

3 - By working with short contact time, higher yields with respect to desired products can be obtained, since the role of secondary reactions may be less significant.

4 - The purity of the products is higher than obtained in homogeneous catalysis.

TYPES OF REACTORS

Catalytic processes with ion exchangers can be performed both discontinuously, in a batch operation, or continuously. In the batch operation the reactants are loaded together with small particles of the ion exchanger in a slurry reactor, and then mixed and heated. In this system the catalyst particles are in a suspended state. At the end of the reaction the mixture is decantated or filtered from the solid.

Continuous operations can be performed in slurry or fixed bed reactors. In the former case, depending on the procedure by which the particles are suspended, different types of slurry reactors can be identified. The most simple is the mechanically agitated slurry reactor, in which the catalyst particles are kept in suspension by means of mechanical stirring, as illustrated in fig. 2a.

A more complex equipment is the bubble column slurry reactor in which the catalyst particles are suspensed by gas induced stirring, as it is shown in fig. 2b. This kind of procedure could be particularly apt when one of the reactant or product is present in gas state.

The fixed bed columns are used with particles of larger size. In this case the fluid phase moves through a stationary bed of catalyst particles, as shown in fig.2c. The raw materials are loaded on the exchanger and the end products appear in the eluate.

KINETICS OF ION EXCHANGE CATALYZED REACTIONS

The theory of catalysis of ion exchangers has not yet been well established. However, the counter ions present into the pores of an ion exchanger are mobile and solvated, which is similar to homogeneous solution. In other words the mechanism of a reaction which occurs into the pores of a resin may be essentially the same as the reaction which takes place in a homogeneous system in the presence of dissolved electrolyte. A significant difference between homogeneous catalysis, in an ion exchange resin, is due to the presence in the latter case of diffusional processes. In fact the reactant must diffuse through the pores into the reaction zone, and the products must diffuse out. Then it follows that the heterogeneous catalytic reaction takes place through a stepwise process, as illustrated in fig. 3. Two diffusion steps are needed in order to transfer the reactants from the bulk of the fluid to the active surface of the resin .Also the counter-diffusion of the reaction products from the active surface to the bulk phase must take place.

The diffusion paths may be divided in two parts:
- external diffusion from the bulk fluid to the outer surface of the resin particles.
- internal diffusion from the outer surface towards the internal surface of the porous resin.

Finally the chemical transformation takes places after the interaction of a reactant molecule with a counter ion of the exchanger. The reaction can be, for instance, first order and irreversible, as it happens in sucrose inversion, or in alcohol dehydration. In other cases more complex kinetic laws are obtained as is the case of esterification of organic acids with alcohols.Here in fact, it can be assumed that the reaction occurs between the molecule of the acid, associated with the protons of the catalyst, and an alcohol molecule in the neighbouring pore liquid.

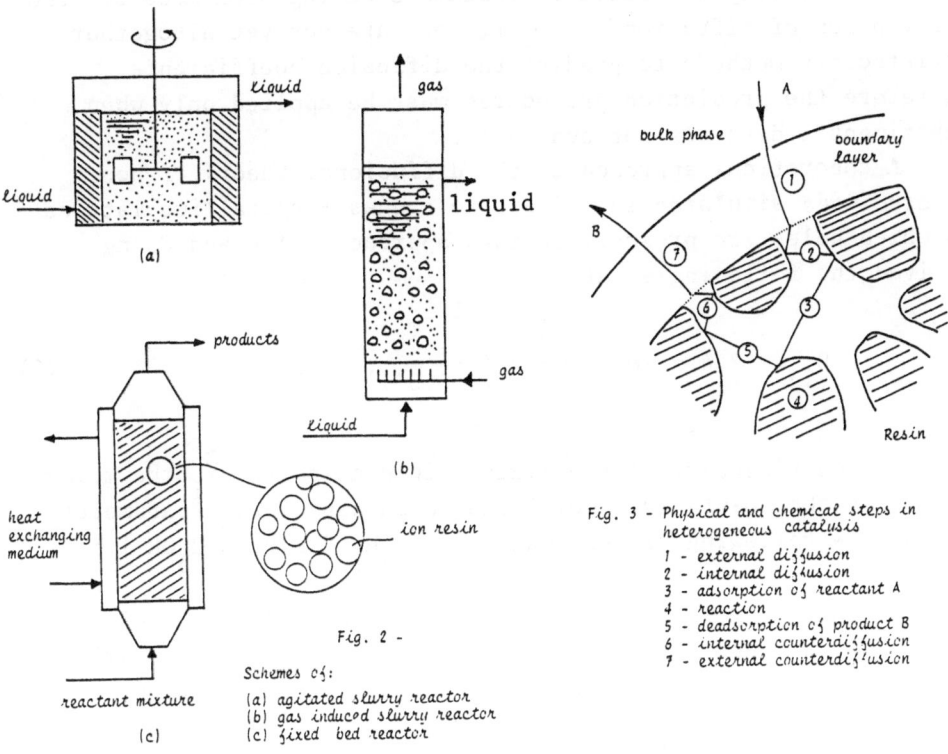

Fig. 2 -

Schemes of:
(a) agitated slurry reactor
(b) gas induced slurry reactor
(c) fixed bed reactor

Fig. 3 - Physical and chemical steps in heterogeneous catalysis
1 - external diffusion
2 - internal diffusion
3 - adsorption of reactant A
4 - reaction
5 - deadsorption of product B
6 - internal counterdiffusion
7 - external counterdiffusion

TRANSPORT PHENOMENA IN LIQUID AND ELECTROLYTIC SOLUTIONS

In a mixture of different components the mass flow of component i is given by [3] :

$$\mathbf{N}_i = C_i \mathbf{u}_i = \underbrace{- D_{im} \nabla C_i}_{\text{diffusional transport}} + \underbrace{\mathbf{u}^* C_i}_{\text{convective transport}} \tag{1}$$

where $\mathbf{u}^*$ is the average molar velocity of the mixture.
The diffusion coefficient D_{im} can be obtained from binary diffusion coefficients D_{ij} as follows:

$$D_{im} = \frac{1 - x_i}{\sum\limits_{j \neq i}^{m} x_j / D_{ij}} \tag{2}$$

where x_i is the molar fraction of component i.

Despite many theoretical treatments having been made for the description of diffusion in liquids, there are not yet altogether satisfactory methods to predict the diffusion coefficients. Therefore the prediction procedures must be applied only when experimental data are not available.

A theoretical approach to the diffusional theory of non electrolytes simulates the liquid to a quasi-crystalline lattice in which holes are present. In this approach Wilke and Chang derived the following equation

$$D_{ij} = 7.4 \times 10^{-8} \frac{(\phi M_j)^{1/2} T}{\mu \tilde{V}_i^{0.6}} \quad (cm^2/s) \tag{3}$$

being μ the viscosity of the mixture in centipoise, $\tilde{V}_i$ the molar volume of the solute (cm^3/mole) and ϕ an association parameter having the following values: water 2.6, CH_3OH 1.9, C_2H_5OH 1.5, benzene 1.

In an electrolytic solution there is an electric potential ψ, as it happens in the pores of an ion exchange resin, producing transference of ions. For this reason equation (3) must be generalized by superposing to the normal diffusion $(J_i)_{diff}$ an electrochemical contribution $(J_i)_{el}$:

$$J_i = (J_i)_{diff} + (J_i)_{el} = - D_{im} (\nabla C_i + Z_i C_i \frac{F}{RT} \nabla \psi) \tag{4}$$

The preceding relation is known as Nerst-Plank equation. Its application to the internal diffusion in ion exchange resins implies the introduction of two additional assumptions, that are respectively:

$$\sum_i Z_i \bar{C}_i = constant, \qquad\qquad electroneutrality \tag{5}$$

$$\sum_i Z_i J_i = 0 \qquad\qquad absence\ of\ electric\ current$$

$\bar{C}_i$ indicates the ion concentration in the interior of the exchanger.

By combining equation (4) and (5) it follows that for a given ion i, an interdiffusion coefficient can be defined with respect to another ion j, in terms of both gradients present in equation (4), according to the following expression [2] :

$$D_{ij} = \frac{D_i D_j (Z_i^2 \bar{C}_i + Z_j^2 \bar{C}_j)}{Z_i \bar{C}_i D_i + Z_j \bar{C}_j D_j} \tag{6}$$

The value of the diffusion coefficient of a component i in the pores of a particle is actually less than the one obtained by means of the preceding equations. In fact if the pores are simulated to an array of cylinders parallel to the diffusion flux, only a fraction ε_p of the flux with no solid present would occur in the solid itself. Besides the length of the tortuos diffusion path in the pores is higher than the distance along a straight line, and the pores are irregularly shaped [4] . By allowing for both the mentioned factors an effective diffusion coefficient can be expressed as follows:

$$D_{eij} = \frac{D_{ij} \varepsilon_p}{\tau} \tag{7}$$

τ is a factor which makes allowance for both the effects of varying the diffusion direction and the pore cross section. It is called tortuosity factor. According to a geometrical model which visualizes the pores as cylinders of fixed diameters which intersect any plane at an average angle of 45°, τ is equal to 2.

EXTERNAL DIFFUSION: ROLE OF HYDRODYNAMICAL FACTORS

The mass flow of a reactive component from the bulk fluid to the surface of particles is usually expressed in terms of a mass transfer coefficient k_c, defined as follows:

$$N_i = k_c (C_i - C_{si}) \quad \text{(moles/time unit surface)} \quad (8)$$

C_i is the concentration of component i in the ambient liquid, and C_{is} is the corresponding value at the external surface of the particles.

A dimensional analysis brings to the following kind of correlation between the mass transfer coefficient, the physical properties of liquid and its hydrodynamical conditions:

$$Sh = f (Re, Sc) \quad (9)$$

The following dimensionless numbers have been introduced:

$$Sh = \frac{k_c d_p}{D} \qquad \text{Sherwood number}$$

$$Sc = \frac{\mu}{\rho D} \qquad \text{Schmidt number}$$

$$Re = \frac{d_p v \rho}{\mu} \qquad \text{Reynolds number}$$

d_p is the diameter of the resin particle.

The correlation of mass transfer data is expressed by means of the following equation:

$$\frac{k_c \rho_\ell}{G} Sc^{0.67} = f (Re) \quad (10)$$

The preceding relationship has been obtained essentially on the basis of dimensionless considerations and empirical analysis of experimental data. It can be justified on the basis of the hydrodynamical boundary layer theory, which describes the behaviour of a fluid stream that flows through a particle bed [5] (fig. 4).

This approach brings to the following equation:

$$\frac{k_c \tilde{\rho}}{v} Sc^{2/3} = \alpha Re^{-1/2} = f(Re) \quad (11)$$

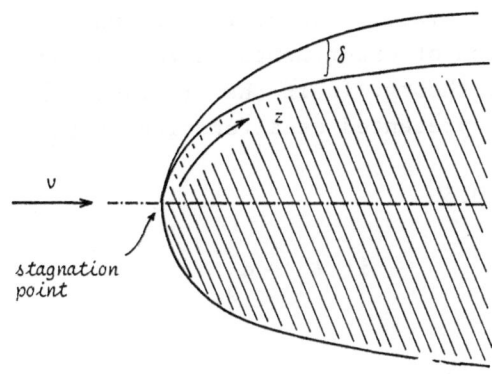

Fig. 4 - *Boundary layer formation in a two dimensional catalytic surface.*

α being a numerical factor which reflects the geometrical characteristics of the particle. The preceding equation, similar to (10) , justifies the dependence of the mass transfer coefficient on the physical properties of the fluid, that can be obtained on the basis of a simple dimensional analysis.

Finally it is important to stress that also if the preceding analysis has been made by neglecting the influence of surface chemical reaction on the external diffusion process, it yields very well approximated values of the mass transfer coefficients.

MASS TRANSFER IN AGITATED SYSTEMS

In a slurry reactor it is necessary to keep the entire solid mass suspended. Different relations have been proposed to calculate the minimum agitation to ensure complete suspension of the catalyst. To explain the mechanism of complete suspension of particles in a flat bottomed vessel, it has been assumed that the suspension is due to eddies of a certain critical scale. In fact the eddies with lower size than the critical value do not have the necessary energy, while the eddies of larger scale have a lower frequency and a lower probability of suspending the particles. This analysis yields the following correlation [6]

$$N_m \div \frac{\mu_\ell^{0.17} (\rho_s - \rho_\ell)^{0.42} d_p^{0.14} w'^{0.125}}{\rho_\ell^{0.58} d_I^{0.89}} \qquad (12)$$

Many investigations have been made on liquid-solid mass transfer processes in agitated reactors, including the presence of ion exchange resins. The correlation of experimental data has been attempted by using the Kolmogoroff theory. In it the liquid-solid mass transfer coefficient can be represented as a function of the Reynolds number, defined as:

$$Re = \left(\frac{e \, d_p^4 \, \rho_\ell^3}{\mu_\ell^3}\right)^{0.5} \qquad if \ \ell_c \gg d_p$$

$$Re = \left(\frac{e \, d_p^4 \, \rho_\ell^3}{\mu_\ell^3}\right)^{1/3} \qquad if \ \ell_c \ll d_p$$

where ℓ_c is the eddy size. It depends only on the energy dissipation rate per unit mass of the slurry e and on kinematic viscosity, as follows:

$$\ell_c = \left(\frac{\mu_\ell^3}{\rho_\ell^3 \, e}\right)^{0.25}$$

The energy supplied per unit mass of slurry depends, in its turn, on power consumption as:

$$e = P /\rho_\ell V_\ell \qquad (13)$$

The application of the preceding theory allows the following correlation to be made [7] :

$$\frac{k_s \, d_p}{D} = 2 + 0.44 \left(\frac{d_p \, \rho_\ell \, u_c}{\mu_\ell}\right)^{0.5} \left(\frac{\mu_\ell}{\rho_\ell D}\right)^{0.38} \qquad (14)$$

u_c is a characteristic velocity which can be predicted from the physical properties of the system and from its dynamical conditions.

Other correlations have been proposed and a comprehensive review of them has been recently given [8] .

ROLE OF INTRAPARTICLE DIFFUSION

Slow intraparticle diffusion may reduce the overall rate, particularly if the reactant molecules are large, and thus they have a small mobility in the resin. In this case the catalyst is not fully utilized since the active ions in the particle centers are not contacted by the reactants and then they remain essentially unused .

The mathematical analysis of the mentioned problem must account both the intraparticle diffusion and chemical reaction [9] . The analysis is relatively simple in the case of a spherical particle with an irreversible first order reaction, with rate equal: $R = kc$. For non uniform and irregularly shaped particles the obtained solution can be still applied as a reasonable approximation if the particle radius is taken as the radius of the equivalent sphere. This is a sphere with the same surface to volume ratio as the actual ion-exchange material.

a) Stationary solution

Let us consider, first of all, the more simple case in which the catalytic reaction is occurring in stationary conditions, as it happens in a continuous process . The concentration profile of the reactant inside the particle can be obtained by integrating the following differential equation, which reflects the stationary material balance of component i, inside the particle.

For a spherical particle, in dimensionless form it can be written as

$$\frac{1}{\rho^2} \frac{d}{d\rho} \left(\rho^2 \frac{du}{d\rho} \right) = \phi^2 u \quad , \quad \text{in } 0 < \rho < 1 \tag{15}$$

while the boundary conditions are:

$$\frac{du}{d\rho} = 0 \qquad , \quad \text{at } \rho = 0$$

$$\text{Bi } (1-u) = \frac{du}{d\rho} \qquad \text{at } \rho = 1 \tag{16}$$

being:

$$u = C/C_f$$

$$\rho = r/R_p$$

$$\phi = R_p \sqrt{k/D} \qquad \text{Thiele modulus}$$

$$Bi = k_c R_p/D \qquad \text{Biot number}$$

The solution is:

$$u(\rho) = \frac{\sinh(\phi\rho)}{\rho\{\sinh\phi + \dfrac{1}{Bi}(\phi\cosh\phi - \sinh\phi)\}} \qquad (17)$$

It is advisable to express the influence of internal diffusion in a catalytic reaction by means of the effectiveness of the catalyst, defined as:

$$\eta = \frac{4\pi\displaystyle\int_{o}^{R_p} r^2\, C\, d\, r}{V_p\, C_f} \qquad (18)$$

It comes out:

$$\eta = \frac{3}{\phi^2}\ \frac{\phi\coth\phi - 1}{1 + \dfrac{1}{Bi}(\phi\coth\phi - 1)} \qquad (19)$$

The previous general equation can be simplified .When $Bi \rightarrow \infty$, that means a negligible influence of the external diffusion:

$$\phi = \frac{3}{\phi}\ (\frac{1}{th\phi} - \frac{1}{\phi}) \qquad (20)$$

Besides if $\phi \rightarrow \infty$, that means a very strong limitation of the rate due to internal diffusion,the preceding equation becomes:

$$\eta \approx \frac{3}{\phi} \qquad (21)$$

It follows that the rate of the overall process can be expressed as:

$$R = \eta\, k\, C_f \qquad (22)$$

where η is given by equation (19).

The preceding treatment concerns first order reactions.Since many catalytic reactions are not in this category a more general analysis must be developed. Often, despite the details of the reaction mechanism, the kinetic equations can be expressed as power law in which the rate depends on the reactant concentration elevated to a reaction order n, that for catalytic reaction is less than one.

In this case a generalized modulus can be introduced:

$$\phi = \frac{R_p}{3} \frac{n+1}{2} \sqrt{\frac{k_v C_s^{n-1}}{D_e}} \qquad (23)$$

It, on the basis of figure 5, gives a good estimate of the effectiveness η.

One example of application of the given concepts is offered by the analysis of the influence of internal diffusion in the reaction of sucrose inversion, catalysed by ion exchange resins [10] :

$$\underset{\text{(sucrose)}}{C_{12}H_{22}O_{11}} + H_2O \overset{H^+}{\rightarrow} \underset{\text{(glucose)}}{C_6H_{12}O_6} + \underset{\text{(fructose)}}{C_6H_{12}O_6}$$

The reaction has been studied kinetically by using a Dowex resin as catalyst, with several different size particles. With the smallest particles, obtained by crushing the original pellets of Dowex, and having a diameter of 0.04 mm, the internal diffusion limitations were absent. Then it has been possible to evaluate directly the Thiele modulus from the observed rate . The data are consistent with a value of the internal diffusion coefficient equal to 2.69×10^{-2} cm^2/s. The preceding analysis stresses the importance of particle size on the overall rate of the process.

b) Non stationary solution

For a single first-order irreversible reaction occurring in a spherical particle, the dimensionless transient mass balance equations are given by:

$$\frac{\partial u}{\partial \tau} = \frac{1}{\rho^2} \frac{\partial}{\partial \rho} \left\{ \rho^2 \frac{\partial u}{\partial \rho} \right\} - \phi^2 u \qquad (24)$$

subject to the initial conditions:

$$u = u^\circ \quad , \quad \rho \in (0,1) \quad , \quad \tau = 0$$

and boundary conditions:

$$\frac{\partial u}{\partial \rho} = 0 \qquad \text{at} \qquad \rho = 0$$

$$\tau > 0 \qquad (25)$$

$$Bi\,(1 - u) = \frac{\partial u}{\partial \rho} \qquad \text{at} \qquad \rho = 1$$

where:

$$u^\circ = C^\circ/C_f \quad \text{being } C^\circ \text{ the initial concentration}$$

$$\tau = t\,D/R_p^2$$

The solution of the problem can be obtained through the finite Fourier transform [11] . After a suitable variable change, in order to have homogeneous boundary conditions, the following expression for the dimensionless concentration can be obtained:

$$u(\rho,\tau) = \frac{Sh\ (\phi\ \rho)}{\rho\left[\ (1 - \frac{1}{Bi}\ )\ Sh\ \phi + (\phi\ Bi)\ Ch\ \phi\right]} + \sum_1^{\infty}\ \psi_n(\rho)$$

$$\left[\gamma_n^o - \frac{\alpha_n}{\lambda_n^2 + \phi^2}\right]\ e^{-(\lambda_n^2+\phi^2)\tau} \tag{26}$$

γ_n^o is the finite Fourier transform of the initial condition of the new variable $\gamma = (u - 1)$:

$$\gamma_n^o = \int_0^1 (u^o - 1)\ \psi_n\ (\rho)\ \rho^2\ d\ \rho \tag{27}$$

and the coefficient α_n is defined as

$$\alpha_n = -\ \phi^2 \int_0^1 \rho^2\ \psi_n\ (\rho)\ d\ \rho \tag{28}$$

λ_n are eigenvalues obtained through the solution of the trascendental equation:

$$\lambda\cos\lambda\ +\ (Bi - 1)\ \sin\lambda\ =\ 0 \tag{29}$$

while the corresponding normalized eigenfunctions are given by:

$$\psi_n(\rho) = \frac{\sin\ (\lambda_n\rho)}{\rho\{\int_0^1 \sin^2\ (\lambda_n\rho)\ d\rho\}^{1/2}} \tag{30}$$

The description of the transient behaviour of a spherical particle, in which diffusion with simultaneous chemical reaction takes place can be expressed in terms of volume–average concentration

$$\bar{u}\ (\tau) = 3 \int_0^1 u\ (\rho,\tau)\ \rho^2\ d\ \rho \tag{31}$$

In the case of $Bi \to \infty$, that is with negligible external diffusion, a simplified version of equation (26) can be derived. By applying to it the volume–average operator (31) , the following expression of average reactant concentration is obtained:

$$\bar{u}(\tau) = 1 + (3/\phi) \quad (Cth\phi - 1/\phi) -$$

$$-6 \sum_{n}^{\infty} \left\{ \exp[- \{(n\pi)^2 + \phi^2\}\tau] \right\} / \{(n\pi)^2 + \phi^2\} \tag{32}$$

KINETICS OF CHEMICAL REACTION

Let us consider now the description of the kinetic behaviour of a reaction catalyzed by ion exchange resins, not limited by diffusional phenomena. In these cases the simple expression of rate law through linear relationships between reactant concentrations in external solution and those in reactive sites in the resins generally fails to fit the experimental data over the entire range of conversions.

A more suitable approach involves the application of Langmuir-Hinshelwood kinetic scheme, which implies the adsorption of the reactants on active sites on the surface of the resin. Basically such an approach assumes that a specific type of sorption occurs. In other words the association of one molecule of reactant with the counter ions of the exchanger, which are essentially closed at sites fixed to the resin skeleton, brings to local concentrations of the adsorbed species, different from the ones present in the pores. Besides, the adsorption model gives rise to a competition in the interaction of different molecules with the reactive sites.

As a first example the case of unimolecular reaction of the adsorbed molecule will be considered. The reaction sequence can be described as follows:

$$A + \sigma \overset{\leftarrow}{\rightarrow} A\sigma \overset{k}{\rightarrow} \text{products} \tag{33}$$

being A the reactant in liquid phase and σ a free active sites. If the Langmuir isotherm is applied to the system the following expression is obtained for the reaction rate:

$$R = \frac{k \, b_A \, C_A}{1 + \sum_i b_i \, C_i} \tag{34}$$

where:

C_i is the concentration of component i in liquid phase

b_i is the equilibrium adsorption constant, or association affinity of component i.

The preceding equation has been applied, for instance, to the description of the kinetics of dehydration of ter-buthyl alcohol, catalyzed by Dowex 50 W (X-2) resin, manufactured by sulfonation of copolymers of styrene with 8% of divinylbenzene [12] . The reaction leads to the formation of olefin and water:

$$CH_3-\underset{\underset{CH_3}{|}}{\overset{\overset{CH_3}{|}}{C}}-OH \quad \to \quad CH_3-C\overset{\diagup CH_2}{\underset{\diagdown CH_3}{}} \quad + \ H_2O$$

In agreement with the previous analysis the expression of the reaction rate is:

$$R = \frac{k\ b_A\ C_A}{1 + b_A C_A + b_{H_2O}\ C_{H_2O}} \tag{35}$$

The preceding equation can be written in linearized form as:

$$\frac{R}{C_A} = \frac{k\ b_A}{1 + b_A\ C^o} - \left(\frac{b_{H_2O} - b_A}{1 + b_A\ C^o}\right) \frac{C_{H_2O}}{C_A}\ R \tag{36}$$

being $C^o \simeq C_A + C_{H_2O}$.

A plot of experimental data, consistent with the preceding equation, is reported in fig. 6

A second example of mechanism of ion-exchange catalyzed reaction, is the one in which a molecule A adsorbed on an active site of the resin, and then associated with a counter ion, reacts with another molecule B, which is present in the neighbouring pore liquid. In this case the reaction sequence is the following:

$$A + \sigma \underset{\leftarrow}{\overset{\rightarrow}{\rightleftharpoons}} A\ \sigma$$

$$A\ \sigma + B \overset{k}{\to}\ products \tag{37}$$

Still applying the Langmuir-Hinshelwood kinetic model, the following expression of the reaction rate is obtained:

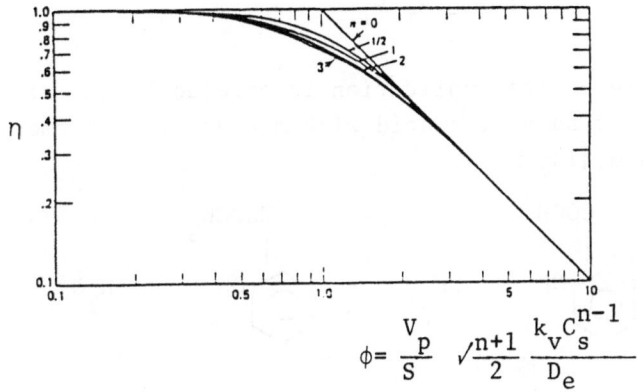

$$\phi = \frac{V_p}{S} \sqrt{\frac{n+1}{2}} \frac{k_v C_s^{n-1}}{D_e}$$

Figure 5-Generalized plot of effectiveness factor for simple order reactions.

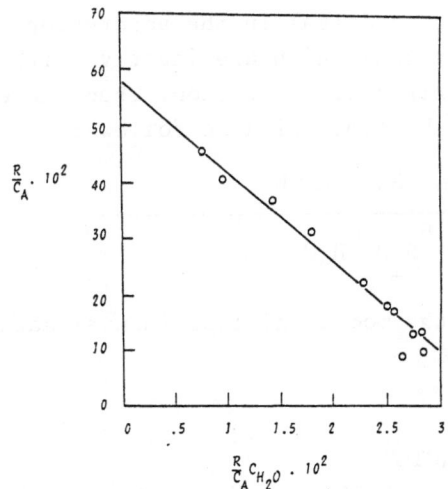

Figure 6-Linearized plot of reaction rate data of terbutyl alcohol dehydration, catalysed by Dowex 50 W(X-2)

$$R = \frac{k \, b_A \, C_A \, C_B}{1 + \sum_i b_i \, C_i} \tag{38}$$

One example of its application is offered by the reaction of esterification of salycilic acid with methanol, performed on Dowex 50-W (X-8) resin [13] :

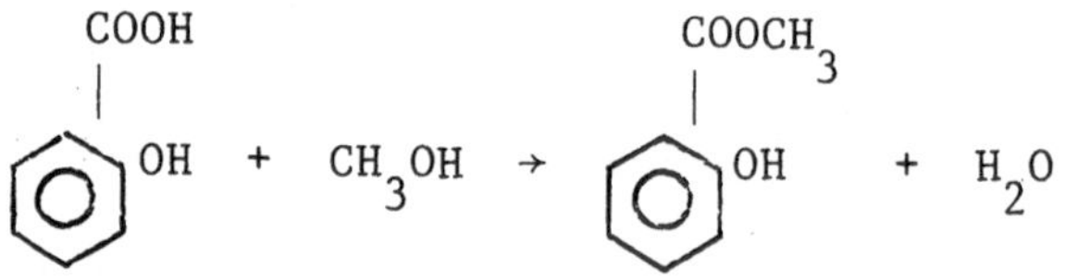

A reasonable mechanistic hypothesis is that the rate limiting step is the reaction between protonated salycilic acid and methanol in solution. Water affects the reaction rate, since it will preferentially associate with the proton, while other species are less strongly adsorbed. It is reasonable to assume that only water competes with salycilic acid (SA) in the adsorption on active sites by forming hydrated protons which are inactive with respect to the catalysis of the esterification reaction. Then the expression of the reaction rate can be simplified as follows:

$$R = \frac{k \, b_A \, C_{SA} \, C_{CH_3OH}}{1 + b_{H_2O} \, C_{H_2O}} \tag{39}$$

A comparison between the model and experimental data is shown in fig. 7.

MODEL FOR A SLURRY REACTOR

A model for the simulation of the behaviour of a continuous slurry reactor will be now derived. In order to give more generality to the treatment the possibility that one of the reactant should be present in gaseous phase will be considered. In fact different reactions catalyzed by ion exchange resins are performed in systems in which three phases can be present, that are: a solid (the catalyst), a liquid and a gas (a reactant). One example is given by the hydration of propylene oxide (PO) to propylene glycol (PG):

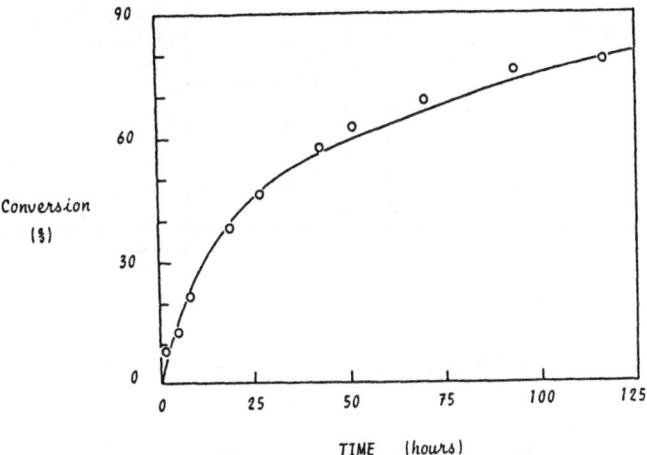

Fig. 7 - Comparison between experimental data and calculated curve of conversion of esterification of salycilic acid with methanol catalyzed by DOWEX 50W (8-X)

$$PO_{(gas)} + H_2O \xrightarrow{\overset{+}{H}} PG_{(liquid)}$$

As a catalyst a resin of acid cross-linked polystyrene beads with SO_3^- as the active group was employed.

The preceding reaction can be represented by the following scheme:

$$A_{(gas)} + B_{(liquid)} \xrightarrow{cat} Products$$

The number of steps necessary for the reaction to occur give [14] :

1. Transport of A from gas phase to liquid bulk
2. Transport of A and B from bulk liquid to particle surface
3. Intraparticle diffusion of A and B within the catalyst with simultaneous chemical reaction.

The behaviour of the concentration profile of component A is illustrated in fig. 8.

The rate of mass transfer of species A from gas phase to the bulk liquid can be written as:

504

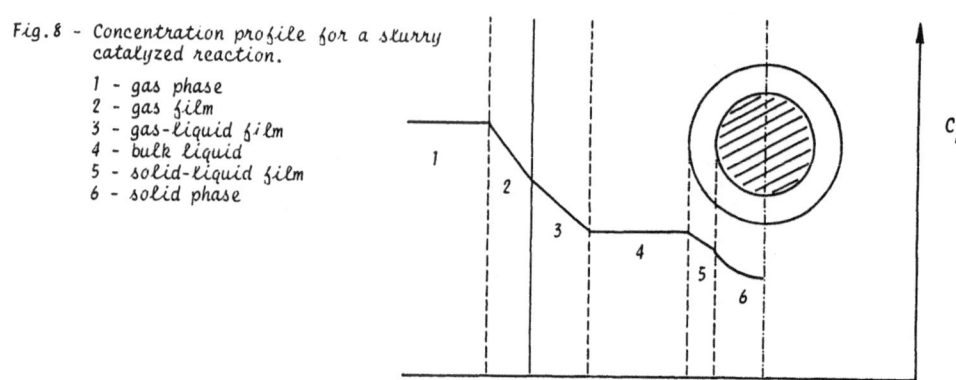

Fig.8 - Concentration profile for a slurry
         catalyzed reaction.
  1 - gas phase
  2 - gas film
  3 - gas-liquid film
  4 - bulk liquid
  5 - solid-liquid film
  6 - solid phase

$$N_A = K_{\ell A} a \left(\frac{C_{gA}}{H_A} - C_{\ell A} \right) \tag{40}$$

where $K_{\ell A}$ is the overall mass transfer coefficient, while H_A is the Henry's solubility coefficient of the gas in the liquid. $K_{\ell A}$ can be expressed in terms of individual gas side and liquid side mass transfer coefficients, as:

$$\frac{1}{K_{\ell A}} = \underbrace{\frac{1}{H_A k_{gA}}}_{} + \underbrace{\frac{1}{k_{\ell A}}}_{} \tag{41}$$

gas side resistence   liquid side resistence

The gas phase in the reactor is assumed to be plug flow, while the liquid phase is assumed completely mixed. At steady state conditions the mass balance of component A yields the following expression:

$$- u_g \frac{dC_{gA}}{dz} = K_{\ell A} a \left(\frac{C_{gA}}{H_A} - C_{\ell A} \right) \tag{42}$$

being u_g the gas velocity and z the axial coordinate of the reactor. The integration of the preceding equation yields the following expression of the concentration of A leaving the reactor:

$$(C_{gA})_{out} = (C_{gA})_{in} e^{-\alpha_A L} + H_A C_{\ell A} (1 - e^{-\alpha_A L}) \tag{43}$$

where:

$$\alpha_A = \frac{K_{\ell A} a}{u_g H_A} \tag{44}$$

and L is the height of the reactor.

The rate of gas adsorbed per unit volume of slurry is:

$$R_A = \frac{H_A Q}{V_L} (1 - e^{-\alpha_A L}) \left[\frac{(C_{gA})_{in}}{H_A} - C_{\ell A} \right] =$$

$$= M_A \left[\frac{(C_{gA})_{in}}{H_A} - C_{\ell A} \right] \tag{45}$$

R_A is also equal to the rate of transport from the liquid bulk to the catalyst surface additioned of the amount of A reacting in the bulk liquid by non catalytic heterogeneous reaction. Then:

$$R_A = k_1 C_{\ell A}^{n_1} + k_s a_p (C_{\ell A} - C_{sA}) \tag{46}$$

Besides the rate of the reaction at the catalyst is given by:

$$k_s a_p (C_{\ell A} - C_{sA}) = w \eta k_2 C_{sA}^{n_2} \tag{47}$$

being η the catalyst effectiveness; the corresponding Thiele modulus is evaluated by equation (23).

By eliminating the unknowns $C_{\ell A}$ and C_{sA} in equations (45) and (46) the following expression of the overall reaction rate is derived:

$$R_A = k_1 \left(\frac{(C_{gA})_{in}}{H_A} - \frac{R_A}{M_A} \right)^{n_1} + \eta w k_2 \left[\frac{(C_{gA})_{in}}{H_A} - R_A \left(\frac{1}{M_A} + \frac{1}{k_s a_p} \right) + \right.$$

$$\tag{48}$$

$$\left. + \frac{k_\ell}{k_s a_p} \left(\frac{(C_{gA})_{in}}{H_A} - \frac{R_A}{M_A} \right)^{n_1} \right]^{n_2}$$

This equation is implicit in R_A and then a numerical procedure must be used to determine such a quantity.

If all the mass transfer resistance are negligible equation (48) can be simplified as follows:

$$R_A = k_\ell \left[\frac{(C_{gA})_{in}}{H_A} - \frac{R_A V}{H_A Q} \right]^{n_1} + w \, k_2 \left[\frac{(C_{gA})_{in}}{H_A} - \frac{R_A V}{H_A Q} \right]^{n_2} \tag{49}$$

From a material balance it comes out:

$$(C_{gA})_{out} = \left((C_{gA})_{in} - \frac{R_A V}{Q} \right) \tag{50}$$

and then it follows that:

$$R_A = k_1 \left[\frac{(C_{gA})_{out}}{H_A} \right]^{n_1} + w \, k_2 \left[\frac{(C_{gA})_{out}}{H_A} \right]^{n_2} \tag{51}$$

One example of application of the described model is the one corresponding to the already mentioned hydration of propylene oxide using a strong acidic cross-linked polystyrene bead resin with SO_3^-, as active group. The employment of an ion exchange catalyst has some advantages over the soluble acidic catalyst from the industrial point of view. Particularly the corrosion of the equipment is minimized and the catalyst can be used repeatedly [15] .

On the basis of a kinetic study of the homogeneous reaction in water, the following mechanism has been proposed for the acid catalyzed hydration of propylene epoxide:

$$CH_3 - \overset{O}{\overset{\diagup\diagdown}{CH - CH_2}} + H^+ \rightarrow CH_3 - \overset{O}{\overset{\diagup\diagdown}{CH - CH_2}} H^+ \rightarrow CH_3 - \overset{\overset{OH}{|}}{CH} - \overset{\overset{OH}{|}}{CH_2}$$

which is consistent with a L-H kinetic equation of the form:

$$R_A = \frac{k\, C_{H^+}\, C_A}{1 + b_A\, C_A} \tag{52}$$

Actually the preceding equation can be approximated as a power law model with a fractional order with respect to A. Experimental analysis revealed that the rate of hydrolysis of propylene can be expressed as

$$R_{po} = k_1 \left(\frac{C_{g.po}}{H_{po}}\right)^{0.433} \tag{53}$$

The heterogeneous reaction has been studied in a slurry reactor and the kinetic data have been interpreted with equation (53). It was found that the intraparticle diffusional resistance can be important. The kinetic data allowed the determination of the values of the effective diffusion of the reactant. From these values a tortuosity factor of three has been determined.

508

NOTATION

a	effective gas-liquid interfacial area per unit volume of slurry
a_p	external area of particles per unit volume of slurry
b_i	equilibrium adsorption constant of component i
C, C_i	molar concentration, molar concentration of component i
C_f	concentration in the bulk fluid phase
C_s	concentration at the solid surface
$C_{gi}, C_{\ell i}$	gas-phase and liquid-phase concentration of component i
$\overline{C}_i$	concentration in the interior of the exchanger
d_I	diameter of the impeller (cm)
d_p	average diameter of the particle
D, D_i	diffusion coefficient of component i
D_{im}	diffusion coefficient of component i in a mixture
e	energy supplied by agitator or gas bubbling per unit mass of slurry
F	Faraday constant
G	mass flow of liquid total cross section normal to the flow
H_i	Henry's law constant of component i, defined as $C_{gi}/C_{\ell i}$
J_i	diffusional mass flow of component i (moles/time surface)
k	reaction rate constant
k_c	mass transfer coefficient (length/time)

$k_{\ell,g,s}$	individual mass transfer coefficient, ℓ = liquid-side, g = gas-side, s = gas-solid
K_i	overall mass transfer coefficient of component i
L	length of a reactor
M_j	molecular weight of component i
n, n_1, n_2	reaction orders
N_i	mass flow of component i (moles/time surface)
N_m	minimum speed agitation for suspension of particles (s^{-1})
P	power consumption for agitation ($g/cm^2/s^3$)
Q	volumetric flow rate of gas (volume/time)
r	radial coordinate
R	reaction rate (moles/time volume)
$\underset{\sim}{R}$	gas constant
R_p	radius of a particle
t	time
T	absolute temperature
$u = C/C_f$	dimensionless concentration
$\boldsymbol{u}_i$	diffusion velocity of component i
$\boldsymbol{u}^*$	average molar velocity
u_g	velocity of a gas
$\tilde{V}_i$	molar volume of component i
V_p	particle volume
V_ℓ	volume of the liquid in the reactor

w	mass of the catalyst per unit volume of the slurry
w'	percentage catalytic loading (g/100 g of solution)
z_i	valence of i-th ion
x_i	mole fraction of component i

Greek Letters

ε_p	internal porosity of a particle
η	effectiveness factor (eq.(18))
μ, μ_ℓ	viscosity coefficient of the liquid (g/cm/s)
$\rho = r/R_p$	dimensionless radial coordinate
ρ_ℓ, ρ_s	density of a liquid, solid
$\tilde{\rho}$	total molar concentration
$\tau = t\, D/R_p^2$	dimensionless time
τ	tortuosity (eq.(7))
ψ	electric field

REFERENCES
1. K. Dorfner, "Ion Exchangers", Ann Arbor Science Publishers, (1972).
2. F. Helfferich, "Ion Exchange" McGraw-Hill Book Company, New York (1962).
3. D.A. Frank-Kamenetskii, "Diffusion and Heat Transfer in Chemical Kinetics" Plenum Press, New York (1969).
4. C.N.Satterfield, "Mass Transfer in Heterogeneous Catalysis", M.I.T. Press, Cambridge, Massachusetts (1970).
5. S. Carrà, "Fundamental Principles in Heterogeneous Catalysis" Nato Advantaced Study Institute, Venice (1971).
6. G.Baldi, R.Conti, E. Alaria, Ind. Eng. Chem. Proc. Des. Dev., 13, 447 (1974).

7. D.M. Levins, J.R. Glastonbury, Chem. Eng. Sci., $\underline{27}$, 537 (1972).

8. P.A. Ramachandran, R.V. Chaudhari, "Three phase Reactors" Gordon and Breach, New York (1983).

9. R. Aris, "The Mathematical Theory of Diffusion and Reaction in Permeable Catalysts", Clarendon Press-Oxford (1975).

10. E.R. Gilliland, E.R. Bixler, J.E. O'Connel, Ind. Eng. Chem. Fund., $\underline{10}$, 185 (1971).

11. M. Morbidelli, A. Servida, G. Storti, S. Carrà, Chem. Eng. Sci., $\underline{37}$, 1645 (1982).

12. V.J. Frilette, E.B.Mower, M.K. Rubin, J. of Catalysis, $\underline{3}$, 25 (1964).

13. M.B. Bochner, S.M. Gerber, W.R. Vieth, A.J.Roger , Ind. Eng. Chem. Fund., $\underline{4}$, 314 (1965).

14. R.V. Chaudhari, P.A.Ramachandran, AIChE Journal, $\underline{26}$, 177 (1980).

15. R. Jaganathan, R.V. Chaudhari, P.A. Ramachandran, AIChE Journal, $\underline{30}$, 1 (1984).

ZEOLITES:SOME CATALYTIC APPLICATIONS

F.Ramoa Ribeiro,F.Lemos,C.Henriques and M.F.Ribeiro

Grupo de Estudos de Catálise Heterogénea
Instituto Superior Técnico
1096 Lisboa Codex,Portugal

1. INTRODUCTION

Zeolites have an excellent ion exchange capacity (1), and so they can be used as ion exchangers for the recovery of radioactive isotopes from contaminated waters, for ammonium removal from munici pal waste water and more recently as phosphate substitutes in deter gent builders (2), as fertilizers and as dietary supplements for animals (3). However, the most important application of zeolites is still as catalysts. Their activity and selectivity can be tailored to produce desired reactions, due to remarkable properties, namely the shape selectivity, ion exchange and adsorption capacities and thermal stability.

With increased knowledge of zeolite chemistry and synthesis we are able to design catalysts with specific shape selectivity,having channels of different diameters to control the size of entered molecules.

The different cations introduced by ion exchange will be the ca talytic sites of the reactions: acid protonic sites catalyse isome-rization and cracking reactions, metallic sites catalyse hydrogena-ting reactions, transition metal sites catalyse oxidation and oligo merization reactions, etc..

The thermal stability of the zeolites depends on the type of the structure, the thermal treatments and the nature of the exchanged cations.

Currently, more than forty natural zeolites and a hundred zeo-lites obtained by synthesis are known, although only a few zeolites have attained extensive industrial applications: zeolite Y in cra-cking of hydrocarbons, mordenite in hydroisomerization of n-paraf-

fins and disproportionation of aromatics and the new pentasil family (4) of ZSM-5 and ZSM-11 in the methanol conversion and xylenes
isomerization.

2. ZEOLITE STRUCTURES

Zeolites are crystalline aluminosicates of (most commonly) sodium cations, presenting a very porous structure with pore opening
diameters from 3 to about 10 Angstroms. The basic structural units
are $[AlO_4^-]$ and $[SiO_4]$ tetrahedra linked to each other by sharing
all the oxygens. The microporous structure depends on the different
arrangements of the tetrahedra. Te negative charges beared by $[AlO_4^-]$
tetrahedra are equilibrated by sodium cations which can be exchanged by different cations.

The chemical formula and pore opening diameters of the best known
zeolites (6) used in catalysis are presented in table 1.

TABLE 1

Chemical formula and pore opening diameters
of different zeolites

Zeolite	Chemical formula	Pore opening dia- meters (A°)
Y	Na_{56} Al_{56} Si_{136} O_{384}, 240 H_2O	7.4
Mordenite	Na_8 Al_8 Si_{40} O_{96} , 24 H_2O	6.7 x 7.0 $\leftrightarrow$ 2.9 x 5.7
ZSM-5	Na_n Al_n Si_{96-n} O_{129}, 16 H_2O $n<27$	5.4 x 5.6 $\leftrightarrow$ 5.1 x 5.5
ZSM-11	Na_n Al_n Si_{96-n} O_{129}, 16 H_2O $n<16$	5.1 x 5.5

Figure 1 (a) shows the framework structure of faujasite (X and Y
zeolites). It is a very open structure and each sodalite unit is
linked to four sodalite units by hexagonal prisms.

The black circles represent sodium cations which can be exchanged by other cations.

Figure 1 (b) shows a view of the chanel structure of ZSM-5 sin-
thesized by Mobil, which has a high content in silica. Two types of
channels can be distinctly seen, originating a three-dimensional
structure.

The size of zeolite pores and cages is similar to that of   many
organic molecules and consequently the occurrence of various chemical reactions in the zeolites becomes dependent on the geometry of
their porous structure. There are several petrochemical processes

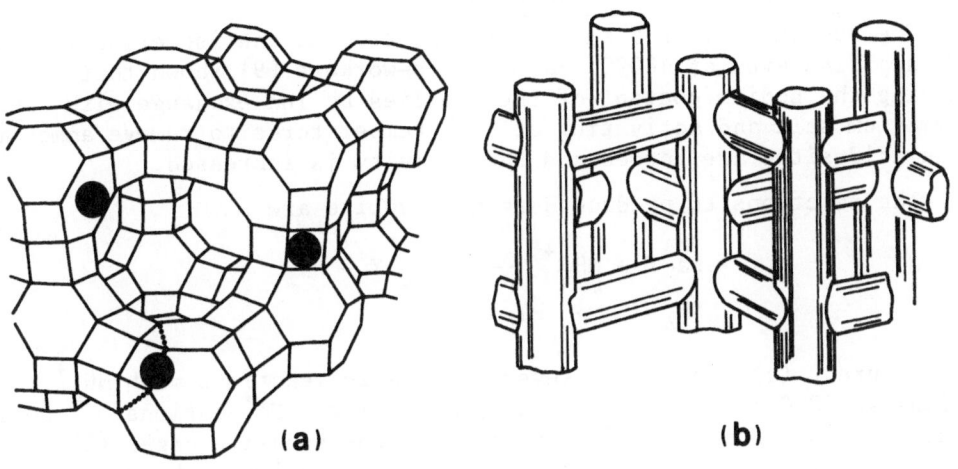

Figure 1 - Structure of: (a) Faujasite (X,Y)
(b) ZSM-5

based on three types of molecular selectivity effects: reactant, product and intermediate molecules (7).

3. CATION EXCHANGE PROPERTIES

The zeolites have cation exchange properties, namely capacity, selectivity and stability, not available in other ion exchangers.

The cation exchange capacity of zeolites is a function of their SiO_2/Al_2O_3 molar ratio and cation form. The selectivities are higher for cations, which easily enter in the zeolite pores. Other variables, such as pH, temperature, type of anions, competing cations and complexing agents can affect the cation exchange performances. However, due to the high stability of zeolite structure, the influence of these last parameters is less complex than with some organic resin in which complex sorption can appear, with changes in their ion exchange performances (8).

Zeolites for use as catalysts are synthesized usually in the sodium form. The sodium cations can be replaced by ion exchange and this process can be repeated to achieve higher degrees of cation replacement. The possibility of exchanging sodium cations by different cations enables us to change considerably the catalytic properties of zeolites and to adjust them to several applications in industrial catalytic processes.

Before mentioning these processes, we will present several examples showing the influence of the ion exchange extent, competing ions, size and nature of exchanged cations.

3.1 Influence of the Ion Exchange Extent

The sodium form of Y zeolite is inactive in the isomerization of
n-paraffins even at 480ºC. Rabo and co-workers (9) shown that reducing the sodium content of the zeolites by ion exchange with an
ammonium salt and activating at high temperatures to remove ammonia
the acid sites are formed and the activity is increased.

The reactions to produce H-form Y zeolite are

$$Na\ Y + NH_4^+ \longrightarrow NH_4\ Y + Na^+$$
$$NH_4\ Y \xrightarrow{\Delta} HY + NH_3$$

Figure 2 (a) shows that several exchange reactions with NH_4^+ cations at 20ºC are not able to displace all the Na^+ cations of the Y
zeolite (10). Some of them localized in the sodalite cages (5) are
not accessible to the exchangeable hydrated cations, whose diameters are larger than the windows (~ 2.2 Aº) of sodalite cages. However at 100ºC, the NH_4^+ - H_2O bond is weakened and consequently the
diffusion of NH_4^+ cation is easier.

Figure 2 (b) shows the influence of the sodium exchange extent
on the acid catalytic activity of Na HY. Since Y zeolite has   a
great variety of sites of different strenght and accessibility, the
activity increases sharply after a threshold relative to exchange
(11). For ZSM-5, which has only one type of sites, activity increases linearly with exchange (12).

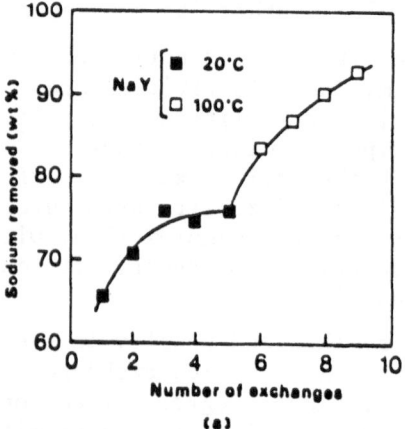

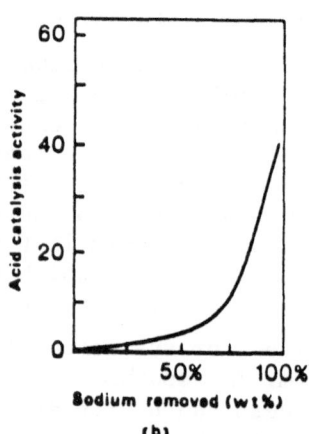

(a)   (b)

Figure 2 (a) Extent of decationization on NaY zeolite as a
function of the number of exchanges at 20 and 100ºC

(b) Acid catalytic activity as a function of sodium
removed in Na HY zeolite

3.2 Influence of Competing Ions

The most common method to prepare zeolites containing metals is the ion exchange technique, followed by several thermal treatments, including reducing under hydrogen.

For the mordenite, if we use a classical ion exchange to introduce a small amount of $[Pt(NH_3)_4^{2+}]$ ions, we will obtain a heterogeneous macroscopic distribution of the metal over the surface. The platinum ions fix on a thin layer over the periphery, due to the exchange rate being higher than the diffusion rate and to an important affinity of the platinum ions for the mordenite possessing a great number of exchangeable sites.

The ion exchange reaction for the NH_4^+ form zeolite is

$$[Pt(NH_3)_4^{2+}]_s + [NH_4^+]_z \rightleftarrows [Pt(NH_3)_4^{2+}]_z + [NH_4^+]_s$$

s - solution
z - zeolite

In order to obtain a homogeneous macroscopic distribution of the metal into the zeolite, we must use the technique of ion exchange with competition (13,14). An excess of competing NH_4^+ ions is introduced, displacing the above equilibrium to the left, which increases the concentration of platinum ions in the solution. Consequently, there is an increase in the rate of diffusion and a migration of the platinum towards the inside region of the zeolite.

Figure 3 shows the competition curve for the NH_4 Mordenite, which represents the fraction of platinum in the solution against the competition.

The competition is defined as the number of the competing ion equivalents added to the solution phase plus the number of competing ion equivalents initially present in support over the number of metal equivalents in solution.

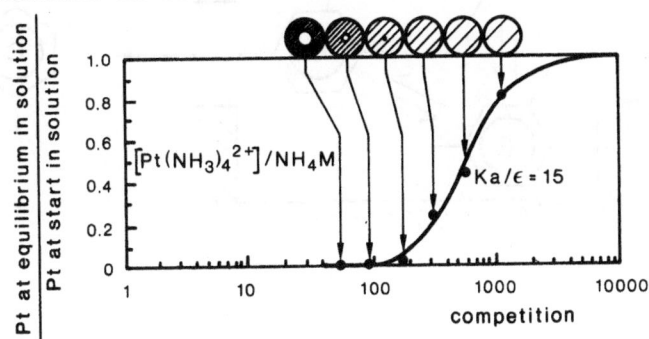

Figure 3 - Competition curve for the NH_4 mordenite
(Experimental conditions: $T = 20°C$, $pH = 7$,
$\frac{V}{P} = \frac{40}{3}$ cm^3 solution g^{-1} $NH_4M; C[Pt(NH_3)_4]^{2+} = 356$ mg l^{-1})

The optimum value of competition from an industrial point of view is ~ 250. In fact we obtain an homogeneous macroscopic distribution of the platinum over the support with simultaneously a very low fraction of metal in solution.

The bifunctional catalysts prepared by competitive ion exchange, present a greater stability and activity compared with identical catalysts, in which the metal was introduced by a classical ion exchange.

3.3 Influence of the Size of Exchanged Cations

The selectivity of aromatic reactions (xylenes, toluene, trimethyl-benzene) on zeolites is very affected by their porous structure.

Figure 4 shows an example of molecular shape selectivity for the products, in which the zeolite structure plays an important role in the selective production of p-xylene (15).

Weisz (16) has considered that in zeolites there exists a configurational diffusion. Due to the microporous structure of zeolites, there is an interaction between the dimensions of reactant molecules and the cages and pores of zeolites.

The introduction by ion exchange of cations with different sizes (Na$^+$ - 0.97 A^o; La^{+++} - 1.016 A^o; Ba^{++} - 1.34 A^o; Cs$^+$ - 1.67 A^o) decreases the pore sizes and the available space in the structural environment of active sites. This will affected in different ways the selectivity of reactions such as isomerization and disproportionation of aromatic compounds on ZSM-5 and mordenite.

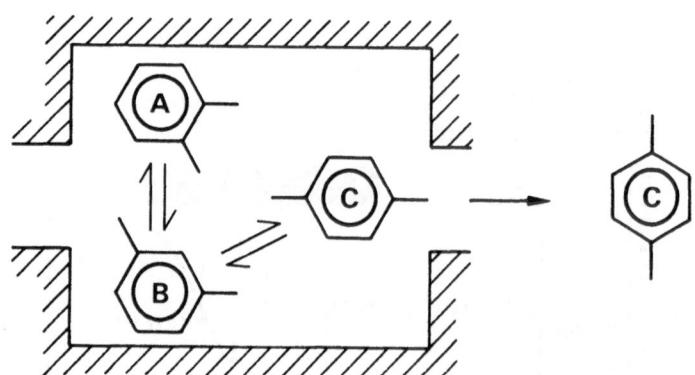

Figure 4 – Product molecules shape selectivity effects in zeolite catalysis (A:o-xylene; B: m-xylene; C:p-xylene)

3.4 Influence of the Nature of the Cation

Zeolites after ion exchange with a proton have acidic properties, but they can present other properties which depend on the nature of the cation.

It is well known that cations can sit in different places of the zeolite framework, thus providing the cation with a choice of coordination simetries. When the cation introduced is a proton,the cation sitting is important only in accessibility but coordination is disregarded. If a transition metal is introduced by ion exchange, thus the zeolite framework can function as a chelating "solid ligand" of high thermal stability or as a "solid solvent" (17).

Transition metal complexes are widely spread as homogeneous or heterogenized catalysts for a great variety of reactions and zeolites which incorporate these metals can become competitive replacements. This field of applications is just now begining, and a review of laboratory works and patents has been recently published (18).

Transition metals exchanged zeolites were found to be active in oxidation, oligomerization, carbonylation and methanation reactions.

4. APPLICATIONS OF EXCHANGED ZEOLITES IN CATALYSIS

Zeolites can be used as catalysts or active catalyst carriers, in which other cations are introduced by ion exchange. They promote several hydrocarbon reactions proceeding through carbonium ion intermediates, such as cracking, hydrocracking, isomerization, disproportionation and alkylation reactions.

The industrial applications of zeolite catalysts are increasing more and more, namely in petroleum refining and petrochemical industry.

4.1 Petroleum Refining

The main application of zeolites as catalysts (more than 95%) is in the catalytic cracking (19). The 1973 oil crisis originated great changes in the refining policy,stimulating the development of processes able to produce more light cuts, starting with heavy crudes (20). Hydrocracking,which use noble metal zeolite catalysts, is one of these processes. Increasing demands for nonleaded gasolines have also promoted hydroisomerization processes.

Catalytic cracking

Catalytic cracking is today one of the most important processes in the refining industry. Originally it made use of silica-aluminas that are slightly acidic, but it received a tremendous boost with the introduction of cation-exchanged zeolites, a fact that led even to a change in the type of reactors (19).

The first kind of zeolites used were HY prepared from the parent NaY by ion exchange with an ammonium salt that was subsequently calcined giving the hydrogen form and releasing ammoniac. These catalysts however showed only a moderate thermal stability. Nowadays, catalytic cracking uses either ultra-stable Y zeolite in the hydrogen form or rare-earth exchanged Y zeolite dispersed in a silica-alumina matrix (21,22).

The ultra-stable HY is prepared by ion-exchange as the original one, but the calcination is performed under wet air. This last step is used to dealuminate the zeolite, conferring enhanced thermal stability and can be replaced by dealumination by acid or chelating reagents (23).

The rare-earth exchange not only assures the presence of acid sites that are produced by hydrolisis of the heavy cation itself, but also, due to the presence of large multi-valent cations, seems to produce an effect similar to the ultra-stabilization (24,25).

These catalysts exhibit a very rapid deactivation due to coke deposition, their residence time being in the order of seconds (26). A large effort is being made to understand the mechanism of such deactivation and to study the effect of the exchange with several cations to prevent or reduce this problem.

Hydrocracking

Hydrocracking is one of the most versatile catalytic refining process. It can be used to obtain light and valuable products, like propane, butanes, light gasoline, diesel oils, jet fuels and middle distillates, from light naphtas, atmospheric and vacuum residue, catalytic cracking recycle oils, coking and visbreaking oils and even deasphalted oils (27).

Hydrocracking requires bifunctional catalysts, exhibiting an acidic function and a hydrogenating - dehydrogenating function, which have to be suitably "balanced", according to the feeds and final desired products (28).

The acidic function is provided by a HY zeolite, in which the sodium ions are replaced by ammonium ions, by successive ion exchanges at higher temperatures.

The hydrogenating - dehydrogenating function is provided by a noble metal, usually palladium, which can be introduced in the zeolite by competitive ion exchange. Zeolite catalysts can operate in the presence of considerable amounts of ammonia and other basic compounds; moreover, their great acidity enhances the resistance of the metallic function to poisoning by sulphur compounds (28).

More recently, new bimetallic zeolite catalysts have been studied (29, 39), in order to increase their stability to the coke deactivation, which will permit the decrease of the reaction temperature and hydrogen pressure (31). The second metal can be introduced into the zeolite also by ion exchange or thermal decomposition of organometallic complexes (32).

Hydroisomerization of C_5/C_6 paraffins

This process converts light gasoline fractions of low octane C_5/C_6 paraffins into higher octane products containing isopentane, methylpentanes and dimethylbutanes.

These isomerization reactions are slightly exothermic, favoured by low temperatures in order to obtain a maximum yield of branched paraffins. The best known process is the Shell Hysomer Process (33), which uses a platinum exchanged acid mordenite of low sodium content, very active at low temperatures. Side reactions, such as hydrocracking and coke formation, are minimized and selectivities for the products of isomerization are increased (34).

4.2 Petrochemical Industry

Recent developments have pushed zeolite catalysts into petrochemical industrial processes. Other applications in this field have been studied but they are not yet commercialized.

Isomerization of aromatic C_8 hydrocarbons

The main industrial interest is in the recovery of para and orto-xylene from C_8 aromatic cut, and various processes with zeolite containing catalysts have been developed to enable the maximization of their yield.

The processes in vapour phase, such as Octafining II (ARCO--Engelhard), Isolene II (TORAY) and Aris (Veb-Leuna-Werke) use a platinum exchanged acid mordenite, and the Mobil process uses a nickel exchanged H-ZSM-5. These processes using bifunctional catalysts convert partially the ethylbenzene of the feedstock.

In the Mobil LTI (Low Temperature Isomerization) process, using a catalyst based on the ZSM-5 (the named zeolite catalysts AP "Aromatic Processing"), the liquid phase conditions prevail and ethylbenzene does not enter into the C_8 equilibrium (35,36).

Disproportionation of toluene

This process converts the less valuable aromatics, such as toluene, into benzene and xylenes, which makes it possible to adjust the production of these more valuable aromatics to the necessities of the commercial market.

The two main industrial processes use zeolite as catalysts. The Toray process (37,38) uses a mordenite exchanged with a non-noble transition metal, which hydrogenates the coke precursors in the presence of hydrogen, increasing the stability of acid mordenites.

More recently, the Mobil L.T.D. (Low temperature Disproportionation) process uses an acid ZSM-5 zeolite (AP Catalysts) with a special treatment. These catalysts present a great resistance to the deactivation by coke, which permits to operate in liquid phase and in the absence of hydrogen.

Alkene oxidation

A (Pd^{2+}, Cu^{2+}) Y zeolite prepared by ion exchange with Cu^{2+} and post-exchanged with Pd^{2+} showed to be highly competitive compared with the classical homogeneous Wacker Process that uses Pd Cl_2 and Cu Cl_2. In fact, the first process (39) converts ethylene to acetaldehyde with a selectivity greater than 90%.

Alkane oxidation

Both Cu^+ and Cu^{2+} exchanged zeolites showed a remarkable selectivity to benzene in the oxidative dehydrogenation of ciclohexane (40).

Cu^{2+} Y zeolite was also found to be active in the oxidation of nitrogen containing molecules (41), whereas Mn^{2+} Y zeolite is active, even at room temperature, in the oxidation of sulphur compounds (42), presenting however a strong deactivation.

Carbonylation of methanol

Rh Na Y zeolite prepared by ion exchange is an active catalyst for this reaction (43) and will perhaps become competitive with the homogeneous rhodium catalyst used by the Monsanto Process for the production of acetic acid.

Methanol to Gasoline (MTG) Process

Nowadays one of the most important new industrial process that take advantage of the zeolites properties, particularly their ion exchange capabilities is the methanol to gasoline (4) which was developed by MOBIL and recently out into action.

This process uses a proton exchanged ZSM-5 zeolite. The exchange is performed directly by diluted mineral acids or using ammonium salts, and is followed, as usual by a calcination under air (44,45).

The process is quite peculiar due to the fact that ZSM-5 is a rather small pore zeolite and thus only hydrocarbons with less than twelve carbon atoms are produced, which corresponds to the end point of gasoline.

The selectivity in gasoline is about 75% and the unleaded research octane number of the gasoline produced is about 90 to 100.

It is clear that this reaction takes advantage of shape selectivity of ZSM-5, and, although it is still somewath controversial, it appears that a new kind of shape selectivity, called molecular traffic control, is present in this reaction (46,47). This feature arises from the fact that ZSM-5 has two kinds of pores: sinusoidal with circular section and linear with eliptical section. Since aromatics and isoparaffins preffer the latter channels, reactants use preferentially the first kind of pores (48).

Fischer-Tropsch Process

Finally we shall refer a process that, although zeolites are not yet commercialized it is probable that they may have an important role in the future, due to the oil crisis and the political need to develope other energetic sources. The Fischer-Tropsch synthesis uses coal as a primary source and is only industrially used in South Africa. One of the greatest drawbacks of this technique is the production of waxes as a by-product when performed with conventional catalysts. Zeolites, due to their shape selectivity produce a non-Schulz- Flory distribution giving a higher gasoline yield (49) although with a still relatively low octane number (~ 75)(44). This process is still in a developing stage (46)but iron or ruthenium ion exchanged ZSM-5 catalysts (49)seem to be quite promising.

5. CONCLUSION

Cation exchange of zeolites modifies their properties, which has contributed to their great success in catalysis, namely in the refi ning and petrochemical industry.

Considering the increasing number of laboratories working in the search of synthesis of new zeolites and modifications by ion exchan ge in the composition and structure of these solids, we can be sure that the future will bring more and more new applications in catalysis.

REFERENCES

1.Flank, W.H., Adsorption and Ion-Exchange with Synthetic Zeolites (Washington, ACS Symposium Series 135, 1980).
2.Sherman, J. in Zeolites: Science and Technology, F. Ramôa Ribeiro et al. Eds. (The Hague, Martinus Nijhoff Publishers, 1984).
3.Breck, D.W. in Properties and Applications of Zeolites, R.P. Townsend Ed. (London, The Chemical Society, 1979).
4.Chang, C., Hydrocarbons from Methanol (New York, Marcel Dekker, 1983).
5.Breck, D.W., Zeolite Molecular Sieves. Structure, Chemistry and Use (New York, J. Wiley, 1974).
6.Tejada, J., Ph.D. Thesis, Université de Poitiers, France, 1983.
7.Csicsery, S.M. in Zeolite Chemistry and Catalysis, J.A. Rabo,Ed. (Washington, American Chemical Society, 1976).
8.Helfferich, F. Ion Exchange (New York, McGraw-Hill, 1962).
9.Rabo, J.A., Pickert, P.E., Stamires, D.N. and Boyle, J.E., Proc. of 2nd International Congress of Catalysis, (1969) 2055.
10.Ramôa Ribeiro, F., Ph.D. Thesis, Université de Poitiers, France, 1980.
11.Ramôa Ribeiro, F., Marcilly, Ch. and Thomas, G., C.R. Acad. Sc. Paris, 287 C (1978) 431.

524

12. Barthomeuf, D. in Zeolites: Science and Technology, F. Ramõa Ribeiro et al., Eds.(The Hague, Martinus Nijhoff Publishers,1984).
13. Guisnet, M., Cormerais, F., Chen, Y., Perot, G. and E. Freund, Zeolites 4 (1984) 202.
14. Ramõa Ribeiro, F. and Ch. Marcilly. Rev. Inst. Fr. Pet., 34(3) (1979) 405.
15. Derouane, E. in Catalysis by Zeolites, B. Imelik et al., Eds. (Amsterdam, Elsevier Scientific Publishing Company, 1980).
16. Weisz, P.B., Chem. Technol.3 (1973) 498.
17. Naccache, C. in Zeolites: Science and Technology, F. Ramõa Ribeiro et al., Eds., (The Hague, Martinus Nijhoff Publishers, 1984).
18. Maxwell, I.E., Advances in Catalysis 31 (1982) 1.
19. Decroocq, D., Catalytic Cracking of Heavy Petroleum Fractions (Paris, Ed. Technip, 1984).
20. Skinner, R., PD 8 (1), 11th World Petroleum Congress, London, 1983.
21. Maselli, J.M. and A.W. Peters. Catal. Rev.-Sci. Eng. 26 (3-4) (1984) 525.
22. Magee, J.S. and J.J. Blazek in Zeolite Chemistry and Catalysis, J.A. Rabo, Ed. (Washington, American Chemical Society, 1976).
23. McDaniel, C.V. and P.K. Mahor, in Zeolite Chemistry and Catalysis, J.A. Rabo, Ed. (Washington, American Chemical Society, 1976).
24. Collins, D.J. and K.J. Mulrooney. J. Catal. 75(1982) 291.
25. Haynes Jr., H.W., Catal. Rev.-Sci. Eng. 17(2)(1978) 273.
26. Bartholomew, C.H., Chem. Eng. 91(23) (1984) 96.
27. Billon, A., Franck, J.P. and J.P. Perier. Hydroc. Proc. Sept. (1975) 139.
28. Guisnet, M. and M. Perot in Zeolites Science and Technology, F. Ramõa Ribeiro et al., Eds. (The Hague, Martinus Nijhoff Publishers, 1984).
29. Henriques, C., Msc. Thesis, Univ. Técnica, Lisboa, 1984.
30. Henriques, C., Dufresne, P. Marcilly, Ch. and F. Ramõa Ribeiro. Proc. Int. Symp. Zeolite Catalysis, Siófok, Hungary, 1985, Acta Physica et Chemica, Nova Series, 31 (1985) 477.
31. Duir, J.M.. Hydroc. Proc. 46(9) (1967) 127.
32. U.S.Pat. 4,456,775 (June 1984).
33. Hydrocarbon Process 53(9)(1974) 212.
34. Ramõa Ribeiro, F. in Zeolites: Science and Technology, F. Ramõa Ribeiro et al., Eds. (The Hague, Martinus Nijhoff Publishers, 1984).
35. Grandio, P., Schneider, F.H., Schwartz, A.B. and J.J. Wise. Amer. Chem. Soc., Div. Petrol. Chem. Preprints (1971) 16, No.3, B; Oil Gas Journal (1971) 62.
36. Grandio, P. and F.H. Schneider. Oil Gas J. 62 (1979) 69.
37. Iwamura, T., Otani, S. and M. Sato. Bull. of Japan Petrol.Inst. 13 (1971) 116.
38. Hydrocarbon Processing, Nov. 1977.

39. Weissermel, K. and H.J. Arpe. Industrial Organic Chemistry (Weinheim, Verlag Chemie, 1976).

40. Mochida, I., Ikeda, Y., Fujitsu, H. and K. Takeshita. Ind. Eng. Chem. Prod. Res. Dev. 15 (1976) 160.

41. Williamson, W.B., Flentge, D.R. and J.H. Lunsford. J. Catal. 37 (1975) 258.

42. Pearco, J.R. and J.H. Lunsford. J. Coll. Int. Sci. 66 (1978) 33.

43. Yashina, T., Orikasa, Y., Takahashi, N. and N. Hara. J. Catal. 59 (1979) 61.

44. Gabelica, Z. in Zeolites: Science and Technology, F. Ramôa Ribeiro et al. Eds. (The Hague, Martinus Nijhoff Publishers, 1984).

45. Olson, O.H., Haag, W.O. and Lago, R.M. J. Catal. 61 (1980) 390.

46. Vaughan, D.E.W. in Properties and Applications of Zeolites, R.P. Townsend Ed. (London, The Chem. Society, 1979).

47. Derouane, E.G. in Zeolites and Technology, F. Ramôa Ribeiro et al. Eds. (The Hague, Martinus Nijhoff Publishers, 1984).

48. Csicsery, S.M. Zeolites 4 (1984) 202.

49. Jacobs, P.A. in Catalysis by Zeolites, B. Imelik et al. Eds. (Amsterdam, Elsevier, Scientific Publishing Company, 1980).

PART V
ALTERNATIVE PROCESSES

ION EXCHANGE MEMBRANES: PRINCIPLES, PRODUCTION AND PROCESSES

Patrick Meares

Department of Chemical Engineering
University of Exeter,Exeter EX4 4QF,Devon,U.K.

1. HISTORICAL INTRODUCTION

Physiologists have been studying permeability phenomena
for well over a century. The membranes bounding biological cells
were identified as the sites where the transport of metabolites
and other substances was subjected to rate control long before
there was any real understanding of the nature of these membranes.
Two aspects of permeability attracted attention, namely the ability
of cell membranes to display very unequal permeabilities to molecules of rather similar sizes and their ability to establish and
maintain for long periods large concentration differences of
certain vital substances between the phases on either side of them.
Such differential transport properties were noted with non-ionic
solutes and with ions.

In order to gain a better understanding of these phenomena
physiologists and physical chemists began systematic work on membranes, particularly synthetic membranes, under _in vitro_ conditions.
Two types of membranes were of particular importance: porous membranes made frequently from collodion and liquid membranes made
from lipoidal substances.

The differential permeabilities of porous membranes towards
various ions attracted much attention but the development of a
sound theoretical basis for the electrochemistry of porous membranes
took a long time. For, although the importance of charges physically adsorbed on the pore walls was noted in 1914, it was not
until 1930 that their role in several otherwise hard-to-understand
membrane phenomena was made clear by Sollner. His ideas and his
experimental work with porous collodion membranes in which ionic

substances were deliberately incorporated and on liquid membranes
in which oil-soluble ionizable substances were dissolved have
formed the basis of the modern electrochemistry of membranes.
Readers interested in the history of these early developments in
membrane electrochemistry are recommended to read two excelllent
reviews by Sollner [1,2].

These ideas were finally put on a quantitative basis by
Teorell [3] and by Meyer and Sievers [4] in 1935 and 1936 when they
independently put forward the so-called fixed charge theory to
explain membrane potentials. Their approach is still held to be
essentially sound and has formed the starting point for the more
complex models and theories developed up to the present time.

All of these early membranes had high electrical resistances
and could convey only small flux densities without serious ohmic
heating and pH changes at the membrane/solution interfaces.
Although they had no commercial possibilities as mass transporting
membranes, they had served the valuable purpose of encouraging the
development of a sound understanding and quantitative theory of
electromembrane phenomena. Thus it was that when synthetic ion-
exchange materials possessing a high ion-exchange capacity and with
a high mobility of the exchangeable ions became available in about
1950 the theoretical basis for their efficient exploitation already
existed.

2. THE IDEAL MODEL MEMBRANE

The chemistry of the ion-exchange membranes in large scale
use closely parallels the chemistry of ion-exchange resins in bead
form but with certain differences which will be described later.
In general, the degrees of crosslinking of membranes are somewhat
lower than those of resins. Thus when a membrane is immersed in
an aqueous solution the extent of swelling is larger than that of
a typical bead resin, specific water uptakes of 30-70% by volume
are common.

When the volume fraction of water exceeds 20-25% there will
be some continuous, though tortuous, aqueous-filled pathways
through the membrane along which ions can be transported relatively
freely. Thus, although the unswollen polymeric matrix of the membrane does not contain fixed pores, other than those randomly located volume elements where the density is lower than average due
to the poor packing of the polymer chains with their bulky ionogenic
side groups, the swollen membrane can be regarded as containing
aqueous filled interconnecting pores. Thus the application of
theoretical principles developed for porous charged membranes to
ion-exchange resin membranes is defensible. The analogy quickly
breaks down when one tries to characterize the membrane in terms

of a stated pore size and it is physically more meaningful to
regard the membrane as a pair of interpenetrating networks: the
physically or chemically interlinked polymeric matrix and the
aqueous micro-phase filling the interstices in that matrix [5].

It will be realized that the amount and distribution of the
absorbed water in the membrane is of great importance in determining
its behaviour. The amount of water taken up is controlled in part
by the hydrophobic-hydrophilic balance of the polymeric matrix
and the density and distribution of chemical crosslinks, chain
entanglements and, perhaps, crystallites. The ionizable groups are
certainly hydrophilic but the polymeric backbone may be very
hydrophobic as, for example, in the perfluorohydrocarbon based
membranes such as Nafion or it may be relatively hydrophilic as in
 the early phenol-formaldehyde condensate membranes such as Zeo-
Karb 315. In the former case the water tends to concentrate in
molecular clusters around the ionic groups and to avoid the exposed
hydrophobic backbones which, consequently, are also clustered
together as in hydrophobic bonding. In the latter case, the water
is more nearly uniformly distributed in the swollen polymer.

In all except the most hydrophobic membranes chemical crosslinks are introduced to restrict swelling or, in case of a high
fixed ion content, to prevent dissolution of the membrane. The
distribution of these crosslinks is another important factor determining the properties of the membrane. If the crosslinks are
introduced when the membrane is swollen either by water or an
organic solvent the final structure resembles that familiar in
solvent-modified or isoporous bead resins whereas if the crosslinks
are introduced into the unswollen polymer a more nearly uniform
homogeneous structure on the molecular scale results. One may
think of the clustered or the solvent-modified membrane matrixes
as giving rise in the swollen state to coarse-grained interpenetrating networks and the homogeneously crosslinked kind as favour-
ing fine-grained networks.

The first water molecules which enter the membrane hydrate
the ions present and so lower the electrostatic free energy in the
region of the ionic groups. The driving force is primarily
enthalpic but the ion pairs are enabled to dissociate so liberating
the counterions. Further water molecules then enter for entropic
reasons. They increase the volume available to the counterions
(c.f. osmotic pressure) and they increase the configurational entropy of the polymer chains. They exert a plasticizing effect on the
chain motions which permits the polymer segments to take part,
albeit a limited part, in the micro-Brownian motion of the system.
This increases the mobilities or diffusion coefficients of the
counterions in the matrix by several orders of magnitude compared
with that in the dry state of the membrane.

The ultimate degree of swelling is reached when the increase in entropy when more water enters is just offset by the elastic work required to stretch the network (i.e. reduce its configurational entropy) against the restraining effect of the crosslinks and the interference with the hydrogen bonding of the water that results when more swelling forces the water to contact hydrophobic groups in the chains.

At the present time no quantitative methods exist which permit the full characterization of the distribution of ions, water and matrix within an ion-exchange membrane, although the industrial importance of perfluorinated cation permeable membranes has stimulated rapid progress towards that objective [6]. Thus a membrane is frequently characterized by its overall degree of swelling and its total fixed ion content or equivalent weight. One may therefore imagine an idealized membrane in which these components are uniformly distributed on the molecular scale. Then one may seek to express the properties of the membrane in terms of single, constant values at equilibrium for its ion and water concentrations and its internal electric potential. A problem arises when expressing its equilibrium thermodynamic properties in choosing a suitable standard state, because the dielectric constant of the fully swollen membrane may be very different from that of water and it is meaningless to talk of bringing the membrane matrix to a state of infinite dilution in water.

Bearing in mind the close association of the ions and water in the membrane, it is physically more reasonable to consider the molal concentration of the ions, i.e. mole per kg of imbibed water, and to use the normal standard state in aqueous solution as a basis for defining activities. The expression of the transport properties then requires two additional parameters; the average tortuosity of the ionic pathways in the membrane, which has a special value on the ideal assumption of uniform distribution of all components [7], and the degree of association, in the form of ion pairs in the sense used by Bjerrum, of the counterions and fixed charges. This quantity is not well defined but has been discussed and studied widely in connection with polyelectrolyte solutions [8].

An alternative approach which treats the membrane material as a porous quasi-crystal with many Frenkel defects i.e. interstitial counterions, has been outlined by Nikolaev [9] but although conceptually enlightening it cannot readily be made operational.

The problem can be avoided by introducing one further experimental parameter of a kinetic kind; for example the mobility of the counterions, measured with tracers or in an electrical conductance experiment when the membrane is in equilibrium with only a very dilute solution of electrolyte. This parameter can then be used to estimate the degree of ion pairing in an ideal membrane [10].

Interestingly, the degree of dissociation of the fixed charges is not very dependent on the concentration of the solution in which the membrane is equilibrated and this behaviour resembles also the dissociation of counterions from micelles of ionic surfactants [11].

3. THE TMS FIXED CHARGE THEORY

The theory introduced by Teorell, and Meyer and Sievers was designed to explain certain observations in membrane conductances and membrane potentials. An important feature of the theory is its assumption that the rates at which ions and water are transported between the phases on opposite sides of the membrane are governed by processes occurring within the membrane. Thus the external phases are regarded as homogeneous and with no concentration or composition polarization close to the membrane/solution interfaces i.e. the permeability of the stagnant boundary layer of solution at each interface towards the flux under study is regarded as very large compared with the permeability of the membrane. Secondly, the assumption requires that there is no significant free energy barrier restricting the flows of the substances studied through the membrane/solution interfaces. Thus although the existence of net fluxes implies that the system is not at equilibrium, nevertheless it is assumed that the conditions just inside and just outside the interfaces can be interrelated by the laws of thermodynamic equilibrium. In electrolytic systems such an interfacial equilibrium permits differences in the ionic concentrations and electric potential on either side of the interface.

According to the fixed charge theory, therefore, the potential difference recorded by a pair of electrodes immersed in the solutions on either side of the membrane is made up of the vector sum of the potentials of the electrodes themselves, the two equilibrium interfacial potential differences and a diffusion potential generated by the different concentrations and mobilities of the ions within the membrane. The idealized membrane model described here, and used more or less implicitly by Teorell and by Meyer and Sievers, makes it appropriate to describe the interfacial potentials by means of the Donnan equilibrium and the diffusion potential through the Nernst-Planck equation, although originally the Henderson equation was used for this purpose.

3.1 The Donnan Potential and Electrolyte Uptake

Each interface may be regarded as a membrane separating an external solution from the mixture of ions, water and polymer inside the membrane. While water, the counterions and co-ions (small ions of the same charge sign as the fixed charges) can cross the interface, the fixed charges are bound to the polymer and play the same role as the large impermeant ions on one side of the membrane considered in the Donnan membrane equilibrium of classical colloid chemistry.

Charged ionic double layers develop on each side of the interface leading to a potential difference between the internal and external phases which are regarded as electrostatically neutral in the bulk beyond the range (a few nm) of the inter-facial effects. Equilibrium is reached when the chemical potentials of the neutral combination of counterions and co-ions are equal inside and outside the membrane. This assumption of local electroneutrality is negligibly disturbed by the existence of the diffusion potential inside the membrane.

The interfacial Donnan equilibrium is important on two counts. It gives an expression for the interfacial potential differences and it enables the concentrations of counterions and co-ions just inside the membrane to be calculated from the electr-olyte concentration outside. These concentrations are needed in order to calculate the contribution of the internal diffusion potential to the observed membrane potential.

The Donnan equilibrium in this connection has been discussed several times by the present author [12,13,14], as well as by others, and only the essential results are repeated here.

Ionic molal concentrations inside and outside the membrane are related by

$$m_n^{\nu_n} m_g^{\nu_g} = m^{\nu_n}(m + M/\nu_n\nu_g)^{\nu_g} = (\gamma_\pm'/\gamma_\pm)^{\nu} m'^{\nu} \qquad (1)$$

where the symbols have the meanings given in the table at the end. In deducing this equation the same reference state was chosen for the solutions inside and outside the membrane. The choice has the effect of fixing the ratio of activity coefficients. However it does little to help in attaching a numerical value to $\gamma_\pm'/\gamma_\pm$ without which eqn (1) cannot be used. The interionic attraction theory is also of little value in estimating this ratio on account of the very high ionic strength and polyelectrolyte nature of the solution inside the membrane.

There is a further difficulty in applying eqn (1). M is the molal concentration of fixed charges. On the one hand M may be taken as the concentration of fixed charge groups deter-

mined by chemical analysis, in which case the activity, and hence
the activity coefficient, of the counterions will be apparently
lowered because no specific account has been taken of the fact
that at any instant a substantial number of fixed charges are
closely associated with counterions, probably in the form of
solvent-separated ion pairs, and so are effectively neutralized.
Alternatively, M may be regarded as the concentration of dissociated (i.e. unpaired) fixed charges but then its value is not
well defined and can be found only in some empirical way, as
mentioned previously.

Qualitatively, eqn (1) correctly describes experimental
data on the ionic equilibrium between an ion-exchange membrane
and a solution of a single electrolyte. Thus the concentration
of co-ions in the membrane is lower than in the solution (the
Donnan exclusion effect) and the higher the valence of the co-
ions the more effectively they are excluded. Conversely, the
higher the valence of the counterions the less effective the co-
ion exclusion and the more dilute the bathing solution the greater
the ratio of the external and internal concentrations of the co-
ions.

When experimental ion uptake data are represented by plotting
$1/\nu$ times the logarithm of the left side of eqn (1) against
log m' a straight line is normally obtained. This suggests that
the ratio of the mean ionic activity coefficients is constant and
is often quite close to unity. However the slope of the line
should also be unity but it is found to be lower than unity and
a characteristic of the particular membrane. This observation
cannot be easily accounted for by adjusting the value of M to
allow for ion pairing. It has been explained by Glueckauf [15,16]
as an indication of the failure of the real membrane to conform
to the ideal model in which the fixed charges are assumed to be
uniformly distributed. The form of eqn(1) is such that any local
fluctuation in the fixed ion concentration will lead to an uptake
of co-ions greater than predicted and the effect is more pronounced
the more dilute the solution phase.

Experimental uptake data can be represented by the empirical
equation

$$m^{\nu_n}(m + M/\nu_n\nu_g)^{\nu_g} = (\alpha m')^{\beta\nu} \tag{2}$$

where α is a constant typical of the electrolyte and β a
constant typical of the membrane. The smaller β, the greater
the non-uniformity of the fixed ion distribution.

If the fixed ion distribution is subject to local variations

so also must be the electric potential, but such variations are
concerned with micro-volume elements determined probably by the
random distribution of the crosslinks or other junction points
in the polymer matrix. On the macroscopic scale, an interfacial
Donnan potential is meaningful and, by considering the equilibrium
distribution of the co-ions, it can be calculated from

$$\Psi_D = (RT/z_nF) \ln(\alpha m'/m) \tag{3}$$

The practical conclusion that must be drawn from this section
is that although the basic physical chemistry of the uptake of
co-ions by a membrane is well understood, even in the case of an
idealized model membrane the uptake cannot be predicted with
confidence but two or three measured values are sufficient to
derive the whole of the uptake curve through the empirical equation
(2). It must be remembered that the co-ion uptake is a vital
factor in determining the selectivity of a membrane and hence the
transport numbers and current efficiency in a separation process
so the discussion here has a more than academic importance.

3.2 Counterion Selectivity and Membrane Discrimination

While some processes that use ion-exchange membranes depend
mainly on the ability of the membranes to transport counterions
and exclude co-ions, others depend upon their ability to discriminate between one type of counterion and another. Such discrim-
ination may arise in part from the different mobilities of the
counterions in the membrane but of at least as great importance is
the difference in the distribution of a pair of counterions between
the membrane, viewed as an ion-exchange resin, and the solution in
contact with it at the interface.

When the two counterions have different charge numbers the
phenomenon of valency selectivity, i.e. the resin shows a preference for the counterions of higher valence, is well known and understood [17]. The effect may be pictured as due to the attractive
effect of the Donnan potential acting more strongly on the ions
of higher charge.

When the counterions have the same valency, the ideal solution
approach predicts that the mole fractions of the ions in the solution and membrane in contact are equal and no specific differences
are predicted between different pairs of counterions with the same
or with unequal valencies. These predictions are contrary to experience and a satisfactory treatment of ion exchange requires the
inclusion of activity coefficients in the solution and membrane
phases with values specific to the individual ions and combined
with a selectivity coefficient which is usually composition and
concentration dependent [18].

The complex nature of these counterion distribution functions has resulted in the quantitative formulation of ion fluxes in mixed counterion systems lagging behind that in single electrolyte systems. In the special case of ion-selective electrodes however, where co-ions can often be regarded as totally excluded, the theory of bi-ionic potentials and electrode selectivity has been carefully studied [19].

3.3 The Nernst-Planck Flux Equation

In principle one may write for each mobile species in the membrane an equation of the form

flux = mobility x concentration x force

The discussion so far has concerned only the term "concentration".

The mobilities may be measured experimentally by using radio-tracers [20] or by measuring electrical conductance and transport numbers [21]. Both methods have their advantages and disadvantages which cannot be covered in detail here. Whereas the concentrations are scarcely dependent on temperature, mobilities are strongly temperature dependent and in many membranes, which are relatively more swollen than common bead resins, the temperature coefficients are not very different from those in aqueous solution.

In general the mobilities of the co-ions are not very dependent on either the solution concentration or the nature of the counterions. They are lower than in free solution due to the tortuosity of the diffusion pathways in the membrane. The ratio between co-ion diffusion coefficients in the membrane and in solution is often close to $[v_w/(2 - v_w)]^2$ [7,20].

For the counterions, the mobilities appear to be an increasing function of concentration but this effect may be to some extent an artefact of the method of measurement and calculation if they do not take account of the association of counterions and fixed charges [10]. In cases where there is a mixture of counterions the mobilities are strong functions of their mole fractions but this may well be due mainly to the stronger association of the preferred counterion species with the fixed charges rather than to a real change in the mobilities of the freely dissociated ions with change of their mole fraction [22].

It is evident from the remarks above that when a membrane is used in a practical process such that its opposite faces are in contact with different solutions, the mobilities of the ions in the membrane may vary considerably with distance through the membrane in a way determined by the, usually unknown, concentration profiles in the membrane.

An ion in a membrane may be subjected to several forces; the most important are an electric field and the virtual force arising from a concentration gradient. If these two only are included and are regarded as being linearly superposable the flux equation obtained is the so-called Nernst-Planck equation

$$J_i = -u_i c_i \ (RT \ d \ln c_i + z_i F \ d \psi)/dx \tag{4}$$

The symbols are defined in the table at the end. Note that, because the flux density is expressed per unit area of membrane and x is the linear coordinate through the membrane, concentrations on the molal scale can no longer be used and must be replaced by molar (i.e. per unit volume) concentrations.

When equations like (4) are written for each mobile ionic species and combined with the conditions (a) that the sum of the products of the ion fluxes and charges must equal the net electric current and (b) that at moderate current densities local electrical neutrality is maintained, the potential gradient can be eliminated and an expression found relating the ion fluxes to the concentrations and current.

The result is a differential equation which has to be integrated between the particular boundary concentrations in the membrane. The variation of the mobilities with concentration and hence with distance x creates a problem here. Frequently this difficulty is ignored and an average mobility for each ionic species is used instead. The error thus introduced is probably not serious for the case of a single electrolyte transferred across a concentration difference but less is known about cases involving the interchange of a number of counterions.

Instead of eliminating the potential, the flux equations can be combined to eliminate the fluxes. Integration then gives the potential, which, in the case of zero current, is the diffusion potential inside the membrane. On adding this to the two interfacial Donnan potentials, the total membrane potential, which was the target of the original fixed charge theory, is obtained.

A number of comparisons have been made between membrane potentials calculated in this way and those measured. The agreements obtained have varied from moderately good to poor [23]. The reasons for these discrepancies lie partly with the factors already discussed and partly with an important factor omitted from eqn (4).

3.4 Convection and Coupling of Flows

The modern theories of non-equilibrium thermodynamics teach that, sufficiently close to equilibrium for the concepts of thermodynamics to be meaningful, fluxes are linearly related to gradients of electrochemical potential. This suggests that in certain circumstances eqn (4) may need to be augmented with a term dealing with a pressure gradient although in most electrochemical operations with ion-exchange membranes the effects of a pressure gradient are small and may be neglected. Furthermore it appears that the logarithmic concentration gradient should be replaced by a logarithmic activity gradient. In the case of a single electrolyte, this correction is not important [24] and this appears to be true also when there is a mixture of counterions of equal valence [25]. Mixed counterions of unequal valence have not been studied systematically from this viewpoint.

There is a far more serious problem concerning eqn (4). The Nernst-Planck equation was developed to describe phenomena in free aqueous solutions where there is no net volume flow. When a swollen membrane separates two solutions it is likely that there will be a net flow of volume through it due in part to the volumes of the transported ions, which should probably be regarded as hydrated, and in part to a flow of solvent driven by osmotic and electro-osmotic forces.

It would not be appropriate to embark here on a discussion of osmotic flow in an incompletely permselective gel membrane; it is important to note however that this flow shows highly anomalous characteristics and its study and explanation has contributed much to understanding the general behaviour of such membranes [1,26]. For our purpose the question that must be faced is "relative to what frame of reference are the ion fluxes in eqn (4) expressed?"

The original fixed charge theory assumed that the frame of reference was the stationary membrane. Later workers [7,27] treated the fluxes of eqn(4) as being expressed relative to a frame moving in the membrane with the average velocity of the volume flow of solvent and ions. In reality, although the ions are intimately associated with the water in the membrane, the counterions, especially, undergo also strong interaction with the charges fixed to the matrix. Thus eqn (4) can be appropriately augmented by a convection term to become

$$J_i = -u_i c_i (RT\, d\ln c_i + z_i F\, d\phi)/dx + \theta_i J_v c_i \qquad (5)$$

where the volume flow J_v has been multiplied by a coupling coefficient θ_i which expresses the tightness of the coupling between the flows of ions i and the water.

That such coupling is significant is proved by the observation that when an electric cirrent is passed through an ion-exchange membrane , a flow of water, sometimes reaching several tens of moles per Faraday and normally in the direction of the counterion current, is observed. By measuring this electro-osmotic flow together with the tracer and electrical mobilities of the ions, it is possible to estimate the value of θ_i [28]. As might be expected, θ_i turns out to be close to unity for the co-ions which avoid the matrix and fixed charges, and it is a good deal less than unity for the counterions which interact strongly with the matrix charges. The situation is more complex than this simple statement suggests and it has been found that if one sets θ_i at 0.5, eqn (5) forms the basis of a reasonably quantitative treatment of fluxes and potentials in homogeneous ion-exchange membranes that correspond fairly well with the idealized model membrane under discussion. There may even be a theoretical basis for the value 0.5 [29].

It may be concluded that there is now a correct understanding of the behaviour of ions under various imposed thermodynamic forces in ion-exchange membranes but understanding of the behaviour of the absorbed water is less complete. There is no satisfactory theory that enables J_v to be quantitatively predicted in all but the simplest cases. This state of affairs means that one can propose how to use ion-exchange membranes to perform separations of practical importance with a good deal of certainty that the expected separation will take place. It is not possible however to predict the degree of separation and its energy requirements and one can speculate only qualitatively on the membrane character-istics required to optimize both of these factors .

3.5 Non-Equilibrium Thermodynamic Representation of Membrane Transport

Theoreticians and experimentalists have devoted a good deal of effort towards improving the situation by adapting the formalism of non-equilibrium thermodynamics to the representation of membrane processes [30]. To do this requires not only a reformulation of the flux equations, it involves also the introduction of many more coefficients to describe the interactions between the fluxes of all pairs of components. The amount of experimental work required to determine these coefficients is too large for this type of refine-ment to become regularly used in the assessment of the practicality of separation processes. Nevertheless such research has contr-ibuted much towards a refined understanding of the molecular situation and events within an ion-exchange membrane[31,32]. Such studies have emphasized that departures from the behaviour predicted for idealized models and by simplified theories are due more to the complex structures of real membranes than to the approximations inherent in the equations derived from eqn (5).

All commercially available membranes have structures more complex than the ideal homogeneous gel discussed above and attempts to tailor membranes to meet the requirements of particular processes

are leading to increasingly complex structures being devized.
Thus it appears that, for the foreseeable future, those wishing to
apply ion-exchange membranes in industry will have to rely upon a
partial characterization based upon a few measured structural
factors including swelling, exchange capacity and conductance
combined with the augmented version of the fixed charge model
equations to give a semi-quantitative interpretation of membrane
performance.

4. METHODS OF PRODUCTION OF ION-EXCHANGE MEMBRANES

A wide variety of materials, including aluminosilicates,
zeolites, hydroxides of heavy metals,etc as well as tissues of
biological origin are known to have ion-exchange properties
and can be formed into membranes. About twenty years ago a
good deal of work was carried out on membranes made by incorporating these inorganic materials into an inert binder but,
perhaps because of the low mobilities of ions in them, such
membranes have not achieved commercial importance.

The membranes of practical interest, currently manufactured
on a considerable scale, are all formed from synthetic polymeric,
ion-exchange materials with high ion-exchange capacities and high
ionic mobilities. In confining this discussion to such membranes,
the author is consciously omitting membranes made by supporting
a liquid ion-exchanger in a porous plastic film such as polypropylene or absorbing it in a nonporous one such as polyvinyl
chloride. These liquid membranes have a currently small but
growing importance in ion-selective electrodes and in the extraction of certain valuable metals in the mineral processing and
industrial waste treatment fields because liquid-exchange media
can show high specificities towards particular ions.

One of the primary factors determining the properties of a
synthetic polymeric, ion-exchange membrane is the chemical nature
of the polymeric matrix i.e. whether the material is hydrophobic
or hydrophilic, whether the chains are stiff or flexible and
whether the polymer is non-crystalline or tends to form crystallites.
The other primary factor is the functional or ionic group covalently bound to the polymeric material. Such groups may be cationic
or anionic and may be strongly or weakly dissociated. They therefore determine whether the membrane is anion or cation selective
and how this selectivity varies with ambient pH.

Secondary factors determined by the method of manufacture
of the membrane control its morphology and have therefore a very
important influence on its swelling behaviour and on its transport
properties.

Similar factors are significant in the manufacture of ion-exchange beads and the basic chemistry of beads and membranes have, therefore, a great deal in common but the balance of properties desirable in beads and membranes are different and so there are differences of emphasis in the processes used to make them.

The author has twice recently reviewed this subject [13,14] so the same material is not repeated here. Instead, a few general comments are made about desirable features of membrane structure and these are followed by a short review of the methods available for characterizing these structures.

When in use, ion-exchange membranes are clamped, usually in batches separated either by spacers or electrodes, in large cell frameworks, commonly one metre square or larger. Such a cell may frequently contain several hundred square metres of membranes in the order of one or two tenths of a millimetre in thickness. The cost of the membranes is high and so also is the cost of the labour and lost production which are incurred if the cell has to be dismantled to replace a faulty membrane. It follows that mechanical strength, toughness, dimensional stability and long lifetime under operating conditions are properties of paramount importance in membranes. They are achieved by making the correct choice of matrix materials and the correct control of matrix morphology through the method of manufacture. Thus the choice and control of the matrix structure is far more important in membrane technology than it is in resin bead technology where replacement is easy and cheapness is often one of the most significant factors in selecting a resin.

The membrane manufacturer has to consider how to meet these structural requirements while providing the main electrochemical properties of high conductance i.e. low energy dissipation in electrically driven processes which are the main users of ion-exchange membranes, and high selectivity as between counterions and co-ions i.e. high current efficiency.

Two additional electrochemical factors are also very important. Normally three or four moles of water per Faraday are transferred through membranes, essentially as water of hydration of the ions, in addition unbound water is dragged by the ions i.e. electro-osmotically transported. This electro-osmotic flow is usually undesirable; its extent is strongly influenced by matrix morphology and this has to be taken into account by the manufacturer. Ion-exchange membranes, like bead resins, display different affinities for different counterions. Such selectivities are the essential "raison d'être" of bead resins; they are frequently a source of problems with membranes.

The problem stems from the fact that ions which are strongly taken up by membranes tend to move only slowly inside them. During electric current flow these slow moving ions are accumulated in the membrane from even trace amounts in solution and to extents far exceeding the ion-exchange equilibrium uptake. As a result the electrical resistance of the membrane increases and then its capacity to convey the more mobile and less strongly bound ions is reduced.

These strongly bound ions may be inorganic ions of high valence and polarizability which permeate right through the membrane and "poison" it by blocking the exchange sites or they may be organic ions, usually univalent and often surfactants, which are strongly adsorbed in the membrane close to the entry surface perhaps to such an extent that the fixed charges are more than neutralized and a highly resistant layer of opposite charge sign is created. An effect known as fouling.

By suitable modification of the matrix structure during manufacture the membrane can be made relatively resistant to fouling but at the expense of increasing the electro-osmotic water flux. Other methods of reducing fouling by organic ions, which involve chemical modification of the functional groups at the membrane surface, are preferable [14].

In order to meet the mechanical and physical requirements of strength and dimensional stability, it has been found desirable to achieve some degree of separation at the micro-level within the membrane. It is necessary also for the swelling of the membrane to be almost independent of the concentration and nature of the electrolyte in the solution being treated and of the temperature. The simple, randomly-crosslinked polystyrene + divinyl benzene networks used to carry the fixed ions in typical bead resins do not meet these requirements at all satisfactorily and it has been found necessary to adopt other structures and devices.

In many, though not all, membranes an improvement in dimensional stability is achieved by forming the resin on a woven or non-woven reinforcement usually made from hydrophobic fibres. The presence of this reinforcing cloth or mesh certainly contributes to the tensile strength and to tear and burst resistance but it cannot resist large changes of osmotic swelling of the active resin which becomes detached from the reinforcement and it obstructs the electric current flow and so increases the resistance per unit area.

Several entirely different manufacturing formulae have been found to minimize the tendency of membranes to swell and shrink with changes in environment. The materials they produce have one common factor : at some level a degree of separation is built in between that part of the material which provides strength and stability and the part which provides the electrochemical properties.

The nature of this separation of functions is obvious in heterogeneous membranes made by dispersing finely powdered bead resins at high loadings in an inert continuous matrix of a binder such as polyethylene or polyvinyl chloride-co- vinyl acetate. Here the rather high swelling and shrinking of the individual ion-exchange particles leads them to debond from the binder with impairment of selectivity. Such heterogeneous membranes are now used mainly to protect the electrodes in electrophoretic painting of automobiles.

In so-called homogeneous membranes the separation is less obvious but nevertheless real. Thus in membranes made by funct-ionalizing the styrene residues in styrene + butadiene block co-polymers, the butadiene chain segments aggregate or micellize to provide a permanent network of multi-functional tie points supporting the structure. In graft co-polymer membranes, a hydrophobic base polymer, such as polytetrafluoroethylene, is activated by γ- irradiation and a hydrophilic polymer or one that can be functionalized by chemical attack is caused to form poly-meric graft chains on to the base material which remains in place and can accommodate to the swelling on hydration of the active graft chains but continues to restrain the whole structure. In the membranes now used in the chlor-alkali industry tetrafluoro-ethylene and a perfluorinated vinyl ether carrying functional groups are copolymerized. The sequences of tetrafluoroethylene units in the resultant chains are extremely hydrophobic and non-polar. When the vinyl ether residues have been functionalized to produce cation-exchanging groups which draw water into the membrane from the surrounding solution segregation of the tetrafluoroethylene and ionic groups + water occurs to create clusters of the latter dispersed in the former.

In another method, interpenetrating networks have been made of polyvinylidene fluoride and polystyrene. The latter can be functionalized in the usual way while the former maintains the integrity and rigidity of the system.

When membranes are made by sulphonation of polyethylene itself, only the amorphous chain segments are subjected to chem-ical attack and a dimensionally rigid network of crystallites remains.

Membranes made by the paste method are a hybrid between homogeneous and heterogeneous membranes. Powdered polyvinyl chloride and a plasticizer are incorporated into the membrane monomers to form a paste which is spread thinly and then polymerized. The polyvinyl chloride particles act in the same way as a filler in, say, an elastomer to impart toughness and tensile strength and, in the case under discussion, there is a close surface integration between the membrane polymer and the filler particles because the latter are swollen by the former before polymerization takes place.

A summary of and references to the literature dealing with these methods of membrane manufacture have been given elsewhere[14].

It must be evident that the membranes which have been found to behave satisfactorily in industrial scale processes have micro-heterogeneous structures at the molecular level which make them behave rather differently from the uniform gel which is the ideal-ized membrane structure considered in the fixed charge theory. It is this structural difference rather than any inadequacy of the theoretical ion distribution and transport equations that leads to the divergence between theory and experiment when the behaviour of commercially produced membranes is compared with the predictions of the fixed charge theory.

5. STUDY AND INTERPRETATION OF MICROSTRUCTURE

It will only become possible to refine the theoretical treatment of ion and water transport in membranes when means have been devized to describe and determine the microstructure of the polymeric matrix. For a long time the methods available for obtaining information on structure were rather insensitive. They included the analysis of co-ion sorption data to characterize the distribution function of the local fixed charge concentration [16]. The changes in the sorption and mobility of the co-ions observed when the counterions are gradually exchanged by another species for which the membrane has a different selectivity can be interpreted in terms of the distribution of local states in the membrane [22]. The existence of so-called "dead-end pores" i.e. regions in which ions may be held without contributing to the transporting capability of the membrane has been inferred by comparing co-ion diffusion coefficients measured by using radio-tracers in steady permeation experiments and in non-steady time lag or sorption kinetic experiments [9].

Such measurements give only a general indication of the degree of dispersity in the internal distribution of fixed charges in the membrane. For more specific structural information one turns naturally to spectroscopic and radiation scattering exper-iments. Observations on infrared absorption spectra of ion-exchange membranes have been used for many years to obtain inform-ation on the energetic states of water molecules and ions in mem-branes [33]. Such studies are still being intensively pursued and give valuable information on the extents of hydration and degrees of pairing of fixed ions and counterions and on the perturbation of the hydrogen bond structure of the water by the organic polymeric matrix.

During the last few years, the growth in the commercial importance of the new perfluorinated cation-exchange membranes used in chlor-alkali cells has stimulated major developments in the means for carrying out structural studies. Such membranes have a complex morphology which responds to the extreme states of temperature, osmotic and electrical stress to which they are exposed in use. This has major implications for power consumption and current efficiency, and hence for product cost. One hopes that similar methods of study will soon be applied to other, less immediately exciting, membranes such as those used in electrodialysis so that the whole science of ion-exchange membranes can benefit and its forward momentum be maintained.

The study of the structures of perfluorinated ion-exchange membranes has been covered in detail [34] and has recently been excellently reviewed [6]. Among the most informative methods of study used to date have been wide and small angle X-ray scattering which have identified small, normal polymeric crystallites in membranes and a two-phase microstructure made up of domains of non-ionic backbone segments and other domains in which the ionic groups are concentrated. The interpretation of the X-ray data has been confirmed by the response of the observed spacings to changes in counterion type and in degree of hydration. Small angle neutron scattering is most sensitive to the amount and distribution of water molecules in the membrane. Data from this technique has shown that the water molecules are freely mobile and are located predominantly in the regions where the ions are aggregated.

Nuclear magnetic resonance can also be used to study the states of the water protons but the chief value of NMR has been the study of the ^{23}Na spectrum. This has demonstrated the existence of hydration sheaths on sodium ions in a fully hydrated membrane and the formation of contact ion pairs between counterions and fixed charges at low degrees of hydration.

Further studies of ion pairing have been carried out with heavy metal counterions by applying electron spin resonance and Mossbauer spectroscopic techniques. Such work has confirmed the close connection between the degree of hydration of the membrane and the existence of strong pairing of counterions and fixed charges. Even well-hydrated membranes contain many loose, solvent-separated ion pairs of long lifetimes i.e. some tens of nanoseconds.

These methods of investigation and some others, including

electron microscopy and the study of dynamical mechanical relax-
ations, have shown that the perfluorinated membranes contain
cluster regions 5-10 nm in diameter into which fixed charges,
counterions and water have diffused and in which the concentration
of ionic material is far higher than the bulk average taken over
the whole membrane. These clusters are not isolated, they are
interconnected by material containing a few ions and some water
molecules that is subjected to continuous fluctuation through the
local molecular Brownian motion. The clusters and their interconnecting regions constitute a network of ion-permeable pathways
embedded in and interpenetrating a matrix of mainly hydrophobic,
micelle-like regions of partly crystalline polytetrafluoroethylene
segments.

It will be revealing to discover through further work whether
such a segrated and clustered arrangement is unique to this type
of membrane and to see what differences in structure are found in
membranes with more hydrophilic matrixes. The evidence available
so far confirms the view that real membranes diverge considerably
from the ideal homogeneous gel structure.

6. SOME PROCESSES THAT USE ION-EXCHANGE MEMBRANES

The processes described briefly in this section have been
selected to illustrate the wide variety of ways in which ion-
exchange membranes are now being used in industry. The coverage is certainly not complete and new uses are constantly being
researched. It seems certain that in the medium term future
such membranes will become of increasing value in the chemical
processing, food and biotechnology fields because they are clean
and non-polluting, permit separation of reactants, products and
waste streams, and can be made very efficient in their use of
energy. Recent progress has been aided too by improvements in
membrane quality, especially in the enhanced lifetimes now attainable at high temperatures (60-80°C) where electrical resistances
are low. The most modern membranes also have excellent resistance
to acids, alkalis and oxidants.

The solutions to which membranes are exposed in these practical processes are usually rather complex and this makes it difficult to interpret the technological observations in terms of the
theory of transport in membranes at its present level of development. Such processes, and the development of membranes specially
tailored to meet their demands, have therefore to be optimized by
a combination of intelligent guesses based on the theoretical principles, empiricism and experience.

6.1 Electrodialytic Desalination and Demineralization

By far the largest use, in terms of tonnage production and
membrane area employed, of ion-exchange membranes is in the desalination of raw brackish water to produce water suitable for
domestic and industrial use. More than 500 such plants are in
use around the world, one of the largest having a capacity of
six million gallons per day. Many detailed accounts of electrodialysis have been published [see e.g. 35,36].

Many problems have had to be faced and solved including
concentration polarization, cell corrosion, electrical leakage
and other extramembrane factors. The principal membrane problems
have been fouling of the membranes, particularly the anion
selective membranes by organic anions in the feed water, and
scaling where either the solubility limit of salts such as calcium
sulphate and calcium carbonate are exceeded in the concentrate
compartments or, where a high pH has developed due to the use of an
excessive current density, $Mg(OH)_2$ or $Ca(OH)_2$ may precipitate
on the cation permeable membranes. Particularly damaging is the
entry of $(MgOH)^+$ ions in the membranes followed by internal
precipitation of hydroxides.

Among the methods used to combat these problems have been
the development of fouling resistant membranes, mentioned earlier,
and the development of the electrodialysis reversal (EDR) process.
In this, the cells are built with a plane of symmetry and a cation
permeable membrane at each end of the stack. By reversing at
intervals the electric current polarity and simultaneously inter-
changing the flows through the feed and product compartments,
the fouling layers deposited in one cycle are displaced again in
the next. This refinement of the electrodialysis process is the
subject of conflicting claims [35] but has apparently achieved
a market success for its suppliers, Ionics Inc.

The procedure of desalting by electrodialysis is straight-
forwardly adaptable to treat process streams other than raw water.
Thus cheese whey [37], blood and molasses are demineralized by
electrodialysis. The cell design has to be adapted to handle
fluids of high viscosity and the high organic concentration in the
solutions to be treated, particularly in whey and blood which
contain proteins that migrate electrophoretically at most pH values,
make it necessary to use the surface modified membranes developed
to combat fouling. If this is not done, the cell resistance rises
rapidly during operation and membrane lifetime is unacceptably
short.

In these applications the electro-osmotic transfer of water
with the ions has the beneficial effect of further concentrating
the demineralized macromolecular product. Whey and blood treatment

by electrodialysis can be beneficially combined with ultrafiltration of the product to separate macromolecular components such as proteins from micromolecular components such as lactose.

Electrodialysis produces two output streams, one is more concentrated than the feed and the other is more dilute. In the schemes described above the dilute stream is the product. In other processes the concentrated stream is the product. They include the concentration of waste streams in order to reduce the volume requiring to be processed or stored and to recover and recycle valuable materials such as nickel and copper from exhausted electroplating baths.

The most important use in this category is the concentration of sea water in the manufacture of table salt [38]. There are several particular problems to be faced in this application. Both streams are relatively concentrated electrolyte solutions thus, to maintain a high current efficiency, co-ion exclusion has to be especially good. This calls for membranes with high exchange capacity. The product concentration is required to be about 4M i.e. about one mole of salt per fourteen moles of water. Since water is electro-osmotically and osmotically transferred into the concentrated stream through both the cation and the anion selective membranes bounding it, the electro-osmotic transference numbers of the membranes must be kept low by devizing membranes with restricted swelling despite their high exchange capacity. Osmotic water transfer is kept low by using relatively thick membranes and this factor helps also to check the back diffusion from the concentrated to the dilute compartments.

Ordinary concentration polarization is not too great a problem in salt production because of the relatively concentrated streams used but sea water contains appreciable amounts of divalent ions Ca^{2+}, Mg^{2+} and SO_4^{2-}. Because these ions move slowly in the membranes they are accumulated during current flow creating a state of composition polarization. Not only do these divalent ions contaminate the product, they can bring about the precipitation of $CaSO_4$ and $MgSO_4$ in the concentrate. To avoid this problem special membranes have been developed with a thin layer of oppositely charged material on the entry face of each membrane i.e. a polycation on the cation selective and a polyanion on the anion selective membrane [39]. The manufacture of table salt from sea water is now a major industrial operation in Japan, Korea and Taiwan that uses this electrodialytic pre-concentration step.

When a membrane is exposed to a mixture of counterions and a current is passed through it, the relative amounts of the counterions transported differ in general from their relative amounts in the feed. Although this phenomenon posed a problem in the sea water concentration process it can also form the basis of a separation procedure. The reason for the difference in the ion ratios is two-fold. The amounts of the counterions taken up by the membrane differ on account of the ion-exchange selectivity. The mobilities of the counterions in the membrane also differ. If the ions have equal valencies, they are subjected to equal electrical forces and the ratio of their fluxes is given by the ratio of the products of their concentrations and mobilities in the membrane i.e.

$$J_A/J_B = c_A u_A / c_B u_B = K_B^A c_A' u_A / c_B' u_B \tag{6}$$

holds to a first approximation.

Unfortunately the ions preferentially taken up by the membrane interact the more strongly with the matrix and so move relatively more slowly within it. Thus the two ratios c_A/c_B and u_A/u_B tend to change in opposite senses reducing the separation obtainable.

The direction of the separation is normally dominated by the ion exchange selectivity and is therefore strongly dependent on the chemistry of the membrane the choice of which offers a means of optimizing the results. Some exploratory work has been done on the separation of anions in this way [40] and such a process may also have potential in hydrometallurgical separations particularly where complex ions can be involved.

6.2 Oxidation and Reduction in Membrane Cells

In electrodialysis and related processes a stack of anion selective and cation selective membranes is set up. In the simplest cases an alternating arrangement of membranes is used, but for more complex separations other sequences may be preferable. Current passing electrodes are arranged at the ends of the stack so that the same current passes through all membranes in series and care is taken to ensure that the fluid streams passing through the electrode compartments do not mingle with and contaminate the process streams.

There exists a different class of processes in which the oxidation and reduction reactions occurring at the electrodes contribute directly to the formation of the product. In such processes each pair of electrodes is separated by only a small number of membranes, usually one, two or three. Depending upon the nature and scale of the process, the cells may be built with a monopolar design or, by making each electrode act as a cathode at one side and anode at the other, a bipolar cell can be assembled. This consists of many sets of bipolar electrodes, ion-exchange membranes and fluid compartmants held between a single pair of end electrodes connected to the external power supply.

The most important process of this kind, judged in terms of

the scale of operation, is the electrolysis of brine to produce
chlorine, hydrogen and concentrated alkali [41]. Here each pair of
electrodes is separated by a single cation-selective membrane.
Limitations of space preclude the inclusion here of a proper description of the anode, cathode, membrane and cell design. The
subject has been exhaustively covered in many recent books and
conference proceedings [e.g. 34, 41]. The membranes used are of
the perfluorinated kind with usually a mixture of sulphonate and
carboxylate fixed ions.

In operation, chloride ions are oxidized to chlorine which
is evolved at the anode, sodium ions pass through the membrane
and combine with hydroxyl ions in the cathode compartment formed
by reduction of water to hydrogen gas and hydroxyl ions at the cathode. The process is very simple in concept but has required the
solution of many difficult technological problems to make it cost
effective in competition with the long-established mercury cells.
Such technological advances have been made possible only in recent
years through the developments in membrane chemistry that led first
to the sulphonated Nafion membranes and then to the introduction
of carboxylate membranes which have a higher cation selectivity in
the face of the immense concentrations of hydroxyl co-ions.

Other membrane-mediated oxidation-reduction processes take
place under less aggressive conditions and can use more conventional ion-exchange membranes.

Many industrial processes involve either the neutralization
of an acid or base and give rise to a troublesome salt as a by-
product or they require the release of an acid or base from a salt
which is achieved by the addition of a mineral acid or base. An
example of the former is the production of large amounts of almost
valueless sodium sulphate in the reconstituted cellulose industry
and in the strippers used to remove oxides of sulphur from flue gases.
Examples of the latter are the production of citric and boric acids
from sodium citrate and borax respectively and the production of
tetramethylammonium hydroxide from tetramethylammonium chloride.

An electrolytic membrane cell can be used to regenerate
sulphuric acid and alkali from sodium sulphate which can then be
reused in the process. The basic process is exactly analogous to
brine electrolysis but with sulphate anions and at the lower concentrations used the gas evolved at the anode is oxygen. The conversion to sulphuric acid goes via $NaHSO_4$ and it turns out to be
helpful to use a four compartment cell with three cation-selective
membranes between the electrodes. The feed solution passes through
the cell twice, first through the membrane bounded compartment
nearer to the cathode where conversion mainly to bisulphate takes
place and then through the adjacent compartment nearer to the anode
where the hydrogen ions entering from the anode compartment replace

the remaining sodium ions. The electrode compartments are fed essentially with water which is split to form the gases evolved, oxygen at the anode and hydrogen at the cathode, and alkali is recovered from the cathode stream.

The production of citric acid is precisely analogous except that the feed solution is sodium citrate.

It may be noted that in these and related examples oxidation and reduction are used to create hydrogen and hydroxyl ions which are then substituted for the cations or anions originally in the feed which is not therefore itself oxidized or reduced.

Hydrogen and hydroxyl ions can be generated directly from water without the evolution of gases by using bipolar ion-exchange membranes [14]. The use of such membranes would enable these ion substitution processes to be carried out in a stack of many membranes and compartments with only a single pair of electrodes as in a conventional electrodialysis cell. Bipolar membranes are not yet a widely available commercial product so that the employment of this more efficient method of acid and base generation must await further improvements in membrane chemistry.

In certain other processes the feed solids are directly reduced at the cathode by discharging hydrogen ions. A cation selective membrane in between anode and cathode allows continuous entry of hydrogen ions from the anode compartment while protecting the product from back oxidation by the oxygen evolution at the anode. Two examples will illustrate this class of process[42]. They are the reduction of uranyl to uranous ions in uranium refining according to the reaction

$$UO_2Cl_2 + 2HCl + 2H^+ - 2e = UCl_4 + 2H_2O$$

and the dimerization and hydrogenation of acrylonitrile to adiponitrile used in the manufacture of Nylon 6,6. The reaction scheme is

$$2CH_2 = CHCN + 2H^+ - 2e = NC(CH_2)_4CN$$

Several other electro-organic syntheses are carried out in membrane cells; they include hydroquinone, aniline and terephthalic acid manufacture among many others. The list could be expanded much further if membranes could be used in non-aqueous ionizing solvents and some research along these lines has already been done.

6.3 Electro-osmotic Separations

The electro-osmotic drag force exerted by the counterions conveys not only water but also non-ionic solutes present in the feed solution. In fact organic solutes with molecules small enough to pass through the membrane matrix are relatively concentrated into the membrane. This can be regarded as a consequence of their exhibiting mildly surface active properties at the internal contacts between the hydrophobic matrix segments and water. Thus the concentration of the organic solute in the electro-osmotic stream is often higher than in the feed. The extent of the difference is a function of the tightness, i.e. degree of crosslinking and network homogeneity of the membrane. The organic molecules are relatively more retarded by tight membranes.

The electro-osmotic transfer of organic solutes creates a problem in the electrodialytic demineralization of molasses when it causes a loss of product. It can be expoited to advantage in whey processing to separate lactose, which can pass through the membranes, from whey proteins which are too large to pass.

The cell is built with an alternation of tight and loose matrix cation-selective membranes and with a tight membrane at each end. Whey is fed to the compartments with a loose membrane on the cathodic side, a salt solution is fed to the other compartments. Lactose is transported through the loose membranes with the electro-osmotic stream flowing towards the cathode. Since the tight membranes largely block the further transfer of lactose it accumulates in the stream from the compartments with tight membranes on the cathodic side. The whey proteins cannot pass through the loose or tight membranes and remain in the effluent from the whey feed compartments.

There are not many examples of this process in commercial operation but it is extensively used in small scale separations in the analytical and clinical fields because it is much faster than normal dialysis.

6.4 Donnan Dialysis

The permselectivity of anion and cation exchange membranes caused by the Donnan exclusion can be used to concentrate ions present at only low concentration in a sample solution to some higher concentration in another solution of smaller volume. This process can be useful in analysis provided a method exists specific to the ion in question and not interfered with by some other species of ions of the same charge. An anion selective membrane is used to concentrate anions and a cation selective membrane to concentrate cations [44].

To explain this process, ideal solutions will be assumed together with the total exclusion of co-ions by the membrane and zero osmotic flow of water. The initial system then evolves to an equilibrium state. In practice, because these ideal assumptions are not met completely, the separation achieved passes through a maximum and the system then slowly runs down to uniform concentrations in both solution phases. Therefore Donnan dialysis is permitted to proceed for an optimum time and then the product removed.

Consider sample ions A and inert ions B and let the sample be introduced on side ' and the other solution on side ". At equilibrium

$$c_A'/c_B' = c_A''/c_B'' \tag{7}$$

holds. If the sample contained initially n_A^o moles and the other solution n_B^o moles, stoichiometric considerations ensure that, at equilibrium

$$n_A^o = n_A' + n_A'', \; n_B^o = n_B' + n_B'' \; , \; n_A'' = n_B' \tag{8}$$

Eqn (7) thus reduces to

$$(n_A^o - n_A'')/n_A'' = n_A''/(n_B^o - n_A'') \tag{9}$$

on the assumption of zero volume flow; whence

$$n_A^o/n_A'' = 1 + n_A^o/n_B^o \tag{10}$$

Provided $n_B^o \gg n_A^o$, most of A is transferred from the sample side to side ". When $V' > V''$ the final concentration, $c_A'' = n_A''/V''$, exceeds the sample concentration n_A^o/V'. Thus the sample volume should be large and the dialysate relatively small in volume and initially concentrated in B. Membranes are chosen that meet as nearly as possible the criteria of zero co-ion and osmotic fluxes. The tubular forms of Nafion have been found particularly well adapted for use in Donnan dialysis.

6.5 Solid Polymeric Electrolytes (SPE)

An electrochemical power source consists of two electrodes which can transfer electrons to or from an external circuit, two chemical reactants which may be the electrodes themselves or are in direct contact with them, such as PbO_2 in lead-acid accumulators, or can be fed to them, as in fuel cells, and an electrolyte placed between the two electrodes maintaining an ionic conduction pathway but separating the electrodes from direct reaction with one another.

There are clear advantages in cell construction in having a solid rather than a liquid electrolyte in the cell. Hydrated ion-exchange membranes offer the possibility of fulfilling this role provided they can meet the requirements of high chemical and thermal stability, especially towards oxidation, high conductivity and selectivity, high mechanical strength without brittleness and low permeability to gases.

Much preliminary work was done with conventional cation-exchange membranes but commercial success in this field was delayed by various inadequacies in the available materials. Now it has been found that the perfluorinated cation-selective materials, especially Nafion, meet the criteria set out above [45,46,47].

The technology is important in electrochemical power sources but can also be adapted to perform a number of the electrolytic processes described earlier. In the most modern chlor-alkali cells porous electrodes are coated with catalyst and directly bonded to the cation exchange membrane which functions therefore as a solid polymeric electrolyte. Of especial importance here is the development of satisfactory methods for bonding the catalyst, usually mixed oxides, to the membrane surface.

Limitation of space does not permit a proper account of this relatively new and rapidly expanding use of ion-exchange membrane materials to be given here but this review would have been incomplete if it had not drawn attention to the SPE field.

TABLE OF SYMBOLS

c_i	molar concentration of ions i
F	Faraday's number
J_i , J_V	flux density of ions i, volume
K_B^A	ion exchange selectivity coefficient
M	molal concentration of fixed charges
m	molal concentration of sorbed electrolyte
m_g , m_n	molal concentration of counterions, co-ions
n_i , n_i^0	number of moles of i at equilibrium, initially
R	the gas constant
T	absolute temperature
u_i	mobility of ions i in the membrane
V	volume of solution
v_w	volume fraction of water in the membrane
x	distance through the membrane
z_i , z_n	charge number, with sign, of ions i, of co-ions
α	ratio of activity coefficients
β	empirical constant of membrane matrix
$\gamma_\pm$	mean molal activity coefficient
θ_i	coupling coefficient between ions i and water flux
ν	number of ions per mole of electrolyte
ν_g, ν_n	number of counterions, co-ions, per mole of electrolyte
ψ	electrical potential
ψ_D	Donnan interfacial potential difference

Unprimed symbols refer to the membrane phase, single primed and double primed symbols refer to solution phases.

REFERENCES

1. Sollner, K. The early developments of the electrochemistry of polymer membranes. in E. Sélègny (ed) "Charged and Reactive Polymers" Vol 3 Pt I, Reidel, Dordrecht, 1976 pp 3-55.

2. Sollner, K. The basic electrochemistry of liquid membranes. in J. N. Sherwood et. al. (eds) "Diffusion Processes" Gordon and Breach, London, 1971 pp 655-730.

3. Teorell, T. An attempt to formulate a quantitative theory of membrane permeability. Proc. Soc. Exptl. Biol. Med. 33 (1935) 282-285.

4. Meyer, K. H. and Sievers, J-F. La perméabilité des membranes I Théorie de la perméabilité ionique. Helv. Chim. Acta 19 (1936) 649-664.

5. Krause, S. Partial solubility parameter characterization of interpenetrating microphase membranes. in D.R. Lloyd (ed) "Materials Science of Synthetic Membranes" ACS Symp. Ser. 269 American Chemical Society, Washington, 1985 pp 351-363.

6. Kyu, T. Structure and properties of perfluorinated ion-exchange membranes. op. cit. Ref. 5 pp 365-405.

7. Mackie, J. S. and Meares, P. The diffusion of electrolytes in a cation-exchange resin membrane. Proc. Royal Soc. A, 232 (1955) 489-505.

8. Manning, G. S. Limiting laws and counterion condensation in polyelectrolyte solutions I. J. Chem. Phys. 51 (1969) 924-933.

9. Nikolaev, N. I. "Diffusion in Membranes" Khimiya, Moscow, 1980 Ch. 6.

10. Meares, P.The self-diffusion coefficients of anions and cations in a cation-exchange resin membrane. J. Chim. Phys. 55 (1958) 273-279.

11. Anacker, E. W. Cationic surfactants. in E. Jungermann (ed) "Surfactant Series" Vol 4, Marcel Dekker, New York, 1970 Ch 7.

12. Meares, P. The permeability of charged membranes. in H. H. Ussing and N. Thorn (eds) "Transport Mechanisms in Epithelia" Munksgaard, Copenhagen, 1973 pp 51-72.

13. Meares, P. Trends in ion-exchange membrane science and technology. in D. S. Flett (ed) "Ion Exchange Membranes" Ellis Horwood, Chichester, 1983 Ch. 1.

14. Meares, P. Ion-exchange membranes. in L. Liberti and F. G. Helfferich (eds) "Mass Transfer and Kinetics of Ion Exchange" Nijhoff, The Hague, 1983 pp 329-366.

15. Glueckauf, E. and Watts, R. E. The Donnan Law and its application to ion exchange polymers. Proc. Royal Soc. A, 268 (1962) 339-349.

16. Glueckauf, E. A new approach to ion exchange polymers. Proc. Royal Soc. A, 268 (1962) 350-370.

17. Helfferich, F. G. "Ion Exchange" McGraw Hill,New York, 1962 pp 156-158.

18. Meares, P. and Thain, J. F. The thermodynamics of cation exchange VI. J. Phys. Chem. 72 (1968) 2789-2797.

19. Koryta, J. "Ion-Selective Electrodes" Cambridge University Press, Cambridge, 1975 Ch. 2.
20. Meares, P. Transport in ion-exchange polymers. in J. Crank and G. S. Park (eds) "Diffusion in Polymers" Academic Press, London, 1968 Ch. 10.
21. Lakshminarayaniah, N. "Transport Phenomena in Membranes" Academic Press, New York, 1969 Ch. 5.
22. McHardy, W. J., Meares, P. and Thain J. F. Diffusion of radio-tracer ions in a cation-exchange membrane. J. Electrochem. Soc. 116 (1969) 920-928.
23 Lakshminarayaniah, N. op. cit. Ref. 21 Ch. 4.
24. Meares, P. and Ussing, H. H. The fluxes of sodium and chloride ions across a cation-exchange resin membrane I. Trans. Faraday Soc. 55 (1959) 142-155.
25. Mackay, D. and Meares, P. Ion exchange across a cationic membrane in dilute solutions. Kolloid Z. 171 (1960) 139-149.
26. Schlögl, R. On the theory of anomalous osmosis. Z. Phys. Chem. N.F. 3 (1955) 73-102.
27. Schlögl, R. and Schödel, U. Behaviour of charged porous membranes during passage of electric current. Z. Phys. Chem. N.F. 5 (1955) 372-397.
28. Meares, P. Coupling of ion and water fluxes in synthetic membranes. J. Membrane Sci. 8 (1981) 295-307.
29. Dickel, G. The Nernst-Planck equation in thermodynamic theories. op. cit. Ref. 14 pp367-393.
30. Schlögl, R. Non-equilibrium thermodynamics - a general framework to describe transport and kinetics in ion exchange. op. cit. Ref. 14 pp 207-212.
31. Meares, P. Some uses for membrane transport coefficients. op. cit. Ref. 1 pp 123-146.
32. Paterson, R. Cameron, R. G. and Burke, I. S. Interpretation of membrane phenomena using irreversible thermodynamics. op. cit. Ref. 1 pp 157-182.
33. Heitner-Wirguin, C. Spectroscopic studies of ion exchangers. in J. A. Marinsky and Y. Marcus (eds) "Ion Exchange and Solvent Extraction" Marcel Dekker, New York, Vol. 7 (1977) Ch 3.
34. Eisenberg, A. and Yeager, H. L. (eds) "Perfluorinated Ionomer Membranes" ACS Symp. Ser. 180, Americam Chemical Society, Washington, 1982.
35. Solt, G. S. Electrodialysis. in P. Meares (ed) "Membrane Separation Processes" Elsevier, Amsterdam, 1976 Ch. 6.
36. Korngold, E. Electrodialysis - membranes and mass transport. in G. Belfort (ed) "Synthetic Membrane Processes" Academic Press, Orlando, 1984 Ch. 6.
37. Ahlgren, R. M. Electromembrane processing of cheese whey. in R. E. Lacey and S. Loeb (eds) "Industrial Processing with Membranes" Wiley-Interscience, New York, 1972 Ch. 4.
38. Nishiwaki, T. Concentration of electrolytes prior to evaporation with an electromembrane process. op. cit. Ref. 37 Ch. 6.

39. Sata, T. Modification of properties of ion-exchange membranes II. J. Coll. Interf. Sci. 44 (1973) 393-406.

40. Hann, R. A., Eyres, R. and Cottier, D. Separation of anions by electrodialysis. op. cit. ref. 13 Ch. 2.

41. Coulter, M. V. "Modern Chlor-Alkali Technology" Ellis Horwood, Chichester, 1980.

42. Seko, M., Miyauchi, H. and Omura, J. Ion-exchange membrane application for electrodialysis, electroreduction and electro-hydrodimerisation. op. cit. Ref. 13 Ch. 12.

43. Zelman, A., Walsh, D., Wayt, H. and Gisser, D. Rapid filtration of non-electrolytes by electro-osmosis. Desalination 25 (1978) 253-261.

44. Kipling, B. General applications of perfluorinated ionomer membranes. op. cit. Ref. 34 Ch. 19.

45. Lee, J. A., Maskell, W. C. and Tye, F. L. Separators and membranes in electrical power sources. op. cit. Ref. 35 Ch. 11.

46. Yeo, R. S., Applications of perfluorosulphonated membranes in fuel cells, electrolyzers and load levelling devices. op. cit. Ref. 34 Ch. 18.

47. Simmrock, K. H., Griesenbeck, E., Jorissen, J. and Rodermund, R Use of perfluorinated cation-exchange membranes in electrolysis processes, particularly in alkali chloride electrolysis. Chem. Ing. Tech. 53 (1981) 10-25.

KINETICS OF METAL EXTRACTION:RATE CONTROLLING STEPS AND
EXPERIMENTAL TECHNIQUES USED TO ESTABLISH A DESIGN EQUATION

E.S. Pérez de Ortiz

Department of Chemical Engineering &
Chemical Technology
Imperial College
London S.W.7

ABSTRACT

A previously established equation for the rate of extraction of a
metal ion from an aqueous into an organic phase is used to
conduct a numerical study of its response to changes in the
concentrations of reactants and mass transfer coefficients.
Results indicate that the rate of extraction can be controlled
exclusively by the chemical reaction (chemical regime) or by a
combination of mass transfer and chemical parameters (mixed
regime). In the mixed regime mass transfer control may occur in
either phase or both depending on the relative concentrations of
reactasnts and the hydrodynamic conditions of the phases. The
hydrodynamic conditions of two laboratory contactors: (i) the
quiescent interface cell and (ii) the single drop, are introduced
into the design equation as mass transfer coefficients to
simulate experimental results obtained with these techniques.
The analysis shows that in order to obtain a design equation
results obtained with the single drop technique are inconclusive
even when the site of the rate controlling reaction is known.

1.INTRODUCTION

During the last two decades the rapid expansion in the industrial
application of solvent extraction of metal ions, usually referred
to as hydrometallurgy, has been followed by the development of
new extractants and an increasing interest in the study of the
kinetics of extraction and its rate controlling steps.
Essentially a problem of mass transfer with multiple chemical
reactions the mechanism of transfer is further complicated by

interfacial phenomena such as the adsorption of species present in the system, and the different structures and degrees of aggregation of some of these species in the bulk. However, simplified models of mass transfer with chemical reaction have helped to explain apparently conflicting experimental results and to establish guide-lines on the experimental techniques and procedures that can eliminate some of the uncertainties in the interpretation of results [1]. Danesi and Chiarizia [2] have published a comprehensive and critical review of proposed mathematical models and experimental techniques used in the study of the kinetics of metal extractions. In this paper a brief overall view of the problem is presented and the effects of the concentrations of the different species on the mass transfer rate controlling steps are analysed in detail for one of the extraction models. The implications of the results in the study of the chemical kinetics is discussed for two of the most widely used experimental techniques: the quiescent interface cell and the single drop.

1.1 Overall reaction

The extraction system consists of an aqueous solution of metal ions, the refinate phase, in contact with an organic phase containing the extractant. At a suitable pH value one of the metal ions reacts with the extractant forming a complex soluble in the organic phase. The reaction is reversible and the metal is re-extracted from the organic phase in the stripping process by an aqueous phase modified to move the reaction equilibrium in the appropriate direction. This modification is usually attained by adjustment of pH of the solution. To avoid reagent losses the solubility of the complexing agent in water must be very low and for an efficient metal separation so should be that of the metal complex. The most commonly encountered overall reaction is of the form:

$$M^{m+} + m\,\overline{HX} \rightleftharpoons \overline{MX_m} + mH^+ \qquad (1)$$

where M denotes the metal, X the anionic part of the complexing agent and bars indicate the species in the organic phase.

1.2 Location of the rate of controlling reactions

Equation (1) is in general the result of several reactions, some taking place in the bulk of the phases, some at the interface. The slowest reaction is rate controlling and it must be either a bulk or an interfacial one. Three main models have been proposed based on different locations of the rate of controlling reactions:

    (i)   Homogeneous reaction in the bulk of the aqueous

phase [3]

(ii) Reaction in a thin aqueous film close to the interface [4]

(iii) Interfacial reaction [1]

Models (ii) and (iii) include mass transfer coefficients in the bulk phases.

1.3 Experimental techniques used in kinetic studies.

Different techniques have been described in the literature for the determination of kinetic data in the two phase liquid-liquid extraction. They fall into three main types:

(a) The single drop [5]. The equipment basically consists of a column containing one of the phases. The second phase is introduced at the top or the bottom in the form of drops of known volume. The drop moves due to the buoyancy forces and the amount of solute transferred from one phase to the other during their contact time is measured. The cross flux per unit of interfacial area can then be calculated.

(b) The quiescent interface cell [6]. The two phases can be stirred independently and the mass flux per unit of interfacial area can be calculated from the variation of the concentration of solute with time.

(c) The stirred tank [7]. One phase is dispersed into the other by the action of the stirrer. The variation of the concentration of solute with time gives the rate of transfer per unit volume. In this case the extension of the interfacial area is not readily known.

Contactors (a) and (b) have the advantage of possessing a well defined interfacial area but type (b) offers in addition the possibility of changing the degree of turbulence of the phases by changing the stirring speed.

1.4 Mathematical model for Extraction with interfacial chemical reaction

The mechanism of transfer of the solute from one phase to the

other may be very complex but it can be reduced to a number of
elementary steps:

(i)   diffusion of the reactants from the bulk of the phases
       to the interface;

(ii)   chemical reaction at the interface;

(iii) diffusion of the reaction products away from the
       interface.

The three steps take place simultaneously and thus mutually
interfere. If the rate of reaction is fast with respect to the
rate of transport of the reactants to the interface, the rate of
formation of the products will depend on the rate of transfer.
In this case the process is diffusion controlled and the system
is said to be in the diffusional regime. On the other hand, if
the reaction is slow with respect to the mass transport rate that
could be attained in the two phases, the extraction rate will be
slowed down by the chemical reaction and the process will be
under chemical control (chemical regime). In this case the mass
transfer coefficients will not affect the rate of extraction. A
third possibility is an interaction of the two mechanisms
resulting in a process where the mass transfer coefficients and
the chemical parameters affect the overall rate.   These
conditions characterize the mixed regime.

Examples of this kind of extraction mechanism are the extraction
of copper by LIX 65N [7], iron (III) by di(2-ethyl-hexyl)
phosphoric acid (DEHPA) [8] and zinc also by DEHPA [9].   The
latter system will be used to study the interaction between
hydrodynamic conditions, concentration of reactants and rate of
extraction. The stoichiometry of the overall reaction is

$$Zn^{2+} + 1.5 \; \overline{H_2A_2} \rightleftharpoons \overline{Zn \, A_2 \cdot HA} + 2H^+ \qquad (2)$$

in which H_2A_2 and HA represent the dimer and monomer of DEHPA
respectively. Figure 1 is a schematic representation of the
concentration profiles near the interface. It is assumed that
the bulk of the two phases is well mixed and the system is in
steady state. At the conditions of the extraction process and at
short contact times only the forward reaction will be
considered.

The rate of interfacial reaction expressed as a flux is given
by [10]:

$$R = k_I \; C_{Zi} \; C_{Oi} C_{Hi}^{-1} \qquad (3)$$

where R is the interfacial rate of reaction, k_I the rate constant and C_{Zi}, C_{Oi} and C_{Hi} the concentrations of zinc, DEHPA and hydrogen ion at the interface.

The fluxes of reactants and products to and from the interface are linked by the stoichiometry of reaction, therefore:

$$R = K_Z(C_Z - C_{Zi}) = \tfrac{1}{2}K_H (C_{Hi} - C_H)$$
$$= 2/3 \, K_O (C_O - C_{Oi}) \tag{4}$$

where K_Z, K_H and K_O are the mass transfer coefficients for zinc, H^+ and DEHPA and C_Z, C_O and C_H are bulk concentrations.

Elimination of the interfacial concentrations between equations (3) and (4) leads to [11]

$$\left[\frac{1}{k_I K_H} - \frac{1.5}{K_O K_Z}\right] R^2 + \left[\frac{1}{2k_I} C_H + \frac{1}{K_Z} C_O + \frac{1.5}{K_O} C_O\right] R - C_O C_Z = 0 \tag{5}$$

Equation (5) shows how the overall transfer rate is governed by the mass transfer coefficients of the species and the chemical reaction through equation (3). It also indicates that at low values of R the first term of the equation can be neglected and if in addition the mass transfer coefficients are of a higher order of magnitude than the rate constant k_I a set of concentrations can be found that make the second and third terms in the second parenthesis negligible. Under these conditions equation (5) reduces to:

$$R_{ChR} = k_I C_Z C_O C_H^{-1} \tag{6}$$

Equation (6) is the same as equation (3) but the concentrations are now bulk concentrations. Equation (6) represents the chemical regime and equation (5) the mixed regime.

2. NUMERICAL STUDY OF THE MODEL

The effect of mass transfer coefficients and relative concentrations of reactants on the rate of transfer can be studied by solving equation (5) for different values of the parameters involved in it. Although the instrinsic diffusivities of Zn^{+2} and H^+ are different, electric neutrality must be maintained. This can only be achieved if $K_Z = K_H$, as indicated by equation (4).

The value of k_I, used in the calculations is 4.27×10^{-7} ms^{-1} as determined by Ajawin et al. [10].

2.1 Effect of concentrations on the mass transfer controlling step

Several sets of concentrations were selected so that the rate of chemical reaction in the chemical regime as given by equation (6)

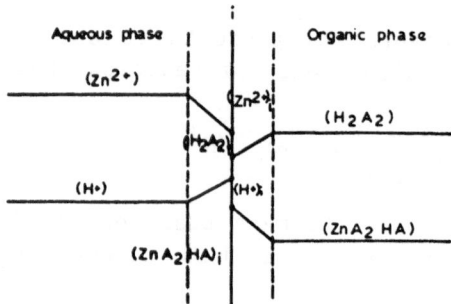

Figure 1.   Concentration profiles at the interface during
extraction.

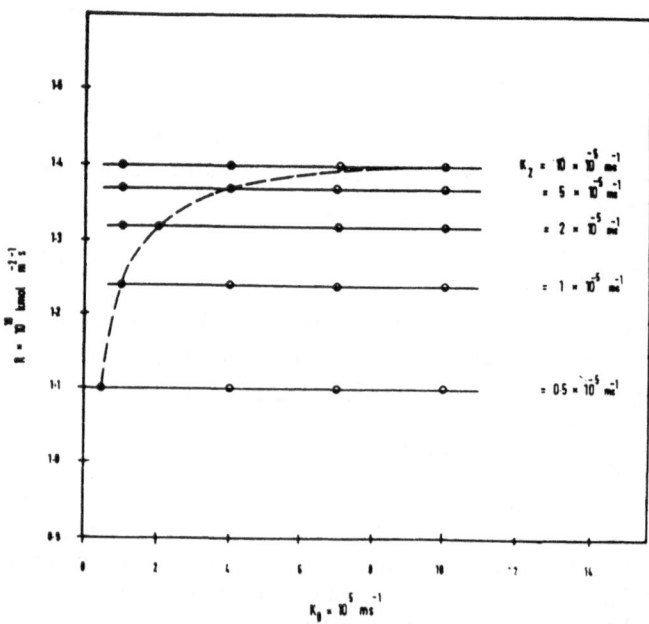

Figure 2.   Variation of the interfacial flux, R, with mass
transfer coefficients C_O= 0.01, C_Z= 0.0001 and
C_H= 0.003 M.

was constant and equation (5) was solved for a wide range of values of the mass transfer coefficients including those encountered in a quiescent interface cell [11]. Figure 2 shows results for $C_O=0.01$, $C_Z=0.0001$ and $C_H=0.003$ kmol m^{-3}. The curves simulate experiments at constant aqueous phase mass transfer coefficient and different values of K_O. This could be achieved in a Lewis type cell by keeping the stirring speed in the aqueous phase constant and varying that in the organic phase. As expected for a case with high excess of organic reactant, the mass transfer control is in the aqueous phase and completely independent of the hydrodynamic conditions in the organic phase. The dotted line shows the curve that would be obtained if the two mass transfer coefficients were changed simultaneously as is usually the case in experimental research with a quiescent interface cell.

Results for C_Z C_O are shown in Figure 3. Here the concentrations are $C_O=0.0001$, $C_Z=0.01$ and $C_H=0.003$. The mass transfer control is now in the organic phase since for a given value of K_O the rate of extraction is independent of K_Z. An increase in K_Z leads to a higher rate of extraction plateau, thus showing the dependence of R on K_O. As in Fig.2 the dotted line indicates results for $K_Z = K_O$.

When C_O and C_Z are of the same order of magnitude the mass transfer control is shared by the two phases as shown in Fig. 4. The concentrations are $C_O = C_Z = 0.001$ $C_H = 0.003$ and now for each value of K_O there is a region in which R increases with K_Z. Here it can be seen that as K_O increases the interval of values of K in which R increases, widens. It is also interesting to note that in this case the value of R reached at $K_O = K_Z = 10^{-5}$ ms^{-1} is much closer to the value corresponding to the chemical regime that in any of the two previous cases. This can be more easily seen in Fig. 5 where the curves of R vs $K_Z = K_O$ for the three sets of concentrations are shown together. These results have implications in the study of the chemical kinetics of extraction as discussed in the following sections.

2.2 Effect of Hydrodynamic Conditions on the Rate of Extraction.

In order to establish an equation of the type of equation (5) experimental results should provide unequivocal results on the orders of the reaction with respect to the relevant species, the effect of mass transfer coefficients on the extraction rate and the locale of the rate controlling reaction. These requirements advocate techniques that allow changes in interfacial area and either have very well described hydrodynamic conditions or give a wide range of values of the mass transfer coefficients. The three basic experimental techniques mentioned before have

566

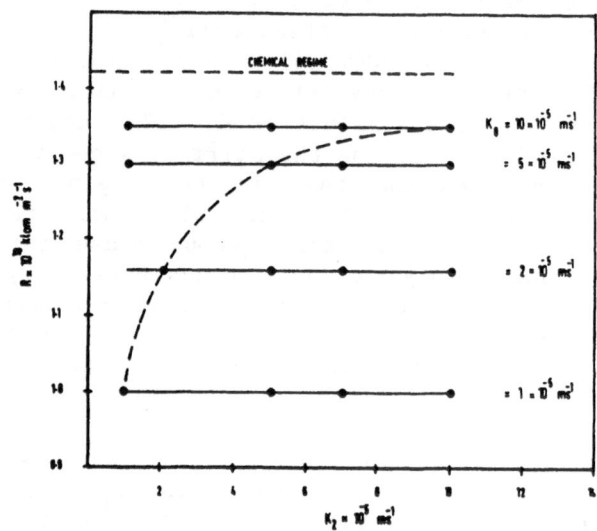

Figure 3. Variations of the interfacial flux, R, with mass transfer coefficients $C_O = 0.0001$, $C_Z = 0.01$ and $C_H = 0.003$ M.

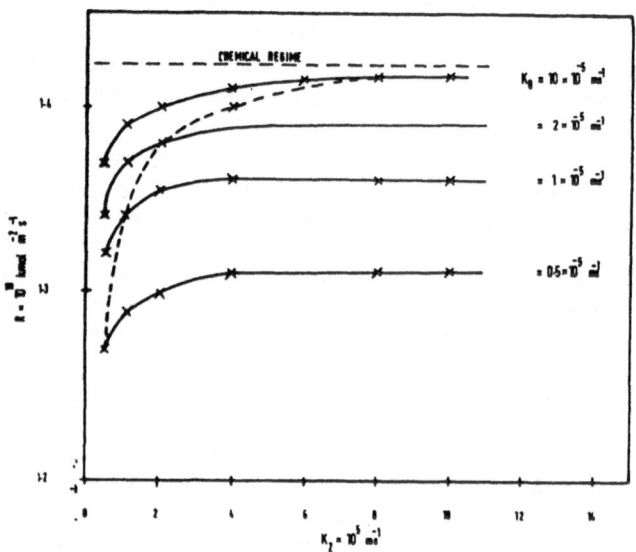

Figure 4. Variation of the interfacial flux, R, with mass transfer coefficients $C_O = C_Z = 0.001$ and $C_H = 0.003$ M.

different hydrodynamic conditions and in some cases not well defined.

In the following two sections the hydrodynamic conditions in a quiescent interface cell and in the single drop contactor represented by mass transfer coefficients are introduced in equation 5 and the simulated 'experimental' results are analysed with the view of helping to understand the limitations of each technique and suggest procedures that may overcome some of their drawbacks.

2.2.1 Kinetic studies in a quiescent interface stirred cell

Assuming that the locale of the rate controlling reaction is known the procedure usually followed to determine the order of the reaction in a quiescent interface cell can be summarized as follows:

(i)   determination of a "plateau" region where the rate of extraction is independent of all mass transfer coefficients (chemical regime);

(ii)   measurement of the variations of R_{ChR} with reactant concentrations at the hydrodynamic conditions of the "plateau" and for values of R_{ChR} that do not exceed those in (i). This step leads to the determination of the orders of the reaction.

The conditions of step (ii) are set to ensure that the system is under chemical rate control in the ranges of concentrations in which the orders of the reaction are investigated. However Fig. 5 shows that for the same value of R_{ChR} and the hydrodynamic conditions of the plateau in curve (3), curves (1) and (2) are in the mixed regime due to the fact that in both cases the concentration of one of the reactants is substantially lower than the other thus requiring a larger mass transfer coefficient to achieve the flux required for the system to be under chemical control.   Therefore in order to eliminate the possibility of mixed control either the relative values of the reactant concentrations should be kept fairly constant or the stirring speed changed to check that the system is still on a plateau.

2.2.2 The single drop technique

In the single drop technique the hydrodynamic conditions of the continuous and disperse phases are determined by the physical properties of the system and it is not possible to change the mass transfer coefficient of either phase in the way shown in Figs 2 to 5. Therefore the use of this technique in chemical kinetic studies requires a mathematical description of the model

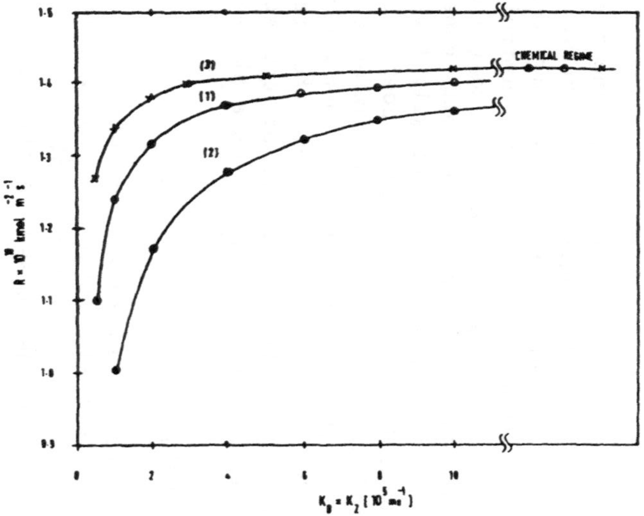

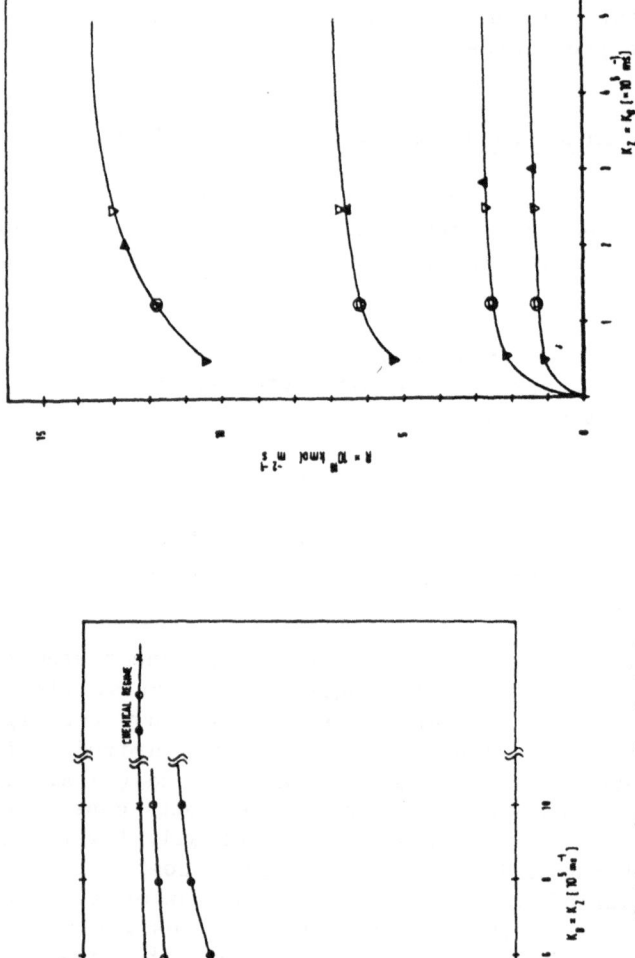

Figure 5. Variation of the interfacial flux, R, with mass transfer coefficients. $C_H = 0.003$, (1) $C_O = 0.01$, $C_Z = 0.0001$; (2) $C_O = 0.0001$, $C_Z = 0.01$ and (3) $C_O = C_Z = 0.001$ M.

Figure 6. Comparison between simulated fluxes for a quiescent interface cell and a single drop. $C_O = 0.01$, $C_H = 0.003$ M, $\sigma = 20 \times 10^{-3}$ Nm^{-1}.

▽ falling oscillating drop; ▼ falling circulating drop; ▲ rising circulating drop, ⊙ oscillating and circulating drops, $\sigma = 5 \times 10^{-3}$ Nm^{-1}.

that includes momentum and diffusion equations. This approach is mathematically complicated and requires a good knowledge of the hydrodynamics of the drop.

Nitsch et al [12] compared experimental results obtained in a quiescent interface cell with the falling drop technique in the extraction of zinc by dithizone. At conditions of continuous phase mass transfer control they found that the falling drop extraction flux corresponded to that in the cell at a stirring speed of 1100rpm. This equivalent stirring speed value was nearly twice the maximum speed that could be used in the cell without deforming the interface. However this parameter is not likely to be an invariant but depend on the hydrodynamic characteristics of the drop and the physical properties of the system. According to their hydrodynamic behaviour drops can be classified as:

(i)   noncirculating

(ii) nonoscillating with internal circulation

(iii) oscillating.

Each of these types of drops has different values of both the disperse and the continuous phase mass transfer coefficient. Expressions based on the correlation of experimental data have been reviewed by Skelland [13]. They show that the power values of mass transfer coefficients occur in noncirculting drops and the highest in systems with oscillating drops The following analysis of the influence of the hydrodynamic drop behaviour on the rate of extraction will be conducted for nonoscillating drops with internal circulation and oscillating drops. The corresponding mass transfer coefficients, given in table 1, were calculated with correlations for single drops given by Skelland [13] for typical values of physical properties, drop diameter and drop velocity.

Table 1.   Mass transfer coefficients for single drops.

Phase	$K \times 10^5 \; [ms^{-1}]$	$\sigma \times 10^3 \; [Nm^{-1}]$
Non oscillating circulating drops (NOCD)	0.6	20
	1	5
Oscillating drops	2	20
	1.2	5
Continuous (NOCD)	4	

As can be seen in table 1 the disperse phase coefficient is a function of the interfacial tension. This is not the case in stirred cells where they only depend on the Reynolds and Schmidt numbers [14], so it can be expected that the equivalent stirring speed of a drop as proposed by Nitsch may not be the same for systems with different values of σ .

Fig.6 shows the variation of interfacial flux with mass transfer coefficients for a system with aqueous phase control. The different curves represent different values of metal ion concentration. Also indicated are the stirring speeds N in the aqueous phase that give the corresponding values of K_Z as determined by Ajawin et al [11] for a Nitsch's cell. In order to have mass transfer control in the disperse phase the drop must be the aqueous phase, i.e. it is a falling drop system and therefore equivalent to that of Nitsch's. Results obtained for oscillating drops in the system with higher interfacial tension correspond to a stirring speed of 625 rpm. However for nonoscillating drops the extraction rate is well below that and equivalent to the one obtained in the cell at 125 rpm. It should be pointed out that the minimum value of N at which homogeneous concentrations in the bulk phases can be attained in the cell is 200 rpm. For the system with lower interfacial tension nonoscillating and oscillating drops give roughly the same values of R and they lie in between those obtained for the system of higher interfacial tension. It can however be argued that if the low interfacial tension is due to the presence of adsorbed extractant, as is usually the case, the drops are more likely to behave as solid spheres, i.e. they will be noncirculating in which case their mass transfer coefficient will be substantially lower thus leading to a lower rate of extraction.

If the experimental technique used is the rising drop the rate controlling step is in the continuous phase. Here it can be seen that there is an equivalent stirring speed at each metal ion concentration. This effect is more easily observed in the case of mass transfer control by both phases in Fig 7. For non oscillating circulating drops of the system with higher interfacial tension the equivalent stirring speed decreases on increasing C_Z for the rising drop and decreases with decreasing C_Z for the falling drop. This is due to the relative change in the concentration of reactants leading to a change in the share of overall mass transfer control exerted by each phase.

The implications of these results become clear in Fig 8. Here the extraction fluxes for the rising and falling drop from Fig.7 are plotted against the concentration of metal ion (curves 1 and 2). As the reaction is first order with respect to C_Z a slope of 1 should be obtained in the absence of mass transfer control. This is not the case in the interval of concentrations of Fig.8. Even more the apparent order of the reaction as given by the

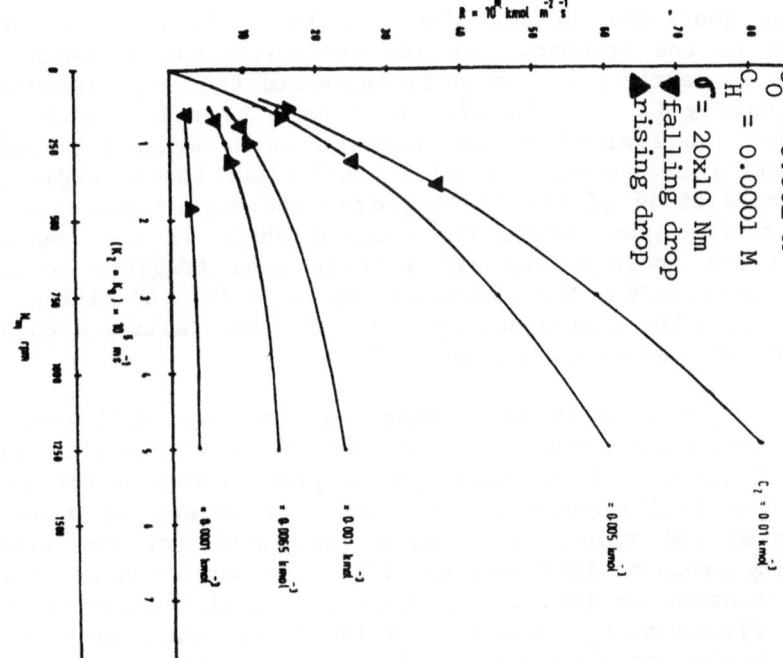

Figure 7. Comparison of fluxes for a quiescent interface cell and a single circulating drop.

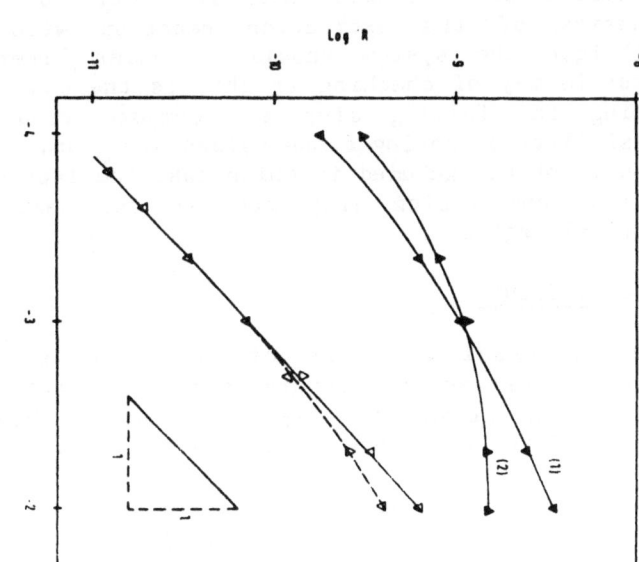

Figure 8. Variations of the interfacial flux R, with zinc concentration for a single circulating drop. $C_O = 0.01$, $C_H = 0.003$ M; ▼falling drop,▲rising drop; $C_O = 0.0001$: ▽falling drop,△rising drop.

slopes of the curves is not the same for the two techniques. This is due to the fact that the mass transfer coefficient in the disperse phase is, in the case of this example, lower than that in the continuous phase. When the drop is falling the continuous phase is the organic. At low concentrations of metal ion the slope is nearly 1 but as it is increased the slope decreases with increasing C_Z and the effect of mass transfer control becomes clear. However since an increase in C_Z makes the system more sensitive to the mass transfer coefficient in the organic phase, the conditions of the falling drop are more favourable than in the rising drop, where the organic phase is now the disperse phase and consequently has a lower mass transfer coefficient. This explains why the slopes of the curve for the rising drop are lower than the corresponding values for the falling drop over the whole interval of C_Z values.

If the system is in the chemical regime the falling and rising drop should give the same extraction fluxes. For this purpose a set of concentrations that give a plateau region for both mass transfer coefficients in the range of values of table 1 was selected and values of R were calculated for the rising and falling drops at different metal ion concentrations. Results are also plotted in Fig.8. In this case both techniques give the same fluxes at $C_Z = 0.001$. At higher concentrations the rising drop begins to give lower values of R. This is an indication that the system is no longer under chemical control and the slope of the curves will not give the true order of reaction. It can therefore be concluded that in order to study the chemical kinetics of the extraction reaction with the single drop technique the system should be under kinetic control and a possible way of checking if this is the case is to use both the rising and falling drop and compare results. However the possibility of having close values of K_0 and K_Z in a given system should not be excluded in which case the extraction rates for the falling and rising drop will be the same even outside the chemical regime.

3. CONCLUSIONS

The numerical analysis of the effects of mass transfer coefficients and concentrations of reactants in the rate controlling steps in a system of known chemical kinetics and locale of the reactions leads to the following conclusions:

(i) in the mixed regime the dependence of the interfacial flux on reactant concentrations does not usually follow the order of the chemical reaction. Therefore results obtained under these conditions give apparent orders of reaction that change with the hydrodynamic conditions of the phases;

(ii) a system in the chemical regime may move to the mixed regime when the concentration of extractants are changed so that the rate of extraction increases. This is due to the increased rate of chemical reaction. But it can also occur even when there is a decrease in the rate of chemical reaction if the ratio of reactant concentrations is such that one of the phases begins to exert mass transfer control;

(iii) an equivalent stirring speed for the single drop technique is constant for different metal ion concentrations only if mass transfer control takes place in the phase but its value changes with the hydrodynamic conditions of the drop; under conditions in which the mass transfer control is shared by the two phases the 'equivalent stirring speed' varies with the concentration of the different species.

These conclusions confirm that in order to establish a design equation the contactor used should have well described hydrodynamic conditions and provide a wide range of mass transfer coefficients so that the conditions for the chemical regime can be established. In this regard the single drop technique, although widely used in kinetic studies poses more difficulties in the interpretation of results than the quiescent interface cell.

REFERENCES

1. Pérez de Ortiz, E.S. Cox, M. and Flett,D., CIM Spec. vol.21, 198 (1979).
2. Danesi, P.R. and Chiarizia, R, CRC vol.10,1 (1980).
3. McClelland,B.E. and Freiser,H., Analytical Chemistry, 361 (12), 2263,1964.
4. Hughes, M.A. and Rod, V., Hydrometallurgy,12, 267 (1984)
5. Whewell, R.J., Hughes, M.A., and Hanson C., J Inorg. Nuclear Chemistry., 37, 2303, 1975.
6. Lewis, J.B., Chem.Eng.Sci., 3, 248 (1954).
7. Flett, D.S., Okuhara, D.N. and Spink, D.R., J.Inorg. Nucl.Chem., 35, 2471 (1973).
8. Roddy, J.W., Coleman, C.F., and Arai, S., J.Inorg. Nucl.Chem., 33, 1099, (1971)
9. Ajawin, L.A., Pérez de Ortiz, E.S, and Sawistowski,H., Int. Solvent Extraction Conf., 1980, Liege, Belgium.
10. Ajawin, Pérez de Ortiz, E.S., and Sawistowski, H., Chem. Eng.Res.Des., 61, 62 (1983).
11. Ajawin, L.A.,Demetriou, J, Pérez de Ortiz, E.S., and Sawistowski, H., I. Chem.E. Symposium Series No.88,

574

185 (1985).
12. Nitsch,W., and Kruis, B., J.Inorg. Nucl.Chem.,40, 857 (1978.
13. Skelland, A.H.P. Diffusional Mass Transfer, John Wiley, 1974.
14. Austin, L.J., and Sawistowski, H., I.Chem.E.Symposium Series No.26,3, 1967.

NOTATION

C	concentration	$kmol\ m^{-3}$
K	mass transfer coefficient	ms^{-1}
k_I	interfacial rate constant of chemical reaction	ms^{-1}
N	stirring speed	rpm
R	interfacial flux of metal	$kmol\ m^{-2}\ s^{-1}$
σ	interfacial tension	$N\ s^{-1}$

SUBSCRIPTS

ChR	chemical reaction regime
H	hydrogen ion
i	interfacial value
O	extractant
Z	Zinc ion

THE SURFACTANT LIQUID MEMBRANE:APPLICATIONS TO METAL EXTRACTION AND POLLUTION CONTROL

E.S. Pérez de Ortiz
Department of Chemical Engineering &
Chemical Technology
Imperial College

The liquid surfactant membrane process (LMP) was originally developed by N.Li and co-workers for the separation of organic components from liquid mixtures [1]. In this particular case, as in many other application of the LMP, this novel process is based on the principles of solvent extraction. The "liquid membrane" is the continuous phase of an emulsion which is dispersed into the liquid feed (or external phase). The formulation of the liquid membrane and the droplets of the emulsion, or internal phase, is such that the species to be separated are extracted by the membrane but once there, equilibrium favours the internal phase. Hence the extraction process takes place at the interface between the feed and the membrane phase, while the stripping process occurs at the interface between the membrane and the internal droplets. The solute then does not accumulate to a great extent in the membrane and extraction ceases when the concentrations of the internal and external phases reach equilibrium values. The fact that this equilibrium constant may be several times higher than that between the external and membrane phases gives the LMP a higher extractive capacity. As a result this process requires lower solvent inventory and fewer separation stages thus leading to potential savings in capital and operation costs.

A literature survey reveals four main areas where investigation on the application of the LMP is in progress

- Hydrometallurgy
- Waste water treatment
- Separation of hydrocarbons
- Biochemistry and biomedicine

The first two will be discussed in more detail.

HYDROMETALLURGY

One of the industrial processes widely used in the extraction of metal ions from leach liquours is conventional solvent extraction (CSX). This process consists of two operations: extraction and stripping. The metal ion is extracted from the aqueous solution by chemical reaction with a specific reactant present in the disperse organic phase. A schematic drawing of the system is shown in figure 1 (a). For economic reasons the reactant must be insoluble in the aqueous phase and since ions are insoluble in non-polar organic solvents the reaction usually takes place at the aqueous-organic interface. A typical reaction is:

$$M^{n+} + \overline{n\ RH} \rightleftharpoons \overline{R_n M} + n\ H^+ \qquad (1)$$

where M^{n+} is the metal ion of valance n, RH the extractant, $R_n M$ the metal complex and the bars indicate species in the organic phase. The concentration-based equilibrium constant, K_c, of reaction (1) is:

$$K_c = [R_n M] [H^+]^n / [M^{n+}][RH]^n \qquad (2)$$

where the square brackets indicate concentrations.

Equation (2) shows that by changing the concentration of hydrogen ion reaction (1) can be shifted either to the right or to the left. Thus the separation of the metal ions from the leach liquour is conducted at low values of H^+ but its recovery from the organic phase is performed by contact with an acidic aqueous phase (stripping operation) as shown in figure 1(b).

In the LMP extraction and stripping take place simultaneously in the same contactor. The acidic solution is emulsified into the organic phase, which now also contains an emulsion stabilizer, and the emulsion is dispersed in the aqueous feed (figure 1(c)).

If despite the presence of the stabilizer, the mechanism of reaction continues being represented by reaction (1), the same equilibrium constant as given by equation (2) will apply at both the external and internal interface of the globule shown in figure 1(c). At the extracting surface, the low concentration of complex will shift the reaction to the right, whereas at the stripping surface, where a large excess of hydrogen ion is present, it will be pushed very far to the left so that the process is practically controlled by kinetics through its whole duration. However the high concentration of surfactant required to stabilize the emulsion (values range from 0.6 to 5%), which under normal conditions would introduce an interfacial resistance to mass transfer, in general, reduce the rate constant of interfacial chemical reactions, thus decreasing the rate of extraction. Exceptions to this behaviour have been reported in the conventional extraction of copper [2] and zinc [3,4] in which some ionic surfactants have enhanced the extraction rate per unit interfacial area. But so far none of these surfactants stabilize the type of emulsion required by the LMP [4].

Most of the work on metal extraction has been conducted in stirred tanks where the interfacial area is usually unknown and a comparison between LMP and CSX does not reveal the effect of the surfactant on the interfacial rate of extraction. For this reason Wongswan et al. [5] conducted their work on the extraction of zinc by di (2-ethylhexyl) phosphoric acid (DEPHA) in a glass spray column where the interfacial area could be measured by direct photography. Their results are given in figure 2. All data were obtained at the same flowrate of disperse phase and in the case of the LMP the emulsion has an aqueous to organic volume ratio of 1. Results are also presented for conventional extraction in the presence of different concentrations of surfactant, in this case Span 80. The lowest value of extraction rate per unit of continue-dispersed interfacial area correspond to LMP. However if the results are expressed in terms of extraction rate per unit volume of organic phase, as in table 1, the LMP has the same performance as CSX for the extraction operation with the advantage that the stripping operation takes place simultaneously.

Table 1. Rate of extraction of Zn by CSX and LMP

	Conventional solvent extraction		Liquid membrane
	Clean interface	3% Span 80	
DEPHA concentration (M)	0.15	0.15	0.15
Initial concentration of Zn in the feed (ppm)	200	200	200
Interfacial area per unit volume of continuous phase (m^{-1})	8	17	11
Extraction rate per unit interfacial area ($kmol\ m^{-2}\ s^{-1}$)	1.4×10^{-7}	0.6×10^{-7}	0.5×10^{-7}

Extraction rate per unit volume of organic phase (kmol m^{-3})			
	11.2×10^{-7}	10.2×10^{-7}	11×10^{-7}

A pilot plant for uranium extraction by the LMP has been in trial operation in Florida, USA [6]. Metals which have been investigated in laboratory scale include copper, cobalt, nickel, uranium and zinc. A recent review [7] gives a summary of the current state of the art as well as the types of problems encountered in the operation of the LMP.

Liquid surfactant membrane processes will become accepted when they offer economic advantages over CSX. A cost comparison has been conducted for the extraction of uranium and copper [6] and the results are presented in tables 2 and 3.

Table 2. Estimated cost of copper recovery from ore leachates.

	CSX	LM
Copper recovered (ktonnes/year)	36	36
Stages	5	1
Plant investment, M $	13	8
Organic inventory, M $	2	1
Direct operating cost c /ob	1.8	1.7

Table 3. Estimated cost of uranium recovery from phosphoric acid.

	CSX	LM
Uranium recovered (kg/year)	145	145
Stages	8	3
Plant investment, M $	38	38
Organic inventory, M $	1.6	0.2
Direct operating cost, $/kg	63	40

It will be noticed that in copper recovery the savings came mainly from plant invetment, whereas in uranium recovery from the reduction in organic losses. These two studies give an encouraging perspective of the commercial application of LMP in hydrometallurgy.

WASTE WATER TREATMENT

The selection of treatment of industrial wastes is based on a combination of legal and economic factors. The range of process options lays between two extreme approaches: recovery processes and treatment without recovery (direct treatment). Some typical types of industrial waste, their source and industrially used treatments are given in table 4.

Table 4. Industrial wastes.

Nature of wastes	Source	Treatment methods
Phenolic	Coal and coal product processing	Activated carbon. Ion exchange. Chemical oxidation. Solvent extraction. Coagulation and precipitation. Biological treatment.
Heavy metals,	Metal treatment,	pH control. Redox control
chromates, cyanides, phosphates	leather treatment	Reverse osmosis and electrodialysis. Ion exchange. Evaporation.
Nitrogenous: nitrate, nitrite ammonia	Explosives manufacture, fertilizer processing, coke ovens, refinery waste.	

In all the cases listed in the table the decision between direct treatment and recovery depends on economic considerations. The threshold limit for economic recovery changes with feedstock prices thus giving incentive for the development of flexible processes.

N. Li and co-workers have applied the liquid membrane technology to waste water treatment both at laboratory and pilot plant scale [6]. Some examples are discussed in the following sections.

(a) Treatment of phenolic wastes

This is an example of the diffusion mechanism or membrane permeation. Since undissociated phenol has an appreciable solubility in organic solvents, the liquid membrane does not need to contain a reactant. The internal phase is an aqueous solution of sodium hydroxide. Thus phenol diffuses through the membrane and reacts on the internal phase to form an oil insoluble anion. Studies conducted by Exxon in mixer-settlers under batch conditions [8] have shown that phenol concentration can be reduced from 2000 ppm to less than 10 ppm in about 5 minutes.

580

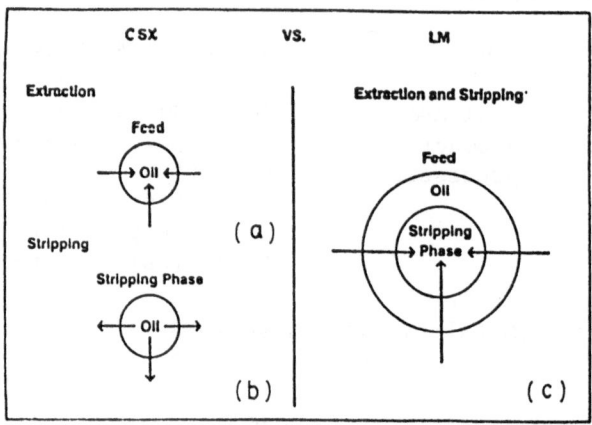

Figure 1- CSX versus LM

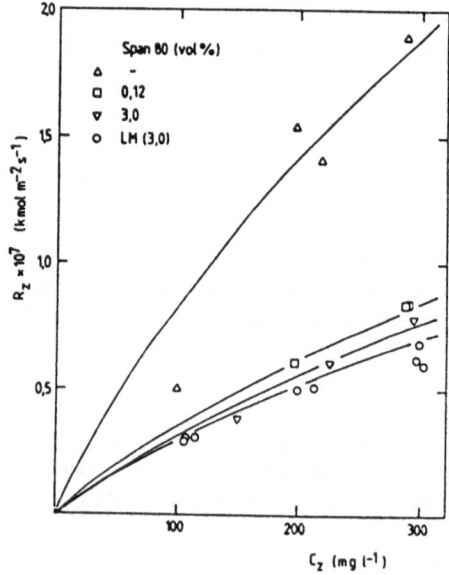

Figure 2- Extraction of zinc byDEPHA

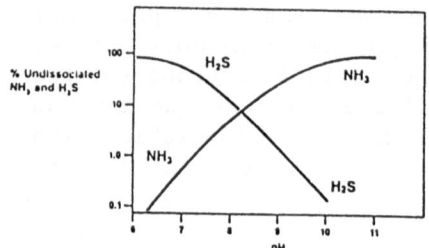

Figure 3- Effect of pH on NH_3 and H_2S dissociation

(b) Separation of ammonia and hydrogen sulphide

Absorption of both ammonia and hydrogen sulphide from refinery gas streams produces a waste stream known as sour water. The conventional process for removing both species is steam stripping followed by combustion of the gases. In this way no recovery is possible.

The main problem in this treatment is caused by pH fluctuations. The dissociation of the ammonium sulphide in solution is highly dependent on pH. Figure 3 shows the effect of pH on NH3 and H2S dissociation. As H2S is stripped off the pH value of the solution increases leading to a decrease of the amount of undissociated H2S. At high pH values the percentage of free ammonia increases and becomes easy to strip. This decreases the pH and the situation reverses to the conditions where H2S stripping is favoured. These pH oscillations make the separation difficult. Large towers and high steam consumption are required to achieve the threshold values of H2S and NH3 fixed by regulations.

To overcome this difficulty a process combining stripping and liquid membranes has been proposed [9]. An emulsion containing sulphuric acid solution as internal phase is fed into the stripping tower along with the sour water and steam or gas (figure 4). The H2S is steam stripped while free ammonia is extracted by the liqid membrane. Figure 5 shows the experimental results obtained by stripping, LMP and the process that combines the two technologies. The combination process is the most effective.

(c) Metal waste treatment

The transport of heavy metals across the liquid membrane barrier is another example of facilitated transport, i.e. extraction with chemical reaction. A typical case is the separation of chromates which is achieved by membranes containing organic amines [10]. The internal phase is an alkaline solution. The extraction of chromate occurs at the external interface by amine neutralization. The organic soluble salt diffuses through the membrane and the amine is regenerated by chemical reaction at the internal interface. Other heavy metals including Cd and Hg have been successfully extracted from dilute solutions [6]. The Takuma Co. of Japan has evaluated liquid membrane processes for heavy metals removal in a continuous pilot plant [10,11]. They found that the process is economically attractive. Furthermore there seem to be no alternates for achieving the low levels of metal required by Japanese law.

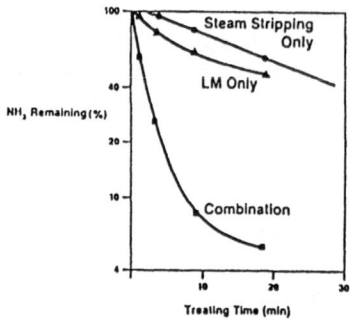

Figure 4- NH$_3$ removal from sour water

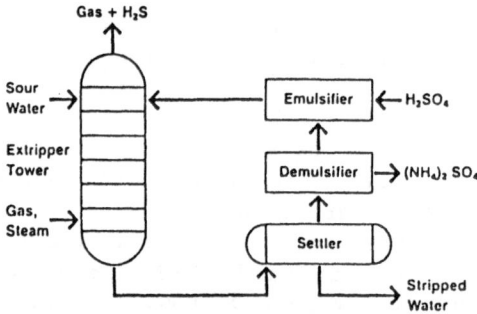

Figure 5-Combination of LM and steam stripping

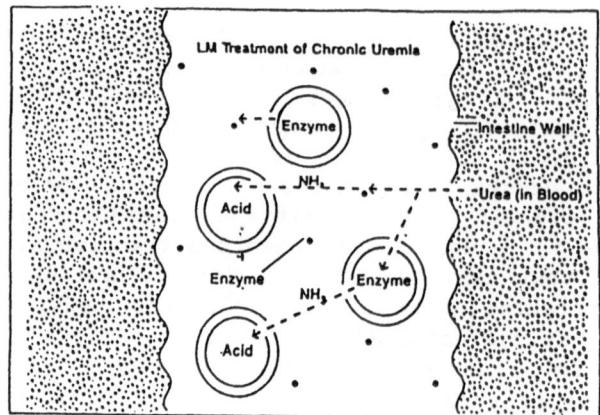

Figure 6- LM treatment of chronic uremia.

SEPARATION OF HYDROCARBONS

Examples of the systems studied include toluene-heptane, hexane-benzene, styrene-ethylbenzene, isomers of xylenes [12 - 16]. A process that combines distillation and liquid membrane extraction has been proposed by Cahn and Li [14] to deal with azeotropic distillation.

BIOCHEMICAL AND BIOMEDICAL APPLICATIONS

An interesting example involving the use of mixed emulsion is the treatment proposed for chronic uremia [17]. The patient drinks a mixture of two types of emulsions. One has urease, an enzyme, encapsulated and is purposely made not so strong so it can rupture in the intestine releasing the enzyme as shown in figure 6. The other emulsion contains an organic acid in the internal phase and is very stable in the gastro-intestinal environment. When the kidney fails and urea reaches the intestine it can either permeate into the enzyme-containing membrane or meet the freed enzyme. In either case, urea wil be decomposed by the enzymatic reaction and the ammonia formed will be removed by the second emulsion and neutralised by the organic acid.

Other applications that have been investigated with promising results include blood oxygenation and treatment of drug overdose.

REFERENCES

1. Li, N.N., Ind. Eng. Chemn. Process Des. Dev., 10 (2) 215, 1971.
2. Yagodin, G.A., Ivakhno, S.Yu. and Tarasov, V.V., Proc. of ISEC 80,3, 15B (1980).
3. Nitsch, N. and Roth,K., Colloid and Polymer Sci., 256, 1182 (1978).
4. Breysse, J. and Pérez de Ortiz., E.S., CHISA 81, Prague (1981).
5. Wongswan, S., Pérez de Ortiz, E.S. and Sawistowski, H., Proc. Hydrometallurgy 81, Soc. Chem. Ind. (1981).
6. Proceedings of the seminar on "Liquid membrane applications in waste-water treatment and metal recovery" held at UMIST, England (1980).
7. Melling, J., Report LR 330 (ME), Warren Spring Laboratory, 1979.
8. Cahn, R.P. and Li, N.N., Separation Science, 9, 505 (1974).
9. Cahn, R.P., Li, N.N. and Minday, R.M., Envir. Sci. Tech., 12, 1051 (1978).
10. Kitakowa, F., Nishikawa, Y., Frankenfeld, J.W. and Li, N.N., Envir. Sci. Tech., 11, 602 (1977).

584

11. Frankenfeld, J.W. and Li, N.N., Recent Developments in Separation Science, Vol. 3 CRC Press, Cleveland, 1977.
12. Alessi P., Kikic, I. and Orlandini-Visalberghi, M., Chem. Eng. J. (Lausanne) 19 (3) 221 (1980).
13. Cahn, R.P. and Li, N.N., "Membrane Separation Processes", Meares, P., ed., 327 (1976).
14. Cahn, R.P. and Li, N.N., J. Membr.Sci., 1 (2) 129 (1976).
15. Kikic, I., Alessi, P. and Orlandini-Visalberghi,M., I.Chem. E. Symp. Ser., 54, 139 (1978).
16. Shah, N.D. and Owens, T.C., Ind. Eng. Chem. Prod. Res. Dev., 11, 58 (1972).
17. May, S.W. and Li, N.N., Biomed. Appl. Immobilized Enzymes Proteins, 1 171 (1977).

REACTION REVERSIBILITY IN BATCH AND CONTINUOUS EXTRACTORS USING EMULSION LIQUID MEMBRANES

R.Baird,D.Reed and A.Bunge

Department of Chemical Engineering and Petroleum Refining
Colorado School of Mines,Golden,Colorado 80401 USA

INTRODUCTION

Emulsion liquid membranes (ELMs) were invented by Li |1|,who first applied this technique to the separation of hydrocarbons.Since then,emulsion liquid membranes have been applied to other separations. These include recovery and purification of metal ions,removal of phenol from waste water,and biological applications such as an artificial kidney.Review articles discuss these separations and ELM uses in more detail |2,3| .

Emulsion liquid membranes(ELMs),shown schematically in Figure 1, are double emulsions.The solute to be extracted from the external bulk phase transfers through the membrane phase to the internal emulsion. Figure 1 illustrates the ELM system assuming that aqueous external and internal phases are separated by an oil membrane.External and internal phases need not be aqueous nor both the same phase.The only

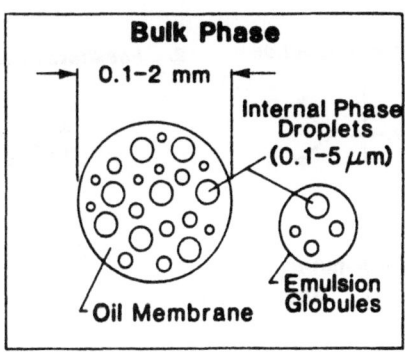

Figure 1- Schematic diagram of emulsion liquid membranes

requirement is that the membrane liquid be immiscible with both the internal and external phases.For example,the membrane could be wate between two hydrocarbons liquids.A gas external phase is also possi ble.

If the transferring solute is not soluble in the membrane liquic (for example,metal ions in an oil),a complexing agent is added to piggyback the solute from the external to internal phases.Carrier species are chosen so that they remain trapped in the membrane phase Often the interaction between carrier and solute is chemically specific permitting selective removal of a single solute from a mixed stream |2,3|.

Reactions with either the solute or the carrier species are used to promote extensive solute extraction.The two cases are diagramed in Figure 2. In the case of amine extraction,the internal phase is acidic and reacts to produce the oil-insoluble amine salt.In the second case,a complexing species carries copper ions,exchanging them for hydrogen ions.An excess of hydrogen ions in the internal phase will assure nearly complete removal of copper ions.

Emulsion extractions are well suited to toxic chemical removal. Unlike conventional extraction techniques,ELM processes can almost completely remove even low solute concentrations from feed streams. The solute is transferred to a much smaller volume solution which can be treated further if required.Solute reaction in the internal phase might be used to combine both separation and detoxification in a single step.

The emulsion configuration could also be used in a reaction-only mode.A homogeneous catalyst which is insoluble in the membrane phase would be trapped in the internal phase.The solute could transfer through the membrane liquid to the catalyst.The reacted solute can then transfer back through the membrane phase.This reactor configuration allows homogeneous catalytic systems without catalyst loss. Encapsulation of enzymes or whole cells as proposed by Li and others exemplify this use of ELMs |4,5|.

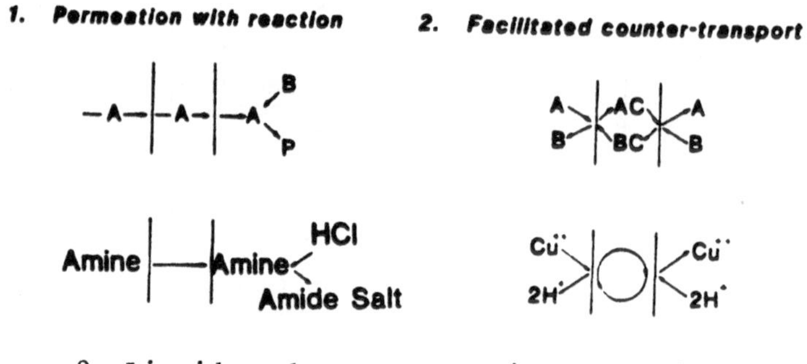

Figure 2- Liquid membrane extraction mechanisms.

In this paper we present results of batch extraction experiments of a series of amines.Extraction rates of both single species and mixtures were measured.These results demonstrate that the reaction equilibrium constant for amine dissociation is an important parameter affecting extraction rates.Experimental results are compared with two predictive models: one which includes reaction equilibrium, and a second which does not.

While batch extractions are useful for laboratory studies,they would probably not be used in larger scale operations.A flow extraction system like that schematized in Figure 3 is more likely.We also present predictions of flow extractor performance as a function of emulsion residence time.

BATCH EXTRACTION THEORY

Recent mathematical descriptions of ELM extraction which involve solute (A) reaction with an internal phase reagent (B)

$$A + B \underset{K}{\overset{\rightarrow}{\leftarrow}} P \tag{1}$$

have followed one of two approaches.The first approach,described by Ho et al.|6| ,assumes that the solute removed from the bulk phase diffuses through the membrane globule to a reaction front,where it is removed instantaneously and irreversibly by reaction with the internal reagent.That is,the equilibrium constant for reaction (1) is assumed to be infinitely large.In turn,the reaction front progresses towards the center as internal reagent is consumed.We refer to this approach as the advancing front model.

An alternate procedure,taken by Bunge and Noble |7,8|,incorporates reaction equilibrium into a description of the rate controlling membrane transport processes.This reversible reaction model predicts interdependent solute,reagent,and product concentrations.

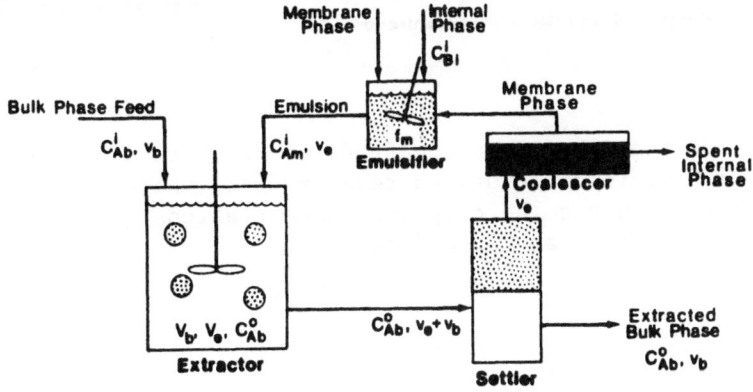

Figure 3- Schematic diagram of continuous flow extraction using emulsion liquid membranes.

Basic assumptions in both models include: (1) membrane/bulk and membrane/internal phases are immiscible, (2) local phase equilibriu[m] between membrane and internal phases, (3) no internal circulation i[n] the globule, (4) unchanging mean globule size, (5) mass transfer is controlled by globule diffusion, (6) internal droplets are solute sinks with finite capacity, (7) internal phase reaction of solute is instantaneous, and (8) no coalescense and redistribution of globules.

The presence of surfactants will restrict globule circulation. Mass transfer films external to the globule will be insignificant whenever diffusion through the membrane phase is substantially slowe[r] than through the external bulk phase. If assumption (8) was not vali[d] and globules did coalesce and break,then undesirable leakage of inte[r]nal droplets into the bulk phase would result. Consequently,preferre[d] globule formulations minimize breakage and recoalescense,thereby ass[u]ring applicability of assumption (8).

The advancing front model depends on two dimensionless parameters, ε and $E/3$,while the reversible reaction model requires specification of four dimensionless groups.These groups are listed and described physically in Table 1.The notable difference between the

Table 1- Characteristic dimensionless parameters for the two model[s]

Reversible Reaction Model

Four Dimensionless Parameters

$$\sigma_1 = f_m \left(\frac{1-f_b}{f_b}\right) K_{bm} = \frac{\text{membrane phase capacity for A}}{\text{bulk phase capacity for A}}$$

$$\sigma_2 = (1-f_m)\left(\frac{1-f_b}{f_b}\right) \frac{K_{bm}}{K_{im}} = \frac{\text{internal phase capacity for A}}{\text{bulk phase capacity for A}}$$

$$\sigma_3 = KC_{Bi}^0 = \text{dimensionless original B concentration}$$

$$\sigma_4 = \frac{KK_{bm}}{K_{im}} C_{Ab}^0 = \text{dimensionless original A concentration}$$

Advancing Front Model

Two Dimensionless Parameters

$$\frac{E}{3} = \frac{\sigma_2 \sigma_3}{\sigma_4} = \frac{\text{original mass of B}}{\text{original mass of A}}$$

$$\epsilon = \frac{3}{E}(\sigma_2 + \sigma_1) =$$

$$= \frac{\text{globule capacity for unreacted A}}{\text{A converted to P when all B reacts}}$$

two models is that the advancing front approach considers only the original mass ratio of solute to reagent ($E/3$),while the reversible reaction view requires the original concentrations of solute and reagent be specified (as σ_3 and σ_4).

Details of model developments can be found elsewhere |6-7|.Here we will only discuss results computed using these two theories.Computationally,the advancing front model (algebraic equations) is preferred over the reversible reaction model which requires a numerical solution of the nonlinear partial differential equation.The parameters, σ_3 and σ_4 provide a quantitative measure of the applicability of the advancing front model.For batch extractions when both σ_3 and

σ_4 are large or when only short dimensionless times are considered, the simpler algebraic description is adequate and preferred.

BATCH EXPERIMENTS

Experiments were designed and operated to test the effect of different reaction equilibrium constants.Table 2 summarizes the constants for phase and reaction equilibria for the four amines studied. The experimental apparatus consisted of a stirred,two-liter reaction vessel fitted with a 4-mm spout on the bottom edge for sampling.Filter paper glued over the sample port allows passage of water but not emulsion drops.All experiments were carried out at 25 C.

Water-in-oil emulsions are prepared by blending 100 grams of oil (Exxon's Solvent 100 Neutral,an isoparaffinic,middle distillate having a molecular weight of 384-404), 3.1 grams of a nonionic-polyamine surfactant (Paranox 106,approximate molecular weight of 1000), and 70 ml of 0.1 N HCl solution.The viscosity of the oil phase was determined to be 0.00425 Ns/m^2 at 22 C using a Cannon-Fenske Routine viscometer.This water-in-oil emulsion is then added to 1 liter of water and stirred,causing small emulsion globules to form.After the globule size has stabilized,a small amount of concentrated amine solution is added to the bulk phase.Samples of the continuous phase are periodically collected and the amine concentration determined by UV absorption.The stirring globules are photographed and analyzed to determine an average globule radius from the Sauter mean diameter |9|.

Table 2 - Equilibrium constants.

	K	K_{bm}
ANILINE	39,450	1.7
m-TOLUIDINE	53,700	3.9
p-TOLUIDINE	120,200	3.9
4-CHLOROANILINE	14,100	5.8

EXPERIMENTAL RESULTS AND DISCUSSIONS

Results of batch extraction experiments for initial amine concentrations of 0.0025 M and 0.0011 M are presented in Figures 4-7. Dimensionless bulk concentration of solute $\phi_b = C_{Ab}/C^0_{Ab}$,is plotted as a function of dimensionless time, $\tau = \bar{D}_{eff} t/R$. The effective mean diffusion coefficient,$\bar{D}_{eff}$,incorporates contributions from the membrane phase and the reacted and unreacted species in the internal phase |7-9|.The symbols in Figures 4-7 represent experimental data; the solid curves,the reversible reaction model; the broken curves, the advancing front model.

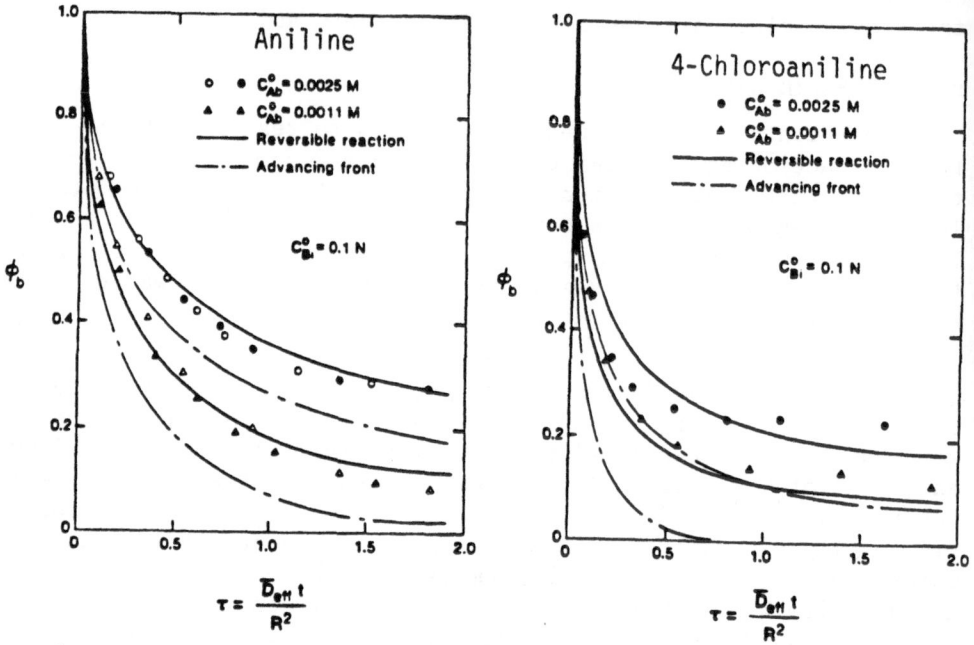

Figures 4-5 :Batch extraction of aniline and 4-chloroaniline.

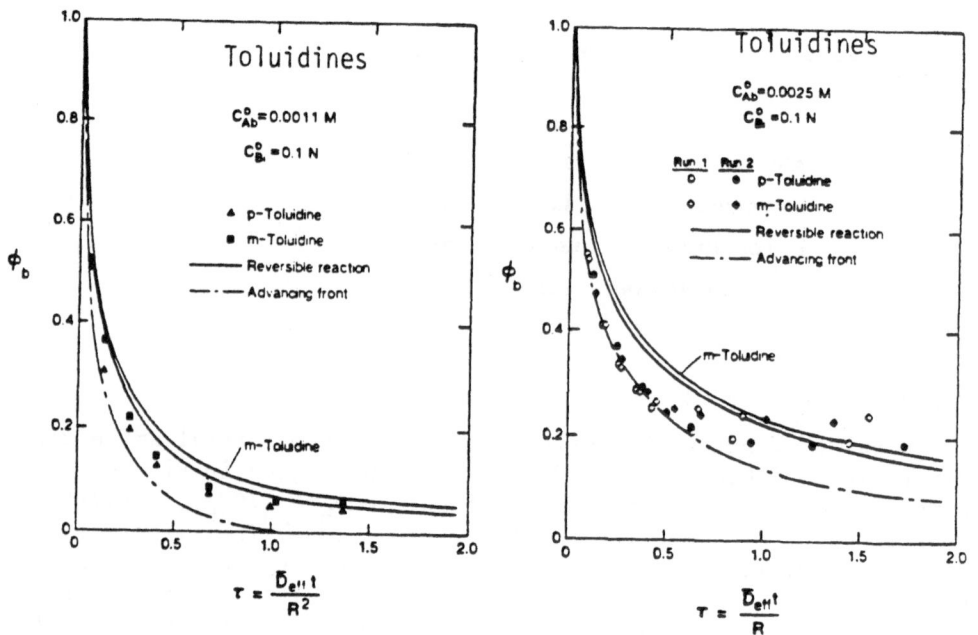

Figures 6-7 :Batch extraction of m- and p-toluidine.

The rate of solute extraction depends on both phase and reaction equilibria.For example,the extraction of m- or p-toluidine exceeds that for aniline since toluidines are more soluble and reactive than aniline.Comparing Figures 4 and 5 for aniline and 4- chloroaniline, we see that the larger solubility of 4-chloroaniline is partially offset by a smaller reaction constant which depresses both the rate and extent of extraction.For fixed internal reagent concentration (0.1 N HCl in these experiments),the dimensionless extraction rate increases as the initial solute concentration decreases.To achieve an equivalent fractional extraction,solute from the higher bulk phase concentration must diffuse further into the globule.

Besides the operating parameters (phase volumes,initial reagent and solute concentrations,and globule radius),comparison of experimental results with model calculations requires specification of the reaction and phase equilibria as well as $\bar{D}_{eff}$.Since the advancing front theory only considers diffusion through a completely reacted shell,the effective diffusion coefficient used to nondimensionalize time must include diffusion of unreacted solute through both the membrane (D_m) and internal (D_i) phases.Solute diffusivities for the membrane and internal phases are estimated using the Wilke-Chang correlation|10|.Ho and coworkers |6| used the Jefferson-Witzell--Sibbett |11,12| parallel-series equation to describe diffusion through such a composite media.

The appropriate diffusion coefficient for the reversible reaction model must account also for the diffusion of the reaction product,P. The concentration of P,estimated from the local concentrations of A and B,is incorporated into the Jefferson-Witzell-Sibbett equation to approximate the concentration dependent,reaction-enhanced effective diffusion coefficient.The mean value,$\bar{D}_{eff}$,is calculated by averaging between the minimum and maximum solute concentrations (between 0 and $K_{bm}C_{Ab}^{o}$).

The distribution coefficients,K_{im} and K_{bm},are assumed to be equal and are measured using the emulsion configuration with neutral water rather than HCl in the internal phase.The distribution coefficient measured this way is somewhat larger than is found by mixing oil and water phases directly,presumably because the solute adsorbs at the membrane/internal phase interface.This apparent effect is presently being studied.If interfacial adsorption does occur,the emulsion-based K_{bm} is actually a lumped parameter which treats interfacial adsorption as an enhanced solubility by the membrane phase.Experimental results and more discussion of this phenomenon are given in reference 8.

The agreement between the experimental results and the reversible reaction model is good.For these experimental conditions,the advancing front theory,shown in Figures 4-7 as the dashed curves, adequately describes only the short time results.For the 0.1 N HCl internal phase and the volume ratios used in these experiments,there is sufficient acid to completely react with all of the amine in the 0.0011 M solutions.Consequently,the advancing front model predicts a final amine concentration of zero which is never observed experimentally.

Of the four amines studied,the advancing front predictions for 4-chloroaniline,shown in Figure 5,are the poorest.Based on a high membrane phase solubility,the advancing front model begins to overpredict the extraction rates from the start because the low reaction equilibrium constant significantly retards the actual extraction rate.In contrast,by accounting for both phase and reaction equilibria, the reversible reaction model is able to predict these experimental measurements more closely.

Figures 8 and 9 show dimensionless concentrations for mixtures of aniline and p-toluidine as a function of dimensionless time.Three mixtures (25,50 and 75 percent aniline) at two constant total amine concentrations (0.001 N and 0.002 N) were prepared and the concentration of both amines measured.The dimensionless concentration,defined as the concentration of species j divided by the initial concentration of species j,is plotted as a function of dimensionless time, $\bar{D}_{eff,mix}t/R^2$. The reaction-adjusted effective diffusivity of the mixture is estimated using average properties $|8|$.Identical symbols are used for aniline and p-toluidine concentrations of the same mixture;open symbols designate the concentration of toluidine,while solid symbols specify aniline.Model predictions for extraction from these mixtures of solutes are also plotted.The reversible reaction predictions for the 25 and 75 percent mixtures are shown as the broken and solid curves,respectively.The advancing front model predicts the same curves for all mixtures which are designated with dashes.

Analysis of the results presented in these two figures reveals that aniline extraction is retarded by the presence of p-toluidine. The single component extraction studies show that the dimensionless solute concentration drops more quickly for solutions of lower initial concentrations.However,in these studies,the dimensionless concentrations of aniline and p-toluidine extract at nearly the same rate for all mixtures.This means that the more soluble p-toluidine is able to penetrate the emulsion globule more quickly than the aniline and partially consume the available reagent.The aniline must therefore compete with the p-toluidine for the available reagent, which significantly slows its extraction rate.

MULTICOMPONENT EXTRACTION OF DISSIMILAR SPECIES

The component extraction rates from binary mixtures can be strongly influenced by mixture composition when the species have greatly dissimilar phase and reaction equilibria.Figure 10 shows calculated results when the more irreversible component j (indicated by the larger K_j or σ_{4j}) has the smaller phase equilibria (indicated by the smaller K_{bmj} or σ_{1j}).In this case the more soluble component (species 2) initially extracts much faster than component 1.

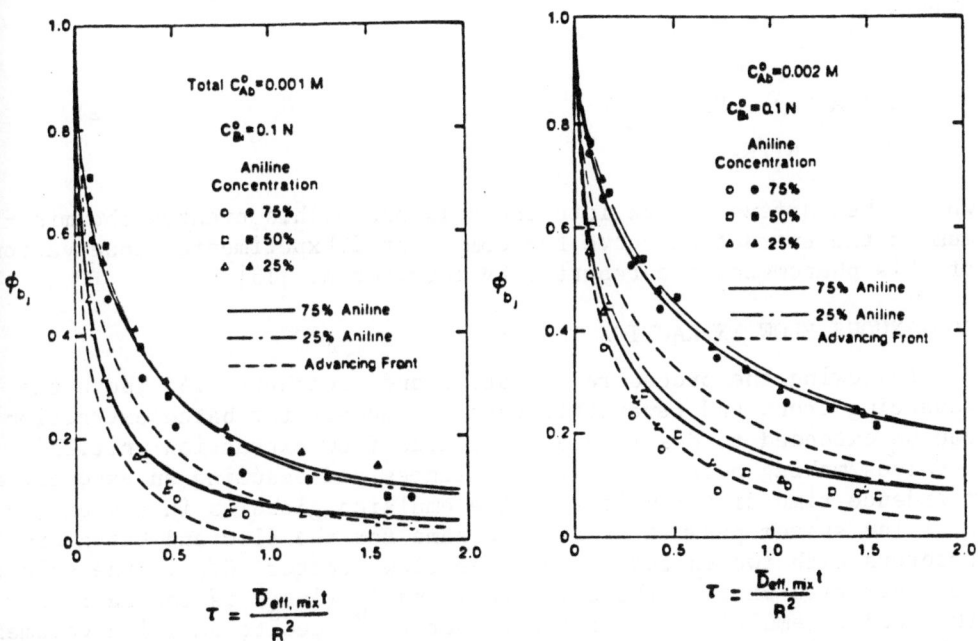

Figures 8-9 : Batch extraction of amines from a binary mixture
of aniline and p-toluidine.

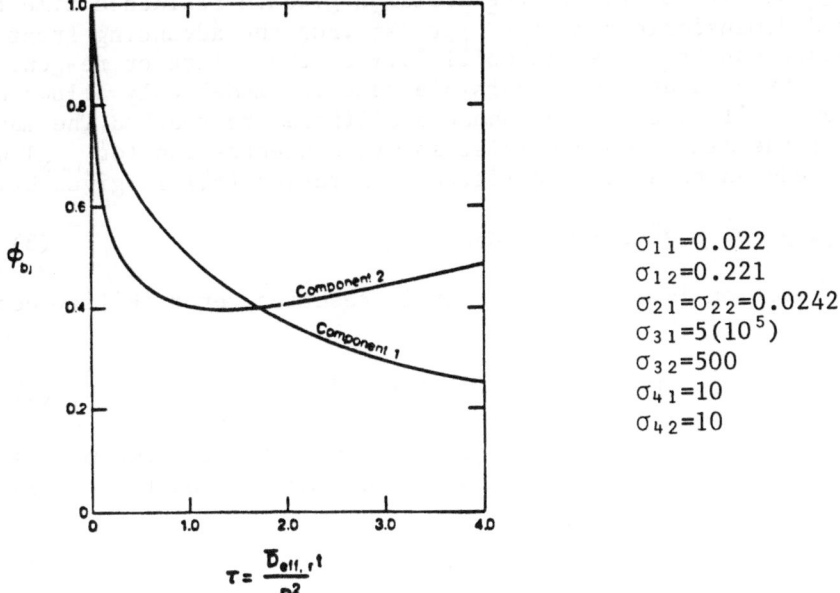

$$\sigma_{11}=0.022$$
$$\sigma_{12}=0.221$$
$$\sigma_{21}=\sigma_{22}=0.0242$$
$$\sigma_{31}=5(10^5)$$
$$\sigma_{32}=500$$
$$\sigma_{41}=10$$
$$\sigma_{42}=10$$

Figure 10-Multicomponent model prediction of extraction when the
more soluble component (1) has the more reversible
reaction

However,as extraction of the less soluble component (species 1) proceeds,it begins to force the reaction product of component 2 to generate A_2

$$P_2 + A_1 \overset{\rightarrow}{\leftarrow} P_1 + A_2 \qquad (2)$$
$$K_1/K_2$$

which then diffuses back into the bulk phase.This creates the minimum in the extraction curve for component 2.Experimental observation of this phenomenon are reported by Terry et al.|13|.

CONTINUOUS FLOW EXTRACTION THEORY

Following the procedure of Hatton and coworkers |14| ,both the advancing front and reversible reaction models for batch extraction can be extended to describe a continuous flow extraction unit.This involves making overall material balances and assuming an exponential residence time distribution for the emulsion globules in the extractor.The system parameters are the same for the flow and batch extractors with two exceptions.For the flow reactor, C^o_{Ab} is the solute concentration exiting the extractor, and $f_b/(1-f_b)$ is the ratio of the feed to emulsion flow rates rather than feed to emulsion volumes |15|.The solute concentration into the extractor,C^i_{Ab},is rendered non-dimensionalized as σ^i_4.

The fraction of maximum possible solute extraction,$F_e=(\sigma^i_4 -\sigma^0_4)/$ $/(\sigma^i_{4,max}-\sigma^0_4)$ is related to the emulsion globule residence time t_e rendered dimensionless as $\tau_e=\bar{D}_{eff}t_e/R^2$ .For the advancing front model,solute can be extracted until all of the solute or reagent is consumed.By contrast,the reversible reaction model only allows extraction until reaction and phase equilibria are reached.The maximum value of the dimensionless inlet solute concentration ($\sigma^i_{4,max}$) which can be reduced to a given outlet concentration (σ^0_4) is given by:

$$\sigma^i_{4,max} -\sigma^0_4 = \sigma^0_4(\sigma_1+ \sigma_2)+ \sigma_2\sigma_3 \qquad (3)$$

if the reaction is irreversible or if reaction reversibility contributes:

$$\sigma^i_{4,max} - \sigma^0_4= \sigma^0_4(\sigma_1+\sigma_2)+\sigma_2\sigma_3\sigma^0_4/(1 +\sigma^0_4) \qquad (4)$$

Table 3 shows typical values for ($\sigma^i_{4,max}-\sigma^0_4$) for the two models.If solute extraction results from only the complete reaction of reagent,

$$\sigma^i_{4,max} - \sigma^0_4 \simeq \sigma_2\sigma_3 \qquad (5)$$

The pseudo steady state solution to the advancing front model requires condition (5) to hold.

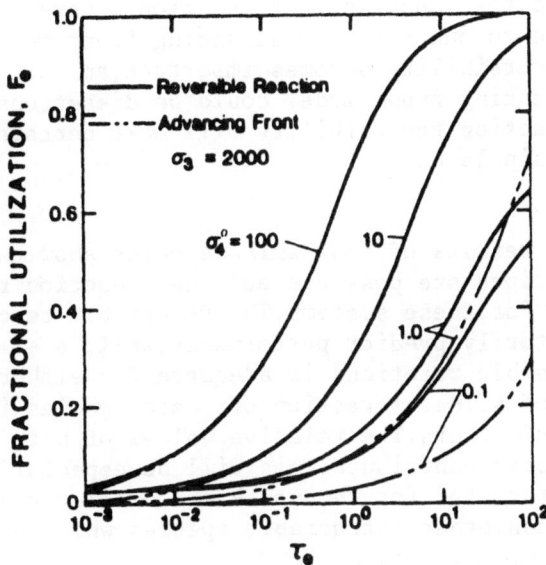

Figure 11- Fractional utilization of emulsion globules as a function of τ_e (f_m=0.6 , K_{bm}=1 , v_b/v_e= 9 and 19).

Table 3- Maximum globule capacity

Maximum Globule Capacity,
$(\sigma^i_{4,max} - \sigma^o_4)$

σ^o_4	Reversible Reaction	Advancing Front	
	Total	Total	Reaction Only
0.1	8.09	88.89	88.88
1.0	44.56	88.99	88.88
10	81.92	89.99	88.88
100	99.12	99.99	88.88

σ_3= 2000 v_b/v_e = 9 f_m= 0.6 K_{bm}= 1.0

From Eqs.(3) and (4) we see that the deviation of (σ_4^0+1) $/\sigma_4^0$ from one measures the importance of reaction reversibility.When thi quantity is close to one then the advancing front model is adequate When reaction reversibility becomes important,an extractor design based on the advancing front model could be disastrously undersized In some cases,reaction reversibility will make certain extractions impossible in a single unit.

CONCLUSIONS

Experimental results of four amine species show that extraction to low concentrations are possible and that reaction reversibility can be important for these systems.The reversible reaction model is able to satisfactorily predict performance,while a simpler model assuming irreversible reactions is adequate for either short times or large values of K.The extraction of each species is influenced by the presence of others.The relative values of both the reaction and phase equilibria contribute.This will be especially problematic when the solute targeted for removal has a smaller reaction equilibrium constant than other extractable species which might be present.

Finally,the extension of batch models to continuous flow reactors is described.Sample calculations indicate that the advancing front model adequately describes extractor performance when $(\sigma_4^0+1)/$ σ_4^0 is nearly one.When feed concentrations become smaller,reversibility is important and must be included in calculations(Figure 11).

ACKNOWLEDGEMENTS

This work was supported in part by the U.S.Environmental Protection Agency through grant number R811247-01-0.Research assistant support was provided through a cooperative agreement (NB83RAH 30001) with the National Bureau of Standards.The authors have benefitted from conversations with and encouragement by Dr. Richard D.Noble at the Chemical Engineering Center of the National Bureau of Standards in Boulder,Colorado.

NOTATION

A - solute species which is being extracted
B - reagent species soluble in internal phase
C_{Ab} - solute concentration in the bulk external phase, mol/dm^3
C_{Ab}^0 - original solute concentration in the bulk external phase (batch extractor); solute concentration of the exiting bulk phase (continuous extractor)
C_{Bi} - original reagent concentration in the internal phase (mol/dm^3)
$\bar{D}_{eff}$ - mean effective diffusion coefficient in a reacting emulsion (see reference 7), cm^2/s
D_i - solute diffusivity in the internal phase, cm^2/s
D^i - solute diffusivity in the membrane phase, cm^2/s
f_b^m - bulk phase fraction of total volume (batch extractor); bulk phase fraction of total volumetric flowrate (flow extractor)
F_e - fractional utilization of emulsion capacity
j - species
f_m - membrane volume fraction of emulsion phase
K^m - reaction equilibrium constant, mol/dm^3
K_{bm} - solute distribution constant between the bulk and membrane phases
K_{im} - solute distribution constant between the internal and membrane phases.
M - molarity, mol/dm^3
P - reaction product soluble in internal phase
R - average radius of emulsion globules determined from Sauter mean diameter, mm
t - time, s
t_e - emulsion residence time in continuous flow reactor, s
v - volumetric flowrate, m^3/s
V - volume, m^3

Greek symbols
τ - dimensionless time = $\bar{D}_{eff} t/R^2$
ϕ - dimensionless solute concentration in the bulk external phase, C_{Ab}/C_{Ab}^0

Superscript
i - inlet
o - outlet

Subscript
b - bulk
e - emulsion
max - maximum
mix - mixture
r - reference

REFERENCES

1. N.Li.Separating Hydrocarbons with Liquid Membranes. U.S.Patent 3,410,794 (1968)
2. Halwachs,W. and R.Schugerl. The Liquid Membrane Technique -A Promising Extraction Process. Int.Chem.Eng. 20(1980)519
3. Way,J.,Noble,R.,Flynn,T. and E.Sloan. Liquid Membrane Transport: A Survey. J.Membrane Sci. 12(1982)239
4. Mohan,R. and N.Li .Reductions and Separations of Nitrate and Nitrite by Liquid Membrane -Encapsulated Enzymes. Biotech.Bioeng 16(1974)513
5. Mohan,R. and N.Li. Nitrate and Nitrite Reduction by Liquid Membrane-Encapsulated Whole Cells. Biotech.Bioeng. 17(1975)1137.
6. Ho,S.,Hatton,T.,Lighfoot,E. and N.Li. Batch Extraction with Liquid Surfactant Membranes:A Diffusion Controlled Model. AIChEJ 28(1982)662
7. Bunge,A. and R.Noble. A Diffusion Model for Reversible Consumptic in Emulsion Liquid Membranes. J.Membrane Sci. 21(1984)55
8. Baird,R.,Bunge,A. and R.Noble. Batch Extraction of Amines Using Emulsion Liquid Membranes:Importance of Reaction Reversibility. AIChE Summer Meeting,Seattle,WA.,Aug.25-28,No 27e(1985)
9. R.Baird. An Experimental Study of Amine Extraction Using Emulsion Liquid Membranes. MSc Thesis.Colorado School of Mines,Golden,CO. (1985)
10. Reid,R.,Prausnitz,J. and T.Sherwood. The Properties of Gases and Liquids (McGraw Hill,New York,1977)
11. J.Crank. The Mathematics of Diffusion (Clarendon Press,Oxford, 1977)
12. Jefferson,T.,Witzell,D. and W.Sibbett. Thermal Conductivity of Graphite-Silicone Oil and Graphite-Water Suspension.Ind.Eng.Chem. 50(1958) 1589
13. Terry,R.,Li,N. and W.Ho .Extraction of Phenolic Compounds and Organic Acids by Liquid Membranes . J.Membrane Sci.10(1982)305
14. Hatton,T.,Lighfoot,E.,Cahn,R. and N.Li .An Internal Recycle Mixer for Solvent Extraction Mass Transfer Characterization with Liquid Surfactant Membranes. Ind.Eng.Chem. Fundam.22(1983)27
15. Reed,D.,Bunge,A. and R.Noble. Influence of Reversible Consumption on Continuous Flow Extraction by Emulsion Liquid Membranes.No 48a presented at AIChE Summer Meeting,Seattle,WA,Aug. 25-28(1985)

FIXED BED CEMENTATION EXPERIMENTS

M.Berteigne,G.Grevillot and D.Tondeur

Laboratoire des Sciences du Génie Chimique du CNRS
1,rue Grandville -54000 Nancy,France

INTRODUCTION

The precipitation of a metal from a solution of one of its salts by another metal is known as "cementation" although "contact reduction" or "metal displacement reaction" are used. Cementation is a very old technique which has been known in China since the Sung's dynasty (1000 before J.C.). Nowadays it is widely used in hydrometallurgy either for extraction or purification. Classic examples of industrial importance include : copper extraction from process streams by cementation on iron, purification of electrolytic zinc sulfate solutions by cementation of copper and cadmium with zinc powder, gold cementation from cyanide solutions with zinc dust.

The general reaction between the cation A^{n+} and a metal B giving the metal A and cation B^{n+} is analogous to an ion-exchange reaction :

$$A^{n+} + B^o \rightleftharpoons A^o + B^{n+} \tag{1}$$

The driving force of the reaction is the difference ΔE of the electrochemical potentials of the two half reactions, reduction of A and oxydation of B, given by :

$$\Delta E = E^o_A - E^o_B + \frac{RT}{nF} \mathrm{Ln} \frac{(A^{n+})}{(B^{n+})} \tag{2}$$

where E^o are the normal reduction potentials.

If ΔE is positive, reaction (1) proceeds to the right side. The equilibrium constant :

$$K = \frac{(B^{n+})}{(A^{n+})} \tag{3}$$

can be calculated from (2) by putting $\Delta E = 0$:

$$K = \exp \left| + \frac{nF(E_A^O - E_B^O)}{RT} \right| \qquad (4)$$

The constants calculated in this way are usualy very high, for exampl 10^{26} for the system Cu/Fe, indicating that the reaction is practicall irreversible.

This is the thermodynamic condition for the feasibility of a ceme tation reaction and as such gives no information at all regarding the rate of the reaction. The main steps which occur during the reaction are shown on Figure 1 :

1. Transport of A^{n+} ions from the bulk of the solution to the interface

2. Adsorption and incorporation of atoms A into a crystal lattice

3. Transfer of n electrons from the anodic sites to the cathodic sites

4. Release of B^{n+} ions into the solution

5. Transport of B^{n+} ions to the bulk of solution through the deposit layer

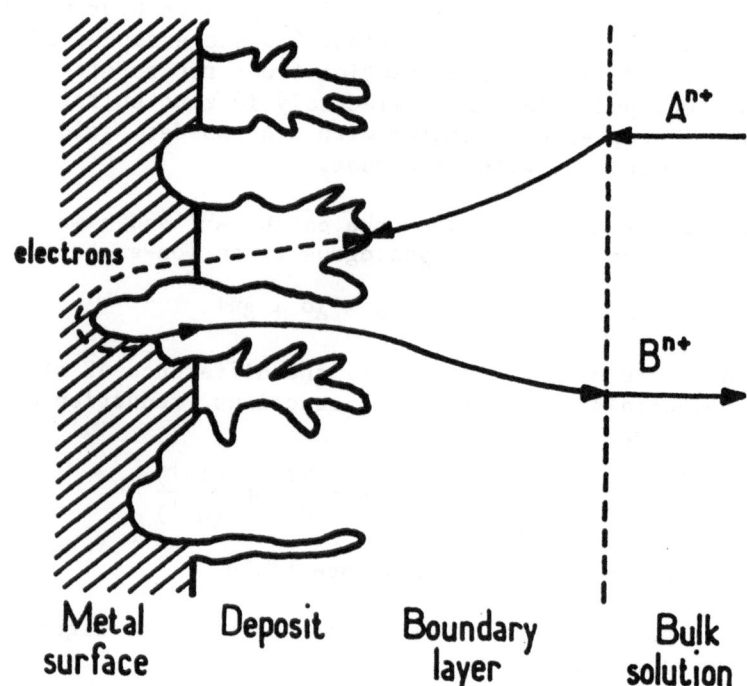

Fig. 1 - Cementation reaction steps.

For most cementation systems, the overall reaction rate is controlled by steps 1 and 5, that is by diffusion through the boundary layer. Almost without exception, results conform to a first-order kinetics, at least in the initial period of the reaction. Thus in a batch reactor containing a volume V of reactant solution of concentration C, the rate at which the reactant disappears from the solution is given by :

$$\frac{dC}{dt} = - k \frac{A}{V} C \qquad (5)$$

where k is the first-order rate constant and A the active surface area of the solid.

The complexity of the phenomenon lies in part in the fact that the interface is continuously modified by the simultaneous deposition and dissolution. An analysis of the morphology of the deposit is essential for a good knowledge of the cementation kinetics.

SURFACE DEPOSIT EFFECTS

The overall cementation reaction comprises an anodic electrodissolution and a cathodic electrodeposition. As for common electrolytic deposits, the morphology of the cement depends on a number of variables (1) : current density (cations flux), activity of depositing ion, temperature, agitation, presence of addition agents (such as arsenic, or antimony), structure of the substrate metal ... Broadly, two classes of deposit may be distinguished : smoothe deposits and dendritic deposits. The formation of smooth deposits is favoured by high agitation and low cations flux (low activity of the depositing ion, low temperature). The reverse conditions favour the formation of dendritic deposits.

The morphology of the deposit is important from two points of view :

- in industrial practice, where a metal powder is used as reagent, a dendritic or particulate deposit is preferred because it can easily be removed from the surface of the metal by agitation and collisions between powder particles ; thus the area for dissolution of the metal is always available and the reaction can proceed up to complete consumption of the initial solid metal. A tight coherent deposit may lead to the inhibition of the reaction as a result of the coating of the particles,

- in interpretation of cementation rate data (2) : a dendritic growth of the deposit increases the cathodic area and also the turbulence near the active surface. These two effects lead to an enhancement of the reaction rate which is generally observed after an initial period.

KINETICS

Experimental Methods

It has been said previously that cementation is a process controlled by kinetics rather than equilibrium. The first step in any kinetic study must be to distinguish between a diffusion or a chemically controlled rate.

In the case of control by diffusion, the rate depends essentially on the sample geometry and stirring. Two main types of experimental geometry are generally used in cementation (1) : the rotating disk and the rotating cylinder. For the rotating disk, the flow over the surface is laminar up to quite high Reynolds numbers (Re = $r^2\omega/\nu \sim 10^5$). Under this condition, the expression for the mass transfer coefficient for a smooth disk has been established by Levich (1) as :

$$k = 0.62 \; D^{2/3} \; \nu^{-1/6} \; \omega^{1/2} \qquad (6)$$

D = diffusion coefficient of reactant
ν = kinematic viscosity
ω = angular velocity

This theoretical relationship has been used successfully by several authors. Often, it is simply the linear dependance of k on $\omega^{1/2}$ which is used as a test of diffusion control. In general, the equation is valid only for the initial period of the reaction, because, as the reaction progresses, deposit accumulates and gives some surface roughness which in turn causes a significant perturbation to laminar flow.

The rotating cylinder has the great advantage that all points on the surface have equal linear velocity and should be equally accessible. On the other hand the transition to turbulent flow occurs at very low Reynolds number (~ 10). For turbulent flow and a smooth cylinder, Gabe and Robinson (1) have proposed the following expression for the mass transfer coefficient :

$$k = 0.079 \; u^{2/3} \; D^{2/3} \; \nu^{-1/3} \; d^{-1/3} \qquad (7)$$

u = peripheral velocity
d = cylinder diameter

A complication arises in the analysis of cementation data for the rotating cylinder due to the end effect : there is a significantly greater deposition at the ends of the exposed section probably because extra-turbulence take place in these zones.

An other type of geometry, suspended particles, is of considerably greater interest than the rotating disk or the rotating cylinder, because cementation on an industrial scale is usually accomplished in stirred reactors with metal powder. However most of the published results are concerned with the size distribution of the cement particles rather than the kinetics of the reaction.

Results

The crude data which are obtained using the above experimental methods are the concentrations of the ions in solution as a function

f time. This permits the calculation of the product kA from Equation
. In the initial period of the reaction, that is as long as the
mount of deposit is not too large, the surface area A can be consi-
ered as constant and the calculated value of k can be tested by use
f the above relationships.

Generally, a good agreement is obtained. From the temperature de-
endence of the reaction rate constant, the apparent activation energy
an be calculated : this energy generally lies between 2 and 6 kcal/
ole (3). It is thus concluded that, for most cementation reactions,
he rate is controlled by the mass transfer through the boundary
ayer of the liquid phase. After the "initial period" of the reaction,
he rate is enhanced in almost all cases (4,5). This is due to the
increase in the cathodic area but probably also the increase in the
mass transfer coefficient k in relation to the morphology of the deposit as pointed out above. This question has received no definitive
answer to date.

INDUSTRIAL TECHNIQUES

The most common equipment for the extraction of copper from leach
solutions is the launder. The solution can be introduced either by
allowing it to flow with gravity at the upper end of the launder or
by injecting it into the mass of scrap iron by means of nozzles.
Grids can be positionned above the bottom to permit the copper cement
to fall to the bottom where it can be recovered. The launders may be
operated batchwise or continuously.

A more compact system used for copper extraction consists in a
down pointing cone feeds at the top with scrap iron. The upper part
of the cone is a screen. The solution is injected at the bottom of
the system and directed tangentially by means of nozzles. The swirling
action washes the copper from the iron surfaces and the copper particles are carried upwards into a reduced velocity zone created by the
larger diameter where they can pass through the screen. The precipitation cone is a continuously operated unit.

Stirred tank reactors are used for the purification of zinc electrolytes. This type of reactor makes easier the control of the reaction by pH and temperature adjustment and addition of reagents such
as antimony or arsenic.

FIXED BED CEMENTATION EXPERIMENTS

The following experiments in fixed bed have been performed to examine to what extent cementation can be compared to ion exchange.

The experimental set-up (Figure 2) is that commonly used for fixed
bed ion exchange except that the feed is maintained under oxygen-free
conditions by a flow of nitrogen in order to prevent oxydation reactions. The column is 10 cm high and has an internal diameter of
0.9 cm. All the experiments have been conducted at 20°C. The flowrate
is about 30 vol/vol/h.

604

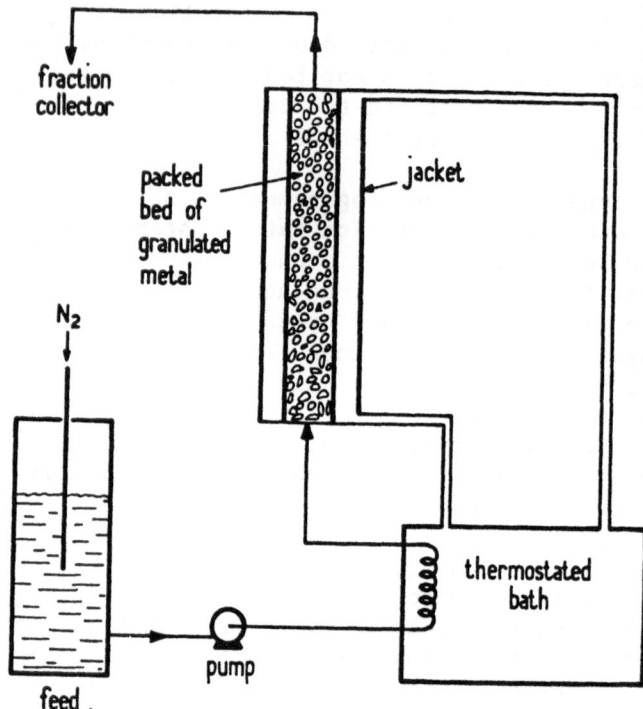

Fig. 2 - Experimental set-up for fixed-bed cementation.

The first cementation system studied here is copper-aluminum shot. The overall reaction for this system is represented by :

$$3 \ Cu^{2+} + 2 \ Al^0 \rightleftharpoons 2 \ Al^{3+} + 3 \ Cu^0$$

The most common commercial method of recovering copper from process streams is by cementation on scrap iron. The main disadvantage of this process is an excess consumption of iron by side reactions with H^+. Aluminum is an attractive alternative to iron because of its world-wide availability in the form of beer and beverage cans. Previous studies have shown that the cementation reaction is extremely slow unless a certain minimum amount of chloride ion is present in the solution in order to break down the surface oxyde layer (6,7). In the present experiments, the copper sulfate feed solution contains no chloride ions but instead a 0.1 N sodium chloride solution was fed into the column just before the start of the experiment itself. Figure 3 shows the copper concentration in the effluent relative to the feed. An   induction period is observed, the copper concentration decreasing in the first liter of effluent. This induction period was observed by other authors for this system. In addition, it has been shown that the period is shortened by increasing the chloride concentration in the solution and so the induction period was supposed to be the time taken for the oxyde film to be ruptured. The small peak betwenn 2 and 3 liters is rather abnormal. This can be attributed to channeling in the column decreasing the contact time of the fluid. Indeed, small bubbles of hydrogen due to the reaction of aluminum

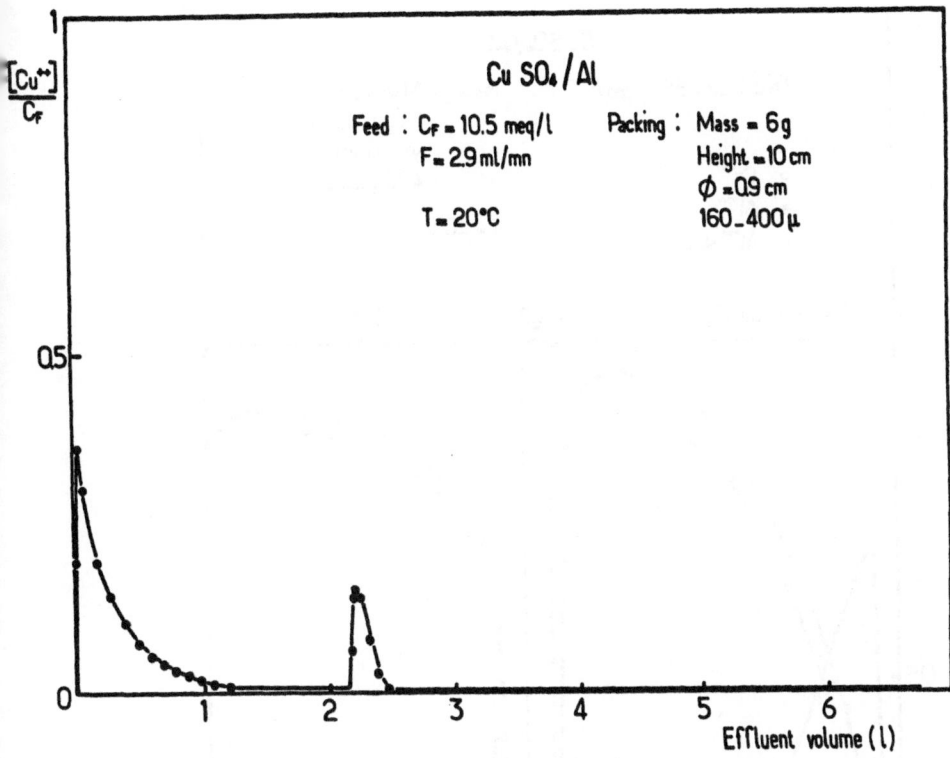

Fig. 3 - Fixed bed cementation of copper on aluminum.

with water, appears in the bed as the reaction proceeds ; some movements and coalescence of the bubbles cause disturbances of the fluid flow. After this peak, the removal of copper from the solution is good, the copper concentration being less than 0.1 meq/l, or 3 ppm. The experiments has ended itself after the percolation of about seven liters of feed because the bed was clogged by the copper deposit between the aluminum particles. The amount of copper deposit deduced from the concentrations of the feed and of the effluent is 72 meq, whereas the initial amount of aluminum in the bed is 667 meq in terms of Al^{3+}. Therefore only 11 % of the aluminum has reacted. However, it could be seen that less than the first third of the bed was concerned by the reaction, and consequently the local yield is higher. The reacted part was a lump which was difficult to remove from the column. Evidently this is a serious problem if fixed bed cementation were considered on a large scale.

Figure 4 shows the results of an experiment with a higher feed concentration : 55.3 meq/l, and with a period with feed modification during which deionized water was flowed through the column instead of the copper sulfate solution. Let us examine first the part of the figure before feed modification. Compared to the previous experiment the induction period is here characterized by a higher initial copper concentration in the effluent, and a shorter duration. As a result the copper leakage during this induction period is about the same in

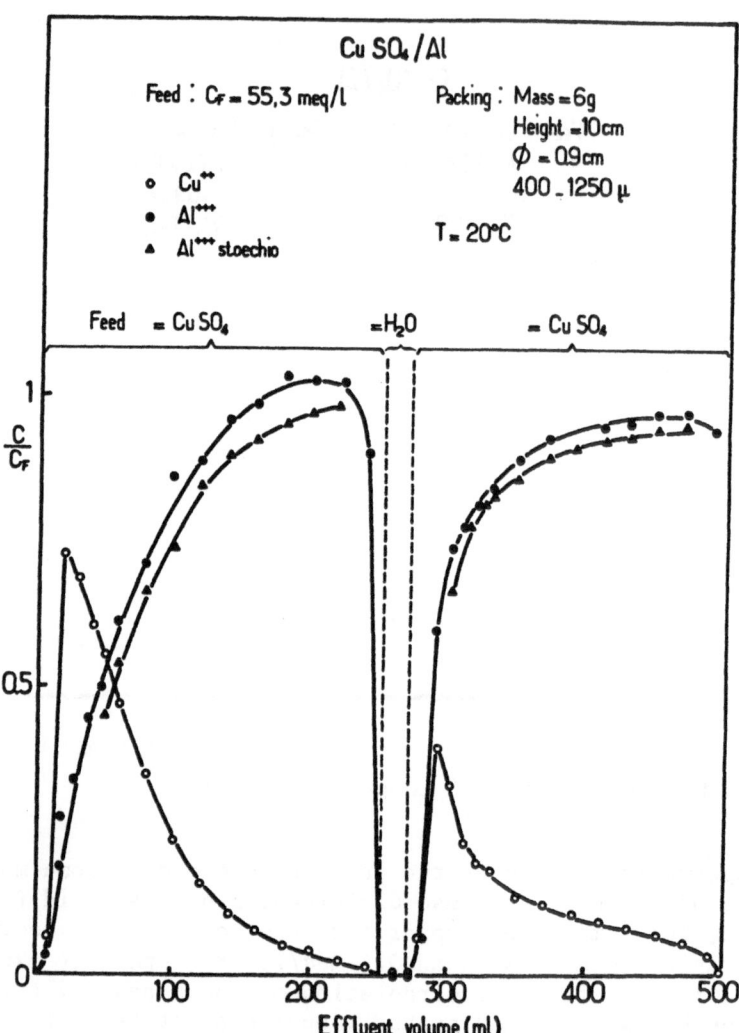

Figure 4-Fixed bed cementation of copper on aluminum with feed interuption.

the two experiments, 2-3 meq. This seems to agree with the idea put forward by other authors that the generally observed enhancement of the cementation rate is correlated to a critical mass of deposit. The aluminum concentration is a bit higher than the stoechiometric one, calculated from the copper concentration and the feed concentration. This excess dissolution is in relation with the observed hydrogen bubbles mentionned above, and can be attributed to the following side reaction :

$$2 \text{ Al}^0 + 6 \text{ H}^+ \rightleftharpoons 2 \text{ Al}^{3+} + 3 \text{ H}_2$$

The right hand side of Figure 4 shows the evolution of the concenrations after a short period of pure water feed. Again an induction period is observed. An explanation is that an oxyde film has formed during the flow of water (not deaereted), and is removed slowly after. It is possible also that the reaction continues partly on new sites of the bed.

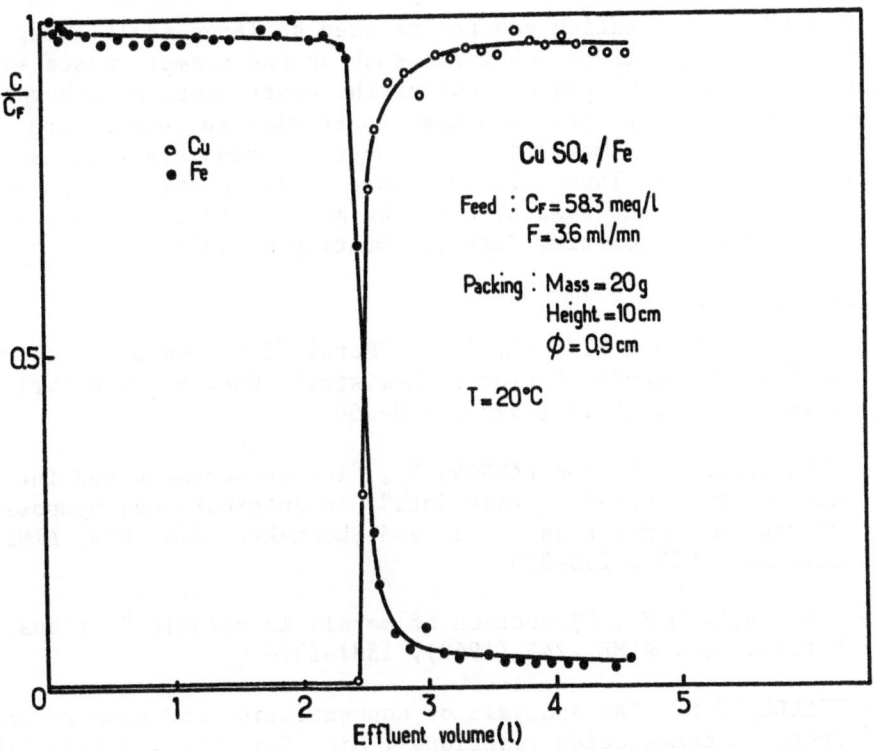

Fig. 5 - Fixed bed cementation of copper on iron powder (50-150 μ)

The second system used is copper-iron powder. The main reaction is :

$$\text{Cu}^{++} + \text{Fe}^0 \rightleftharpoons \text{Fe}^{2+} + \text{Cu}^0$$

The results are shown on Figure 5. Here there is no induction period. The effluent is immediately almost free of copper and the reaction is stoichiometric. After about 2.5 1, the copper concentration rises to the feed concentration whereas the iron concentration falls near to a low value. This very sharp wave is like a stable wave obtained in binary ion exchange when the equilibrium is favorable to the ion which fixes in the column. The stoichiometric volume in binary ion exchange corresponds to the complete removal of the ion occuping initially all the sites of the bed. An analogous volume could be defined in cementation, corresponding to the complete dissolution of the metal initially in the column. However surface deposit effects can lead to some trouble. In the present cementation experiment, the volume at which the wave exits the column corresponds to a copper deposit of 147 meq that is significantly less than the initial amoun of iron (717 meq on Fe^{2+} basis). Probably the deposit coats progress vely the powder particles and the reaction is inhibited by the inacc sibility to the fluid of the anodic area.

CONCLUSION

Fixed bed cementation can not be used at large scale except perhaps for very dilute solutions such as for example waste water containing precious metals. However the experiments at laboratory scale are easy to perform and seem to be rich in information. In addition, they used metal in the powder or granulate form as in industrial practice. Thus, if a reliable interpretation of phenomena can be obtained, the results could be more directly applicable than that obtained on rotating disk or rotating cylinder.

REFERENCES

1 - POWER, G.P. and RITCHIE, I.M., "Metal displacement reactions", in"Modern aspects of electrochemistry", Conway and Bockris Eds, Plenum, N.Y., n° 11 (1975), 199-250

2 - STRICKLAND, P.H. and LAWSON, F., "The measurement and interpretation of cementation rate data", in International Symposium on Hydrometallurgy, Evans, D.J. and Shoemaker, R.S. Eds, AIME, Chicago (1973), 293-330

3 - WADSWORTH, M.E., "Reduction of metals in solution", Trans. Metall. Soc. AIME, 245 (1969), 1381-1394

4 - MILLER, J.D., "An analysis of concentration and temperature effects in cementation reactions", Min. Sci. Eng., 5 (3), (1973), 242-254

5 - STRICKLAND, P.H. and LAWSON, F., "The cementation of metals from dilute aqueous solution", Proc. Aust. Inst. Min. Met., 237 (1971), 71-79

6 - MacKINNON, D.J. and INGRAHAM, T.R., "Copper cementation on aluminum canning sheet", Can. Metall. Quarterly, 10 (3), (1971), 197-201

7 - ANNAMALAI, V. and MURR, L.E., "Effects of the source of chloride ion and surface corrosion patterns on the kinetics of the copper-aluminum cementation system", Hydrometallurgy, 3 (1978), 249-263

LIST OF PARTICIPANTS AND LECTURERS

Angola
Anabela Leitão-Universidade de Angola-POBox 1756-Luanda
Belgium
P.Van Cutsem-Facultés Universitaires Namur-rue de Bruxelles 61-
    B-5000 Namur
Canada
A.Tombalakian-Laurentian University-Sudbury,Ontario P3E 2C6
Denmark
E.Basby-Asnaes Power Station -4400 Kalundborg
Finland
H.Heikkila-Finish Sugar Co -SF 02460 Kantvik
France
Y.Frere -CRM/CNRS-6 rue Boussingault -67083 Strasbourg
P.Gramain-CRM/CNRS- 6 rue Boussingault-67083 Strasbourg
J.Debas-Societe Rexim- 80400 HAM 33
M.Bailly-LSGC/CNRS-1 rue Grandville-54042 Nancy
G.Grevillot-LSGC/CNRS- 1 rue Grandvilee-54042 Nancy
R.Nicoud-LSGC/CNRS-1 rue Grandvilee-54042 Nancy
A.Gorius-LSGC/CNRS-1 rue Grandville-54042 Nancy
D.Tondeur-LSGC/CNRS- 1 rue Grandville-54042 Nancy
P.Grammont-Duolite International-BP 48-02301 Chauny
Germany
H.Schneider-Dow Chemical GmbH -D 2160 State Butzfleth
K.Geckcler-Universitat Tubingen -D 7400 Tubingen
M.Guilhem-Dow Chemical Europe- D 7587 Rheinmunster
R.Sievers-Dow Chemical Europe- D 7587 Rheinmunster
R.Bartz-Dow Chemical Europe-D 7587 Rheinmunster
W.Hoell-Nuclear Research Center-POBox 3640-D 7500 Karlsrhue
M.Grote-Universitat Paderborn- D 470 Paderborn
Greece
K.Matis-University of Thessaloniki- KT 54006 Thessaloniki
Israel
C.Heitner-Wirguin-Hebrew University-Jerusalem 91904
P.Kaushansky-The Weizman Institute of Science -76100 Rehovot
E.Kvaalen- IMI Institute for Research and Development-Haifa 32002
A.Warshawsky-The Weizman Institute of Science- 76100 Rehovot
Italy
L.Rizutti-Universita di Palermo-90128 Palermo
V.Augugliaro-Universita di Palermo-90128 Palermo
C.Sarzanini-Universita di Torino- V.P.Giuria 5-Torino
A.Baradel- Eniricerce-20097 San Donato Milanese-Milan
F.Evangelista-Universita degli Studi degli Aquila Abruzzi- 67100
    L'Aquila
G.Storti-Politecnico di Milano-Piazza Leonardo da Vinci 32-Milano
A.Masi-Politecnico di Milano-Piazza Leonardo da Vinci 32-Milano
S.Carra-Politecnico di Milano-Piazza Leonardo da Vinci 32-Milano
Japan
M.Yokomizo -Japan Organo Co -Tokyo 113

610

Netherlands
J.Liou- TNO Institute of Applied Chemistry-POBox 108-3700 AC Zeist
A.Kusters- Delair-4841 BM Prinsenbeek
J.Wesselingh-Delft Institute of Technology-2628 BL Delft
Norway
H.Brattebo- SINTEF - 7034 Trondheim
T.Barth- University of Bergen - N 5000 Bergen
Portugal
A.Rodrigues-University of Porto - 4099 Porto
C.Costa-University of Porto -4099 Porto
F.R.Ribeiro- Instituto Superior Tecnico-Lisbon
J.L.Figueiredo- University of Porto- 4099 Porto
F.Lemos -Instituto Superior Tecnico -1096 Lisboa
C.Henriques-Instituto Superior Tecnico- 1096 Lisboa
M.Filipa Ribeiro- Instituto Superior Tecnico- 1096 Lisboa
J.M.Loureiro - Faculty of Engineering-University of Porto-4099 Porto
R.M.Ferreira- Department Chemical Engineering-University of Coimbra-
    Coimbra
Elisa Ramalho-Instituto Superior Engenharia Porto-Porto
Rui M. Ganho- The New University of Lisbon -Monte da Caparica
F.Delmas- LNETI/DMM - 1699 Lisboa
Luis S.Xavier- CNP-Companhia Nacional de Petroquimica -Sines
Orlando Valdez - CNP-Companhia Nacional de Petroquimica- Sines
Amélia Lemos -Instituto Superior Tecnico- 1096 Lisboa
Madalena Ribeiro- E.S.Garcia do Orta- Porto
Manuela Brotas- Faculty of Sciences-University Lisbon- Lisbon
Filipe Freire-Instituto Superior Tecnico- 1096 Lisboa
Spain
M.Diaz- Universidad del Pais Vasco -Lejona-Apartado 644-Bilbao
M.Lopez Alvarez-Facultad de Ciencias- Oviedo
G.Una -Universidad de Santiago de Compostela- Santiago
J.Pampin-Universidad de Santiago de Compostela-Santiago
R.Caeiro-Universidad de Santiago de Compostela- Santiago
South Africa
B.Hendry-University of Capetown -Rondebosch 7700 Capetown
Turkey
A.Basaran -METU-Environmental Engineering-Ankara
G.Surucu- METU-Environmental Engineering-Ankara
United Kingdom
C.Baylis-Coventry Polytechnic -Coventry CV5 6NQ
M.Hudson- University of Reading -Whiteknights-Reading RG6 2AD
M.Streat-Imperial College London -London SW7 2BY
S.Ortiz-Imperial College London· London SW7 2BY
P.Meares-University of Exeter-Exeter EX4 4QF
USA
Jim Joseph-University of Pennsylvania - Philadelphia,PA 19104
A.Bunge-Colorado School of Mines-Golden,Colorado 80401
G.Klein- University of California,Berkeley-Richmond,CA 94804
F.Helfferich-Penn State University-University Park,PA 16802
A.Myers-University of Pennsylvania-Philadelphia,PA 19104
P.Wankat-Purdue University -West Lafayette,Indiana 47907

INDEX